MW00453122

Food Policy

Looking Forward from the Past

Food Policy

Looking Forward from the Past

Janel Obenchain
Arlene Spark

CRC Press
Taylor & Francis Group
Boca Raton   London   New York

CRC Press is an imprint of the
Taylor & Francis Group, an **informa** business

CRC Press
Taylor & Francis Group
6000 Broken Sound Parkway NW, Suite 300
Boca Raton, FL 33487-2742

© 2016 by Taylor & Francis Group, LLC
CRC Press is an imprint of Taylor & Francis Group, an Informa business

No claim to original U.S. Government works

Printed on acid-free paper
Version Date: 20150619

International Standard Book Number-13: 978-1-4398-8024-1 (Hardback)

Visit the Taylor & Francis Web site at
http://www.taylorandfrancis.com

and the CRC Press Web site at
http://www.crcpress.com

Printed and bound in Great Britain by
TJ International Ltd, Padstow, Cornwall

Aaron Johnston
1974–2010

He would have advised against this project,
but without him I had no reason not to proceed.

A. Daniel Ochs, MD
1928–2013

Husband, father, brother, uncle, opa,
polymath, and Arlene's hero.

Contents

Foreword

Everyone, it seems, is an expert on food. Eaters know what they like and dislike when they taste food. Parents know what is good for their kids, if not always how to get the little darlings to try the kale salad or grilled tilapia. Food companies, of course, know exactly what their customers want—and how to persuade them to find their most profitable products irresistible. And many politicians are certain, for example, that cutting back on food benefits for poor people will motivate them to work more or that subsidizing corn, sugar, and soy producers will protect American agriculture.

So with all these experts, why have my colleagues Janel Obenchain and Arlene Spark written this comprehensive overview of what is known about food and food policy? Well, it turns out, the authors tell us, that more than one kind of expertise is needed to create the food policies and programs that can prevent diet-related disease, reduce hunger and food insecurity, protect American farmers and food workers, reduce the environmental damage that our food production system causes, and make healthy food accessible to all sectors of our population. Furthermore, the authors note, policy based on evidence is more likely to achieve its goals than policy based only on values and beliefs.

What Obenchain and Spark offer in *Food Policy: Looking Forward from the Past* is an interdisciplinary synthesis of recent scholarship on food and food policy. Using the lenses of history, economics, nutrition, public policy, and others, they complicate the certainty of self-proclaimed experts. Their goal, they tell us, is not to provide answers but to raise questions—questions that will lead both experts and nonexperts to the evidence that can inform food policy. The global food system that shapes who eats what and where is a complex organism whose anatomy and physiology have only recently been studied systematically. To understand and change this organism will require a solid foundation in this emerging knowledge.

Food Policy: Looking Forward from the Past provides students, practicing professionals, activists, and concerned citizens with an accessible guide to the recent history, politics, and future of food and food policy. It provides a starting point for those who need to know more, those who have come to appreciate the limits of their expertise, and those who are committed to shaping food policies, environments, and systems that can better meet the world's food needs. Its readers will be better equipped to take on these challenges.

Nicholas Freudenberg
Distinguished Professor of Public Health
City University of New York School of Public Health and Hunter College
Director, New York City Food Policy Center at Hunter College

Preface

In the mid-1980s, Mr. Meinke asked his Wayzata Junior High School social studies class what we thought about the proliferation of food brands on the grocery store shelves as an example of competition in the market. A classmate had a particularly long-winded response about how the numerous products for sale reflected the success of capitalism in the marketplace through providing this many choices to the consumer. As an avid label reader even at that tender age, I had a different opinion. I thought the enticing variety was illusory, as all those different brands were just owned by a few larger companies. We had fewer true distinct choices than appeared. But, we ran out of time, and I did not get to answer the question.

In an alternate universe, I answer that question. An illuminated path appears before me, a guide to a future career dedicated to critical inquiry about our food and food system—a path leading to a lawyer, a food policy lawyer, an activist lawyer, an activist, a food policy activist, a philosopher of food and water ethics, an emergency food supplier, a nutrition/consumption epidemiologist, and a farmer raising pigs.

In this universe, those illuminated runway lights stayed dark. Arlene Spark, however, is eponymic, and indeed, a catalyst. She reminded me of the question, and gave me the opportunity to answer.

Janel Obenchain
San Jose, California

I started working in *food policy* long before I knew what it was. A year out of college, I found myself teaching home economics (sewing, in particular) to incarcerated men and women who were recovering from heroin addiction. Sewing turned out to be a great form of occupational and group therapy for the young men and women in my classes. They wore with pride the colorful dashikis they made. And they talked about things that mattered to them, much like the women in a quilting bee would talk among themselves as they sewed. It seemed to me that introducing cooking would further enhance the program by providing the participants with shopping, budgeting, and food preparation skills that would be useful to them once they left the program. So I lobbied the administration for cooking classes. An empty room in the basement was converted into a kitchen, a retiring cookbook author donated his pots, I secured donations from food companies, and we launched a pilot project that proved so popular that it became a line item in the agency's budget.

Years later, I convinced the medical school where I worked at the time that it would be more useful for future doctors to learn about preventive nutrition (medical jargon for *eating well*) than memorizing the classical vitamin deficiency diseases. Another policy change!

As an educator, it is not surprising that the changes I was able to catalyze took place in educational settings. At Hunter, while teaching all manner of nutrition courses, I realized that conspicuous by its absence from our BS, MS, and MPH nutrition curricula was an examination of how nutrition tenets of the past century shaped where we are in food policy today. Thus, with encouragement from Hunter Distinguished Professor of Public Health Nicholas Freudenberg, I created our first course in food policy. The course is required for CUNY SPH students pursuing the MPH in nutrition and the DPH with a concentration in nutrition. The prerequisite for our food policy course is 30 MPH credits, meaning that students must be at least two-thirds of the way through the MPH degree in order to be eligible to enroll in the course. As a teacher, I am proud to say that the work of some of our former food policy students appears in several chapters of this book.

Arlene Spark
Demarest, New Jersey

Acknowledgments

I was unable to stay in New York City, and unable to return to New York City, each time losing the opportunity to continue on for a doctor of public health (DrPH) degree at the Hunter School of Public Health. In lieu of being able to study for that degree, I studied for my own through the writing of this book.

I now dedicate this book to all my friends and family, whom I dearly hope will be willing to revitalize our friendships and connections after too many times of me saying I can't ... because, book. I owe a particular debt for the kind offers of house-sitting gigs and places to stow myself away to write all over the Bay Area, in Mountain View, Los Altos, San Jose, Oakland, and San Francisco; and far beyond the Bay in Montana and Minnesota. Each proved absolutely invaluable and indispensable at different times for different reasons.

Being of Finnish-American immigrant descent, I inherited the lingering traces of a culture of farmers dependent on the joint efforts of a small close-knit community to survive in a foreign land. Barn-raising? I am there. Jam-making? Yes, please. Fencing club construction? Sign me up. Help writing a book? Of course.

However, the lingering culture of group participation also leads to group recognition, and thus, no names will be named. I am indebted to all of you. I am grateful to have all of you in my life. You have all participated in this end result of me, and the book, one way or another.

And you all know who you are: my family, my long-time fencing family and friends, my former colleagues and nonfencing friends, my Irish family, my adopted Finnish family, and most particularly due to the timing of this project, my Finnish fencing lad.

Janel Obenchain
San Jose, California

This book would not have been possible without my colleagues, graduate students, friends, and family, all of whom made me who I am today. In no particular order, I want to thank Philip Alcabes, who, with my late husband, taught me all I need to know about Marx (the other one, Groucho). Jack Caravanos has been steadfast and loyal. Susan Klitzman has morphed from colleague to friend and now she is my senior associate dean. What is most important is that she is my friend. Two unique groups have made my life richer, both personally and professionally. The AcaDames is a tightly knit unit of four female professors from different fields who meet intermittently for food, laughter, and moral support. Marilyn Auerbach, Elaine Rosen, and Dava Waltzman have been there for me, and I for them, for 15 years. City University of New York (CUNY) Food and Nutrition Network (FANN) was originally comprised of everyone at CUNY who taught courses in food and nutrition. When not meeting in person, we communicated through a listserv. Our goal was to help others by such activities as sharing teaching strategies and CVs of prospective adjunct faculty. Among us, we have four dietetic internships and a dietetic tech program and degrees from associate through doctorate, so we never ran out of things to discuss. CUNY FANN gave me confidence. I hope all the other members value the group as much as I do.

I have been blessed by almost 35 years of students teaching me. A few of them are credited in this book, but most of them go unsung. This book is my gift to them. Two students I thank in particular. Janel Obenchain was a food policy student of mine. Nothing could make me prouder than her name as first author of this book. Lauren Dinour was a masters and doctoral student whose input into this book you may not see. But it is there. Both Lauren and Janel are coauthors of my last book (I promised my family!), which will be published later this year.

I have lived in the same house on the same dirt road for 35 years. My neighbors are my second family. We are what is now known as "aging in place." Mary Hamilton and Jennifer Otto, in particular, have been stalwart friends who watch out for me. Alas, Emily Mikulewicz moved a few years ago to be nearer to her children. I have a large, loving, and supportive family. My brothers Bob and Mark Workman have always believed in my abilities. Sisters- and brothers-in-law Jayne Workman, Carol Ochs, Brian Wachs, and Michael Ochs are much more than in-laws. They are my friends. Eva Haberman is my late husband's sister. She knew my husband longer than anyone else so I continue to learn about him through her. To Harrison, Justin, and Alexander, I am Bubbe. I touch the future through them. My most ardent supporters are my sister, Barbara Wachs, and daughter, Danielle Bier. The special bond among the three of us is, by far, my biggest source of strength.

I am challenged every day by my colleagues who make up Hunter's small yet vocal nutrition program—Khursheed Navder, May May Leung, Ming-Chin Yeh, Ann Gaba, and Sikha Bhaduri. You can tell from our names that our grandparents came from different parts of the world. Nevertheless, it is not our ethnicities that make us so different—it is our ages. I am a generation older than they are. (You know who is the least computer-savvy.) It is precisely our age difference that made me realize I remember things they have only read about, such as the seven food groups, the very first edition of the Dietary Guidelines, food stamps before they became electronic benefits cards, and school lunches that were almost everything advised against by the U.S. Department of Health and Human Services (HHS). Thus, the genesis of this book is the confluence of past and present food and nutrition policy. Janel and I examined how the past gave rise to the present and, by this effort, hope to help readers determine the future. If there are any errors in this book, the fault is ours.

Arlene Spark
Demarest, New Jersey

We cannot thank enough the students of Arlene's food policy class. We wanted to present their work. Unable to do so, we can only highlight here some of the contributors to this effort: Lauren Dinour, DPH, RDN; Jeanine Kopaska-Broek, MPH; Megan Lent, MPH; Jenna Mandel-Ricci, MPH; Matthew Nulty, MPH, RDN; Angelica Santana, MPH, RDN; Lia Wallon, MPH; Andrea Wilcox, MPH, RDN; and Craig Willingham, MPH. And of course, the talented CRC Press team: Stephen Zollo, Marsha Pronin, Tara Nieuwesteeg, and Syed Mohamad Shajahan, who so kindly guided us through to completion.

Arlene and Janel

Authors

Janel Obenchain majored in philosophy at Northwestern University. If she had been smarter at the time, she would have continued on to a PhD in philosophy. Pragmatically, she thought law school might afford better employment opportunities. And it would have, too, if she had gone to law school. Fortunately, she was waitlisted and then rejected by Stanford, which allowed her to instead participate in the far riskier lifestyle of attempting to qualify for the 1996 Olympic team in women's épée and working in numerous dot-com startups in the San Francisco Bay Area. Unfortunately, she never made it to the top of either of these two pyramid schemes.

The details between then and her arrival at the public health program at Hunter College, CUNY, are not worth repeating. While residing in New York, she studied community nutrition, including that of her own neighborhood. She worked for United Way NYC as a site evaluation consultant for food pantries and soup kitchens and, with a team of professors that included Arlene, developed a policy agenda for the New York City Food and Fitness Partnership.

Arlene Spark attended City College of New York (CCNY) and Columbia University Teachers College. After majoring in English at CCNY, she worked for a year as a narcotics caseworker for the New York City Department of Social Services and then for three years as a home economics teacher at the Narcotics Addiction Control Commission (NAAC). The commission awarded her a paid educational leave of absence for an MS in public health nutrition. She subsequently earned an EdM in community nutrition and began teaching at the college level—and has been teaching at the college level ever since. A United States Public Health Traineeship allowed her to return to school to complete a doctorate in nutrition education.

Dr. Spark's career in nutrition includes 12 years of clinical practice and teaching in the Departments of Pediatrics and Community & Preventive Medicine at New York Medical College. At the American Health Foundation, she was a co-PI on an National Heart, Lung, and Blood Institute (NHLBI) intervention study in pediatric preventive cardiology, and she served as a co-I on an NHLBI study of the elderly in the Department of Community Medicine at Mt. Sinai School of Medicine (now the Ichan School of Medicine at Mt. Sinai). Since 1998, Dr. Spark has taught at CUNY. She is professor in the masters and doctoral programs in the CUNY School of Public Health. She has a New York State permanent teaching license in home economics (it was through home economics that she discovered nutrition). She was in the first cohort of registered dietitians (RDs) to become board certified in pediatric nutrition and in the second cohort of RDs to become a fellow of the American Dietetic Association (now the Academy of Nutrition and Dietetics). Dr. Spark is the author of *Nutrition in Public Health: Principles, Policies, and Practice* (2007). She is completing the second edition of the book with two of her former graduate students, Lauren Drucker, DPH, RD, and Janel Obenchain, MPH.

Introduction

1

The genesis of this book is the confluence of past and present food and nutrition policy. Each generation of people interested in the topics of food and nutrition start with a certain grounding and basis in the knowledge then currently agreed on. The elder author of this book struggled with seven food groups in school, the younger author learned about four. Those four food groups are now so entrenched in our dietary guidance that we chose to reflect that organization in this book; with a chapter dedicated each to Milk, Meat, Grains, and, last but not least, Fruits and Vegetables.

Although this book is not meant to be a nutrition textbook, some of the chapters have more detailed nutrition information than others; those interested in food policy may gain some exposure to nutrition perspectives and nutrition students may gain a perspective on food policy issues. We aim to illustrate how the past gave rise to the present. Today, we may struggle to perceive why people did not immediately embrace the benefits of pasteurized milk. Illuminating how current policies arise from choices made (or abandoned) informs us that just like knowledge, food policies are mutable, and subject to our change.

Building from definitions, this chapter introduces certain concepts and themes that are helpful when examining controversies or implications of food policy in the United States. This chapter also highlights useful information such as a short description of the federal agencies primarily involved in US food policy, the key role of the Commerce Clause and conceptual questions examining American values around the right to know, the nature of risk, and the balancing of economic interests. But a note of caution that although these brief explorations are meant to inform your reading of this book, they are not intended to parallel the remaining chapters; this introduction is not outlining the remainder of the book. We start by asking questions.

WHAT IS POLICY?

The Oxford Dictionary defines policy as "a course or principle of action adopted or proposed by an organization or individual."[1] The American Heritage Dictionary defines policy differently: "a plan or course of action, as of a government, political party, or business, intended to influence and determine decisions, actions, and other matters."[2] The American Heritage Dictionary leaves out "individual." Definitions are important, as differences in definition can impact policy implementation. For example, consider the implications of the idea that policies are—or are not—able to be made by individuals.

Combining elements from these two definitions leads us to a third definition: policy is created or adopted by governments, institutions and individuals to guide choices, decisions, and actions.

WHAT IS FOOD?

To quote Potter Stewart, an Associate Justice of the US Supreme Court: "I shall not today attempt further to define the kinds of material I understand to be embraced within that short-hand description; and perhaps I could never succeed in intelligibly doing so. But I know it when I see it."[3] Although Justice Stewart's famous quotation refers to pornography, this statement is equally applicable to food. We all eat. Each of us considers the particular items that we prefer to consume to be food. If pressed to define with specificity what we consider to be food, we may find doing so very difficult—but we know food when we see it. Or when we eat it.

Dictionaries do not agree on the definition of "food." One dictionary definition highlights the components of food as the defining characteristic: "consisting essentially of protein, carbohydrate, and fat."[4] Another dictionary uses a more holistic concept of food as a "nourishing substance."[5] Both of these definitions, however, also refer to the role of food as providing energy and promoting or sustaining growth and life.

Even though we usually "know what food is when we see it," it is still useful to carefully think about the implications, such as value judgments, that may arise from defining food. Differing ideas of "food" might lead to differing guidelines and decisions about food policy. For example, the "food as components" definition above elicits an underlying concept of food emphasizing nutrition, whereas the "nourishing substance" definition hints at a more evocative cultural context of the role of food. Another way to frame this idea: *Do you consider soda to be food?*

Sugars are carbohydrates, a source of energy. If viewing the definition of food through a nutrition component-based lens, soda is food. The act of defining food— or alternatively, rejecting the act of defining food—can significantly contribute to the implementation of food policy. Underlying philosophies about the role and nature of food are evident in government federal assistance programs. For example, soda cannot be purchased with Women, Infant and Children Supplemental Nutrition Program (WIC) benefits. However, soda can be purchased with Supplemental Nutrition Assistance Program (SNAP) benefits. *Should a government food assistance program allow all foods, even unhealthy ones? What if the program goal is to address hunger? What if the program goal is really just income transfer?*

The SNAP program website has a detailed explanation of why SNAP benefits may be used to purchase soda.[6] The explanation revolves around how food is defined, or in reality, not defined, in the Food and Nutrition Act of 2008 (FNA).[7] The FNA defines food in a self-referential, we "know it when we see it" style by making only specific exclusions. Soda is not specifically excluded, and neither are other non-nutritious items, and therefore these items are considered "food" and can be bought with SNAP benefits. See Chapter 6, "Federal Food Assistance," for further information on SNAP, soda, and federal food assistance programs.

WHAT IS FOOD POLICY?

Food policy may be as simple as an individual's personal guideline of "What foods do I strive to choose to eat?" However, although individuals may have a guiding policy for their choices, they will still be constrained by resource availability and by decisions others (including corporations and governments) have made as to what one should eat. In Chapter 6, "Federal Food Assistance," you will read about specific decisions the government has made as to what we should eat while using benefits from one food assistance program (WIC) and conversely, a complete refusal by the government to decide what we should eat while using benefits from another federal food assistance program (SNAP). *Is it appropriate for the government to make decisions about what we should eat? If so, which decisions?*

The government promulgates food policy through laws, through federal agency regulation implementing laws, and through action plans with stated goals and objectives (such as Healthy People) and through guiding statements (*Dietary Guidelines for Americans*). Some, like food activist and food policy council expert Mark Winne, author of *Closing the Food Gap*, 2008, argue the government creates food policy even through inaction.[8] When the government fails to take action, the result may be *de facto* food policy manifested through decisions made by other organizations, such as corporations.

Food policy can also be promulgated by local and regional governments, states, or even set by an institution. Think of the numerous things we do with food. We produce and process food, doing so in considerable quantity and with differences in quality; we package, label, distribute, protect, store, access, purchase, prepare, eat, and, in the end, we dispose of excess or spoiled food. These activities constitute our "food system." A food policy is any guideline, rule, practice, or regulation that affects how we do any of these activities.

As a shortcut, a definition of food policy might simply refer to things that impact the food system.[8] However, in the United States today, most people have limited engagement with the entire cycle of our food system. It is impossible for most of us to control food access and intake from "seed to table," and instead we participate mostly at the end of this cycle by making purchasing decisions about what to eat, and disposing of what we do not eat, often as trash but sometimes as compost. Describing food policy operating at the level of "food system" is accurate but minimizes the fact that policies carry unstated values impacting individuals within the system—both human and animal. Tim Lang, Professor of Food Policy at City University London's Centre for Food Policy, humanizes food systems and food policy by simply asking who eats what, when, how, and with what impact.[9]

WHO EATS WHAT, WHEN, HOW, AND WITH WHAT IMPACT: FOOD POLICY IN THE UNITED STATES

Examples of food policy in the United States include

- The eligibility standards for participation in food assistance programs (*who eats*)
- Dietary guidance and subsidization of agricultural production (*what*)
- Food safety processing and preparation regulations (*when*)
- Taxes, ordinances, zoning, and licensing requirements or incentives for food-related businesses (*how and with what impact*)

Other policies that are not specifically about food can also affect who eats what, when, how, and with what impact. For example, historically, the limits of transportation affected who ate what and when. Access to fresh foods throughout the United States was greatly improved by the signing of the Federal-Aid Highway Act by Eisenhower in 1956 and the subsequent development of the Interstate Highway System. No transportation-related laws or regulations directly address the issues of food access or agricultural transport. Although not a distinct guiding directive of "food policy," federally supported changes in infrastructure nonetheless had a distinct impact on our food system and diet, including the development of fast food restaurants along the highways as well as greater access to fruits in nongrowing regions.[10]

MAJOR FEDERAL AGENCIES INVOLVED WITH FOOD POLICY

The Department of Health and Human Services (HHS) is one of the two major players in federal food policy, the other being the United States Department of Agriculture (USDA). The USDA's mission broadly covers agriculture and related concerns, such as natural resources, rural development, and nutrition. The HHS is charged with protecting the health of Americans, and is home of the Food and Drug Administration (FDA) and the Centers for Disease Control and Prevention (CDC). The FDA has primary oversight of food safety, food labeling, and veterinary drugs. The CDC conducts research that informs policy, rulemaking, and the public. The USDA and the HHS jointly develop and publish *The Dietary Guidelines for Americans* every five years, and the HHS also develops 10-year health objectives for the nation through the Healthy People initiative. For the most part, the discussion in this book involves these two major players. *Beyond the USDA*, a publication of the Institute for Agriculture and Trade Policy, provides an excellent overview of other federal agencies involved with food policy, which we will recommend you to review instead of repeating the information here.[10]

We turn to forecast some themes discussed in our chapters, starting with agricultural policies.

AGRICULTURAL POLICIES: "FARM" OR "FOOD"?

2012 was a year of some importance to food policy in the United States, as 2012 was the 150th anniversary of the USDA—as well as the purported renewal year for the omnibus "Farm Bill" legislation. Agricultural policy in the United States, and thus, food policy, is greatly driven by the colloquially known "Farm Bill" which is up for renewal every five years. The renewal deadline for the 2008 Farm Bill (officially the Food, Conservation and Energy Act of 2008) was September 30, 2012, although Congress failed to authorize renewal at that time. By January 2013, Congress managed to implement a short-term extension for many of the programs, and agreement on the Farm Bill (officially the Agricultural Act of 2014) was finally reached on January 3, 2014.

The mechanisms of federal agricultural support affect what food is available, in what quantity, at what price, and who produces the food, and the impact on the American food supply and diet can be far-reaching. Journalist/author Michael Pollan has encouraged the adoption of the term "Food Bill" instead of "Farm Bill" in an effort to emphasize the immediate importance of this legislation to a society for which farming has become an abstraction.[11]

Paradoxically, even though the USDA is in charge of agriculture and also authors the Dietary Guidelines, our agricultural production subsidies do not always support the recommendations of our dietary guidelines. For example, the United States does not produce enough fruits and vegetables for each of us to be able to eat the 2010 Dietary Guidelines recommended quantities of fruits and vegetables. Per capita, we have only six cups of fruit available to apply toward the weekly recommendation of 14 cups of fruit and just over 11 cups of vegetables available to apply toward the weekly recommendation of just over 24 cups of vegetables.[12] *The USDA is involved in promulgating both agricultural policy and dietary guidelines. What impact might this have on food policy?*

In the late 1970s, then Senator George McGovern (D-SD) strongly connected food production and processing with medical care, thus pushing for a nutritional direction for agricultural policy. His Congressional Committee on Nutrition and Human Needs produced the *Dietary Goals for the American People*, which called out items that Americans should increase, such as fruits and vegetables, and items to decrease, including the specific statement "reduce consumption of meat." Under food industry pressure, the second edition of the Dietary Goals softened the "reduce consumption of meat" language. After the second edition of the Dietary Goals, the torch was passed to the USDA and the HHS for the first production of the *Dietary Guidelines for Americans*. As the reader reads about our agricultural policies and our dietary guidance, he or she can make his or her own assessment as to if we have a nutritional direction for agricultural policy.

FEDERAL REGULATION OF FOOD PRODUCTION: THE COMMERCE CLAUSE

Generally, today we have come to accept that government in the United States—local, state, and federal—can regulate food production as a social good. Arguments today concerning the Farm Bill are generally over the contents of the Farm Bill (and funding allocation), not as to if Congress has the power to regulate food production, but that was not always the case.[13] During the Great Depression, the New Deal era legislation reflected efforts by the federal government to assert more control over the failing economy through regulating food production (see Chapter 2, "Agricultural Policies"). However, the Supreme Court disagreed that the federal government had that power. In *United States v. Butler (297 U.S. 1 1936)* the Supreme Court struck down the first Farm Bill, the Agricultural Adjustment Act of 1933. The court stated that a statutory plan to control and regulate agricultural production was an infringement of Tenth Amendment state rights by the federal government.[14]

As the Great Depression lingered on, however, the underlying tenants of the New Deal Legislation gained more traction. Eight years later, in 1941, in a case concerning quotas and penalties on wheat production, the Supreme Court held that the Commerce Clause, which gave Congress the power to regulate commerce among the states, meant that the federal government had the power to regulate food production.[15]

The application of the interstate Commerce Clause is significant for food policy in the United States. Although much of our food now crosses state lines as part of interstate commerce, this was not always the case, and in the early 1940s, even less so. *When food was a much more local system than today, how could the Supreme Court argue that the interstate Commerce Clause provided authority for the federal regulation of food production?*

The Supreme Court's decision was grounded in the idea that even small, private consumption decisions by individuals could, if replicated and aggregated throughout society, create an excessive cost to the public as a whole.

FOOD POLICY AND PUBLIC HEALTH: SAFETY AND LABELING

Even though we may take the information on our labels for granted now, the role of the federal government in food safety and consumer protection was also highly contentious. The federal government began to accept a role of responsibility in food safety in 1906 with the Meat Inspection Act and the Pure Food and Drug Act, but only after a long run-up period. About 190 different proposals of legislation on this topic had been introduced between January 1879 and 1906.[16] Although the 1906 Acts initiated some measures of consumer protection, it also helped industry, as these acts provided a veil of perceived safety that had been drawn aside by Upton Sinclair's *The Jungle*, an expose of working conditions in the meat packing industry. Following Sinclair's book, domestic markets for American meat weakened and sales fell sharply and confidence had to be restored internationally and domestically in order to recoup sales.[17]

Today, at least 15 federal agencies[18] are involved in food safety; the ones most critically connected to the consumer are the USDA, the FDA, the CDC, and the Environmental Protection Agency (EPA). None of these agencies have food safety as a primary mission; although the FDA's food safety authority was expanded under the Food Safety Modernization Act of 2011 (FSMA).

The FDA is the major food policy player when it comes to labeling. In addition to regulating the marketing claims and statements made on packages, the FDA also regulates the Nutrition Box and ingredient list on labels, which is discussed in Chapter 4, "Nutrition Information Policies." The FDA's authority over ingredients extends from the label into the contents of the food, by setting food identity standards. Food identity standards provide some protection to the consumer; if purchasing a bottle labeled ketchup, the consumer is safe to assume that, unless otherwise stated, the contents are as expected, made from tomatoes. The FDA also oversees the safety of food additives. Substances added to food are subject to approval by the FDA prior to the product being sold on the market unless the substance is generally recognized as safe (GRAS). These topics are briefly explored in this book, but deserve further inquiry by students of food policy.

Food product labeling illustrates a trend in increasing federal regulation over food and increasing customer demands for information about their purchases. In addition to nutrition information, consumers may rely on packaging information to assess other values they hold dear when selecting food, such as "organic" or "humane," as discussed in Chapter 8, "Meat," or "healthy," as discussed in Chapter 9, "Grains." In the absence of informative labeling, the consumer must unknowingly accept any risk of consumption, which may be hard to swallow when there are very little perceived benefits being offered in return. Thus, activism has played a large role in pushing for more informative labeling.

FOOD POLICY AND "AMERICAN" VALUES

Food policy can come into conflict with cherished "American" values, such as free speech. Many of the tax, zoning and ordinance-based policies discussed in Chapter 12, "Nanny State," are aimed at changing the food environment to promote healthier living. Some of these policies directly limit consumer access or exposure to unhealthy foods, like the moratorium on fast food establishments in Southeast Los Angeles. Limiting consumer access or restricting marketing to children could be considered restrictions on commercial speech, which does have some protection under the First Amendment of the United States Constitution. The Supreme Court has also ruled that this protection in turn prohibits compulsory speech. The 1943 decision *West Virginia Board of Education v. Barnette* determined that the First Amendment prohibits the government from compelling speech. However, courts have upheld laws that compelled companies to disclose facts when relevant to a legitimate government interest.

When New York City moved to require chain restaurants to publish calorie information, the regulation was fought by the industry under the grounds that it infringed free speech. The lawsuit against the New York City calorie labeling law argued that providing calorie information was compelling an opinion statement, and thus fell under the prohibition of compulsory speech. However, the courts determined that calorie information was a fact, and the provision of this information therefore was a legitimate governmental requirement that did not constrict commercial speech.[19] *What do you think? Could calorie information actually be an opinion?*

Another value that comes into discussion is that of individual liberty and self-autonomy. Those who oppose food regulations might argue the efforts to regulate our interactions with food are "nanny-esque" or even "anti-American." In Chapter 12, "Nanny State," we discuss this tension between regulations and liberty through a brief discussion of John Stuart Mill's "harm principle," namely *That the only purpose for which power can be rightfully exercised over any member of a civilized community, against his will, is to prevent harm to others.*[20]

Governmental regulation of the production of food has become generally accepted as an appropriate form of food policy. However, government regulation of consumption remains contentious, and can be quite controversial. Regulatory efforts at modifying our food consumption rather than production rely on extending public health's traditional sphere of improving common goods (i.e., water) into the realm of our food system. We will make a case later in this book that food is a social (i.e., common) good.

Public health initiatives, through the effort to change the behavior of many people, may result in a shift of a cultural norm. A cultural or social norm can also be created through changing our physical environment or our food environment such that no decision about our behavioral choices need be made. For example, for the most part, we no longer make a decision about buying pasteurized milk. We just buy milk. When it comes to regulating our food consumption choices in light of public health efforts to address chronic health risks, we have accepted regulatory efforts at taxes, educational information such as calorie information on menus, and physical modifications of food characteristics that make our choice healthier without any effort on our part, such as pasteurization.

However, as far the authors are aware, beyond war-time rationing, we have accepted government regulation limiting the amount we may choose to consume only for food served to children at school, and even this may still be contentious. In the case of moderating the amounts of our consumption, corporate self-regulation may be more acceptable to an American public sensitive to infringement of individual liberties by government.

FOOD POLICY: ECONOMIC INTEREST

Food policy can be in reality, an economic policy. The 2009 Recovery Act increased SNAP benefits and loosened eligibility temporarily. This is an overt example of a "food policy" that, as simply a transfer of money without nutritional guidelines, is really an economic policy.

Food policy has many intersections with economic interest. Government at any level may be interested in shaping food policy that changes the choices available to consumers, but food manufacturers also play a role in what choices are available to consumers. Consumer desires may change over time due to increased information, trends in tastes, or interest in other values associated with food, such as local production as well as costs. The information disseminated in the Dietary Guidelines may cause the creation of new food products as well as increased consumption of foods promoted for nutritional value.

Similarly, food safety concerns may impact choices available to consumers. Some restrictions may be promoted through the federal government (such as the requirement that milk must be pasteurized to be sold across state lines), but others may simply be the result of decisions made by the food producer or retailer regarding the burden of risk, including the economic burden. Food retailers have announced that they would stop selling sprouts due to the high likelihood of contamination.[21] Farmers may choose to stop growing items such as cantaloupes, with some dropping out of the market, or conversely, coming together as a group to create a safer (and branded) product.[22] Even though the retailers and producers have no desire to cause illness, these are at heart decisions of economic interest.

Money may be a key consideration in food policy that impacts "who eats what." Be aware of how programs, research, and policies are promoted and funded. Think back to the personal and private decisions made about what foods you prefer to consume. *Is it fair and reasonable that these decisions are made public if such decisions impact the public good?*

What responsibility do we have, if any, if our food consumption decisions are publically funded? The average American does not have a healthy diet as measured against US Dietary Guidance. *Should we expect more from those using food assistance programs than from the average American?*

Our federal food policy, through agricultural subsidies and other mechanisms, promotes the availability of some types of foods at a low cost, including "fast-food" or ready-made foods available to eat outside of the home and without preparation. We are able to consume inexpensive food in part because our federal food policy supports cheap food for the animals that we butcher for meat. In the long run, who receives the economic benefit of the low cost food? Who is bearing the risk—the economic risk; the health risk? *Is there a greater cost to our food production than what the producer is taking on?*

Production in any industry can create external costs; these are costs that are borne by society instead of solely by the producer. Examples of external costs could be taxpayer subsidies, health care costs, the use of limited resources such as water, and the environmental costs of cleaning up damage. There has been controversy over large employers, such as Walmart, hiring employees for full-time work, and then encouraging them to apply for SNAP benefits. In other words, the wages offered were low-enough that someone working a full-time job still qualified to receive federal food assistance benefits. *Is it appropriate for a retailer to protect profits by transferring costs to taxpayers?*

As a society, we take on the external costs of our agricultural production, and our industrial meat production, in particular, may not be sustainable in light of these external costs. "Meatless Monday," a public health awareness campaign reborn in 2003 at Johns Hopkins, is drawing on historical US government war-effort campaigns asking the public to do their part by eating less of certain items.[23] This campaign encourages people simply to eat less meat, not to become vegetarians. Should this idea be supported through federal food policy? *Would reducing our meat consumption benefit the public good?*

Another social mechanism that supports inexpensive food is limitations on how food-service workers are paid and if they receive health care and sick-pay benefits. Our inexpensive food can become costly if we consider the potential public health and food safety ramifications, not just for the food-service worker, but also for those being served. Income is considered an important social determinant of health. Federal law does not currently require employers to provide paid leave benefits to workers; and those who are undocumented, part-time employees, or minimum wage earners are not likely to have access to paid leave from their employers.[24] In the case of foodborne disease outbreaks, the income of the food service worker may become a determinant of health, not just of the worker's, but also for those who receive the services of the food-worker who cannot afford to stay home when ill.

CONCLUSION

This introduction has posed numerous questions, and additional ones will appear in the text. This book is not intended to answer these questions. Our goal instead is to inspire you to ask more questions. If something intrigues you, go learn more! Instead of presenting learning objectives, we suggest that when starting to read a chapter, ask yourself if there is something you want to learn from the chapter. If your question or idea was not addressed, follow it up. Similarly, although many texts offer recommendations for further reading, we decided not to do so, because food is "hot" right now, and any suggestions we might make will be quickly superseded. There are numerous interesting articles, books, blog discussions, and websites. Look at the endnotes cited when you find an idea of interest; search the internet for the reference. *See what else comes up. Critically evaluate our sources, thoughts, and questions. Become your own expert and agent of change.*

REFERENCES

1. The Oxford Dictionaries. Available at: http://oxforddictionaries.com/definition/english/policy, accessed September 5, 2014.
2. The American Heritage Dictionary. Available at: http://www.ahdictionary.com/word/search.html?q=policy, accessed September 5, 2014.
3. *Jacobellis v. Ohio*, 378 US 184, 187, 1964.

4. Merriam-Webster Online Dictionary. Available at http://www.merriam-webster.com/dictionary/food, accessed September 5, 2014.

5. Dictionary.com. Available at http://dictionary.reference.com/browse/food?s=t, accessed September 5, 2014.

6. United States Department of Agriculture, Food and Nutrition Service, SNAP. Available at http://www.fns.usda.gov/snap/eligible-food-items, last updated July 17, 2014.

7. United States Department of Agriculture, Food and Nutrition Service, SNAP, The Food and Nutrition Act of 2008, as amended through PL 112—240. Definitions, Section 3 (k).

8. Harper A et al. Food policy councils; Lessons learned. Institute for Food and Development Policy, 2009; 1–63.

9. Lang T. Food control or food democracy? Re-engaging nutrition with society and the environment. *Public Health Nutr.* 2005;8(6a):730–737.

10. Gosselin M. Beyond the USDA: How other government agencies can support a healthier, more sustainable food system. February 3, 2010, Institute for Agriculture and Trade Policy. Available at: http://www.iatp.org/documents/beyond-the-usda-how-other-government-agencies- can-support-a-healthier-more-sustainable-foo, accessed September 14, 2014.

11. Pollan M. You are what you grow, *New York Times Magazine*, April 22, 2007. Available at http://www.nytimes.com/2007/04/22/magazine/22wwlnlede.t.html?pagewanted=all&_moc.semityn.www, accessed September 12, 2012.

12. Palma MA, Jetter KM. Will the 2010 *Dietary guidelines for Americans* be any more effective for consumers?. *Choices: The Magazine of Food, Farm and Resource Issues*, 2012;1:27.

13. Peck A. Revisiting the original 'tea party': The historical roots of regulating food consumption in America. *University of Missouri-Kansas City Law Rev.* 2011;80.

14. The National Agricultural Law Center, Farm Bills, www.nationalaglawcenter.org/farm-bills, accessed November 16, 2014.

15. Wickard, *Secretary of Agriculture et al. v. Fillburn*, 317 U.S. 111 (US 1942), accessed August 17, 2013. Available at http://www.law.cornell.edu/supremecourt/text/317/111#ZO-317_US_111n12.

16. Regier CC. The struggle for federal food and drugs legislation. *Law Contemp. Prob.* 1933;1:3–15.

17. Barkan ID. Industry invites regulation: The passage of the pure food and drug act of 1906. *Am. J. Public Health.* 1985;75:1.

18. Johnson R. The Federal Food Safety System: A Primer, Congressional Research Service, December 15, 2010.

19. Farley TA et al. New York City's fight over calorie labeling. *Health Affairs*, November/December 2009;28:w1098–w1109.

20. Mill JS. *On Liberty*, 3rd edition. New York: Dover, 2002.

21. Weise E. Kroger stores stops selling sprouts as too dangerous. *USA TODAY*, October 20, 2012. Available at http://www.usatoday.com/story/news/2012/10/19/kroger-bans-sprouts-too-dangerous/1645147/, accessed October 4, 2013.

22. Wyatt K. *Colorado Cantaloupes Return; Growers Push Safety.* The Associated Press, July 12, 2012. Available at http://bigstory.ap.org/article/colorado-cantaloupes-return-growers-push-safety, accessed October 4, 2013.

23. Meatless Monday, History. Available at http://www.meatlessmonday.com/about-us/history/, accessed December 16, 2013.

24. Siqueira CE et al. Effects of social, economic, and labor policies on occupational health disparities. *Am. J. Ind. Med.* May 2014; 57(5):557–572.

United States Agricultural Policies*

2

> *The Agricultural Department, under the supervision*
> *of its present energetic and faithful head, is rapidly*
> *commending itself to the great and vital interest it*
> *was created to advance. It is precisely the people's*
> *Department, in which they feel more directly concerned*
> *that in any other. I commend it to the continued attention*
> *and fostering care of Congress.*
> President Abraham Lincoln
> *May 15, 1862*[1]

INTRODUCTION

Understanding current food policy in the United States requires at least a modicum of familiarity with the history of the various Farm Bills that have steered necessary funds to the nation's food and agriculture programs. Historical events such as the Great Depression, both World Wars, the Dust Bowl, climate change, and food activism in support of various food and agriculture programs have all contributed to our current policies regarding food and nutrition. This chapter introduces the United States Department of Agriculture, the cabinet-level department that is most closely aligned with food and agriculture, provides a history of the Farm Bill from its earliest days, presents an overview of the Agricultural Safety Net of commodity crops, crop subsidies, and crop insurance, and concludes with a look at policy and agrarian values.

The United States Department of Agriculture (informally, the "Agriculture Department" or "USDA") is the U.S. federal executive department responsible for developing and implementing national policy on farming, agriculture, and food. The department was formed in 1862 by President Abraham Lincoln and received Cabinet status in

---

* Portions of this chapter were published previously in Morland, K.B. (ed.), *Local Food Environment: Food Access in America*, Boca Raton, FL: CRC Press, 2014. Copyright 2014 from *Local Food Environment: Food Access in America* by K.B. Morland. Reproduced by permission of Taylor and Francis Group, LLC, a division of Informa plc.

1889 under President Grover Cleveland. The current USDA mission and vision statements are, respectively: "We provide leadership on food, agriculture, natural resources, rural development, nutrition, and related issues based on sound public policy, the best available science, and efficient management" with the aim "to expand economic opportunity through innovation, helping rural America to thrive; to promote agriculture production sustainability that better nourishes Americans while also helping feed others throughout the world; and to preserve and conserve our Nation's natural resources through restored forests, improved watersheds, and health private working lands."[2] The USDA is headed by a Secretary whose nomination by the President must be approved by Congress.

THE FARM BILL, THEN AND NOW

The United States supports its agricultural sector through a variety of programs. The primary legal framework for agricultural policy is established through what is known as the "Farm Bill," a periodic process of developing omnibus legislation—a giant piece of multi-year legislation that deals with a multitude of subjects and programs. Congress drafts, debates, and passes a Farm Bill every 5–7 years. In legal terms, functions of the bill include amending some and suspending many provisions of laws enacted through previous Farm Bills; reauthorizing, amending, or repealing provisions of preceding temporary agricultural acts; and establishing new policy provisions for a limited time into the future.[3]

Actually, the term "Farm Bill" is a misnomer.[4] In addition to farming, the legislation involves (in alphabetical order) ecology, economics, employment, energy, nutrition, research, and trade. Food activists and many others feel that the legislation should be called the "Food and Farm Bill" since most of the money involved is earmarked for domestic antihunger nutrition programs.

Since 1965, each bill has been given a new name (Table 2.1). The name of each Farm Bill signifies both what is important in the bill and what the authors want the public to think is important. For example, the title of the 2008 Farm Bill—the Food, Conservation, and Energy Act—signals that the legislation contains provisions that are important to consumers, "conservation" calls attention to the importance of the environment, and "energy" points to concerns about high fuel prices and its effect on food prices. Interestingly, the 2014 Farm Bill has reverted to the older naming convention, and is known as the "Agricultural Act of 2014."

The Farm Bill became omnibus legislation in 1973, with the inclusion of nutrition program assistance.[3] Starting in 1973, farms bills have included titles for major issue areas. The order and number of Farm Bill titles varies from year to year. Titles are added (or eliminated) as topics become crucial to the food and farm economy. *Nutrition* appeared for the first time in 1973[3] and *horticulture* in 2008, along with four other new titles. In 2014, there were 12 titles, down from 15 titles in 2008, but still more than the 10 titles of 2002. *What do you think should be emphasized by title in the next Farm Bill?* Compare some titles from the past to think about looking forward (Table 2.2).

The 2008 Farm Bill was set to expire in 2012. The 112th Congress (2011–2012) did not complete the legislation in time, requiring new bills to be introduced in the 113th

TABLE 2.1   U.S. Farms Bills, 1933 till present

YEAR	PUBLIC LAW NUMBER[a]	FARM BILL NAME
1933	73–10, 48 Stat. 31	Agricultural Adjustment Act of 1933
1938	75–430, 52 Stat. 31	Agricultural Adjustment Act of 1938
1948	80–897, 62 Stat. 1247	Agricultural Act of 1948
1949	81–439, 63 Stat. 1051	Agricultural Act of 1949
1954	83–690	Agricultural Act of 1954
1956	84–540	Agriculture Act of 1956
1965	89–321	Food and Agricultural Act of 1965
1970	91–524	Agricultural Act of 1970
1973	93–86	Agricultural and Consumer Protection Act of 1973
1977	95–113	Food and Agriculture Act of 1977
1981	97–98	Agriculture and Food Act of 1981
1985	99–198	Food Security Act of 1985
1990	101–624	Food, Agriculture, Conservation, and Trade Act of 1990
1996	104–127	Federal Agriculture Improvement and Reform Act 1996
2002	107–171	Farm Security and Rural Investment Act of 2002
2008	110–246	Food, Conservation, and Energy Act of 2008
2014	113–79	Agricultural Act of 2014

[a] The full text of each bill is available at the National Agricultural Law Center, http://www.national-aglawcenter.org/farmbills, accessed September 13, 2014.

Congress (2012–2014). Due to the complexities of the omnibus legislation, not all programs under the Farm Bill are affected in the same manner by an expiration date. Some programs require reauthorization or they will cease to function. Other programs, such as nutrition assistance programs, may keep operating without reauthorization as long as appropriations are granted. Farm commodity programs, however, would expire and revert back to 1940s era permanent law, but other agricultural support programs, like crop insurance, have permanent authority and reauthorization is not required.[3]

Some provisions that expired in September, 2012 were extended, but without funding. Some programs have mandatory funding, but others do not, and could be authorized but not funded.

Mandatory funding means that their continuation is virtually guaranteed. On the other hand, programs with discretionary funding survive or fall at the hands of the Appropriations Committee. Although programs with discretionary funding may be authorized in the Farm Bill, authorization is separate from funding. Such programs are paid for separately in annual appropriations bills, and funding may not be appropriated.

In brief, this is how a new Farm Bill is drafted:

- Discussions about the Farm Bill begin in the House and Senate Agricultural Committees, where hearings are held to review current policies, introduce possible changes to the legislation, and determine funding strategies for any revisions. Generally, these discussions take place two years before the deadline for reauthorization. Simultaneously, representatives from Congress hold meetings with trade organizations, farm lobbyists, and civil society

TABLE 2.2   Farm Bill titles, 2002, 2008, and 2014

2002 FARM BILL		2008 FARM BILL		2014 FARM BILL	
TITLE NO.	TITLE	TITLE NO.	TITLE	TITLE NO.	TITLE
I	Commodity Programs	I	Commodity Programs	I	Commodities
II	Conservation	II	Conservation	II	Conservation
III	Trade	III	Trade	III	Trade
IV	Nutrition Programs	IV	Nutrition	IV	Nutrition
V	Credit	V	Credit	V	Credit
VI	Rural Development	VI	Rural Development	VI	Rural Development
VII	Research and Related Matters	VII	Research	VII	Research, Extension, and Related Matters
VIII	Forestry	VIII	Forestry	VIII	Forestry
IX	Energy	IX	Energy	IX	Energy
X	Miscellaneous (crop insurance, disaster aid, animal health and welfare, specialty crops [fruits and vegetables], organic, farmers markets, civil rights, etc.)	X	Horticulture and Organic Agriculture	X	Horticulture
		XI	Livestock	XI	Crop Insurance
		XII	Crop Insurance	XII	Miscellaneous (livestock, socially disadvantaged/ limited resource producers and more)
		XIII	Commodity Futures		
		IV	Miscellaneous		
		XV	Trade and Taxes		

organizations that have interests in the contents of the Farm Bill. Lawmakers also conduct regional and local town hall meetings to gather input about the Farm Bill from their constituents and stakeholders.
- While Congress holds its discussions, the White House and the USDA set out their own plans for revisions to the Farm Bill. Officials from each agency

conduct research, talk to key stakeholders, and generate proposals for legislation or policy briefings to submit to Congress when the bill comes to table. Congress has the final say on the content of the Farm Bill, but these proposals and briefings can play an important role in shaping congressional negotiations, and this is where lobbying comes in on the part of food activists.

• Drafting of the Farm Bill begins after the final round of hearings in the Agricultural Committees. A final version of the bill is produced by both Senate and House Agricultural Committees, and is sent to the House and Senate floors for the next-to-the last round of discussions. A joint version of the Farm Bill is produced, voted on by both houses of Congress, and sent to the president for signature to pass the bill into law. Once passed, the bill moves into appropriations—a process that determines how much money each Farm Bill program receives. See Box 2.1 Individuals and groups that contribute to the Farm Bill.

The current 2014 Farm Bill is the latest chapter in the legislation's history, which dates back to its origin during the Great Depression. The Congressional Budget Office (CBO) estimated at the time of enactment that the 2014 Farm Bill would incur $489 billion of mandatory spending from FY2014 to FY2018. Most of the money, 80% of the $489 billion, is earmarked for the Supplemental Nutrition Assistance Program (SNAP), 13% for farm commodity support and crop insurance, 6% for agricultural conservation, and the remaining 1% for all other Farm Bill titles (trade, horticulture, energy, rural development, research, and forestry).[3]

BOX 2.1   INDIVIDUALS AND GROUPS THAT CONTRIBUTE TO THE FARM BILL

Many individuals and organizations contribute to the Farm Bill, including members of government and special interest groups, which span a broad range of issues. For instance, advocacy organizations play an important role, such as the Center for Rural Affairs, the Environmental Working Group, and the Food Research and Action Center, among others.

Individuals, consumer groups, and professional organizations have a role to play in this process by contacting their Representatives or Senators and telling them what they want, such as a bill that rewards farmers for taking care of the land, putting fresh, healthy food in our schools and neighborhoods, helping young people get into farming, and restoring fairness in the marketplace.

Key industry groups that have a stake in the Farm Bill include the American Farm Bureau Federation, the National Corn Growers Association, and the International Dairy Foods Association. In 2013, agribusiness lobbyists spent $150 million lobbying Congress to get their points across.

OpenSecrets.org, Agribusiness. Available at http://www.opensecrets .org/industries/indus.php?Ind=A, accessed September 14, 2014.

To appreciate the roots of the Farm Bill, it is instructive to examine the economic climate of the United States and indeed the world in the 1930s when the first Farm Bill was written.

The Great Depression

The Great Depression was a severe worldwide economic depression in the decade preceding World War II. It was the longest, most widespread, and deepest depression of the twentieth century. In the 1920s, production surpluses aggravated unemployment and lack of consumer buying power. The Depression originated in the United States, after the fall in stock prices that began in September 1929, and became worldwide news with the stock market crash of October 29, 1929 (known as Black Tuesday). The timing of the Great Depression varied across nations, but in most countries it started in 1930 and lasted about 10 years. The Depression had devastating effects in countries rich and poor. Personal income, tax revenue, profits, and prices dropped, while international trade plunged by more than 50%. Unemployment in the United States rose to 25%, and in some countries reached 33%. People in the United States and internationally could not afford to buy the crops and animals produced by U.S. farms, and drought made it increasingly difficult for farmers to plant and harvest their crops. As a result, many farmers lost their farms, and many moved out of the agricultural Great Plains states (Colorado, Kansas, Montana, Nebraska, New Mexico, North Dakota, Oklahoma, South Dakota, Texas, and Wyoming) in search of any kind of work they could find. Many became migrant farm laborers on the West Coast. The impact of the Great Depression on the national psyche was severe, and is reflected in the music, writing, and art of the time (see Box 2.2).

The Farm Bill was originally developed as an emergency measure for farmers during the Great Depression. When Franklin D. Roosevelt (FDR) became president, U.S. farmers were producing prodigious amounts of corn. The glut on the market led to an increased supply that caused a decrease in demand, resulting in decreased profits for the farmers.

**BOX 2.2   AWARD-WINNING ARTISTS DEPICT THE
GREAT DEPRESSION AND THE DUST BOWL**

The Pulitzer Prize-winning novel, *The Grapes of Wrath* (1939), by Nobel Prize winner John Steinbeck (1902–1968), depicts a family of share-croppers who were driven from their land due to the dust storms of the Dust Bowl during the Depression. Arguably the most influential folk musician of the first half of the twentieth century, who was known for his ballads of the Dust Bowl and the Great Depression, Woody Guthrie (1912–1967) was an unemployed worker who left Oklahoma for California in 1935. Recently, Emmy Award-winning film maker Kenneth "Ken" Burns (1953–) directed a 3-hour documentary *The Dust Bowl* (2012) that recounts the impact of the Dust Bowl on the United States during the Great Depression.

Supply and demand

At this point, it is appropriate to summarize the law of supply and demand:

1. When demand increases and supply remains unchanged, a shortage occurs, leading to a higher price.
2. When demand decreases and supply remains unchanged, a surplus occurs, leading to a lower price.
3. When demand remains unchanged and supply increases, a surplus occurs, leading to a lower price.
4. When demand remains unchanged and supply decreases, a shortage occurs, leading to a higher price.

Thus, the third law of supply and demand explains why farmers who produced enormous crops suffered financially. To remedy the situation, FDR enacted a series of economic programs known as "The New Deal." Box 2.3 lists some New Deal programs.

The Agricultural Adjustment Act of 1933, regarded as the first Farm Bill, was one of the first pieces of "New Deal" legislation. The Agricultural Adjustment Act of 1933 restricted agricultural production so commodity prices would rise (the fourth law of supply and demand). Farmers were paid a subsidy to restrict their output. This subsidy was paid to farmers through a tax levied exclusively on companies that processed farm products.

The U.S. Supreme Court ruled the act unconstitutional in 1936 because the tax structure used to fund the program was deemed an over-reach of the powers of the federal government. Congress promptly replaced it with the Soil Conservation and Domestic Allotment Act, which encouraged conservation by paying benefits for planting soil-building crops such as alfalfa that would not be sold on the market, instead of such staple crops as corn and wheat. On signing the Soil Conservation and Domestic Allotment Act in 1936, FDR said "The new law has three major objectives which are inseparably and of necessity linked with the national welfare. The first of these aims is conservation of the soil itself through wise and proper land use. The second purpose is the reestablishment and maintenance of farm income at fair levels so that the great gains made by agriculture in the past three years can be preserved and national recovery can continue. The third major objective is the protection of consumers by assuring adequate supplies of food and fiber now and in the future."[5]

BOX 2.3   FDR's "NEW DEAL" SPAWNED AN ALPHABET SOUP OF ACRONYMS

AAA	Agricultural Adjustment Administration
CCC	Civilian Conservation Corps
CWA	Civil Works Administration
FDIC	Federal Deposit Insurance Corporation
FSCC	Food Surplus Commodities Program
FSP	Food Stamp Program (experimental)
TVA	Tennessee Valley Authority
WPA	Works Progress Administration

Two key initiatives arose from the first Agricultural Act, namely the Agricultural Adjustment Administration (AAA) and the Food Surplus Commodities Corporation (FSCC). The AAA implemented a "domestic allotment" plan that subsidized producers of basic commodities for cutting their output. Its goal was the restoration of prices paid to farmers for their goods to a level equal in purchasing power to that of 1909–1914, which was a period of relative prosperity.

Previously known as the Federal Surplus Relief Corporation, the Food Surplus Commodities Corporation (FSCC) was established in 1935 with AAA funding to re-establish farmers' purchasing power by supporting agricultural exports and encouraging domestic consumption of surpluses.

Five years later, the Agricultural Adjustment Act of 1938 instituted the recurring 5-year Farm Bill concept, and created price supports for corn, cotton, and wheat, later expanded to include soybeans, other grains, and dairy. The 1938 Act empowered the AAA in years of plenty to make loans to farmers on staple crop yields and to store the surplus produce, which could then be tapped in years of low yield. Largely excluded were fruits and vegetables, as well as livestock. Farmers' cash income doubled between 1932 and 1936, but even so, it took the enormous demands of World War II to reduce the accumulated farm surpluses and to increase farm income significantly.

Reduction of Agricultural Excess: Commodity Distribution and Trade

Commodity distribution: Food stamps

From 1939 to 1943, an experimental Food Stamp Program was in place as an innovative way to respond to severe hunger caused by the Depression while simultaneously supporting farmers by buying their excess crops. The project, called the "Food Stamp Plan," preceded the current major food assistance program (which began again as Food Stamps in the 1960s, until 2008 when the program was retitled SNAP). Residents who received public financial assistance (aka "welfare") purchased booklets of orange colored stamps at a value equal to what they normally spent on food each month, or at least $4 per month per family member. Orange stamps could be used to purchase "any food usually sold in a grocery store," as well as "household articles usually bought in grocery stores, such as starch, soap, matches, etc.," but could not purchase alcohol, tobacco, or food eaten at stores. (Today, SNAP may not be used to purchase nonfood items.) For every $1 in orange stamps purchased, 50 additional cents of blue stamps were received, at no cost. Blue stamps were used for purchasing commodity surplus foods. Grocers posted lists of the allowed surplus foods, adding fresh produce as it came into season.[6] Thus, $1 would purchase a dollar's worth of any food plus another 50 cents' worth of commodities. Figure 2.1 shows the food stamps in 25 cent denominations.

In other words, government subsidized food purchased by those in need, as long as those purchases helped to reduce the nation's agricultural surplus. From the inception of the food stamp program, government had a say in what could be purchased, and because surpluses change with the natural cycle of the seasons, the bonus commodity list was updated monthly. In general, commodity products included dairy products,

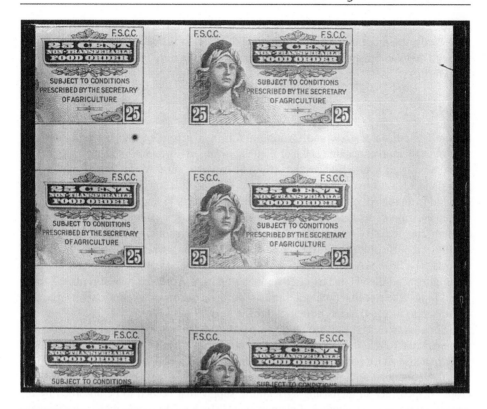

FIGURE 2.1 1939 surplus food stamps in 25 cent denominations. "First food stamp" photograph. (From Harris and Ewing, Library of Congress Prints and Photographs Division, Washington, DC, reproduction number: LC-DIG-hec-26518 (digital file from original negative), http://www.loc.gov/pictures/item/hec2009013216/, accessed May 5, 2015.)

eggs, flour, and seasonal fruits and vegetables. Thus the food stamp program helped alleviate hunger and raise farming income. *Today, SNAP helps alleviate hunger. Does it help raise farming income?*

Some nutrition assistance programs still specifically help to reduce agricultural surplus. Commodity donation programs that supported the post-Depression farm economy were precursors to the National School Lunch Program.[7] The Emergency Food Assistance Program (TEFAP), Commodity Supplemental Food Program (CSFP), and the Food Distribution Program on Indian Reservations (FDPIR) are nutrition assistance programs that still benefit from USDA commodity foods as well as USDA's donation of bonus commodities, which USDA purchases based on agricultural producers' identification of surplus goods or need for price support. Federal nutrition assistance programs are discussed in further detail in Chapter 6, "Federal Food Assistance."

As part of the 2008 Farm Bill, the name of the Food Stamp program was changed to Supplemental Nutrition Assistance Program (SNAP). SNAP represents the program's focus on nutrition and putting healthy food within reach for low-income consumers.[6] Today, the former "food stamps" have been replaced with an electronic benefits transfer system (EBT) that allows state welfare departments to issue benefits via a magnetically encoded payment card. Once farmers' markets were allowed to accept purchases with EBT cards, the program returned full circle to FDR's original vision of matching supply with demand by allowing struggling families to purchase fresh, healthy foods from local farmers.

Trade

As the United States grew and farming and infrastructure expanded west, supplying the domestic market became cheaper than exporting to Europe. Nonetheless, American farmers enjoyed a small, but steady, flow of exports, until World War II made shipping dangerous. Convoy protection from the U.S. Navy allowed trade to continue and demand from the Allies grew. When the war ended in 1945, the United States provided monetary support to help rebuild European economies by removing trade barriers, modernizing industry, and making Europe prosperous again through the Marshall Plan (1948–1952, officially, the European Recovery Program), which continued the flow of exports. The flow of American food, made inexpensive through the Marshall Plan, may have influenced dietary habits elsewhere as imported American wheat replaced traditional foods like rice and potatoes,[8] and encouraged continued export markets. Exports continued to increase, remaining high for the next 40 years.

Since American farm productivity paralleled the increased demand, American farmers came to rely on exports to prop up prices and farm incomes.[9] U.S. agricultural trade increases have paralleled gross farm income.[10]

The continued and ever-expanding subsidies for corn, soy, and other "commodity crops" helped fuel an increasing agricultural supply that needed to be absorbed by the market, giving rise to a number of phenomena: (1) Increased exports, which disadvantage farmers in countries that cannot produce their own crops as cheaply. (2) Corn is fermented into ethanol (the residual grains are fed to animals). (3) Most corn grown by American farmers is not eaten by people, but rather is fed to animals in feedlots and livestock warehouses, thus effectively lowering the cost of animal protein in the United States. (4) Corn is used to produce sweeteners, such as high fructose corn syrup (HFCS), which has primarily replaced sucrose (table sugar) in the U.S. food industry. In addition to corn subsidies, other factors that have resulted in the proliferation and ubiquity of HFCS include governmental production quotas on domestic sugar and a tariff on imported sugar,[11] which combine to raise the price of sucrose to levels above those of the rest of the world, thus making HFCS cheaper for many sweetener applications (see also Chapter 9, "Grains"). In sum, we have become a nation with a preference for meats and items containing HFCS because these food products are comparatively cheap. Corn subsidies have encouraged the popularity of hamburgers and soft drinks in the United States.

THE AGRICULTURAL SAFETY NET

Do federal agricultural subsidies and insurance continue to provide a safety net for farmers and agriculture? Have the original goals of these food safety net programs become so subverted that they are wasting the taxpayers' money and diverting funds from social-responsible initiatives while endangering small- and medium-sized farming and sabotaging our health? In the next section, we attempt to answer these questions by examining commodities, subsidies, crop insurance, and trends of consolidation in agribusiness.

Commodities, Subsidies, and Crop Insurance

What is referred to as *commodity food crops* includes wheat, feed grains (grain used as fodder, such as maize or corn, sorghum, barley, and oats), milk, rice, peanuts, sugar, and oilseeds such as soybeans. An agricultural subsidy is governmental assistance paid to farmers and agribusinesses to supplement their income, manage the supply of agricultural commodities, and influence the cost and supply of such commodities. In general, the 2014 Farm Bill continues the transition from traditional farm price and income support to risk management; that is, crop insurance is the primary support for farmers instead of payment programs. The federal government heavily subsidizes crop insurance, thus crop subsidies and crop insurance are referred to as *farm safety net programs*.[12,13]

Here's how the system works. Crop insurance is purchased by agricultural producers, including farmers, ranchers, and others to protect themselves against the loss of their crops due to natural disasters such as hail, drought, floods, or the loss of revenue due to declines in the prices of agricultural commodities. Crop insurance was authorized in 1938 as part of the first Farm Bill to help agriculture recover from the combined effects of the Great Depression and the Dust Bowl. Four crops—corn, cotton, soybeans, and wheat—now account for three-quarters of total acres enrolled in crop insurance. But the program is not limited to crops. Relatively new or pilot programs protect livestock and dairy producers from losses due to price declines. The government pays, on average, 60% of the premium for coverage and the farmer pays the rest. Over the past decade, crop insurance has become *the most important* crop subsidy program in the United States. In 2013, crop insurance accounted for about three-fifths of all budgeted outlays for farm subsidies.

The government programs are covering much more than pure crop failure due to disaster. If the crop insurance is too good, farmers are incentivized to take risks they probably should not. Also, some question why we are funding insurance agents to induce farmers to buy heavily subsidized crop insurance, instead of consolidating all government programs. It is beyond the scope of this chapter to examine the details of crop insurance in full; however, interested readers will want to pursue for themselves answers to these questions: *What is the average payout compared to the premium? Is the risk of the insurance borne by the taxpayer? How does the public benefit from taxpayer risk management subsidies? Has the transition to crop insurance as a solution been influenced by insurance lobbying?*[14]

In total, insurance policies are available for more than 100 crops (including coverage on a variety of fruit trees, nursery crops, pasture, rangeland, and forage). Many specialty crop producers depend on crop insurance as the only "safety net" for their operation, unlike field crop producers, who are also eligible for farm commodity program payments. Crop insurance covers about 75% of total area for selected specialty crops.

While farmers who are insured may be more inclined to embrace some sustainable forms of agriculture, such as using fewer chemicals against pests, government-subsidized insurance has also encouraged corn farmers to take unforeseen risks. Because crop insurance reduces financial risk, insured farmers may engage in risky practices, such as failing to rotate the kinds of crops planted or planting on marginally productive land that could be conserved or used for grazing. Indeed, heavily subsidized agricultural operations are associated with plowing up environmentally sensitive wetlands and grasslands. For example, more than 23 million acres of grass and wetlands were plowed for cash crops like corn and soybean from 2008 through 2011; the land losses were greatest where crop insurance subsidies were largest.[15]

In addition to threatening wildlife and the environment, these subsidies also take a heavy toll on American taxpayers. A decade ago, USDA paid on average 30% of the cost of crop insurance premiums. In 2000, to encourage more farmers to participate in crop insurance, Congress dramatically increased the premium payments. The payments shot up from less than $2 billion in 2001 to $7.4 billion in 2011. Today, USDA pays, on average, 62% of farmers' premium subsidies and $1.3 billion to insurance companies and agents. The cost of this program is soon expected to be $9 billion.[15]

Economists and others who champion sustainable agriculture contend that the existing farm safety net programs should be reformed. The lobbying group, American Farmland Trust,[16] is the only national conservation organization dedicated to protecting farmland, promoting sound farming practices, and keeping farmers on the land. The Trust argues that a modern farm safety net needs crop insurance and crop subsidies that are *complementary rather than duplicative*, which would save taxpayers' significant amounts of money. Further, they believe that farmers should only receive assistance if unavoidable losses are suffered, which would do away with subsidies for not planting.[16]

The Environmental Working Group (EWG)[17] is a nonprofit 501(3)(c) organization that publishes a database of agricultural subsidies and their recipients. Their lobbying arm, the EWG Action Fund, advocates for Farm Bill reform in the form of decreased disaster payments and subsidies for commodity crops, and increased funding for nutrition programs, conservation, specialty crops (i.e., fruits and vegetables), and "organic" agriculture. EWG points out that the $9 billion a year noted above overwhelmingly flows to the largest and financially lucrative farm businesses. Unlike other farm subsidies, crop insurance subsidies are not subject to means testing or payment limits and farmers are not required to adopt basic environmental protections in exchange for premium support from the taxpayer. There are no limits on the size of insurance subsidies, and the USDA is prohibited from disclosing which growers receive them—whether billionaires or members of Congress. The EWG advocates making conservation compliance a condition of receiving crop insurance subsidies, placing limits on who can receive subsidies and how much they can receive, and rejecting cuts to voluntary conservation programs.[17]

Impact of Subsidies and Crop Insurance

Artificially low prices may result from commodities that are subsidized. Since the 1970s when Earl Butz was Secretary of Agriculture (1971–1976), U.S. agriculture has operated under a "cheap food" policy that was spurred by production of a few commodity

BOX 2.4   EARL BUTZ, THE 18TH SECRETARY OF AGRICULTURE

In an interview in *King Corn* (2007), Earl Butz said that prior to his being Secretary of Agriculture (1971–1976), farmers were paid not to produce, "...one of the stupidest things we ever did, I think." The policy shifts enacted under Butz came about when oil shortages and inflation pushed up food prices—provoking widespread hunger abroad, and in the United States, the rise of major agribusiness corporations and the declining financial stability of small family farms. His policies have been associated with starting the rise of corn production, large commercial farms, and the abundance of corn in American diets. In the *King Corn* interview, Butz claims that the increased agricultural production in the United States consequent to his policies has driven food prices down now so that we only pay about 17% of our incomes on food, whereas previous generations paid twice that amount. Consequently, we now live in an "age of plenty." He acknowledged that his policies led to large-scale farming, created by farmers who took out huge bank loans in response to his repeated exhortation to "get big or get out." That practice was responsible for thousands of smaller farms foreclosing during the farm financial crisis of the 1980s.

King Corn, *directed by Aaron Woolf.*
Distributed by Balcony Releasing, 2007.

crops. What has changed since "cheap food" agricultural policies were enacted is that obesity has replaced hunger as the more prevalent nutritional problem. Box 2.4 has more details on Earl Butz.

Because government subsidizes the production of corn, soybeans, and other commodity crops, farmers are incentivized to grow prodigious quantities of these commodities. We do not directly consume much of these commodity crops; only about 1% of U.S.-produced corn is for direct human consumption. The vast majority is turned into biofuels, fed to livestock, or processed into ingredients used in industrial food processing, such as HFCS or hydrogenated vegetable oils.[18] Subsidies have resulted in market distortions, leading to the following means of using up the overabundance of commodity corn and soybeans.

- Corn processed into HFCS has since the 1970s become a ubiquitous ingredient in American food products due to its low cost, ease of production, and widespread availability. High fructose corn syrup is exactly "as described on the tin"; it is corn syrup that is higher in fructose than regular corn syrup. Regular corn syrup consists almost entirely of glucose. Glucose, fructose, and sucrose, among others, are types of sugars. Table sugar is sucrose, which is made up of a molecule of glucose bonded to a molecule of fructose (in other words, half glucose and half fructose). The fructose, which is almost twice as sweet as glucose, is what distinguishes sugar from other carbohydrate-rich foods such as bread or potatoes that break down upon digestion to glucose alone. The more fructose in a substance, the sweeter it will be. High-fructose corn syrup is made by converting a portion of the glucose found in corn syrup to fructose through enzymatic isomerization, most commonly in a ratio that turns the syrup into 55% fructose and 45% glucose.

- *Health consequences*: Although increased fructose consumption in many developed nations coincides with the increased prevalence of obesity, this association does not prove causation. However, fructose metabolism in the liver does favor lipogenesis (fat production), and a number of studies have shown that fructose consumption induces hyperlipidemia (elevated blood levels of fat) and particularly increases the kind of fat known as triglycerides after meals. These effects may be pronounced in people with a condition known as *metabolic syndrome* (existing hyperlipidemia and insulin resistance).[19]

- Soybeans ground up into meal, with the meal going to feed cows and the liquid skimmed off and turned into fat-based additives like trans fatty acid-rich partially hydrogenated vegetable oils.

- *Health consequences*: Trans fatty acids have been shown to be atherogenic—they increase the risk of coronary heart disease by raising levels of low-density lipoproteins and lowering levels of high-density lipoproteins. Starting in 2007, New York City banned *trans fats* in restaurants, bakeries, and other food-service establishments, the first major city to institute such legislation. Within 5 years, the average *trans fat* content of fast-food restaurant meals dropped from about 3.0 g to 0.5 g.[20]

- Corn fed to cattle confined in concentrated animal feeding operations (CAFOs), rather than allowing the cattle to graze at pasture. In this industrial system of meat production, meat animals are raised where their mobility is restricted. They are fed a high-calorie, grain-based diet, often supplemented with antibiotics and hormones.

- *Health consequences*: The meat from grain-fed cattle has a less healthful nutritional profile; meat from grass-fed cattle is lower in overall fat, which may reduce the risk of some types of cancer, and contains higher levels of certain omega-3 fatty acids. Meat and milk from grass-fed animals tends to have, in particular, alpha-linolenic acid (ALA) and conjugated linoleic acid (CLA). The presence of ALA in the pasture plants leads to the development of other potentially healthful fatty acids in the cow's body such as EPA (eicosapentaenoic acid) and DHA (docosahexaenoic acid). The evidence supporting the health benefits are mixed; the data are stronger for some fatty acids than for others. The strongest evidence, encompassing animal studies as well as experimental and observational studies of humans, supports the effects of EPA/DHA on reducing the risk of heart disease. ALA also appears to reduce the risk of fatal and acute heart attacks. Finally, animal research on CLA has shown many positive effects on heart disease, cancer, and the immune system, but these results have yet to be duplicated in human studies.[21]

The obesity epidemic in the United States has many causes, among them snacking on foods with little or no redeeming nutritional value and eating at fast food restaurants at a frequency and quantity that makes the non-nutritious calories insidious in our diet. On average, one-quarter of adults' total daily calories are consumed at snacking occasions. However, for some adults, snacks provide an ever-greater proportion of daily calorie intake. For nearly 1 in 6 adults (16%), over 40% of their total daily calories are obtained from foods and beverages they report as being consumed as snacks.[22]

Frequent fast food consumption has been shown to contribute to weight gain. Data from the 2007–2010 National Health and Nutrition Examination Survey (NHANES) indicate that fast food accounts for an average of about 11.3% of adult Americans' calorie intake, which is a decrease from 12.8% over 2003 to 2006.[23] More than one-quarter (28%) of adults have fast food two or more times a week,[24] and 40% of high school students consume fast food on any given day.[25]

What snack foods and fast foods have in common are the raw ingredients—beef, corn, soy, and wheat—crops that are heavily subsidized by the U.S. government. Thus, cheap ingredients depress the cost of foods made from corn and soy, leading to a glut of nutritionally poor but extremely palatable low-priced food available for us to eat.

Here's Michael Pollan's succinct description of the loop: "When yields rise, the market is flooded with grain, and its price collapses. As a result, there is a surfeit of cheap calories that clever marketers sooner or later will figure out a way to induce us to consume.... Cheap corn, the dubious legacy of Earl Butz, is truly the building block of the 'fast-food nation....' The political challenge now is to rewrite those rules, to develop a new set of agricultural policies that don't subsidize overproduction—and overeating."[25]

Writing the Rules

The USDA is not inclined to fund research critical of its own activities. Since most agricultural policy research is funded by the USDA, there is little incentive for government to interfere with income-producing regulations.[26]

Furthermore, in recent years, farm subsidies have remained high even in times of record farm profits. Even with the 2012 drought and other climate concerns, American farmers have recently had unparalleled success. U.S. agricultural exports exceeded $136 billion in 2011 and economists anticipate that exports will continue to grow at above historical averages over the next decade. Both agricultural exports and net farm income are at record levels, while farm debt has been cut in half since the 1980s.[27] In 2012, the overall value of both crop and livestock agricultural sales was the highest ever recorded, with crop sales exceeding livestock sales for only the second time, the other being 1974.[28] *Why does agricultural policy continue to support the market distortions commented on earlier when farming income is high?*

Agricultural policy has favored farmers over the course of U.S. history because, in part, farmers enjoy favorable proportional political representation in Congress. The Senate grants more power per person to inhabitants of rural states, because there are two Senators per state, no matter how small the population of that state is. Also, the 435-member House of Representatives is reapportioned only once every 10 years (following the Constitutionally mandated decennial census) and farmers are often left with greater proportional power until the next reapportionment because population tends to shift from rural to urban areas. The U.S. Census reports a decrease in rural population and increase in urban population decennially from 1790 to 2010, from 5.1% of the total population to 80.7%. The migration of labor geographically, out of rural areas, and occupationally, out of farm jobs, is one of the most pervasive features of the agricultural transformation and economic growth from small farms to agribusiness.

The Agriculture Establishment

For 150 years, food producers have worked intensely, if not intimately, with the House of Representatives and Senate Agricultural Committees and the USDA to develop what is informally known as the *agriculture establishment*. This establishment has united in a way to secure federal policies and legislation related to land use and food distribution that support the food industry. This has been done in several ways. First, many members of the Congressional Agricultural Committees are from the Great Plains farm states. Second, committee members serve long terms, sometimes decades. Third, committee members are sometimes replaced with former food industry lobbyists and, conversely, some committee members become lobbyists themselves. As Marion Nestle says, "Today's public servant is tomorrow's lobbyist." The Congressional Agricultural Committees are important targets for the food industry lobby because it is legal to lobby for committee members, whereas it is not permitted for USDA members to interact with lobbyists.[29]

Food Industry Lobbying

Lobbying is a practice whereby companies hire people to influence political action (i.e., policies and legislation). In general, lobbyists offer expertise, make campaign contributions, provide perquisite (perks) such as lavish meals, and use other methods of persuasion to influence politicians to pass laws that support the companies that hire them. Note that lobbying is not technically or legally considered a form of bribery, although at times, there seems to be a fine line between the two practices. Lobbying refers to the act of trying to influence members of a legislative body to vote in favor of the lobbyist, whereas acts of bribery are defined as cases where cash or property is offered in exchange for a specific influence that tips in the briber's favor. Almost all major U.S. companies have some sort of active lobby, and the corporations within the food industry are no exception. As agribusiness, the interests of agricultural corporations remain highly represented; agribusiness spent $151.8 million on lobbying in 2013.[30]

Examples of money paid to lobby by different types of companies within the food industry are presented in Table 2.3.

Who Benefits?

Do all farms and farmers benefit equally from Farm Bill support? Traditionally, Farm Bills have not provided livestock and poultry producers with price and income support programs

TABLE 2.3 Lobbying expenditures for selected agribusiness companies: 1998–2010

COMPANY	TYPE OF BUSINESS	AMOUNT
Walmart	Retailer	$33,245,000
Kraft Foods	Food company	$13,155,000
Smithfield Foods	Food producer/processor	$9,635,000
Archer Daniels Midland	Food processor	$6,270,000
Tyson Foods	Food producer	$13,876,285
Dole Food Company	Food producer/distributor	$3,375,000
Monsanto	Seed producer	$46,522,720

Source: Adapted from Food & Water Watch, *Farm Bill 101*, January 2012. Available at http://www.foodandaterwatch.org/tools-and-resources/farm-bill-101, accessed May 4, 2014.

like those for the commodity crops, such as grains and oilseeds. Instead, the livestock and poultry industries have looked to government for leadership in such areas as reassuring buyers that their products are safe, of high quality, and free from pests and diseases.

The 2008 Farm Bill was the first to include a title that specifically covered livestock and poultry issues (Title XI), but livestock has been relegated back to a "miscellaneous" title in the 2014 Farm Bill. Some groups favor more government oversight of the industry in order to address competition issues. Other groups, such as the National Cattleman's Beef Association, oppose separate livestock titles in order to minimize government's oversight of their industry.[31] There are other areas of the Farm Bill that concern livestock and poultry producers. For example, government authorizes conservation programs that take croplands out of production, and government also provides incentives to shift corn to fuel use, both of which increase the price of corn as a feed ingredient (the first law of supply and demand). Other Farm Bill "conservation" programs, such as the Environmental Quality Incentives Program (EQIP), support building the actual infrastructure for CAFOs to process their waste.[32]

Overall, the historic Farm Bill policy mechanisms of price and income support have traditionally heavily favored commodity crops rather than livestock. The largest farms reap more profit from subsidies. Small farms do survive on the subsidies. According to *Food Fight: The Citizens Guide to the Next Food and Farm Bill*,[33] in 2003 small farms earned an average net income of $30,000 a year from farming, more than half of which came from subsidy payments. Furthermore, in addition to farm income and subsidies, most of these small farm households supplemented their income from off-farm employment. Without subsidies they would perish. Conversely, the megafarms turn a profit from subsidies. These large commercial farms earn over $250,000 per year, and are flourishing.

U.S. agriculture production has continued to become increasingly concentrated. Just 4% of all U.S. farms, namely those with more than $1 million in sales, produced two-thirds of the total value of U.S. agricultural products in 2012.[34] Five years prior, in 2007, the top 4% produced just under 60%, and 10 years prior, in 2002, the top 4% produced under 50% of the total value of U.S. agricultural products.[29] Farms with more than $5 million in sales make up less than one-half of 1% of the total number of farms in the United States, but produce almost one-third of agricultural products. *How did some farms become so large? Why are they able to profit from a subsidy system that was originally designed as a safety net, not a cash cow?*

Consolidation of Farms

Conducted every 5 years by USDA's National Agricultural Statistics Service (NASS), the *Census of Agriculture* is a complete count of the U.S. farms and ranches and the people who operate them. The Census looks at land use and ownership, operator characteristics, production practices, income and expenditures, and other topics. It provides the only source of uniform, comprehensive agricultural data for every county in the nation.

NASS defines a farm as any place from which $1000 or more of agricultural products were, or normally would be, produced and sold during the Census year. Although the details from the 2012 census have not been fully released at the time of this writing, the 2007 Census tells a story about the increasing concentration of wealth in U.S. agriculture.[35] In 1935, there were 6.8 million farms in the United States with an average size of 155 acres. In 2008 there were a third as many farms (2.2 million), which were on average almost three times larger (418 acres); 144,000 farms produced three-quarters of the value of U.S. agricultural output in 2002. By 2007, just 5 years later, it took only 125,000 farms to produce the same amount.

The percentage of Americans who live on a farm diminished from nearly 25% during the Great Depression to about 2% now. Of that population, less than 1% claim farming as an occupation, and even less work full-time on the farm.[35]

With the loss of families operating farms as their main source of income, and with the advent of technology that replaces farm workers and is hugely productive but costly, and the growth and consolidation of farms to provide efficiency, the farm sector has been converted to a manufacturing model predicated on providing its consumers with the lowest cost product. That low cost can only be guaranteed by *economies of scale*—the cost advantages that an enterprise obtains due to the size of the operation, allowing a producer's average cost per unit to fall as the scale of output is increased. The expensive technology costs are amortized when spread over a maximum number of acres or animals. Thus, big farms reap more profits per acre than small ones. Additionally, these factory farms are structured as complex family corporations configured to circumvent subsidy laws that limit how much a single farming operation can receive.

When it comes to livestock, animal feed is the livestock and poultry producers' single largest input (expense). In the past, farms raised both livestock and the grain to feed them. Small farms may still do so, but again, given the manufacturing model, a larger livestock producer is not likely to do so. Because grain is heavily government-subsidized, livestock producers do not need to grow their own grain to contain costs. The cheap inputs of low-cost feed (and other external costs discussed in Chapter 8, "Meat") keeps meat and chicken prices down, and is in part responsible for the current low meat costs we enjoy as consumers.

Currently, corn is produced through a system of farming known as *monoculture*, an unsustainable agricultural practice that maximizes yield by concentrating on a single crop. Intensive monoculture depletes soil and leaves it vulnerable to erosion. Herbicides and insecticides that harm wildlife can pose human health risks as well. Biodiversity in and near monoculture fields is negatively impacted, as populations of birds and beneficial insects decline.[36] See Box 2.5 for more on bees.

BOX 2.5 BEES

Bees and other pollinators provide an indisputable value to our food system; the estimated value in the United States alone is $16 billion annually. Many of the crops produced in the United States rely on managed and wild bee pollination, and much of these are fruit and vegetables.

In 2006, commercial beekeepers began reporting severe losses of bees in their colonies—a phenomenon described by scientists as colony collapse disorder (CCD), which is characterized by colony populations that are suddenly lost and an absence of dead bees. Honeybee decline has continued since 2006, although not solely attributable to CCD. The USDA has estimated that the loss of bees over the winter has averaged 30% annually since 2006. Congress increased funding for bee research in the 2008 Farm Bill and again in the 2014 Farm Bill, reauthorizing and expanding "high-priority research and extension initiatives" at USDA. Research efforts include an annual winter survey loss for honeybees and a database of honeybee health.

There are thought to be numerous possible factors influencing bee loss, including parasites, pathogens, diseases, habitat loss, interspecies conflict, pesticides (singularly and as a cumulative effective), crops that contain pesticides due to bioengineering, and bee genetics, as a lack of diversity can result in increased susceptibility and lowered disease resistance.

For commercial bees, bee management issues may be contributory. Some of these issues are evocative of concerns familiar from industrial animal husbandry such as overcrowding, feeding practices, antibiotics and miticides used by beekeepers, confinement and temperature fluctuations, susceptibility to disease, and a diet that is of low nutritional value to the bees (based on the crops the bees are pollinating or the supplements provided by the keepers). In general, when it comes to pollen, honeybees are "omnivorous," collecting from diverse plants, and a colony used for monoculture agricultural pollination may not acquire all needed nutrients. Bees may also feed on plants that are toxic, plants that have been treating with toxic pesticides, and some sugars are toxic. A diversity of plants, even near agricultural areas, is optimal.

In 2013, USDA and the U.S. Environmental Protection Agency (EPA) published a joint report "National Stakeholders Conference on Honey Bee Health." One of the EPA's key findings is a need for *improved nutrition for honeybees*, concluding that federal and state programs should consider land management strategies that maximize available nutritional forage and to protect bees by keeping them away from pesticide-treated fields. The 2014 Farm Bill directs the USDA to consult with the Department of the Interior and EPA about recommendations related to allowing for managed honeybees to forage on National Forest System (NFS) lands.

In February 2014, USDA's Natural Resources Conservation Service (NRCS) announced that it would provide nearly $3 million to help implement conservation practices that provide bees with forage areas while providing other environmental benefits. (See "Bee Informed" about Bee Health and the USDA research efforts at: http://beeinformed.org/about/bip-database/.)

> *Johnson, R., and Corn, M.L., Bee health: Background and issues for Congress, April 9, 2014, Congressional Research Service. Available at http://fas.org/sgp/crs/misc/R43191.pdf, accessed October 26, 2014; Brodschneider, R., and Crailsheim, K., Nutrition and health in honey bees, Apidologie, 2010;41(3):278–294.*

Agribusiness: Consolidation of Livestock, Processors, and Food Production

Ownership and control of the swine industry has become particularly concentrated. Fifty years ago, nearly three million farms in the United States raised hogs. By 2003, only about 3% of that number, or 85,000 farms, raised the same number of hogs. The number of hogs per farm has increased dramatically. In addition, ownership and control are concentrated to specific regions of the country, with Iowa and North Carolina containing the greatest proportion of CAFOs for hog production. Although hog confinement operations began as early as the 1970s in Iowa, the greatest growth was in the mid-1990s. In 1980, Iowa's 80,000 farms averaged 250 hogs per farm. By 2002, there were only 10,000 farms, averaging 1500 hogs per farm. The number of hogs per farm increased more rapidly in Iowa than in the nation as a whole and by 2002 was nearly double the national average.[37] In addition to the inhumane treatment of animals and reliance on monoculture to produce animal feed, there are other consequences of CAFOs, such as antibiotic resistance in humans, negative impacts on the environment, and respiratory problems in workers and nearby residents. The human and environmental impact of CAFOs is examined further in Chapter 8.

In 2000, just four companies, Tyson, Cargill, JBS Swift, and National Beef, were responsible for 80% of the country's beef production. In fifth place was Smithfield Foods, which was then also the top pork producer in the country and, together with Tyson, ConAgra, and Cargill, produced 60% of the pork in the United States at the time.[38]

Concentration of the livestock industry is a global trend. In 2013, Smithfield Foods was purchased by the WH Group, a Chinese company formerly known as Shuanghui International. The WH Group is now the largest pork company in the world, as it owns Smithfield Foods, is a majority shareholder in China's largest meat processing business, and has a 37% shareholding in Campofrio Food Group, the largest pan-European packaged meat products companies.[39]

Archer Daniels Midland (ADM) is the largest corn producer and processor in the world and the leader in manufacturing HFCS, with sales in 2009 reaching $61.7 billion. ADM also processes oilseeds (soybeans, canola, sunflower seeds, and cotton seeds) in North America, Europe, South America, and Asia. ADM mills wheat, and grinds 16% of the world's total cocoa crop, invented textured vegetable protein, and produces ethanol, among many other products and endeavors. Generally, ADM processing plants turn crops into products for human food, animal food, and industrial and energy uses; with multiple products from a single crop, ADM is "extract (ing) the maximum possible value from each seed, kernel or grain."[40]

Most food is manufactured by a small number of companies. The top 10 producers are Associated British Foods, Coca-Cola, General Mills, Kellogg, Mars, Mondelez International (formerly Kraft Foods), Nestlé, PepsiCo, and Unilever. Collectively, these companies generate revenue of more than $1 billion a day. Fewer than 500 companies control 70% of the food choices for 7 billion people on the planet.[41] The companies use a number of brand names to sell their products. Food is a big business, globally, and these businesses reach deeply into our lives. See Chapter 5, "Activism," for a discussion on the Nestlé boycott.

Agribusiness: Food Retailers

Along with consolidation of farms and ranches that produce the foods that enter the U.S. food chain, there is also consolidation of retail food stores that serve as the middlemen between producers (farmers and ranchers), manufacturers, and the consumer. The same law applies to retailers as to producers and manufacturers—economy of scale. The U.S. food retail sector is big business and similar business practices that have evolved in other parts of the food industry are also taking place in the food retail sector, such as the concentration of retailers.

Retailers such as Walmart and Costco have become increasingly significant nontraditional sources of retail food sales. Walmart became the largest U.S. food retailer in 1988, just 12 years after opening its first supercenter that sold food products.[42] The share of food sales going to traditional food retailers fell from 82% in 1998 to 69% in 2003.[43] In response, traditional retailers consolidated through mergers and acquisition to capture their own economies of scale during the late 1990s. Sales by the 20 largest food retailers in the United States totaled $420.8 billion in 2012, constituting 62% of all U.S. grocery sales, a significant increase from 39.2% in 1992. Walmart remains the largest food retailer in the United States, with an estimated $113.2 billion in grocery sales. Kroger, the largest traditional grocery retailer, had sales of $73.9 billion in 2012.[44]

The proliferation of these stores appears oligopolistic. When there are only a few sellers, we call the structure an oligopoly and the sellers are oligopolists, from the Greek root *oligo* meaning "a few." If there were only one seller, the structure would be a monopoly, from the root *mono*. The consumer may see the mergers as opportunities for cheap food as these consolidated food retailers may have the wherewithal to sell products at lower prices. But the trade-off in allowing a small number of corporations to control a large proportion of the food retail market is an environment in which a small number of large stores now control not just prices, but also selection, quality, and the locations of where food can be purchased. Food retailers such as Walmart and Whole Foods take the processing of foods one step further by branding their own food products, giving customers limited brand variability for purchases. Food retailers, like other corporations, are in the business of making a monetary profit and those profits are derived from selling a large volume of goods—regardless of the nutritional values of those goods.

The consolidation of food retailers allows them to become so large and so few that they exert enormous influence over food producers, processors, shippers, and other suppliers to control the food market and food messaging.

All humans have to eat. Unlike other retail markets such as electronics or furniture, people cannot modify their need for calories to survive. But individuals are no longer the targeted customers of the food production industry when the retailer has such expansive market control. Two questions remain: *Where is the government regulation within the food retail sector? Where in the Farm Bill is the title to address the delivery of nonemergency, noncommodity food to Americans?*

2014 Farm Bill Highlights

The 2014 Farm Bill deserves a much closer and detailed analysis than we can provide in a single chapter, but a few highlights by Title need to be mentioned. In general, the 2014 Farm Bill continues the transition from traditional farm price and income support to risk management; that is, crop insurance is the primary support for farmers instead of payment programs. The 2014 Farm Bill titles are summarized further.

Title I, commodities

Commodity crops lost some support, including the elimination of direct payments (payments that were made even though the farmer had not suffered a loss). However, other price programs were retained, such as Price Loss Coverage and Agriculture Risk Coverage, both of which make payments calculated on base acres, not planted acreage. Although these payment programs sound like crop insurance, crop insurance is a separate program, and now the dominant one.[45]

Title II, conservation

Crop insurance is now connected with certain wetland and conservation rules that farmers who receive insurance subsidies must address. Referred to as conservation compliance, this requirement has existed since 1985 but has excluded crop insurance since 1996. However, conservation programs were consolidated to reduce overlap and funding overall reduced by close to $4 billion.[46]

Title III, trade

Food aid law is amended to emphasize improving nutritional quality of food aid and to avoid disruption of local markets. Annual export credit guarantees in the amount of $5.5 billion are available. The USDA is required to propose a reorganization plan for the international trade functions, including creating the position of an Under Secretary of Agriculture for Trade and Foreign Agricultural Affairs who will have multiagency coordination responsibilities.[46]

Title IV, nutrition

One of the interesting differences between the House and Senate proposals was that the House proposed a different schedule for the authorization of the nutrition program than the rest of the Farm Bill, hinting at a possible future separation of the nutrition provisions. *What do you think about that possibility?* Changes in SNAP eligibility were made, but most benefit and eligibility calculations were not impacted. While SNAP funding was decreased by $8 billion over 10 years, funding for the Emergency Food Assistance program (TEFAP) was increased by $205 million over 10 years, $125 of which is front-loaded in the first 5 years.[46]

Title V, credit

The USDA provides direct and guaranteed loans to farmers and ranchers who cannot get commercial loans. Policy changes have been made that appear to give more flexibility to extend credit and favorable terms, with some specific terms to help beginning farmers and facilitate purchases of highly fractionated land in Native reservations.[46] This Farm Bill makes law a microloan (up to $50,000) program that the USDA had been conducting through regulation.

Title VI, rural development

Two rural business programs were consolidated into one, but overall spending seems enhanced. Definitions of "rural" and "rural area" were retained, but the proposal from the Senate to add a category "rural in character" was rejected. Areas deemed rural between 2000 and 2010 will retain that designation until the data is in from the 2020 census, and the population threshold has been increased to 35,000 from 25,000. The Bill is also supporting a 0% interest rate loan for energy efficiency improvements and funding for increased access to broadband, distance learning, and telemedicine service.

Title VII, research

Mandatory spending for research is increased for Organic Agricultural as well as for Specialty Crop Research; $200 million is provides to establish the Foundation for Food and Agriculture Research, a nonprofit corporation that will solicit and accept private donations to award grants for public/private partnerships with USDA scientists.[46] The 15-member board has 7 industry nominated members, and 8 from a National Academy of Sciences list. As Marion Nestle points out in her blog, the board has to raise money while navigating potential conflicts of interest. *Will industry donate money without strings attached?*[46]

Title VIII, forestry

Programs were reauthorized; expired programs or those that never received appropriations were repealed; and provisions address the management of the National Forest System.[47] The Conservation Stewardship Program was enhanced, as well as expanded and strengthened market opportunities, such as prospects for wood in green building markets.[46]

Title IX, energy

Although USDA renewable energy programs include solar, wind, and anaerobic digesters, biofuels have been the primary focus, and in particular cornstarch-based ethanol. The 2008 Farm Bill attempted to shift biofuel policy toward noncorn feedstock such as cellulosic biofuels. The 2014 bill extends most of the 2008 renewable energy provisions.[48] Producers of biomass energy sources can receive matching funds for production; commodities or other already subsidized crops are not eligible (except for Title 1 commodity crop residues) and the crops must be harvested or collected under a conservation or forest stewardship plan.[46]

Title X, horticulture

Title X reauthorizes many provisions supporting specialty crop and organic agriculture, in particular, those relating to marketing and promotion, organic certification, data and information collection, pest and disease control, food safety and quality standards, and local foods.[49] Increased funding was made for several specialty crop programs, such as the Specialty Block Grant Program, plant pest and disease programs, USDA's Market News for specialty crops, the Specialty Crop Research Initiative, and the Fresh Fruit and Vegetable Program. Programs in other titles also benefit specialty crop and organic producers, including conservation, crop insurance, and expanded opportunities for local food systems and beginning farmers and ranchers. The research, nutrition, and rural development titles also contain provisions supporting local food producers.[46]

Title XI, crop insurance

The 2014 Farm Bill increased funding for crop insurance by an additional $5.7 billion over 10 years. Much of this is for a new insurance product for cotton. Crop insurance for specialty crops is expanded and improved. Although more than 100 crops can currently be insured, studies or policies are being required of the USDA concerning additional crops and products, such as alfalfa, sorghum grown for renewable energy, peanuts, rice, food safety and contamination-related losses in specialty crops, catastrophic disease in swine, reduction in the margin between market prices and production costs for catfish, and commercial poultry production against business disruptions caused by integrator bankruptcy (and also catastrophic event coverage). Private-sector index weather insurance is funded as well, which insures against specific weather events, not against actual loss.[46]

Title XII, miscellaneous

This title covers livestock, socially disadvantaged farmers, limited-resource products, and other miscellaneous issues. A production and marketing program for sheep is established. Inspection of catfish is transferred from the FDA to the USDA. The USDA is also required to conduct an economic analysis of its country-of-origin labeling (COOL) rule within 6 months. (Some industry groups had been pushing for repeal of COOL, see Chapter 8, "Meat," for more details.[46]) Additional miscellaneous provisions deserve a much closer look than we can in this chapter, one such being the establishment of a standing agriculture committee on the EPA's Science Advisory board.

AGRICULTURE POLICY AND AGRARIAN VALUES

Despite that many U.S. agricultural policies favor Big Food, some policies and the Farm Bill do reflect an attempt to address consumer values around sustainability, organic and local. This section briefly discusses U.S. organic agriculture policy and Farm Bill support for local food system connections. Additionally, the Farm Bill does have a range of programs around land conservation and rural development that are not addressed here.

Federal Organic Regulations

"Organic" became a U.S. federally regulated standard with the Organic Food Production Act (OFPA), which was Title XXI of the 1990 Farm Bill. The Congress passed OFPA to set the foundation for national standards covering the production and handling of "organic" products that would facilitate marketing and provide consumer confidence in their purchase choice. OFPA authorized USDA to establish the National Organic Program (NOP) to administer the standards and a National Organic Standards Board to advise the NOP. The NOSB advises on which substances should be allowed or prohibited but does not address food safety or nutritional value.[50]

The Food and Agricultural Organization (FAO) of the United Nations provides a long list of the ecological advantages of organic agriculture versus conventional farming. These advantages include sustainability over the long term, protection of soil and water resources, decreased nonrenewable energy use, support of biodiversity and an overall less polluting agricultural system. The hidden costs of agriculture to the environment in terms of natural resource degradation are reduced.[51] The USDA Economic Research Service reports that a number of studies have found enhanced soil tilth, higher biodiversity, higher soil organic matter and productivity, lower energy use, and reduced nutrient pollution with organic production methods when compared with conventional methods.[52]

Organic Standards

According to the USDA Agricultural Marketing Service, the arm of the USDA that implements the organic program, "organic" is a labeling term that indicates that the food or other agricultural product has been produced through approved methods that integrate cultural, biological, and mechanical practices that foster cycling of resources, promote ecological balance, and conserve biodiversity.[53] The USDA organic program is implemented through the marketing arm of the USDA. *Is this a characterization of organic as a consumer driven choice rather than an overall food security or environmental solution?*

For crops, the USDA organic seal signifies that synthetic fertilizers, sewage sludge, irradiation, prohibited pesticides, and genetically modified organisms were not used in production. For livestock, the USDA organic seal signifies that animal health and welfare standards were met, antibiotics or growth hormones were not used, and the animals were provided with 100% organic feed and access to the outdoors.[53] Note that "animal health and welfare standards" are distinct from label claims of "humane treatment" (which are not USDA regulated) and that access to the outdoors does not necessarily entail pastured livestock (see Chapter 8, "Meat," for more on this topic). A multi-ingredient food can carry the USDA organic seal if 95% of the ingredients are organic. If 70–95% of the ingredients are organic, the label can state "made with organic (specified ingredients or food groups)," but cannot carry USDA's organic seal.[54]

Demographics and Marketing Outlets

Once a niche product sold in a limited number of retail outlets, organic foods are now available in a wide variety of venues. According to USDA's Economic Research Service, 2005 was the first year that all 50 states reported having some certified organic farmland.[54] The 2008 Organic Production Survey counted 14,540 organic farms and ranches in the United States, comprising 4.1 million acres of land, which is only about 1% of total U.S. cropland.[55] Of those farms, three-quarters were USDA certified and the rest were exempt from certification due to small volume of sales (less than $5000), although such farms must still follow production, labeling, and record-keeping certifications. The top 10 states with organic farms as of 2008 are listed in Table 2.4. The organic tabulation from the 2012 census of agriculture was not yet released at the time of this writing.

In 2008, most (nearly 83%) of U.S. organic sales were to wholesale markets, while direct-to-retail sales comprised 10.6% including 5.3% to conventional supermarkets and 3.5% to natural food stores (both cooperatives and supermarkets). The remaining 6.8% were direct to consumers, including 2.4% on-site (e.g., farm stands and you-pick operations), 1.9% via farmers' markets, and 1% via community-supported agriculture (CSA). Almost one-half (44%) of producers sold their organic products locally (within

TABLE 2.4   States with the largest number of organic farms in 2008

STATE	NO. OF FARMS
California	2714
Wisconsin	1222
Washington	887
New York	827
Oregon	657
Pennsylvania	586
Minnesota	550
Ohio	547
Iowa	518
Vermont	467

Source: Adapted from USDA, National Agricultural Statistics Service, The 2008 Organic Production Survey. Available at http://www.agcensus.usda.gov/Publications/2007/Online_Highlights/Fact_Sheets/Practices/organics.pdf, accessed September 14, 2014.

100 miles of the farm), almost one-third (30%) sold their products 100–500 miles from the farm (regional), one-quarter reported (24%) selling nationally (more 500 miles from the farm), and only 2% reported selling internationally.

Federal Support for Organics in the 2014 Farm Bill

The growth in demand and interest in organic agriculture is reflected in Farm Bill trends. Financial support for organic agriculture has continued to increase from about $20 million in the 2002 Farm Bill, to just over $100 million in the 2008 Farm Bill (when organic production appeared in its very own Farm Bill Title), to over $160 million in the 2014 Farm Bill.[52] The 2014 Farm Bill expands cost-share assistance for organic certification (funding has more than doubled since 2008), expands organic research funding, improves organic crop insurance by increasing the maximum price producers may elect to reflect the premium organic crops command, and strengthens regulation enforcement which supports marketing and consumer confidence. Interestingly, the number of organic certified farmers in the United States has leveled off. The increase in funding for the organic certification cost-share program means that a substantial portion of the fees can be covered for smaller operations, and this may help attract small-scale producers.[52]

Improved price election makes crop insurance more attractive. Organic producers were first able to get crop insurance under the 2000 Farm Bill, but, due to lack of data, had to pay 4% more in premiums and could only insure for the "conventional" price instead of the premium organic could command. The 5% upcharge was dropped at the beginning of 2014.[56]

Recognizing the inextricability of organic agriculture and environmental stewardship, the Farm Bill provisions for organic farming research speak to contemporary land use policy concerns. In 2008, it was climate change, as Congress cited the need for research on the potential of organic farming to capture atmospheric carbon and store it in the soil. In 2014, it is water, as Congress encourages research on weed and pest management for organic farming systems that maintain healthy water resources.[56]

The People's Department

Previous Farm Bills have provided support to rural economic and local food system development. The 2002 bill introduced the Farmers Market Promotion Program and expanded the Value-Added Producer Grant program to include organic and grass fed. The 2008 bill provided mandatory funding for the Farmers Market Promotion Program and the Specialty Crop Block Grant program, and added new rural economic programs aimed at micro-entrepreneurship, as well as local and regional food enterprises loan programs. The Value-Added program expanded to include locally produced and marketed food.[57]

Although the 2014 Farm Bill made lesser strides in rural economic programs, local and regional food systems continued to see increased attention. The Farmers Market Promotion Program was expanded into the Farmers Market and Local Food Promotion Program, supporting direct farmer-to-consumer marketing channels as well as local and regional enterprises that process, distribute, aggregate, or store locally or regionally produced food products. The bill also provides $30 million in annual mandatory funding, triple the amount this program received during the final years of the 2008 Farm Bill. The Community Food Projects program, which supports the development of community-based food projects in low-income communities, was given $9 million in mandatory funding, which was nearly double the funding level in the 2008 Farm Bill.[57]

The incorporation of the Healthy Food Financing Initiative (HFFI) into the Farm Bill is a powerful step toward creating equitable and sustainable access to fresh and healthy foods across America. Modeled on the successful Pennsylvania Fresh Food Financing Initiative (see Chapter 12, "Nanny State") HFFI provides grants and loans to retailers of healthy foods to encourage entry in underserved areas. Support of regional food systems and locally grown foods is one of the priority categories listed for funding. Although authorized to receive up to $125 million in appropriated funds, whether HFFI actually receives funding will be determined by future annual agriculture appropriations bills.[57]

In addition to the intention of Congress to support local food system development manifest in the Farm Bill, the USDA has initiated various other programs. For example, HFFI originated as a joint program by the USDA and other agencies before garnering permanent status through Congress in the 2014 Farm Bill. The 2014 Farm Bill also explicitly supports farm-to-school initiatives by including two pilot programs, although, like HFFI, these programs do not have mandatory funding.

Other initiatives from the USDA include The People's Garden and Know Your Farmer, Know Your Food (KYF2). The People's Garden was named in honor of President Lincoln's description of the USDA as the People's Department. This program was started in 2009 and challenged USDA employees to create gardens at USDA facilities, but blossomed into a collaborative effort with over 700 local and national organizations to establish community and school gardens that provide a community benefit, are run collaboratively, and are sustainable.[58] Know Your Farmer, Know Your Food (KYF2)[59] aims to strengthen USDA's support for local and regional food systems by ensuring cross-collaboration within the USDA; the program is run by a USDA taskforce representing every agency within the department. Through KYF2, USDA integrates programs and policies that

- Stimulate food—and agriculturally based community economic development.
- Foster new opportunities for farmers and ranchers.
- Promote locally—and regionally produced and processed foods.
- Cultivate healthy eating habits and educated, empowered consumers.
- Expand access to affordable fresh and local food.
- Demonstrate the connection between food, agriculture, community and the environment.

How do these programs make the USDA once again a "people's department"? What else can the USDA do?

CONCLUSION

In 1977, then Senator George McGovern (D-SD) strongly connected food production and processing with medical care, thus pushing for a nutritional direction for agricultural policy. He stated, "Food production and processing is America's number one industry and medical care ranks number three. Nutrition is the common link. Nutrition is a spectrum which runs from food production at one end to health at the other. By recognizing this connection… [hopefully] … nutrition will become a major priority of this Nation's agricultural policy" (p. xxii).[60]

Food policy is important to all of us. We all eat. We all have values around food, and these values differ. Some may question why SNAP makes up the largest portion of the USDA's budget. Some may see our need to affordably feed people to be more important than sustaining small, local farming. Should future Farm Bills incentivize risky agricultural practices? Or will our food supply be nutritiously diminished if our farmers cannot take risks? What balance must we strike between plow and plover? Does the mere inclusion of nutrition assistance programs in the USDA constitute a nutritional direction for our agricultural policy? Looking forward, we will need to face the political challenges of finding the common ground between some of these values and create a new set of agricultural polices that truly consider nutrition.

REFERENCES

1. Rasmussen WD. Lincoln's Agricultural Legacy. United States Department of Agriculture. National Agricultural Library. Available at http://www.nal.usda.gov/lincolns-agricultural-legacy, accessed September 25, 2014.

2. United States Department of Agriculture, About USDA. Mission Statement. Last updated April 15, 2014. Available at http://www.usda.gov/wps/portal/usda/usdahome?navid=MISSION_STATEMENT, accessed September 18, 2014.

3. Johnson R, Monke J. What is the farm bill? Congressional Research Service. July 23, 2014. Available at as.org/sgp/crs/misc/RS22131.pdf, accessed September 16, 2014.

4. USDA Transcript, Release No. 0458.11. Agriculture Secretary Vilsack on Priorities for the 2012 farm bill. Remarks As Delivered. October 24, 2011. Available at http://www.usda.gov/wps/portal/usda/usdamediafb?contentid=2011/10/0458.xml&printable=true&contentidonly=true, accessed September 13, 2014.

5. Roosevelt FD. Statement on signing the Soil Conservation and Domestic Allotment Act. March 1, 1936. Online by Peters G, Woolley JT. *The American Presidency Project.* Available at http://www.presidency.ucsb.edu/ws/?pid=15254, accessed September 13, 2014.

6. Partners for a Hunger-Free Oregon-Ending hunger before it begins. History of Food Stamps and SNAP.2012. Available at http://oregonhunger.org/history-food-stamps-and-snap, accessed September 13, 2014.

7. Gunderson GG. USDA. FNS. Website, The National School Lunch Program: Background and Development, last updated June 17, 2014. Available at http://www.fns.usda.gov/nslp/history_4, accessed September 13, 2014.

8. Ganzel B. Farming in the 1940s: Exports & Imports. Available at http://www.livinghistoryfarm.org/farminginthe40s/money_09.html, accessed September 13, 2014.

9. Ganzel B. Living History Farm, Farming in the 1940s. Available at http://www.livinghistoryfarm.org/farminginthe40s/money_09.html, accessed September 13, 2014.

10. Schnepf R. U.S. Farm Income. Congressional Research Service. 7-5700 R40152. February 28, 2014. Available at http://www.fas.org/sgp/crs/misc/R40152.pdf, accessed December 3, 2012.

11. Groombridge MA. America's Bittersweet Sugar Policy. Executive Summary. CATO Institute. Center for Trade Policy Studies. December 4, 2001. Available at http://www.cato.org/pubs/tbp/tbp-013.pdf, accessed September 14, 2014.

12. USDA. Risk Management Agency. A History of the Crop Insurance Program. Available at http://www.rma.usda.gov/aboutrma/what/history.html, accessed September 14, 2014.

13. Shields DA. Federal Crop Insurance: Background and Issues. Congressional Research Service. 7-5700. R40532. December 13, 2010. Available at http://fas.org/sgp/crs/misc/R40532.pdf, accessed September 14, 2014.

14. Babcock B. Should government subsidize farmers' risk management? *Iowa Ag Rev.* 2009;15:2. Available at http://www.card.iastate.edu/iowa_ag_review/spring_09/article1.aspx, accessed November 17, 2014.

15. Sumner DA, Zulauf C. Economic & Environmental Effects of Agricultural Insurance programs. July 2012. The Council on Food, Agricultural, and Resource Economics (C-FARE). Available at http://ageconsearch.umn.edu/bitstream/156622/2/Sumner-Zulauf_Final.pdf, accessed September 14, 2014.

16. American Farmland Trust, Agenda: 2012: Transforming U.S. Farm Policy for the 21st Century. Available at http://www.farmbillfacts.org/wp-content/uploads/2011/11/Transforming-U.S.-Farm-Policy-for-the-21st-Century.pdf, accessed September 14, 2014.

17. Environmental Working Group. Key Issues, Crop Insurance. Available at http://www.ewg.org/key-issues/farming/crop-insurance, accessed September 14, 2014.

18. Russo M. Apples to twinkies: Comparing federal subsidies of fresh produce and junk food. U.S. PIRG education fund, September 2011. Available at http://www.uspirgedfund.org/sites/pirg/files/reports/Apples-to-Twinkies-web-vUS.pdf, accessed September 14, 2014.

19. Havel PJ. Dietary fructose: Implications for dysregulation of energy homeostasis and lipid/carbohydrate metabolism. *Nutr. Rev.* 2005;63:133–57.

20. Lichtenstein AH. New York City trans fat ban: Improving the default option when purchasing foods prepared outside of the home. *Ann. Intern. Med.* 2010;157:144–5.

21. Clancy K. Greener pastures: How grass-fed beef and milk contribute to healthy eating. Union of Concerned Scientists. Available at http://www.ucsusa.org/food_and_agriculture/solutions/advance-sustainable-agriculture/greener-pastures.htm, accessed July 29, 2014.

22. Sebastian RS, Enns CW, Goldman JD. Snacking patterns of U.S. adults: What we eat in American, NHANES 2007–2008. Food Survey Research Group Dietary Data Brief No. 4. June 2011. USDA. Available at http://www.ars.usda.gov/SP2UserFiles/Place/12355000/pdf/DBrief/4_adult_snacking_0708.pdf, accessed September 14, 2014.

23. Fryar CD, Ervine RB. Caloric intake from fast food among adults: United States, 2007–2010, NCHS Data Bried No. 114, February 2013. Available at http://www.cdc.gov/nchs/data/databriefs/db114.pdf, accessed September 14, 2014.

24. Anderson B, Rafferty AP, Lyon-Callo S, Fussman C, Imes G. Fast-food consumption and obesity among Michigan adults. *Prev. Chronic. Dis.* 2011;8:A71.

25. Bowman SA, Gortmaker SL, Ebbeling CB, Pereira MA, Ludwig DS. Effects of fast-food consumption on energy intake and diet quality among children in a national household survey. *Pediatrics.* 2004;113(1Pt 1):112–118.

26. Pollan M. The way we live now:10-12-03; The (agri)cultural contradictions of obesity. *New York Times.* October 12, 2003. Available at http://www.nytimes.com/2003/10/12/magazine/the-way-we-live-now-10-12-03-the-agri-cultural-contradictions-of-obesity.html?pagewanted=all&src=pm, accessed September 14, 2014.

27. Pasour EC, Jr. Agricultural economics and the state. *Econ. J. Watch.* 2004;1:106–133.

28. Sundell P, Shane M. The 2008-09 Recession and Recovery: Implications for the Growth and Financial Health of U.S. Agriculture. USDA/ WES-1201 May 2012. Available at http://www.ers.usda.gov/media/619162/wrs1201_1_.pdf, accessed September 14, 2014.

29. United States Department of Agriculture, National Agricultural Statistics Service, 2012 Census of Agriculture, last updated May 21, 2014. Available at http://www.agcensus.usda.gov/Publications/2012/Online_Resources/Highlights/Farm_Economics/, accessed September 14, 2014.

30. Nestle M. *Food Politics: How the Food Industry Influences Nutrition and Health,*10th edition, Berkeley: University of California Press, 2013.

31. OpenSecrets.org. Agribusiness. Available at http://www.opensecrets.org/industries/indus.php?Ind=A, accessed September 14, 2014.

32. Chite RM. Previewing the next farm bill. Congressional Research Service.7-5700. R42357. February 15, 2012. Available at http://www.fas.org/sgp/crs/misc/R42357.pdf, accessed September 18, 2014.

33. Imhoff D (ed.). *The CAFO Reader: The Tragedy of Industrial Farm Animals.* Berkeley: University of California Press, 2010.

34. Imhoff D. *Food Fight: The Citizens Guide to the Next Food and Farm Bill.* Healdsburg, CA: Watershed Media, 2012.

35. USDA. National Agricultural Statistics Service. 2007 Census of Agriculture. Available at http://www.agcensus.usda.gov/Publications/2007/Full_Report/usv1.pdf, accessed September 14, 2014.

36. United States Environmental Protection Agency. Ag 101. Demographics. Available at http://www.epa.gov/oecaagct/ag101/demographics.html, accessed September 14, 2014.

37. Union of Concerned Scientists (UCS). Our Failing Food System: Industrial Agriculture, last updated August 30, 2012. Available at http://www.ucsusa.org/food_and_agriculture/our-failing-food-system/industrial-agriculture/, accessed September 14, 2014.

38. Flora JL, Chen QL, Bastian S, Hartmann R. Hog CAFOS and Sustainability: The Impact on Local Development and Water Quality in Iowa. The Iowa Policy Project, 2007. Available at http://www.iowapolicyproject.org/2007docs/071018-cafos.pdf, accessed May 3, 2015.

39. Hendrickson M, Heffernan W. Concentration of agricultural markets, February 2002. Available at http://www.foodcircles.missouri.edu/CRJanuary02.pdf, accessed September 7, 2014.

40. WH Group, Corporate Profile. Available at http://www.wh-group.com/en/about/profile. php, accessed September 7, 2014.

41. Archer Daniels Midland Company, Our Company, Our Role: Understanding ADM and Food. Available at http://www.adm.com/company/Documents/ADM_and_Food_Brochure_ lores.pdf, accessed September 19, 2014.

42. Oxfam, Behind the Brands: Food justice and the 'Big 10' food and beverage companies," February 26, 2013. Available at http://www.oxfam.org/sites/www.oxfam.org/files/bp166-behind-the-brands-260213-en.pdf, accessed September 19, 2014.

43. Food & Water Watch, Consolidation and Buyer Power in the Grocery Industry, Fact Sheet, December 2010. Available at http://documents.foodandwaterwatch.org/doc/ RetailConcentration-web.pdf, accessed September 19, 2014.

44. Leibtag E. The Impact of Big-Box Stores on Retail Food Prices and the Consumer Price Index, Economic Research Report No. (ERR-33), December 2006. Available at http://www. ers.usda.gov/publications/err-economic-research-report/err33.aspx#.VBsATEu4mFI, accessed September 19, 2014.

45. United States Department of Agriculture, Economic Research Service, Retail Trends. Available at http://www.ers.usda.gov/topics/food-markets-prices/retailing-wholesaling/ retail-trends.aspx#.VBsEMEu4mFI, accessed September 19, 2014.

46. Chite R. The 2014 Farm Bill (P.L. 113-79); Summary and Side-by-Side, February 12, 2014. Available at http://www.farmland.org/programs/federal/documents/2014_0213_CRS_Farm BillSummary.pdf, accessed September 19, 2014.

47. Nestle M. Food Politics Blog, August 11, 2014. Available at http://www.foodpolitics. com/2014/08/dan-glickman-heads-board-of-foundation-for-food-and-agriculture-research/, accessed September 19, 2014.

48. American Forest Foundation, Congress Passes Strongest Farm Bill for Forests Yet, February 4, 2014. Available at https://www.forestfoundation.org/congress-passes-strongest-farm-bill-for-forests, accessed September 19, 2014.

49. FarmEnergy.org, 2014 Farm Bill's Energy Title Provisions, March 7, 2014. Available at http://farmenergy.org/news/summary-2014-farm-bills-energy-title-provisions, accessed September 19, 2014.

50. United States Agricultural Department, Agricultural Marketing Service. National Organic Program. Available at http://www.ams.usda.gov/AMSv1.0/nop, accessed September 14, 2014.

51. FAO. Organic Agriculture FAQ, What are the environmental benefits of organic agriculture? Available at http://www.fao.org/organicag/oa-faq/oa-faq6/en/, accessed September 14, 2014.

52. United States Department of Agriculture, Economic Research Service, Agricultural Act of 2014: Highlights and Implication, last updated April 11, 2014. Available at http://www.ers. usda.gov/agricultural-act-of-2014-highlights-and-implications/organic-agriculture.aspx#. VBeLBUu4mFI, accessed September 15, 2014.

53. United States Department of Agriculture. Agricultural Marketing Service. National Organic Program. Available at http://www.ams.usda.gov/AMSv1.0/nop, accessed September 14, 2014.

54. Johnson R. Organic Agriculture in the United States: Program and Policy Issues. Congressional Research Service, updated November 25, 2008. Available at http:// nationalaglawcenter.org/wp-content/uploads/assets/crs/RL31595.pdf, accessed September 16, 2014.

55. Chite RM. Previewing the next farm bill. Congressional Research Service.7-5700. R42357. February 15, 2012. Available at http://www.fas.org/sgp/crs/misc/R42357.pdf, accessed September 14, 2014.

56. United States Department of Agriculture, Risk Management Agency Fact Sheet. Organic Farming Practices, May 2013. Available at http://www.rma.usda.gov/pubs/rme/2013revise dorganicsfactsheet.pdf, accessed September 16, 2014.

57. National Sustainable Agriculture Coalition, 2014, Farm Bill Drilldown: Local and Regional Food Systems, Healthy Food Access, and Rural Development. February 11, 2014. Available at http://sustainableagriculture.net/blog/2014-farmbill-local-rd-organic/, accessed September 18, 2014.

58. United States Department of Agriculture, The People's Garden. Last updated May 29, 2014. Available at http://www.usda.gov/wps/portal/usda/usdahome?navid=GARDEN_RT1&parentnav=PEOPLES_GARDEN&navtype=RT, accessed September 18, 2014.

59. United States Department of Agriculture, Know Your Farmer-Know Your Food (KYF2). Available at http://www.usda.gov/wps/portal/usda/usdahome?navid=KYF_MISSION, accessed September 18, 2014.

60. Select Committee on Nutrition and Human Needs United States Senate. Press Conference, Friday January 14, 1977. George McGovern, Chairman. Preface. *Dietary Goals for the United States*, 2nd edition, Washington, DC: U.S. Government Printing Office, 1977.

61. OpenSecrets.org. Agribusiness. Available at http://www.opensecrets.org/industries/indus.php?Ind=A, accessed September 14, 2014.

62. Harris and Ewing, Library of Congress Prints and Photographs Division, Washington, DC, 20540 USA. http://www.loc.gov/pictures/item/hec2009013216/, accessed May 5, 2015.

63. USDA. National Agricultural Statistics Service. The 2008 Organic Production Survey. Available at http://www.agcensus.usda.gov/Publications/2007/Online_Highlights/Fact_Sheets/Practices/organics.pdf, accessed September 14, 2014.

64. Johnson R, Corn ML. Bee health: Background and issues for Congress. April 9, 2014. Congressional Research Service. Available at http://fas.org/sgp/crs/misc/R43191.pdf, accessed October 26, 2014.

65. Brodschneider R, Crailsheim K. Nutrition and health in honey bees. *Apidologie*. 2010;41(3):278–294.

Dietary Guidance

3

When it comes to eating, what's more useful than a plate?
First Lady Michelle Obama[1]

INTRODUCTION

During the past century, what messages did we get from the U.S. government, the food industry, and health organizations about how we should eat? What is the relationship between population health objectives and dietary guidance? This chapter reviews the early twentieth century focus on foods to prevent deficiencies, war-time food messaging and the developing science of nutrients, the postwar focus on health promotion and nutrients, and the more recent emphasis on dietary patterns. During this time period, our consumer nutrition education messaging changed graphically. We currently discuss dietary guidance in terms of a plate. This messaging around the contents of our plate is also evocative of seeing our food as a totality rather than as just a collection of nutrients. Will our next wave of U.S. dietary recommendations reflect additional concepts of food as a totality, such as sustainability?

U.S. government dietary guidance stems from research about the characteristics of food and on what foods we are actually eating as a population. Chapter 4, "Nutrition Information Policies," discusses how research on dietary status contributes to food policy such as the dietary related health objectives for the population published in Healthy People. This chapter focuses on dietary recommendations, from Food Groups to Dietary Guidelines, how such messages are communicated (see "More Matters"), and the use of colorful graphic icons (such as "MyPlate").

Portions of this chapter are heavily influenced by Davis and Santos' comprehensive review, Dietary Recommendations and How They Have Changed Over Time, in the United States Department of Agriculture (USDA) monograph, *America's Eating Habits: Changes and Consequences* (1999).[2]

Definitions

Throughout this chapter, we refer to several kinds of food- and nutrition-related advice. Let us start by examining some generally accepted meanings for dietary goals, dietary guidelines, food-based dietary guidance, and nutrient intake recommendations.[3]

47

Dietary goals are desirable food or nutrient intakes that support population health. Usually expressed as average national intakes, they are used for long-term planning on the national level.

Dietary guidelines are broad targets for which people can aim. They are sets of advisory statements for government policy or consumer messages that give dietary advice to the population to promote overall nutritional health.

Food-based dietary guidance consists of educational tools, often presented through graphic illustrations of food groupings that support a recommended dietary pattern consistent with dietary guidelines. The Food Guide Pyramid and MyPlate are examples of food-based dietary guidance.

Nutrient intake recommendations are suggestions for the quantity of nutrients or food factors that should be consumed, on average, every day. Nutrient recommendations serve as the basis for food fortification, nutrition labeling, menu planning for groups, and diet planning for individuals. The Recommended Daily Allowances (RDAs) is an example of nutrient intake recommendation.

Pioneering Federal Dietary Guidance: Wilbin Olin Atwater

Wilbin Olin (W.O.) Atwater, PhD (1844–1907), was the first director of the Office of Experiment Stations of the USDA and created the system to measure energy in food by units, known as food calories. Atwater was a pioneer in energy balance research. He built the first direct human calorimeter. Subsequently, using indirect calorimetry, he studied human energy requirements and how energy in food is used for fuel by the human body. See Box 3.1 on calorimetry.

BOX 3.1   CALORIMETRY

Direct calorimetry is the measure of heat, specifically heat loss, as a result of energy expenditure. Indirect calorimetry estimates heat production by measuring the consumption of oxygen and the production of carbon dioxide. Under the duress of aerobic activity, the body takes in oxygen and fuels (carbohydrates, fats, and proteins), converting these to heat, energy (ATP), and by-products, including carbon dioxide, which is then exhaled. In the resting state, the rate of heat production from this oxidative process is equal to the loss of heat into the environment. Measuring heat can be estimated through measuring the consumption of oxygen and the expulsion of carbon dioxide. This can be done in a closed room with a constant supply of fresh air and a gas analyzer measuring oxygen consumed and carbon dioxide produced. The ratio of oxygen to carbon dioxide produced will depend not only on the person's activity but also on the types of fuels used by the body. Carbohydrates, protein, and fat produce different ratios of respiratory gases exhaled by the body. This ratio is known as the respiratory quotient ($RQ = CO_2 : O_2$). Carbohydrates produce an RQ of 1.00, whereas fat produces an

RQ of 0.7. If more carbohydrate is used as fuel, the RQ will be higher. A closed room calorimeter, such as the one in the Smart Foods Centre at the University of Wollongong in Australia, measures a person's ability to burn different fuel as well as their energy expenditure.

University of Wollongong, Food Trials Laboratory, Human Whole Room Calorimeter, http://smah.uow.edu.au/medicine/smartfoods/ foodtrialslab/UOW046888.html, accessed June 6, 2014.

Atwater's research helped to determine the amount and types of food required to maintain a healthy body weight. It was conducted at a time when the major challenge was helping people to get enough energy to meet energy demands in an environment that required high levels of physical activity for the tasks of daily living as well as for work. A century ago, underweight was more the problem than overweight.[4]

Human nutrition activities in USDA, initiated by W. O. Atwater, have always have been linked with the nutritive value of foods, human nutritional requirements, the kinds and amounts of foods that Americans consume relative to their needs, and strategies for improving diets and the food supply.[5] His food composition studies formed the basis for the USDA National Nutrient Database for Standard Reference (NNDSR). The most current release, Release 27, provides the nutrient content of more than 8600 different foods.[6] The NNDSR is the basis for therapeutic diets, the food composition tables in introductory nutrition textbooks, and all online diet and food calorie calculators.

Obscured by Atwater's renown as the founder of food chemistry in the United States, is the fact that in 1894, he published the first Federal dietary recommendations. His advice, published in a farmers' bulletin, recommended providing diets for men based on proportions of protein, carbohydrate, and fat and "mineral matter" (specific vitamins and minerals were not yet discovered).[7]

In *Principles of Nutrition and Nutritive Value of Food* (1902), Atwater advocated variety, proportionality, and moderation; minimizing waste; measuring calories; and an efficient, affordable diet that focused on nutrient-rich foods and less fat, sugar, and starch. He counseled that "unless care is exercised in selecting food, a diet may result which is one-sided or badly balanced—that is, a diet in which either protein or fuel ingredients (carbohydrate and fat) are provided in excess.... The evils of overeating may not be felt at once, but sooner or later, they are sure to appear—perhaps in an excessive amount of fatty tissue, perhaps in general debility, perhaps in actual disease."[8]

Atwater's message contains advice remarkably similar to the themes of variety, proportionality, and moderation promoted in USDA's Food Guide Pyramid (1992–2011) and subsequent dietary plan graphic MyPlate (2011). Atwater's work and that of his colleagues also inspired modern research that examines the cost of different foods relative to their energy and nutritive value and association with diet-related chronic diseases. See Box 3.2 for current research based on this model.

Atwater's work at the USDA on food composition and human nutritional needs set the stage for the development of the first quantitative U.S. food guide (1916), which translated nutritional needs into food intake recommendations.

BOX 3.2   CURRENT RESEARCH BASED ON ATWATER'S MODEL OF CATEGORIZING FOOD BY COST AND NUTRITIONAL VALUE

When Atwater examined the value of foods relative to their protein content, he found the least expensive source of protein was beans, and the most expensive was fruit. A century ago, getting enough protein was a paramount concern, so beans would have been the most economical choice. Today, University of Washington epidemiologist Adam Drenowski, PhD, looks at food relative to the cost of calories, and finds the most expensive foods per calorie are vegetables and fruits, and that nutrients such as dietary fiber, vitamins A, C, D, E, and B12, beta carotene, folate, iron, calcium, potassium, and magnesium that are associated with a lower risk of chronic disease are also associated with higher diet costs, and food factors, such as saturated fats, trans fats, and added sugars that are associated with higher disease risk are associated with lower diet costs. Drenowski and his collaborators conclude that following the current dietary advice to eat more vegetables and fruits might be economically prohibitive for low-income segments of the population. Helping consumers to select affordable as well as nutritious diets ought to be a priority for researchers and health professionals.

Atwater, W.O., Foods: Nutritive Value and Cost, *Washington, DC: Government Printing Office, 1894 (USDA, Farmers' Bulletin No. 23.), http://www.ars.usda.gov/SP2UserFiles/Place/12355000/pdf/hist/ oes_1894_farm_bul_23.pdf, accessed September 12, 2014; Atwater, W.O.,* Principles of Nutrition and Nutritive Value of Food, *Washington, DC: Government Printing Office, 1902 (USDA, Farmers' Bulletin No. 142.), http://openlibrary.org/books/OL24231176M/Principles_of_nutrition_and_ nutritive_value_of_food, accessed September 11, 2014; Drewnowski, A.,* The cost of US foods as related to their nutritive value, *Am. J. Clin. Nutr., 92, 2010, 1181–88; Aggarwal, A., Monsivais, P., Drewnowski, A., Nutrient intakes linked to better health outcomes are associated with higher diet costs in the US, PLoS One, 2012;7(5):e37533, epub 2012 May 25.*

1910s TO THE 1930s: DEFICIENCY PREVENTION

A science-based approach to food guidance began in the United States, a century ago. In 1916, the 20-page brochure *Food for Young Children*,[9] was published by Caroline L. Hunt (1865–1927), a nutrition specialist in USDA's Bureau of Home Economics. It was followed the next year with the 14-page *How to Select Foods: What the Body Needs*,[10] aimed at the general public, providing "a simple method of selecting and combining food materials to provide an adequate, attractive, and economical diet." These two guides promoted five food groups—milk and meat, cereals, vegetables and fruits, fats and fatty foods, and sugars and sugary foods. A third guide, released in 1921, the 27-page booklet, *A Week's Food for an Average Family* suggested amounts of food to purchase for a household with two adults and three young children or four adults, using the same five food groups. In 1923, a revision was published that included households of different compositions.

In 1921, Emma A. Winslow, a lecturer at Teachers College, Columbia University and Home Economics specialist for the Charity Society Organization of New York published *Food Values: How Foods Meet Body Needs*, which presents, for the first time, nutrition information in easy-to-read graphic form. The bulletin is based on Atwater's pioneering publication about the composition of 2600 foods commonly consumed in the United States, a precursor of subsequent food composition manuals produced by the USDA. Although the bulletin is typeset, the graphs are hand-drawn. The 48 charts depict the percent of the then recommended intake of protein, calcium, phosphorous, and iron provided by one pound of each food for a physically active man. The book could be purchased from the USDA for 10 cents.[11] Winslow subsequently pursued a doctorate degree in Economics and Sociology at the University of London. Table 3.1 summarizes these early USDA food guides.

Food Recommendations during World War I

When the United States entered World War I in 1917, the war had already been raging for three years and much of Western Europe was devastated. Many Allies were starving because the war had disrupted transportation and reduced farming productivity since farmers left to fight and were replaced by women, children, and the elderly, who were less equipped for arduous farm work.

In 1917, the newly formed U.S. Food Administration—led by Herbert Hoover before he became president—was charged with a home front campaign of *winning the war with food.* Hoover was convinced that the United States would not need to ration its foodstuffs; enough food for the war effort overseas could be allocated by voluntarily conserving resources, planting vegetable war gardens, and adopting healthy dietary practices. Hoover asked Americans to participate in "Meatless Mondays" and "Wheatless Wednesdays" in order to free up food to send overseas. See Box 3.3 about Meatless Mondays then and now.

Inciting voluntary diet behavior change across the nation was a significant undertaking. To this end, each State appointed a home economics director to create volunteer networks trained to educate women about food substitutions such as using corn meal or potatoes instead of wheat; substituting fish, pork, beans, fish, and cheese for beef; substituting corn and rye and peanut flour for wheat flour; and sweetening with syrups and honey in place of sugar.[12] Numerous posters provided educational and patriotic messages, such as shown in Figure 3.1, a poster recommending the use of potatoes.

Home economists also stressed the importance of increasing consumption of fresh fruits and vegetables (which were too perishable to be shipped to the troops) while generally raising the nation's consciousness to eat less. Posters, such as shown in Figure 3.2, emphasized the need for Americans to eat less in order to send food to help the Allies.

TABLE 3.1 Early UDSA food guides, 1884–1923

YEAR	USDA BULLETIN NO.	TITLE	AUTHOR/S	LENGTH (PAGES)	URL
1884	23	Foods: Nutritive Value and Cost	OW Atwater	30	https://archive.org/details/CAT87201446
1902	Na	Principles of Nutrition and Nutritive Value of Food (rev ed)	OW Atwater	48	https://archive.org/details/CAT31127342
Quantitative Food Guides					
1916	717	Food for Young Children	CL Hunt	20	https://archive.org/details/CAT87206470
1917	808	How to Select Foods I. What the Body Needs	CL Hunt and HW Atwater	14	https://archive.org/details/CAT87202872
1917	817	How to Select Foods II. Cereal Foods	CL Hunt and HW Atwater	23	https://archive.org/details/CAT87202539
1917	824	How to Select Foods III. Foods Rich in Protein	CL Hunt and HW Atwater	19	https://archive.org/details/CAT87202545
1921	975	Food Values: How Foods Meet Body Needs	EA Winslow	37	https://archive.org/details/foodvalueshowfoo975wins
1921	1228	A Week's Food for an Average Family	CL Hunt	27	https://archive.org/details/CAT87203126
1923	1313	Good Proportions in the Diet	CL Hunt	18	http://digital.library.unt.edu/ark:/67531/metadc3473/

BOX 3.3   MEATLESS MONDAYS: 1917 AND 2003

The Monday Campaign, Inc. was founded in 2003 in association with the Johns Hopkins Bloomberg School of Public Health Center for a Livable Future to encourage Americans to make healthy choices at the beginning of every work week. Among the Healthy Monday campaigns are Meatless Mondays, The Kids Cook Monday, Monday 2000, Quit and Stay Quit Monday, Move it Monday, and The Monday Mile. The U.S. Food Administration's "Meatless Mondays" preceded Johns Hopkins' by more than a half century.

The Monday Campaigns, The day all health breaks loose, http://www.mondaycampaigns.org/, accessed September 12, 2014.

Working at both the Federal and State levels, home economists created hundreds of recipes, menus, wrote articles, books and pamphlets, and traveled widely to demonstrate food conservation methods, such as canning and drying. As a result of these efforts, the United States did not legislate mandatory rationing during World War I as it had to when we entered the Second World War.

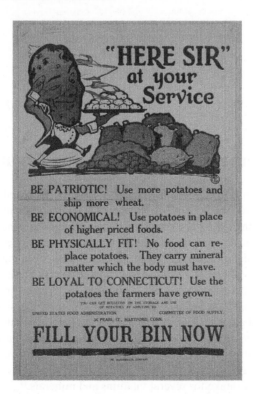

FIGURE 3.1   World War I poster extolling patriotic virtue of consuming potatoes. (With permission from Special Collections, National Agricultural Library.)

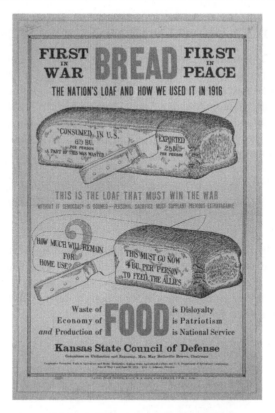

FIGURE 3.2   World War I poster emphasizing eating less to support the war effort. (With permission from Special Collections, National Agricultural Library.)

Food Recommendations during the Great Depression

In 1933, as a response to the Depression that started in 1929 and gripped the United States until we entered the Second World War, the Federal government introduced food plans at four different cost levels. Hazel Stiebeling, a food economist at USDA's Bureau of Home Economics, developed a set of dietary standards designed for maintenance of health. The standards built on and expanded the work of Atwater and Hunt by increasing the number of food groups from five to 12. A full four of these groups were allotted to produce, as the emerging discoveries of vitamins encouraged grouping along certain nutrients: tomatoes and citrus fruits; potatoes and sweet potatoes; leafy green and yellow vegetables; and all other fruits and vegetables. The guide additionally emphasized the importance of balancing "protective" (nutrient-dense) foods, such as milk, fatty fish, vegetables, and fruits with less protective foods and nonprotective foods, such as meat and sweets. Siebeling also used the term "dietary allowances" for quantitative estimates of individual dietary intake for several vitamins and minerals.[13]

The standards developed by Stiebeling were used to design the proposed four food plans and craft a buying guide to help people to shop for food on a budget. The lowest cost plan was identified as a restricted diet for emergency use; the next costly plan was identified as an adequate diet at a minimum cost.[14]

1940s TO THE 1970s: NUTRIENTS FOR HEALTH

From the 1940s through the 1970s, dietary advice focused on getting enough nutrients for health. As noted in the previous section, dietary guidance in the early twentieth century mostly concerned proportions of macronutrients (fat, protein, and carbohydrate). The advances in discovering other characteristics of foods that were essential for animals and people to survive and thrive brought a new depth to the study of nutrition. The American Institute of Nutrition, the first independent society devoted exclusively to nutrition, was founded in 1933.[15] The discovery of vitamins led to a greater emphasis on nutrients within dietary guidance messages, and eventually, the establishment of Recommended Dietary Allowances (RDAs) for the intake of vitamins.

Development of the Recommended Dietary Allowances (1940)

In 1940, the Committee on Food and Nutrition of the National Research Council was established to advise on nutrition for the U.S. population. Unfortunately, the nutritional problem being addressed was not simply that of improving the health of the public. The specific mandate of the committee was to advise on nutrition problems in connection with national defense. This committee developed the first RDAs.[16]

Harriet Stiebeling proposed dietary allowances that were 50% more than the average requirement to account for variability in the population, and thus cover most normal individuals. Based on the combined judgment of nutrition authorities and the best available evidence, the committee developed a table of RDAs listing specific intakes for calories, protein, iron, calcium, and vitamins A, C, D, thiamin (B1), riboflavin (B2), and nicotinic acid (B3), now commonly known as niacin. The RDAs were presented and accepted at the 1941 National Nutrition Conference for Defense. The report also provided a sample dietary pattern and a sample low-cost meal plan for a day (at a daily cost of 32 cents).[16] The inclusion of oleo (margarine) with vitamin A as something to be eaten on toast or bread at every meal is notable.

The USDA's "Basic 7" Food Groups (1943)

In 1943, USDA and the War Food Administration published the National Wartime Nutrition Guide. The civilian population was subject to food rationing to support military efforts in World War II (see Box 3.4). The "Basic 7" food group guide promoted seven food groups (see Figure 3.3) to help in maintaining nutritional standards under

**BOX 3.4   RATIONING AND THE COMMUNITY
GARDENING MOVEMENT**

In 1942, sugar became the first consumer commodity to be rationed, with an allotment of one-half pound per person per week, which was about one-half of normal sugar consumption at the time. Consider that just over 50 years later, our consumption of added sweeteners had doubled to 103 pounds per person per year.[18] Coffee was rationed in 1942 to one pound per person every five weeks, about half of then normal consumption. When comparing diets then and now, one is compelled to ask: *When were we eating better?*

All of these were rationed until the end of the war: processed foods, meats, canned fish, cheese, evaporated milk, fats (butter, oil, margarine, and lard), and baby food. Consumers were limited in purchasing amounts, but food manufacturers and grocers were also proscribed limitations.

Although produce was not rationed during World War II, fresh fruit and vegetables were in short supply, which prompted the government to launch a victory garden campaign. An estimated 20 million victory gardens were planted throughout the US, many by families that may not have gardened before. Gardens were planted in backyards and empty lots; some neighborhoods pooled their resources by planting different kinds of foods and forming cooperatives. Sales of pressure cookers soared, as people started canning the bounty of their gardening efforts. It is estimated that 9–10 million tons were harvested. In general, these efforts were not maintained after 1946, but the gardens did become the predecessors of vegetable gardening movement that is experiencing a resurgence in the United States—not just in the backyards of private homes, but in place of front lawns,[19] in previously vacant lots in communities,[20] on school grounds (the Edible School Yard Project) and the at White House, where, in 2009, the first major garden was planted since Eleanor Roosevelt's Victory Garden during World War II.[21]

rationing conditions which restricted access to food. This model did not suggest eating prescribed numbers of servings from each group each day. Rather, a large number of groups were presented to choose from in case of limited supplies of a certain type of food. For example, if wheat-based foods in group 6 were scarce, the suggestion was to use more potatoes from group 3. This model not only provided a large variety of choices[17] but also operationalized the concept of food group substitutions that had earlier been introduced during the First World War.

The "Basic 7" Reissued (1946)

Following the war, the seven food groups were revised and reissued. This guide suggested numbers of servings needed daily from each food group, but its complexity and lack of specificity regarding serving sizes led to future modification and simplification. The elder author of this book remembers learning about the "Basic 7" in a fourth grade

FIGURE 3.3 National wartime nutrition guide promoting seven food groups. (From U.S. Department of Agriculture.)

class. Although always interested in food and nutrition, tried as she might, she could never remember more than six of the seven groups at one time. Butter, for example, being its own group, would be easy to overlook.

The USDA's "Basic Four" (1956)

Beginning in 1956, freed from the constraints of war, the USDA promoted a pared-down "Basic Four" food groups plan that recommended a minimum number of servings of foods from four food groups to provide the foundation for a nutritionally adequate diet. Many readers will recall the four food groups from school, and you may have already noticed the influence of the four food groups on this book's organization: Milk, Meat, Grains (Cereals and Bread), and Fruits and Vegetables.

Just as the Basic 7 suffered from being too complicated, the Basic Four suffered from being too simplified. As the nation was beginning to segue from a focus on eating enough to not eating too much, the Basic Four failed to provide adequate nutritional guidance regarding fats, sugar, sodium, and maximum number of servings from each food group. It recommended a minimum of two servings from the milk and milk products group and from the group composed of meat, fish, poultry, eggs, dry beans, and nuts. It recommended a minimum of four servings from fruits and vegetables, and four servings of grain products. It did not differentiate whole foods as compared to foods that had been structurally modified through processing.

However, to its credit, the meat and nondairy animal products group also contained dry beans, a highly nutritious while inexpensive food category that was often categorized as carbohydrate-rich (which it is) and, therefore, unfortunately mistakenly grouped with breads and cereals.

Conversely, however, unfortunately, potatoes, corn, and peas are starchy vegetables that are carbohydrate-rich that pass muster as vegetables. The decreasing number of food groups—12 to 7 and then to 4—highlights the difficulty of parsing dietary guidance for cooking and food consumption through the lenses of both botany and nutrition, while simultaneously being easy to understand and remember.

"Food Groups": Industry Influences on Education

In the middle of the twentieth century, the one food industry group that unarguably profited the most from the "foods groups" guidance was the dairy industry, which enjoyed the particular luxury of a food group dedicated solely to their products. The USDA recommendation of a "dairy group" gave the dairy industry a free ride into the hearts and minds of teachers and their charges. The National Dairy Council, established in 1915, provided free food and nutrition education materials to K-12 teachers in the guise of promoting the four food groups. The dairy industry monopolized K-12 nutrition education in this country by providing free lesson plans, masterly scope and sequence charts, handouts, recipes, teaching kits, and other irresistible aids to overworked undercompensated teachers with no to minimal food and nutrition background who were nevertheless assigned to teach about food and nutrition.

The meat and grain industry groups also benefitted from being a major player in the basic four. Representatives of the meat and grain industries develop and widely disseminate food and nutrition education materials although their items are neither as elaborate as those from dairy, nor have they been in the nutrition education business as long.

With all due respects, however, the fruit and vegetable industry finally caught up in the nutrition education business, too. In fact, in what may be the biggest *coup* of them all, fruits and vegetables developed a unique partnership with the National Institutes of Health (NIH) to promote their wares through the first industry-government dietary advice-dispensing partnership. This 1991 partnership is discussed later in this chapter.

In a classic (1980) withering essay, Joan Gussow warned that "… when anyone concerned with the public health is offered an 'educational package' by a food company,

> **BOX 3.5   WHO ARE YOU TO TELL ME WHAT TO EAT?**
>
> Consider the various sources of dietary recommendations and consider how funding sources might impact recommendations. The United States Government is a tax-supported organization (funded by the public at large through taxes); the allocation of such tax monies are potentially influenced by lobbyists and industry funds, as well as by the intended democratic desires of the public at large. The Food and Nutrition Board (FNB) of the Institute of Medicine (IOM), part of the National Academies, which provides scientific guidance on nutrition to policy makers, is a nonprofit organization but private. Current members elect new members, who are distinguished by their contributions of original research, and election is considered one of the highest honors in the professions.[22] Foundations are also nonprofit organizations. In the United States, the designation "nonprofit" means that the business has been granted a tax-exempt status by the Internal Revenue service. Any profits (surplus revenue above expenses) must go back into furthering the goals of the nonprofit, rather than be distributed to owners or shareholders. Chronic disease organizations, such as the American Cancer Society, the American Diabetes Association, and the American Heart Association are examples of nonprofit organizations that may also have a voluntary membership component. Food companies, however, are most likely for-profit organizations, and can be privately held or publically owned by shareholders; in either case potentially beholden to the interests of investors or shareholders.

it is essential to consider whether the public's health is really being served. While such consideration is proceeding, the caveat to hold in mind is an old one: 'He who pays the piper calls the tune.'"[23] See Box 3.5 for additional thoughts on sources of dietary recommendations.

1970s TO THE 1980s: PREVENTION OF DIET-RELATED CHRONIC DISEASES AND CONDITIONS

During the 1970s and 1980s, the direction of diet and health reports and food guides began to shift away from focusing on reducing the risk of nutrient deficiency diseases toward concern about heart disease (and later obesity) and other diet-related chronic conditions associated with excesses of calories, fat, sugar, and sodium.

Dietary Goals for the United States (1977–1978)

In 1968, the United States Senate formed a new committee to address hunger and malnutrition in the United States, although a battle over funding allocation meant the

Committee on Nutrition and Human Needs did not come into fruition until 1969. The existing Senate and House Agricultural committees were not interested in the problem of hunger; perhaps seeing food aid as antithetical to farming interests or perhaps from conceiving of poverty and hunger as a personal problem arising from poor character. In 1968, a member of the House Agricultural committee explained hunger as ignorance and indifference as to what constituted a balanced diet.[24] Hunger may not have been enough to fully snare the attention and cooperation of Congress on dietary goals. However, the possible connection between diet and heart disease did strike a chord—a reported 8 U.S. Senators died in office of heart disease in the 1960s and 1970s.[25] With cardiovascular disease (CVD), the Committee on Nutrition and Human Needs found a foothold in addressing connection between diet and health.

Eight years later, in early 1977, led by Senator George McGovern (D-SD), the Committee released recommendations called the *Dietary Goals for the American People.* The goals provided complementary food and nutrition recommendations that suggested Americans should eat less, specifically, less refined and processed sugars, total fat, saturated fat, and sodium. In order to achieve these nutritional guidelines, the recommendation called out specific food items that should be increased (fruit, vegetables, and whole grains) and those that should be decreased (foods high in salt, high in sugar, high in total fat, and high in animal fat, as well as eggs, butterfat, and high-fat dairy products).[26]

The message to eat less was not well received from the subject sugar, dairy, beef, and egg industries and even McGovern's home state cattle producers called for a recall of the report. Although arguments against the recommendations often entailed concerns about the underlying science, as Marion Nestle points out, the real point of contention was an economic one. Eventually, under intense pressure, the Committee released very shortly thereafter a second edition of the Dietary Goals, which came out stronger on obesity and alcohol, but softened the contentious language and recommendations on salt, egg, and meat consumption.

In particular, "reduce consumption of meat" was changed to "choose meats, poultry, and fish which will reduce saturated fat intake."[27] A clear statement about eating less of a particular food was modified to a more complex recommendation about eating less of foods with particular characteristics, albeit one perhaps more technically accurate. This guidance also influenced a transition within the poultry industry from providing mostly whole birds for sale to providing packaged cuts, in particular the boneless and skinless chicken breast in the early 1980s.[28]

Nonetheless, in the second edition of the *Dietary Goals for the United States* (1977), McGovern strongly connected food production and processing with medical care, thus pushing for a nutritional direction for agricultural policy. He stated in the preface: "Food production and processing is America's number one industry and medical care ranks number three. Nutrition is the common link. Nutrition is a spectrum which runs from food production at one end to health at the other end. By recognizing this connection... [hopefully] ... nutrition will become a major priority of this Nation's agricultural policy.[29]

The second edition's dietary goals and suggested strategies to achieve the goals are reproduced in Table 3.2. This is the first time government articulated dietary recommendations aimed at reducing diet-related chronic disease risk. While the goals and strategies may seem timid by today's standards, the report did usher in the era of food and nutrition guidance targeting postindustrial society chronic diseases. The prefaces for both editions of the

TABLE 3.2 *Dietary goals for the United States, 2nd edition*

US DIETARY GOALS	*STRATEGIES TO ACHIEVE THE GOALS*
To avoid being overweight, consume only as much energy (calories) as is expended; if overweight, decrease energy intake and increase energy expenditure.	
Increase the consumption of complex carbohydrates and "naturally occurring" sugars from about 27% of energy to about 48% of energy intake.	Increase consumption of fruits and vegetables and whole grains.
Reduce the consumption of refined and processed sugars by about 45% to account for about 10% of total energy intake.	Decrease consumption of refined and other processed sugars and foods high in such sugars.
Reduce overall fat consumption from approximately 40% to about 30% of energy intake.	Decrease consumption of foods high in total fats, and partially replace saturated fats, whether obtained from animal or vegetable sources, with poly-unsaturated fat.
Reduce saturated fat consumption to account for about 10% of total energy intake; balance that with poly-unsaturated and mono-unsaturated fats, which should account for about 10% of energy intake each.	Decrease consumption of animal fat, and choose meats, poultry, and fish, which will reduce saturated fat intake.
Reduce cholesterol consumption to about 300 mg a day.	Decrease consumption of butterfat, eggs, and other high cholesterol sources. Some consideration should be given to increase the cholesterol goal for premenopausal women, young children, and the elderly in order to obtain the nutritional benefit of eggs in the diet.
Limit the intake of sodium by reducing intake of salt to about 5 g [*sic*] a day.	Decrease consumption of salt and foods high in salt content.

Dietary Goals for the United States are both well worth reading to better appreciate the groundbreaking nature of this initiative.

USDA's "Hassle-Free Guide" (1979)

Subsequent to the second release of the Dietary Goals, the USDA published the *Hassle-Free Guide to a Better Diet* (1979) to address the problems associated with an excess of energy-rich nutrient-poor food choices. This guide not only recommended minimum servings of foods from the basic four food groups to provide the foundation for a nutritionally adequate diet, but also formally recognized and illustrated a fifth group to be used in moderation, namely, "Fats, Sweets, and Alcohol." This was the first time government dietary recommendations focused on both nutrient adequacy and moderation. This calling out of food groups associated with negative consequences arose from the increasing recognition that avoiding excess intake of some food components may be associated with preventing CVD, type-2 diabetes (at that time referred to as adult onset diabetes), and some forms of cancer.

The USDA, however, did not adopt the Dietary Goals directly, but instead, in conjunction with the Department of Health, Education, and Welfare (now the Department of Health and Human Services), conducted a supporting investigation of the underlying science, resulting in the first issuance of the *Dietary Guidelines for Americans* in 1980.[26] The Dietary Guidelines are discussed later in this chapter.

FOOD-BASED DIETARY GUIDANCE: FROM PYRAMID TO PLATE

The theme of nutrient adequacy and moderation introduced in 1979 was graphically depicted in the 1992 Food Guide Pyramid, which expanded the five food groups to six by separating fruits and vegetables. The Pyramid expressed the recommended servings of each food group: 6 to 11 servings of bread, cereal, rice, and pasta occupied the large base of the pyramid, followed by sectors depicting 3 to 5 servings of vegetables (which included potatoes), 2 to 4 servings of fruits, 2 to 3 servings of dairy products, 2 to 3 servings of meat, poultry, fish, dry beans, eggs, and nuts, and finally fats, oils, and sweets in the pyramid's small apex (to be used sparingly). Inside each sector were several images of representative foods, as well as symbols representing the fat and sugar contents of the foods (Figure 3.4).

In 2005, the USDA updated its guide with MyPyramid, a more abstract design often displayed without images of foods. MyPyramid replaced the hierarchical levels of the Food Guide Pyramid with colorful vertical wedges. Stairs were added with an icon of someone climbing them to represent exercise. The share of the pyramid allotted to grains now only narrowly edged out vegetables and milk, which each were of equal proportions. Fruits were next in size, followed by a narrower wedge for protein and a small sliver for oils. An unmarked white tip represented *discretionary calories* for items such as candy, alcohol, or additional food from other groups (Figure 3.5).

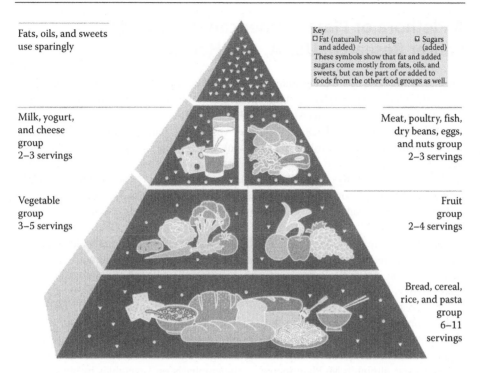

FIGURE 3.4   Food guide pyramid: A guide to daily food choices. (From U.S. Department of Agriculture.)

FIGURE 3.5   USDA's MyPyramid, depicting food groups through the use of colored vertical wedges. (From U.S. Department of Agriculture.)

A Plethora of Plates: American Institute for Cancer Research, USDA, and Harvard

Since 2011, the USDA's food guidance system and icon is MyPlate. The USDA did not invent the plate iconography. Its forerunner was published in 1999 by the American Institute for Cancer Research (AICR). AICR's New American Plate stresses *proportion* (aim for meals that are at least two-thirds of vegetables, fruits, whole grains or beans, and a maximum of one-third animal protein) and appropriate *portion* sizes. USDA's suggested serving sizes are recommended. In terms of number of servings, AICR suggests at least five servings a day of vegetables, 6–8 servings of other plant-based foods, such as whole grain brown rice, barley, quinoa, whole-grain breakfast cereal, oatmeal, and whole-wheat bread, and to limit consumption of red meat to less than 16 ounces per week (Figure 3.6).

MyPlate is the nutrition guide published by the USDA in 2011 to replace the pyramid food guidance system. My Plate consists of a diagram of a plate and glass. The plate is divided into the following categories: fruit, vegetable, grains, and protein. The glass is dairy.

As iconography, MyPlate is a helpful improvement over the pyramid. The use of a plate connects abstract nutrition information to a physical context of how we actually consume food. The visual reminder of proportions that should be on our plate is a simple answer to the question "am I eating enough vegetables?" (Figure 3.7).

However, although MyPlate increases the emphasis on vegetables by showing the vegetable sector as the largest sector (marginally bigger than grains), the underlying food components of the groups remain the same as promulgated in the Pyramid and

FIGURE 3.6   The AICR's new American plate stresses proportion. (From American Institute for Cancer Research.)

FIGURE 3.7 USDA's MyPlate provides a visual reminder of proportions. (From U.S. Department of Agriculture.)

MyPyramid. Starchy vegetables, such as potatoes, corn, lima beans, taro, and cassava are still vegetables. Coloring the vegetable sector of MyPlate green does create a more evocative reminder to eat swiss chard, collard greens, kale, or broccoli as one's vegetables, but technically, bread (as grains) and potatoes (as vegetables) could together comprise the majority of the plate.

Moreover, protein is a component, not a food. Since the "protein" category includes meat, fish, beans—and could also incorporate dairy—why not also have a "carbohydrate" category that includes grains and starchy vegetables? *Or should U.S. dietary guidance continue to include potatoes and other starchy vegetables in the "vegetable" category?*

Recall the discussion in the introduction chapter about how we think about food. Characterizing dietary guidance through the nutritional elements of foods that we group together in botanic and other cultural dietary groupings is a challenge. Potatoes are indeed vegetables, albeit part of the "starchy vegetable" subgroup.[30] And potatoes, as the United States Potato Board reminds us, are indeed a source of vitamin C and potassium, other micronutrients and fiber.[31] However, arguably, the greatest contribution of potatoes to our diet may be as the potentially lowest-cost source of potassium and fiber, as reported in a study partially funded by the Potato Board.[32] When a more varied diet is available and consumed, fruit, spinach, and broccoli, among others, contribute vitamin C and potassium. Then potatoes become, most simply, an excellent source of inexpensive carbohydrate calories.

In 2011, the Obama administration proposed limiting the amount of starchy vegetables served in school lunches to one cup per student per week, as well as banning them from breakfast. The Senate blocked the proposal with an amendment to the (then anticipated 2012) Farm Act prohibiting maximum limits placed on vegetables in school meal programs.[33] In 2014, potatoes once again became "hot," as the potato industry lobbied for inclusion in the list of approved foods in the Special Supplemental Nutrition Program for Women, Infants, and Children (WIC). Both the Senate and the House approved this inclusion into WIC by incorporating it into their own version of the 2015

Fiscal Year Agriculture Appropriations bill.[34] White potatoes have not been included in the WIC package since 2009. The reason is simple—the IOM had determined that the population receiving this benefit is already eating plenty of potatoes. The goal of the WIC program is to encourage consumption of foods which provide critical nutrients that the target population is otherwise less likely to consume. However, some argue that the consumption data on which this exclusion relies upon is now quite old.[35] And, in fact, as of early 2015, potatoes are now being recommended for inclusion in the WIC package over concerns that consumption of starchy vegetables by this population did not meet 2010 Dietary Guidelines.

The Healthy Eating Plate (Harvard School of Public Health and Medical School, 2011) aims to correct the weaknesses they perceive in the USDA's food graphic. The Healthy Eating Plate is divided into four groups: vegetables, fruits, healthy proteins, and whole grains. The graphic includes explanatory text and qualifying statements (healthy and whole). The text describes healthiest choices in the major food groups, provides basic nutrition advice to help to choose a healthy diet, distinguishes between potatoes and nonstarchy vegetables by making it clear that potatoes do not count as a vegetable, addresses fats, and emphasizes water and other noncaloric beverages, rather dairy (Figure 3.8).[36]

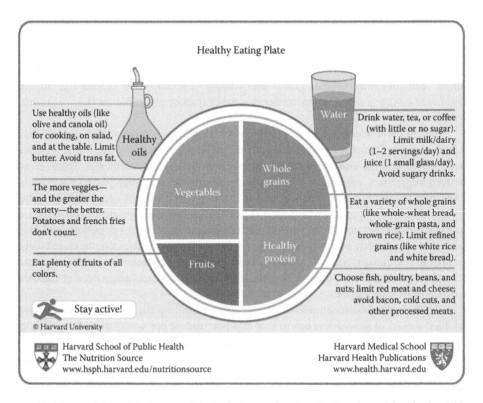

FIGURE 3.8   The Healthy Eating plate includes explanatory text and provides the healthiest choices. (With permission from Harvard University.)

DIETARY GUIDELINES FOR AMERICANS (1980)

The Dietary Guidelines provide a national nutritional policy statement that is used to support federal food assistance and nutrition education objectives. The Dietary Guidelines are a joint effort between the U.S. Department of Health and Human Services and the USDA and have been published every five years since 1980. The Departments voluntarily published the Guidelines through the 1990 edition. Subsequently, the National Nutrition Monitoring and Related Research Act of 1990 (7 U.S.C. 5341) mandated the five-year review and renewal cycle. The initial 1980 publication leveraged the second edition of McGovern's committee Dietary Goals, but starting in 1985, a Dietary Guidelines Advisory Committee (DGAC) consisting of nationally recognized experts in the field of nutrition and health was appointed. The DGAC is charged with reviewing the current scientific and medical knowledge and preparing a report for the Secretaries of HHS and USDA that provides recommendations for the next Dietary Guidelines.[37] Thus, for each cycle, two documents are produced—the DGAC report and the actual Dietary Guidelines.

The Dietary Guidelines evolved over time to make better use of nutrition science and to better communicate the science, and interestingly, increased in length. The first four editions (1980–1995) were small pocket-sized brochures aimed at consumers. In a departure from the first four consumer-focused editions, the fifth edition (2000) was published as a 39-page document that targeted policy-makers as well as consumers. This reflected a move by the government toward providing nutrition professionals who apply dietary guidance in their work with increased information about the science behind the consumer material.

Substantial changes appear in the sixth (2005) and seventh (2010) editions of the Guidelines. The 2005 edition is a 70-page booklet that serves solely as a nontechnical policy document to be a foundation of the work carried out by nutrition educators, nutritionists, and policymakers. The seventh edition (2010), a 95-page booklet, serves as a policy document intended for policymakers to design and carry out nutrition-related programs and for nutrition educators and health professionals, who develop nutrition curricula, teaching tools, and advice for consumers.

A comparison of the Dietary Goals of 1977 and the Dietary Guidelines of 2010 reveal that they are remarkably similar. As illustrated in Table 3.3, over the course of 35 years, the overall theme is to decrease consumption of total and saturated fat, refined carbohydrates and sodium by consuming a more plant-based diet that is richer in fruits, vegetables, whole grains and complex carbohydrates, and to achieve or maintain a healthy weight.

The 2015 Dietary Guidelines has not yet been finalized at the time of writing this chapter, and we cannot compare to see if the overall trend continues in 2015.

While both the sixth and seventh editions of the Dietary Guidelines are based on a search and review of the scientific literature, an even more robust, systematic approach was used to organize and evaluate the science underlying the 2010 Guidelines. Research quality was assessed using evidence-based reviews. The DGAC's scientific conclusions are given an overall grade, selecting from the following: Strong, Moderate, Limited; Expert Opinion Only, and Grade not assignable.

TABLE 3.3   A comparison of the Dietary Goals (1977) and Guidelines (2010)

DIETARY GOALS, 1977	STRATEGIES TO ATTAIN THE DIETARY GOALS	DIETARY GUIDELINES FOR AMERICANS, 2010	STRATEGIES TO MEET THE DIETARY GOALS
Avoid overweight.	Consume only as much energy (calories) as is expended; if overweight, decrease energy intake and increase energy expenditure.	Maintain calorie balance over time to achieve and sustain a healthy weight.	
Increase the consumption of complex carbohydrates and "naturally occurring" sugars from about 27% of energy to about 48% of energy intake.	Increase consumption of fruits and vegetables and whole grains.	Eat more vegetables and fruits. Eat a variety of vegetables, especially beans, peas, and vegetables that are dark green, red, or orange. Get at least half of your grains from whole-grain sources, such as whole-wheat bread or pasta, brown rice, oatmeal, and cracked wheat.	Try to eat fewer foods made with refined grains (almost anything made with white flour or white rice).
Reduce the consumption of refined and processed sugars by about 45% to account for about 10% of total energy intake.	Decrease consumption of refined and other processed sugars and foods high in such sugars.		Eat less foods with added sugars (especially sodas and juices sweetened with sugar or high-fructose corn syrup).
Reduce overall fat consumption from approximately 40% to about 30% of energy intake.	Decrease consumption of foods high in total fats, and partially replace saturated fats, whether obtained from animal or vegetable sources, with poly-unsaturated fat. (Note: The Dietary Goals recommend increasing the carbohydrate content of the diet to 58% of calories, decreasing fat to 30%, while keeping protein constant, changes that would require switching to a more plant-based diet, even if not stated directly as such.)	Use oils whenever possible in place of solid fats like butter, margarine, and shortening. Choose fat-free or low-fat milk and dairy products.	Eat less solid fat (which we get mainly from whole milk, butter, cream, and red meat).

(Continued)

TABLE 3.3 (Continued)   A comparison of the Dietary Goals (1977) and Guidelines (2010)

DIETARY GOALS, 1977	STRATEGIES TO ATTAIN THE DIETARY GOALS	DIETARY GUIDELINES FOR AMERICANS, 2010	STRATEGIES TO MEET THE DIETARY GOALS
Reduce saturated fat consumption to account for about 10% of total energy intake; and balance that with poly-unsaturated and mono-unsaturated fats, which should account for about 10% of energy intake each.	Decrease consumption of animal fat, and choose meats, poultry and fish, which will reduce saturated fat intake.	Choose a variety of protein sources, such as seafood, lean meat and poultry, eggs, beans and peas, and nuts and seeds. The most healthy protein packages are low in saturated fat and contain healthy fats. Eat fish or other seafood more often, in place of red meat.	Consume less than 10% of calories from saturated fats (mainly in red meat and dairy products). Try not to get any trans fats by limiting your intake of foods made with partially hydrogenated oils.
Reduce cholesterol consumption to about 300 mg a day.	Decrease consumption of butterfat, eggs, and other high cholesterol sources. Some consideration should be given to increase the cholesterol goal for premenopausal women, young children, and the elderly to obtain the nutritional benefit of eggs in the diet.		Consume less than 300 mg of cholesterol per day.
Limit the intake of sodium by reducing intake of salt to about 5 g a day (which contains about 2400 mg sodium).	Decrease consumption of salt and foods high in salt content.	Choose a variety of protein sources, such as seafood, lean meat and poultry, eggs, beans and peas, and nuts and seeds. The most healthy protein packages are low in sodium.	Reduce daily sodium intake to less than 2300 mg or to less than 1500 mg if you have high blood pressure, diabetes, or chronic kidney disease, are in your 50s or older, or are African American.

The DGAC's scientific conclusions were derived from an objective analysis of the preponderance and quality of published research conducted by the staff of USDA's Nutrition Evidence Library (NEL). The NEL evaluates, synthesizes, and grades the strength of the evidence to support conclusions. The web-based NEL allowed the Committee to ask and process more questions in a systematic, transparent, and evidence-based way.

For example, the 2010 Committee asked the NEL to determine the relationship between glycemic index (GI) and type-2 diabetes. The same topic was considered by the 2005 DGAC. The conclusions expressed in the 2010 DGAC Report are informed by the evidence compiled for the 2005 DGAC Report, but are based primarily on the NEL evidence gathered from 2000 on. The NEL concluded: "a moderate body of inconsistent evidence supports a relationship between high GI and type-2 diabetes." The overall grade for this conclusion is Moderate.[38]

In addition to reviewing the scientific literature, consumer, professional, and food industry comments from individuals and organizations were solicited, with transcripts of oral and written comments available online. The record is particularly thorough for the 2010 version of the DGA, and no doubt will likely be so for the 2015 version as well. Written public comments were received throughout the Committee's deliberations through an electronic database designed for collecting public comments. This database allowed for the generation of public comment reports as a result of a query by key topic areas. Comments received on and before late April, 2010, were compiled into these reports and shared with all Committee members to be considered before the release of their final report the following month. Three-quarters of the nearly 1000 comments received between October 2008 and April 2010 were posted and are available at www.dietaryguidelines.gov. A general description of the types of comments received and the process used for their collection is described in the Appendix E-5 (Public Comments) of the 2010 DGAC report.

Oral comments were received from organizations and individuals. In general, the input for 2010 provided suggestions for improving the guidelines. For example, on behalf of the United Fresh Produce Association, Lorelei DiSiogra, EdD, RD, recommended replacing such vague messages as "food to encourage" and "make wiser food choices" with concrete guidance. She advocated using "Half Your Plate Should be Fruits and Vegetables" to graphically illustrate the importance of eating more fruits and vegetables at every meal, pointing to NCI and USDA research demonstrating the efficacy of the "Half Your Plate" message.[39]

The online comment process for the DGA 2015 is open at the time this chapter is being finalized. Meetings are being held by webcast, with notebooks of prior meeting, including presentation slides, available online. Comments can be made online on any topic, but some of the subcommittees have posted specific questions for which they are seeking public input.[40]

For the 2015 DGAC, just as for the 2010 DGAC, oral comments were received, some on behalf of organizations that represent food industry organizations. The "revolving door" that is sometimes referenced between government and industry is also manifest in the crafting of the Dietary Guidelines. Dr. Joan Slavin, a former 2010 DGAC member, testified to the 2015 DGAC on behalf of the grain industry coalition group known informally as the Grain Chain.[41]

BOX 3.6   WORKING SUBCOMMITTEES FOR DGAC

2010 DGAC	2015 DGAC
1. Sodium, Potassium, and Water	1. Food and Nutrient Intakes and Health: Current Status and Trends
2. Nutrient Adequacy	2. Dietary Patterns, Foods and Nutrients, and Health Outcomes
3. Energy Balance and Weight	3. Diet and Physical Activity Behavior Change
4. Carbohydrates and Protein	4. Food and Physical Activity Environments
5. Ethanol	5. Food Sustainability and Safety
6. Fatty Acids	
7. Food Safety and Technology	

Oldways, Sneak Peek at the 2015 Dietary Guidelines, January 23, 2014, http://oldwayspt.org/community/blog/sneak-peek-2015-dietary-guidelines, accessed June 9, 2014.

Although lacking information about the 2015 Dietary Guidelines at the time of writing this chapter, some likely trends are still evident. Quantitative recommendations are expected for saturated fats, sodium, and added sugar. The DGAC committee also examined literature on low-calorie sweeteners, as the higher use of low-calorie sweeteners could be one consequence of a recommendation regarding added sugar. The DGAC emphasized replacement and shifts in food intake and overall dietary patterns, rather than solely on reduction.

Sustainability is a certain topic of discussion, given that a subcommittee is assigned to the topic. The Food Sustainability subcommittee is actively seeking public input on approaches and current examples of sustainability for both food systems and food groups.[40] With this idea of sustainability in mind, one question that seems likely to be raised at the food group level is if a fish consumption dietary recommendation is sustainable.

If the recommendation to consume fish is based on a nutrient profile, are there other dietary options available that can also provide those nutrients?

Conversely, there also appears to be a focus on dietary patterns instead of nutrients for the 2015 Dietary Guidelines. See Box 3.6 for a comparison of the working subcommittees of the 2015 DGAC to those of 2010.

Sustainability: What Is a Sustainable Diet and Is It Compatible with Current Dietary Guidelines?

The word "sustainable" is derived from two the Latin words: *sub* (from underneath) and *tenere* (to have, hold, possess), thus sustainable means to "uphold from underneath" or be able to endure. A sustainable diet, therefore, is one that can last for a long time. Like all good diets, it must taste good and be affordable. But to endure over the long run, it must also conserve and regenerate natural resources, reduce greenhouse gas emissions, preserve biodiversity, promote access to healthy food and clean water, and optimal health and well-being for all people.

As recently as 30 years ago, people were not necessarily considering the effect of dietary choices on water use, land use, waste, biodiversity, and the social, ethical, and economic issues that threaten future food security. It was not until 1976, when the concept of sustainability was introduced in a food-related activities guide for teachers[42]—but the concept was not given a name. A decade later, Gussow and Clancy[43] warned that we must adopt a "sustainable diet" to save the planet. They proposed developing dietary guidelines for sustainability to help consumers "make food choices that contribute to the protection of our natural resources." At last, this paradigm had a name.

Dietary guidelines for sustainability should be incorporated in future revisions to food-based dietary guidelines. The sustainability guidelines should include information about labor and trade ethics, the role of packaging within a healthy and sustainable food system, and labeling that provides consumers with unambiguous information about the food, such as its country of origin, how it was raised, harvested, processed, as well as its ingredients and nutritional content.[44] And, of course, guidelines must suggest the foods we should eat, as proposed by Gussow in the first article of the first issue of the journal, *Hunger and Environmental Nutrition.*[45]

- Seasonal fruits and vegetables: dried, canned, or frozen when fresh is unavailable. That fruits are not available in winter should be noted, reminding consumers that vegetables provide the same nutrients as fruits.
- Protein sources: beans and nuts, dry foods shippable without refrigeration. Pasture-raised animal sources of protein; seafood in moderation.

Nutritionally, a sustainable diet would follow the recommendations outlined in many dietary guidelines (eat a healthy, more plant-based diet rich in fruits, vegetables, whole grains and oils from plants) and provide enough additional information for individuals to make choices that contribute to the protection of our natural resources.

The Public Health Concerns behind the Dietary Guidelines

In prior decades, the health concerns driving our dietary recommendations were emphatically those of cardiovascular health. Mortality from CVD had become a significant problem in the United States and other developed countries. By 1950, one out of every three men in the United States developed CVD before reaching the age 60 and it was the leading cause of death. CVD was twice as common as cancer.[46] In 1963, 805 people per 100,000 in the U.S. died due to CVD, stroke and other circulatory diseases, but in 2010, less than 236 per 100,000 in the U.S. died of these causes.[47]

The diet-heart hypothesis proposes a link between dietary fat intake and risk of CVD, and encourages total dietary fat reduction (with an emphasis on saturated fat) without necessarily further specifying the replacement nutrient.[48] In the United States, the overall dietary focus was on reducing the consumption of saturated fat and cholesterol while increasing consumption of complex carbohydrates. This meant suggestions to consume margarine instead of butter, low fat, or skim milk instead of whole milk, and to reduce consumption of eggs, particularly yolks.

The general emphasis on fat reduction also meant a flurry of activity in the food industry to produce new "low fat" food products. One objective of Healthy People 2000 was that there would 5000 new low- and reduced-fat products by 2010. Private industry met this goal by 1995.[49]

Unfortunately, food producers still had to make lower in fat foods palatable, and this usually meant adding sweeteners of some kind. The vision of the heart healthy model diet has led to a "substantial decline" of the percentage in our energy intake attributed to saturated fats, but has also apparently contributed to a compensatory rise in added sugars and refined carbohydrates.[50]

Is it possible that our dietary advice to address one dietary-related health concern helped foster the rise of other such concerns, namely obesity and diabetes?

Gary Taubes, a science journalist, began critically questioning the emphasis on a low-fat diet in 2002, contending that the replacement of fat with carbohydrates (often refined), drove the obesity epidemic, noting Walter Willett's studies and Katherine Flegal's statistics showing a steep rise of obesity in the 1980s that continued in the 1990s, consistent through "all segments of American society."[51]

Furthermore, by 2006, evidence was available from a large randomized control trial study of women that a low (saturated) fat approach, replaced principally by carbohydrates, was not effective in reducing coronary heart disease (CHD), stroke, or CVD incidence.[52] And a 2009 pooled analysis of 11 American and European cohort studies ($n = 344,696$ persons) found no association between replacing saturated fat with carbohydrates and a decreased risk of ischemic heart disease (IHD).[53] Similarly, a 2010 meta analysis of 21 cohort studies ($n = 347,747$ subjects) found no significant association between intake of saturated fat compared with carbohydrates and risk of IHD, stroke, and total cardiovascular events.[54]

Instead of emphasizing a low fat dietary pattern in which carbohydrates replace saturated fats, should our recommendations simply be to substitute polyunstatured fat for saturated fat?

Could this improve our cardiovascular health without leading to other unwanted health consequences?

According to a June 2014 review article, meta analyses have shown that the relationship between consumption of saturated fatty acids on CVD risk is not independent of the replacement nutrient. This review also found that the consumption of animal products was not necessarily associated with increased CVD risk, but consumption of nut and olive oil was associated with reduced CVD risk.[48] In general, the article concluded the overall "matrix" of a food is more important than is fatty acid profile when considering the potential effect of a food on CVD risk.

Thus, current research is emphasizing "matrices" and "dietary patterns" instead of isolated micronutrients. Consider that one of the subcommittees of the 2015 Dietary Guidelines Committee is "Dietary Patterns, Foods and Nutrients, and Health Outcomes." *Did we reach "peak nutrient" emphasis in the late 1990s?* We turn back now to review the development of the dietary reference intakes (DRIs).

Dietary Reference Intakes

The table of RDAs was updated periodically after inception in 1941 to incorporate new scientific knowledge. The 1974 edition clarified that RDAs are the levels of intake of essential nutrients judged to be adequate to meet the known needs of practically all healthy persons. The last update to the RDAs was published in 1989. In 1995, the FNB at the Institute of Medicine of The National Academies (IOM, formerly National Academy of Sciences) decided that a more comprehensive approach was needed, in part because the RDAs did not distinguish well enough between population guidelines and those for individuals. A population guideline or level that is adequate for practically all healthy persons may be too great for most (97%) healthy persons.[55]

Since 1997, the FNB has issued a series of nutrient reference values, collectively referred to as DRIs. Included within the DRIs are the following four reference values: the Estimated Average Requirement (EAR), the Adequate Intake (AI), the RDA, and the Tolerable Upper Intake Level (UL). See Box 3.7 for definitions of these terms. DRI is the overall term for this set of reference values.

The DRIs vary by age and gender and are used to plan and assess nutrient intakes of healthy people in a variety of settings. For example, individuals would look to RDAs or AIs as a goal for *average* daily intake and ULs as an indicator of highest safe intake. EARs would be used to set policies for food supplies for groups and populations.[55]

**BOX 3.7   DEFINITIONS OF THE FOUR
REFERENCE VALUES OF THE DRIs**

- *Estimated Average Requirement (EAR):* Reflects the estimated median requirement and is particularly appropriate for applications related to planning and assessing intakes for groups of persons.
- *Recommended Dietary Allowance (RDA):* Derived from the EAR. Average daily level of intake sufficient to meet the nutrient requirements of nearly all (97–98%) healthy people.
- *Tolerable Upper Intake Level (UL):* As intake increases above the UL, the potential risk of adverse effects may increase. The UL is the highest average daily intake that is likely to pose no risk of adverse effects to almost all individuals in the general population.
- *Adequate Intake (AI):* Established when evidence is insufficient to develop an RDA and is set at a level assumed to ensure nutritional adequacy.
- *Acceptable Macronutrient Distribution Range (AMDR):* An intake range for an energy source associated with reduced risk of chronic disease.

United States Department of Agriculture, Agricultural Research Service, News 1999, Dietary Reference Intakes (DRIs)—New Dietary Guidelines Really Are New!, http://www.ars.usda.gov/Research/ docs.htm?docid=10870, accessed September 12, 2014.

Marion Nestle, for one, has had the temerity to admit that the DRIs are almost impenetrable.[56]

Just when you thought it was complicated enough....note that a nutrient's daily value (DV) that appears on food labels is not necessarily always the same as one's RDA or AI for that nutrient. The DVs were developed by the FDA for the purpose of providing information about the nutrients in the labeled product by serving basis. See Chapter 4, "Nutrition Information Policies," for more details on DVs.

Dietary Patterns

The Harvard group behind the "Healthy Eating Plate" described above recommends a plant-based diet rich in fruits, vegetables, and whole grains. Instead of emphasizing low-fat, the emphasis is on choosing healthier fats, such as olive oil, limiting consumption of saturated fats, and avoiding trans fat. Sugary drinks and salt should also be limited. The overall message includes keeping calories in check in order to avoid weight gain.

Proponents of the "Mediterranean Diet" have long advocated for this style of eating—which is higher in fat content (not lower) than the typical American diet—to reduce the risk and mortality for a number of chronic diseases. Not a specific diet, the Mediterranean eating plan is a collection of eating habits traditionally followed by people living in the areas bordering the Mediterranean Sea. The eating style diet refers to a dietary profile commonly available in the early 1960s in the Mediterranean regions and characterized by a high consumption of fruit, vegetables, legumes, and complex carbohydrates, with a moderate consumption of fish, and the consumption of olive oil as the main source of fats and a low-to-moderate amount of red wine during meals.[57]

In general, researchers are now recommending that the focus of the diet-heart paradigm be shifted away from restricted fat intake and toward reduced consumption of refined carbohydrates. They are recommending a moderately restricted carbohydrate intake but rich in vegetable fat and vegetable protein to improve blood lipid profiles.[50] Benefits of this plant-based, low-carbohydrate diet are likely to arise from higher intake of polyunsaturated fats, fiber, and micronutrients as well as the reduced glycemic load in the dietary pattern.

Criticisms of the Dietary Guidelines

As noted above, as part of the preparation of The Dietary Guidelines, the DGAC releases a report covering their evaluation. The reports of the DGACs and the resulting Dietary Guidelines have never been—and never will be—above reproach, as indicated by the several examples presented here concerning the 2010 DGAC Report and the 2010 edition of the Dietary Guidelines.

The 2010 Report of the DGAC has been criticized for making strong recommendations in light of weak evidence, and of making recommendations that are premature. For example, the report unambiguously states that "healthy diets are high in carbohydrates" yet on the same page it calls for developing and validating carbohydrate assessment methods because "studies of carbohydrates and health outcomes on a macronutrient level are often inconsistent or ambiguous due to inaccurate measures and varying food categorizations and definitions."[58] The report seems to be asking for studies that back up the strong statement "healthy diets are high in carbohydrates."

Note the change in style of language between the 2000 Dietary Guidelines and the 2005 Dietary Guidelines. Starting with 2005, the Dietary Guidelines were produced as policy documents designed for nutrition professionals and policy makers, people who require unambiguous information they can eventually translate into consumer messages. Accordingly, in a major departure from previous editions, the 2005 DG presents quantitative recommendations in terms of cups of food to be eaten, although still referencing number of servings as a comparison. Moreover, the recommended number of servings of fruits and vegetables generally increased, ranging from 4 to 13 servings, depending on calorie level. Four servings would suffice for a diet of 1000 calories but 13 servings would be necessary for a diet at the 3000–3200 calorie level. Expressing this range of 4–13 servings as cups translates as 2 cups to 6½ cups, which seems like a much less daunting goal to achieve. The "cup" unit of measure was felt to provide more precise guidance than the ambiguous "serving."

Although the idea of measuring in cups is more accessible to consumers than servings, arguably dieticians and other nutrition science professionals might prefer an even more unambiguous method of expressing quantities as grams (which measure weight or mass), not cups (ounces and spoons), which measure volume (the amount of space a food takes up).

Concerns have also been expressed regarding the use of the Dietary Guidelines to set standards for federal food programs. For example, the 2010 Dietary Guidelines stated an upper limit of 35% of calories for fat. This cap can counterproductively distort menus, especially if incorrectly applied to individual foods or meals. As the Harvard team points out, applying the fat limit to a side dish of vegetables might exclude broccoli drizzled with olive oil but include mashed potato with butter.[59]

The role of the USDA as coauthor of the Dietary Guidelines has also been criticized. Consider this: *Is it a conflict of interests for the USDA to serve as a lead agency in the development of dietary guidance when their mandate is protecting the interests of agriculture?*

When the USDA was created in 1852, two roles were assigned to it—promoting a sufficient and reliable food supply and advising the public about subjects related to agriculture.[60] The agency's current mission is to "provide leadership on food, agriculture, natural resources, and related issues based on sound public policy, the best available science, and efficient management."

NYU's Nestle has long argued that *food industry influences nutrition and health* (the subtitle of her book) and that it is a conflict of interests for the USDA to develop dietary guidelines because the agency's mandate is to get Americans to eat more (although many should eat less).[56] In concrete terms, according to Harvard's Willett, "... the continued failure to highlight the need to cut back on red meat and limit most dairy products suggests that 'Big Beef' and 'Big Dairy' retain their strong influence within this department. Might it be time for the USDA to recuse itself because of conflicts of interest and get out of the business of dietary advice?"[61] This statement from Willett was published online post the release of the 2010 Dietary Guidelines and is still available as of the fall of 2014.

Research has, of course, continued since the 2010 Dietary Guidelines were published, providing new dietary information. Some research, as a 2014 review article notes, has not found that the need to cut back on red meat to be so clear-cut.[48] Even so, the language used by the USDA in the 2010 Dietary Guidelines is not a strong, clear message when it comes to discussing food that contemporary scientific consensus considered important to reduce, as is evident by the statement to reduce "solid fats" instead of plainly stating reduce butter or red meat.[62] As another example, see the trend of language in the Dietary Guidelines regarding sugar usage in Box 3.8.

There is also a putative weakness in the manner in which background literature is reviewed for the DGA. Nutritionists critical of the NEL review process suggest that NEL reviewers who are affiliated with the Academy of Nutrition and Dietetics may not be objective because their professional organization relies heavily on food industry to underwrite many educational programs for its members.[58] The USDA may thus be relying on the dietetics focus of Academy-trained nutritionists to confirm their own industry-friendly guidelines.

The 2015 Dietary Guidelines are not yet available; yet criticisms (aka concerns) are being leveled that the committee is acting outside the boundaries of its mandate by considering nondietary factors. Interestingly, one of the groups leveling this criticism sees the problem as being that industry is not involved enough—charging that the committee is composed of "nearly all epidemiologists from elite academic institutions with no direct experience in the practical realities of how food is produced and what average Americans may choose to eat."[63] *But where does the data for our food policies come from?*

The data epidemiologists rely on includes dietary information reported by individuals' dietary recall as part of the National Health and Nutrition Examination Survey

---

BOX 3.8   EVOLUTION OF THE RECOMMENDATIONS FOR SUGAR IN THE *DIETARY GUIDELINES FOR AMERICANS*, 1980–2010

- 1980: Avoid too much sugar.
- 1985: Avoid too much sugar.
- 1990: Use sugars only in moderation.
- 1995: Choose a diet moderate in sugars.
- 2000: Choose beverages and foods to moderate your intake of sugars.
- 2005: Choose and prepare foods and beverages with little added sugars or caloric sweeteners, such as amounts suggested by the USDA Food Guide and the DASH eating plan.
- 2010: Limit the amount of added sugars when cooking or eating (using less table sugar); consume fewer and smaller portions of foods and beverages that contain added sugars, such as such grain-based desserts, sodas, and other sugar-sweetened beverages.
- 2015 _____

_____

[fill in the blank].

(NHANES) study and other studies, such as the Nurses' Health Study, which is a large cohort study.[64] NHANES is examined in Chapter 4, "Nutrition Information Policies."

One criticism leveled from the data collection perspective is that cohort studies are observational, and can suggest associations and correlations, but cannot show direct causation. Gary Taubes, the science journalist mentioned above, points out observational studies are a tool for developing hypotheses, but not a tool for directly developing a dietary edict.[65] Conversely, randomized trials are not practical for all research questions; and having a narrow focus that randomized trials are the "gold standard" to the exclusion of other forms of scientific inquiry may diminish opportunities to build scientific knowledge.

CONCLUSION

Establishing dietary guidelines implies having a quantitative (scientific) understanding of what humans need physiologically from our diet as well as what characteristics foods contribute to our diet. At any one time, dietary guidance can only be based on what is known and generally agreed on at the time the dietary guidance is given. Research, whether government or food industry based, may race ahead following differing areas and ideas. New information is constantly available, which individuals may freely adopt, but there can be a lag between knowledge and guidance, particularly at the governmental level, which may contribute to consumer confusion.

Through the Dietary Guidelines, the United States has adopted an evidence-based approach for evaluation nutritional research and promoting dietary advice. Nonetheless, the food industry may still have an undue influence on the development of the guidelines. Recall that MyPlate was developed by USDA to operationalize the dietary advice recommended by the 2010 Dietary Guidelines. In turn, Harvard's Healthy Eating Plate attempts to correct what its developers perceive to be the weaknesses in MyPlate, which they attribute to USDA's mandated support of the food industry.

Even in the absence of industry influence on the actual recommendations, food producers can have a significant impact on how dietary guidance can become our actual diet through the characteristics of the foods we consume. The first Dietary Goals (1977) suggested American should eat less, and specifically less refined and processed sugars, total fat, saturated fat, and sodium. Seeking to profit from consumers looking to follow dietary guidance, the food industry quickly provided us with delicious fat-free and low-fat food products. An undue emphasis on fat as a singular culprit may have contributed to a general increase in the consumption of refined carbohydrates, encouraging a new dietary health problem.

Looking forward, the content and the impact of the 2015 Dietary Guidelines remain to be seen. But given the current trends toward dietary patterns and sustainability, and accepting the fact that based on history and politics the Dietary Guidelines may not come out with strong recommendations to eat less of any particular food item, is it possible that our next dietary guidance theme might simply be Michael Pollan's summation: "Eat less, mostly plants"?

REFERENCES

1. First Lady Michelle Obama. MyPlate Press Conference. Park M. Plate Icon to guide Americans to healthier eating. June 2, 2011. *CNN.* Available at: http://www.cnn.com/2011/HEALTH/06/02/usda.new.food.plate/, accessed November 18, 2014.

2. Davis C, Santos E. Dietary recommendations and how they have changed over time. In Frazao E (ed.). *America's Eating Habits: Changes and Consequences,* Chapter 2. U.S. Department of Agriculture, Economic Research Service, Food and Rural Economics Division. Agriculture Information Bulletin No. 750. May 1999. Available at: http://www.ers.usda.gov/publications/aib-agricultural-information-bulletin/aib750.aspx, accessed June 6, 2014.

3. Preparation and use of food-based dietary guidelines. Report of a Joint FAO/WHO Consultation. WHO Technical Report Series 880. Geneva: World Health Organization, 1998. Available at: http://apps.who.int/iris/bitstream/10665/42051/1/WHO_TRS_880.pdf, accessed October 26, 2014.

4. Darby WJ. Contributions of Atwater and USDA to knowledge of nutrient requirements. *J. Nutr.* 1994;124(9 Suppl):1733S–1737S.

5. Combs GF. Celebration of the past: Nutrition at USDA. *J. Nutr.* 1994;124(9 Suppl): 1728S–1732S.

6. U.S. Department of Agriculture, Agricultural Research Service. 2014. USDA National Nutrient Database for Standard Reference, Release 27. Nutrient Data Laboratory Home Page. Available at: http://www.ars.usda.gov/ba/bhnrc/ndl, accessed September 11, 2014.

7. Atwater WO. *Foods: Nutritive Value and Cost.* Washington, DC: Government Printing Office, 1894. (USDA Farmers Bulletin No. 23). http://www.ars.usda.gov/SP2UserFiles/Place/12355000/pdf/hist/oes_1894_farm_bul_23.pdf. Accessed September 12, 2014.

8. Atwater WO. *Principles of Nutrition and Nutritive Value of Food.* Washington, DC: Government Printing Office, 1902. (USDA, Farmers' Bulletin No. 142.) Available at: http://openlibrary.org/books/OL24231176M/Principles_of_nutrition_and_nutritive_value_of_food, accessed September 11, 2014.

9. Hunt CL. Food for Young Children. 1916. USDA. Farmers' bulletin no. 717. Available at: http://openlibrary.org/books/OL24364099M/Food_for_young_children/, accessed September 12, 2014.

10. Hunt CL, Atwater HW. How to Select Foods: What the Body Needs. 1917. USDA. Farmers' Bulletin 808. Available at: http://archive.org/details/howtoselectfoods00hunt, accessed September 12, 2014.

11. Winslow EA. Food Values: How Foods Meet Body Needs. 121. USDA. Farmers' Bulletin 975. Available at: https://archive.org/details/foodvalueshowfoo975wins, accessed September 12, 2014.

12. Cornell University Albert R. Mann Library, Home economists in World War I. Available at: http://exhibits.mannlib.cornell.edu/meatlesswheatless/meatless-wheatless.php?content=three, accessed September 12, 2014.

13. Stiebeling H. *Food Budgets for Nutrition and Production Programs.* Washington, DC: United States Department of Agriculture, Miscellaneous Publication No 183, 1933.

14. Stiebeling H and Ward M. *Diets at Four Levels of Nutritive Content.* Washington DC: United States Department of Agriculture, Circular No. 296, 1933.

15. Rosenfeld L. Vitamine-vitamin. The early years of discovery. *Clin. Chem.* 1997;43(4): 680–685.

16. National Research Council. *Recommended Dietary Allowances.* Washington, DC: The National Academies Press, 1941. Available at: http://www.nap.edu/catalog.php?record_id=13286, accessed June 8, 2014.

17. Welsh SO, Davis C, Shaw A. *USDA's* Food Guide: Background and Development. USDA. Human Nutrition Information Center. Miscellaneous Publication Number 1514. September 1993. Available at: http://www.cnpp.usda.gov/Publications/MyPyramid/OriginalFood Guide Pyramids/FGP/FGPBackgroundAndDevelopment.pdf/, accessed June 7, 2014.

18. Haley S, Reed J, Lin B-H, Cook A. Sweetener Consumption in the United States: Distribution by Demographic and Product Characteristics. USDA. SSS 243–01, http://www.ers.usda.gov/media/ 326278/sss24301_002.pdf, accessed June 8, 2014.

19. Soler I. *The Edible Front Yard: The Mow-Less, Grow-More Plan for a Beautiful, Bountiful Garden.* Portland, OR: Timber Press, 2011.

20. Lawson LJ. City Bountiful: A Century of Community Gardening in America. Berkeley: University of California Press, 2005.

21. Obama M. *American Grown: The Story of the White House Kitchen Garden and Gardens Across America.* New York: Crown Publishers, 2012.

22. The National Academies, http://www.nationalacademies.org/memarea/index.html, accessed September 12, 2014.

23. Gussow JD. Who pays the piper? *Teachers College Record.* 1980;81(4, summer):448–66.

24. Raphael C, *Investigated Reporting, Muckrakers, Regulators, and the Struggle over Television Documentary.* Illinois: University of Chicago Press, 2005: 47.

25. Aubrey A, Why we got fatter during the fat-free boom, *NPR,* March 28, 2014. Available at: http://www.npr.org/blogs/thesalt/2014/03/28/295332576/why-we-got-fatter-during-the-fat-free-food-boom, accessed September 12, 2014.

26. Dietary Guidelines Advisory Committee, 2010. Report of the Dietary Guidelines Advisory Committee on the Dietary Guidelines for Americans, 2010, Appendix E-4, History of the Dietary Guidelines for Americans. Available at: http://cnpp.usda.gov/Publications/DietaryGuidelines/2010/DGAC/Report/E-Appendix-E-4-History.pdf, accessed September 12, 2014.

27. Nestle M. *Food Politics: How the Food Industry Influences Nutrition and Health.* Berkeley: University of California Press, 10th Edition, 2013:38–42.

28. The National Chicken Council. Available at: http://www.nationalchickencouncil.org/about-the-industry/history/, accessed September 12, 2014.

29. Select Committee on Nutrition and Human Needs United States Senate. George McGovern, Chairman. Preface. *Dietary Goals for the United States,* 2nd Edition. Washington, DC: U.S. Government Printing Office, 1977.

30. United States Department of Agriculture, What Foods are in the vegetable group? Available at: http://www.choosemyplate.gov/food-groups/vegetables.html, accessed June 9, 2014.

31. The United States Potato Board. Available at: http://www.potatogoodness.com/, accessed June 9, 2014.

32. Drewnowksi A, Rehm C. Vegetable cost metrics show that potatoes and beans provide most nutrients per penny. *PLoS One* 8(5):e63277.

33. Pear R, Senate saves the Potato on School Lunch Menus, *The New York Times,* October 18, 2011. Available at: http://www.nytimes.com/2011/10/19/us/politics/potatoes-get-senate-protection-on-school-lunch-menus.html?_r=1&, accessed June 9, 2014.

34. Yeager H, A good week for Big Potato, *The Washington Post,* May 23, 2015. Available at: http://www.washingtonpost.com/blogs/post-politics/wp/2014/05/23/a-good-week-for-big-potato/, accessed June 9, 2014.

35. The National Potato Council, Separating Fact from Fiction in the WIC/Potato debate. Available at: http://nationalpotatocouncil.org/, accessed June 9, 2014.

36. Anon. Healthy Eating Plate vs. USDA's My Plate. The Nutrition Source, Department of Nutrition at the Harvard School of Public Health. Available at: http://www.hsph.harvard.edu/nutritionsource, accessed September 12, 2014.

37. Health.Gov, Dietary Guidelines for Americans, 2015. Available at: http://www.health.gov/dietaryguidelines/2015.asp, accessed September 12, 2014.

38. United States Department of Agriculture, Nutrition Evidence Library, DGAC 2010, What is the relationship between glycemic index or glycemic load and type 2 diabetes? Available at: http://www.nel.gov/conclusion.cfm?conclusion_statement_id=250354, accessed September 12, 2014.
39. DiSogra L, Testimony 2010 Dietary Guidelines Public Meeting, July 8, 2011. Available at: http://www.unitedfresh.org/assets/files/DiSogra%20Dietary%20Guidelines%20Testimony%20July%208%202010.pdf, accessed September 12, 2014.
40. Dietary Guidelines.Gov, 2015 Dietary Guidelines Advisory Committee request for Public Comments, May 2014. Available at: http://www.health.gov/dietaryguidelines/2015DGAC RequestForPublicComments.asp, accessed September 12, 2014.
41. Health.gov, Dietary Guidelines Advisory Committee, Meeting 2: Materials and Presentations. Available at: http://health.gov/dietaryguidelines/2015-binder/meeting2/ OralTestimony.aspx, accessed September 12, 2014.
42. Katz D, Goodwin MT. *Food: Where Nutrition Politics & Culture Meet.* Washington, DC: Center for Science in the Public Interest, 1976.
43. Gussow JD, Clancy KL. Dietary guidelines for sustainability. *J. Nutr. Educ.* 1986;18:1–5.
44. Clonan A, Holdsworth M. The challenges of eating a healthy and sustainable diet. *Am. J. Clin. Nutr.* 2012;93:459–60.
45. Gussow J. Reflections on nutritional health and the environment: The journey to sustainability. *J. Hunger Env. Nutr.* 2006;1:3–25.
46. Framingham Heart Study, Epidemiological Background and Design. Available at: http:// www.framinghamheartstudy.org/about-fhs/background.php, accessed September 12, 2014.
47. National Heart, Lung, and Blood Institute. *Fact Book Fiscal Year 2012.* Bethesda, MD: National Institutes of Health; 2013, accessed September 12, 2014.
48. Michas G, Micha R, Zampelas A. Dietary fats and cardiovascular disease: Putting together the pieces of a complicated puzzle. *Atherosclerosis,* 2014;234:320–328.
49. Peters JC. Combating obesity: Challenges and choices. *Obesity Research.* 2003;11:7S–11S.
50. Hu F. Are refined carbohydrates worse than saturated fat? *Am. J. Clin. Nutr.* 2010;91:1541–42.
51. Taubes G. What if it's all been a big fat lie? July 7, 2002, *The New York Times.* Available at: http://www.nytimes.com/2002/07/07/magazine/what-if-it-s-all-been-a-big-fat-lie.html, accessed September 12, 2014.
52. Howard BV, Van Horn L, Hsia J et al. Low-fat dietary pattern and risk of cardiovascular disease: The Women's Health Initiative Randomized Controlled Dietary Modification Trial. *J. Am. Med. Assoc.* 295 2006, pp. 655–666.
53. Jakobsen MU, O'Reilly EJ, Heitmann BL et al. Major types of dietary fat and risk of coronary heart disease: A pooled analysis of 11 cohort studies. *Am. J. Clin. Nutr.* 2009;89:1425–32.
54. Siri-Tarino PW, Sun Q, Hu FB, Krauss RM. Meta-analysis of prospective cohort studies evaluating the association of saturated fat with cardiovascular disease. *Am. J. Clin. Nutr.* 2010;91:535–46.
55. United States Department of Agriculture, Agricultural Research Service, News 1999, Dietary Reference Intakes (DRIs) – New Dietary Guidelines Really Are New!. Available at: http://www.ars.usda.gov/Research/docs.htm?docid=10870, accessed September 12, 2014.
56. Nestle M. *Food Politics: How the Food Industry Influences Nutrition and Health.* Berkeley: University of California Press, 10th Edition, 2013:306–307.
57. Sofi F, Abbata R, Gensini GF, Casini A. Accruing evidence on benefits of adherence to the Mediterranean diet on health: An updated systematic review and meta-analysis. *Am. J. Clin. Nutr.* 2010;92:1189–96.
58. Hite AH, Feinman RD, Guzman GE, Satin M, Schoenfeld PA, Wood RJ. In the face of contradictory evidence: Report of the Dietary Guidelines for Americans Committee. *Nutrition.* 2010;26:915–924.

59. Nutrition Source, Harvard School of Public Health, New U.S. Dietary Guidelines: Progress, Not Perfection. Available at: http://www.hsph.harvard.edu/nutritionsource/dietary-guidelines-2010/, accessed September 12, 2014.
60. Nestle M. Dietary advice for the 1990s: The political history of the food guide pyramid. *Caduceus.* 1993;9(Winter):136–51.
61. Willett W. MD, DrPH, MPH, Fredrick John Stare Professor of Epidemiology and Nutrition, and chair of the Dept. of Nutrition at Harvard School of Public Health. Nutrition Source, Harvard School of Public Health. Available at: http://www.hsph.harvard.edu/nutritionsource/dietary-guidelines-2010/ accessed September 12, 2014.
62. Nutrition Source, Harvard School of Public Health. Available at: http://www.hsph.harvard.edu/nutritionsource/dietary-guidelines-2010, accessed September 12, 2014.
63. Independent Women's Forum, Two-Dozen Free-market Organizations Voice Concern And Urge Reform of the Dietary Guidelines Advisory Committee, May 28, 2014. Available at: http://www.iwf.org/media/2794073/, accessed September 12, 2014.
64. Nurses' Health Study, http://www.nhs3.org/, accessed September 12, 2014.
65. Swidey N. Walter Willett's Food Fight, July 28, 2013, *The Boston Globe.* Available at: http://www.bostonglobe.com/magazine/2013/07/27/what-eat-harvard-walter-willett-thinks-has-answers/5WL3MIVdzHCN2ypfpFB6WP/story.html, accessed September 12, 2014.
66. University of Wollongong, Food Trials Laboratory, Human Whole Room Calorimeter, http://smah.uow.edu.au/medicine/smartfoods/foodtrialslab/UOW046888.html, accessed June 6, 2014.
67. Drewnowski A. The cost of US foods as related to their nutritive value. *Am. J. Clin. Nutr.* 2010;92:1181–88.
68. Aggarwal A, Monsivais P, Drewnowski A. Nutrient intakes linked to better health outcomes are associated with higher diet costs in the US. *PLoS One.* 2012;7(5):e37533. Epub May 25, 2012.
69. The Monday Campaigns; The day all health breaks loose, http://www.mondaycampaigns.org/, accessed September 12, 2014.

Nutrition Information Policies*

4

*Most importantly, nutrition knowledge will become
a means by which Americans can begin to take
responsibility for maintaining their health and reducing
their risk of illness.*
Senator George McGovern

INTRODUCTION

We concluded Chapter 1 by quoting from Senator George McGovern's preface to the second edition (1977) of the *Dietary Goals for the United States,* and this chapter is introduced with another quotation from the same document. The Dietary Goals themselves are discussed in Chapter 3, and we will not be discussing the content again here. However, the call to action of the Dietary Goals became the groundwork of our policies about nutrition information. If you are not already convinced of the relevance of this document, we will provide an additional quote from the press conference announcing the release of the Dietary Goals: "Our greatest bulwark against the interests that have helped to create the present problems is an informed public."[1]

Food policy, as described in the introduction chapter, covers "who eats what, when, how and with what impact."[2] This chapter centers on a "with what impact" area of food policy, namely that of federal involvement in the relationship between nutrition information and public health.

The United States Department of Agriculture (USDA) is only briefly discussed here as other chapters discuss in detail the USDA's role in *what we eat* (Chapter 2), *how we eat* (Chapter 3), and *who eats* (Chapter 6). To explore "with what impact" we focus on the Department of Health and Human Services (HHS), the home of the Food and Drug Administration (FDA), and the Centers for Disease Control and Prevention (CDC). The FDA plays an overall role in regulating how nutrition information reaches the public

---

* Jenna Mandel-Ricci, MPH, contributed primary research and wrote a draft version of a portion of this chapter.

through food product labeling. Food product labeling is used to illustrate a trend in increasing federal regulation over food and also to exemplify how nutritional labeling helps implement health policy objectives. The CDC contributes ongoing nutrition surveillance which informs our policy and quantifies the "with what impact" the policy may have. The research conducted by the CDC through the National Health and Nutrition Examination Survey (NHANES) contributes an evidence basis for the development of food and public health policy, such as published in the Dietary Guidelines and Healthy People.

United States Department of Agriculture

The United States Department of Agriculture (USDA) is a major government player in United States federal food policy, as agricultural policies and the resulting food supply directly affect our food system. Signed into existence by President Abraham Lincoln in May of 1862, the newly created USDA was charged with the task "to acquire and to diffuse among the people of the United States useful information on subjects connected with agriculture in the most general and comprehensive sense of that word, and to procure, propagate, and distribute among the people new and valuable seeds and plants."[3] However, a commissioner lacking Cabinet stature led this new department, and the USDA did not become a Cabinet department until February of 1889.[4] By 1914, legislation had established land grant universities, state agricultural experiment stations, and cooperative education services. USDA continued to grow through the twentieth century, even spinning off the Food and Drug Administration (FDA) in 1940.

The USDA now has the following Mission Areas: Farm and Foreign Agriculture Services; Food, Nutrition and Consumer Services; Food Safety; Marketing and Regulatory Programs; Natural Resources and Environment; Research, Education and Economics; and Rural Development[5] and an overall mission of "... leadership on food, agriculture, natural resources, rural development, nutrition and related issues based on sound public policy, the best available science, and efficient management."[6] For 2015, 76% of the USDA's budget was earmarked for Nutrition Assistance, 11% on Farm and Commodity programs, 8% on Conservation and Forestry, and 5% on all other areas—including research, which was the original mission of the USDA.[7]

Beyond the USDA

Beyond the USDA is a publication of the Institute for Agriculture and Trade Policy discussing how other government agencies (beyond the USDA!) can contribute to a more sustainable food system, including many other agencies that are not commonly thought of as involved in food policy. As *Beyond the USDA* describes it, the USDA is "titanic, charged with the administration of commodity, conservation and nutrition programs, Cooperative Extension, the soil survey, the development of dietary guidelines ... (and more)."[8] *Beyond the USDA's* description of the USDA as titanic is illustrated by comparing the original 1862 USDA mission, the most recent mission information, and the recent budget allocation, as described above.

Department of Health and Human Services

The Department of Health and Human Services (HHS) is "the principal agency for protecting the health of all Americans."[9] A number of different agencies under HHS are involved with food policy, including the FDA and the Centers for Disease Control and Prevention (CDC). The CDC, among other tasks, conducts the National Health and Nutrition Examination Survey (NHANES). We will return to NHANES later in this chapter with a discussion of Healthy People, the current "operating" policy document of health objectives for the country. Briefly, Healthy People began with a 1979 Surgeon General's report—*Healthy People: The Surgeon General's Report on Health Promotion and Disease Prevention*[10]—and subsequently transformed into a HHS initiative developing health objectives for the nation on a 10-year cycle. Healthy People 2020, released in 2010, is the current "operating document."

Food and Drug Administration

The FDA "is responsible for protecting the public health by assuring the safety, efficacy and security of human and veterinary drugs, biological products, medical devices, our nation's food supply, cosmetics, and products that emit radiation."[11] In general, the FDA is responsible for assuring that foods sold in the United States are safe, wholesome, and properly labeled.[12] For matters in which the FDA needs scientific expertise, the FDA can call on the Institute of Medicine (IOM), the health arm of the National Academy of Sciences chartered by President Lincoln in 1863. The IOM was established in 1970 by Congress to act as a nongovernmental organization to provide unbiased advice.[13]

The FDA has broad-ranging responsibilities to protect the public health through the regulation of food labeling. Although the Nutrition Facts and the ingredient panel are prominent food label characteristics, the FDA's authority over labels also extends to other claims made on the package, such as health claims.

Health claims refer how the consumption of food relates to reducing the risk of a specific disease or health condition. "Adequate calcium throughout life, as part of a well-balanced diet, may reduce the risk of osteoporosis" is an example of a health claim.[14] Although health claims on packaging fall into the "with what impact" intersection between food policy and public health, we have chosen to highlight health claims in Chapter 9, due to the key role of the cereal industry in policy developments. Labeling regarding food additives, GRAS, allergens, and GMOs is discussed in Chapter 13, as is the Food Safety Modernization Act of 2010 (FSMA), which greatly expanded the powers granted to the FDA and further increased the high public visibility of the FDA.

How did the FDA acquire such a primary role in regulating food labeling? The FDA was originally housed under the USDA as the Bureau of Chemistry and engaged primarily in research. With the 1906 Pure Food and Drugs Act, mentioned below and discussed in more detail in Chapter 13, the scope of the bureau expanded to include regulatory functions, including an authority over misbranded foods. To deal with the expanded regulatory requirements stemming from the 1906 Act, the USDA split authority between different departments within the USDA, including the Bureau of

Chemistry. By 1931 the bureau had been renamed the FDA and the research function had been dropped. The initial shared responsibility over the 1906 regulations has left a complex legacy of food safety responsibilities split between the FDA and the USDA today. In 1940 the FDA was moved out of the USDA and housed in the Federal Security Agency, and then in 1953 moved to the Department of Health, Education and Welfare, which is now the Department of Health and Human Services (HHS).[15]

The FDA's developing authority over consumer information on food packaging and labels to include nutritional information is also reflective of the FDA's movement through agencies. As part of HHS, the FDA is now more closely aligned with nutrition as a health science then with the agricultural production sources of food.

FEDERAL REGULATION OF FOOD AND PRODUCT LABELS

Advancements in manufacturing, packaging, and distribution technologies created a burgeoning market for nationally distributed brands by the end of the nineteenth century. In other words, the increasing ability to sell food nationally also encouraged the development of national regulation. Leading manufacturers, who had the most to lose from competitors hawking poor quality imitations, provided industry support for federal regulation of food products, packaging and labeling. Increasing federal oversight over food and food production, however, was not necessarily a political slam-dunk, and Congress struggled and argued for several years over approaches until the publication of *The Jungle,* Upton Sinclair's expose of insanitary and inhumane working conditions in meatpacking plants, caused public outrage and demand for action. Within a year of *The Jungle*'s publication, Congress passed the 1906 Meat Inspection Act and the Pure Food and Drug Act.

The FDA's responsibility to protect public health through federal regulation of misbranded food was first articulated in the 1906 Pure Food and Drug Act. Even the name "Pure Food and Drug Act" codifies a value about food and speaks to the FDA's mandate for safety and labeling. For any consumer to be confident in the wholesomeness and purity of a purchased food product—or at least, to have confidence that they are getting what they believe to be purchasing—basic information must be made available to the consumer.

However, although first articulated in 1906, the effective authority by the FDA over food labeling has been a slow, developing process refined through years of legislation, namely the 1938 Federal Food, Drug and Cosmetic Act, the 1967 Fair Packaging and Labeling Act, the 1990 Nutrition Labeling and Education Act, and, almost 100 years after "Pure Foods," the 2004 Food Allergen Labeling and Consumer Protection Act (discussed in Chapter 13).

The 1906 Pure Food and Drug Act prohibited adulteration of foods and false or misleading labels, but otherwise did not contain any affirmative requirements for food labels, and it failed to give the authority to the FDA to set standards. Thus, although this Act hinted at consumer protection, the information available in a consumer transaction was still very much weighted toward the benefit of the manufacturer. Manufacturers were not required to provide an accurate list of ingredients or of the weight or measure of the contents, and the statute failed to prohibit misleading packaging. Congress passed a further amendment in 1913 requiring manufacturers to list the net quantity of the content. Even so, the FDA continued to struggle with manufacturers over deceptive packaging and packages slackly filled for another 25 years, without receiving much support from the judicial branch, until the 1938 Federal Food, Drug and Cosmetic Act strengthened the FDA's authority over misbranding.[16]

The 1938 Federal Food, Drug and Cosmetic Act (FDCA) provided stronger authority to regulate misbranding, partially through invoking interstate commerce, and also by granting authority to the FDA to set standards of identity and quality. Much of the FDA's authority today still stems from the 1938 FDCA. However, regardless of Congressional intent in increasing federal authority to regulate food packaging, the courts continued to push back and thwart the FDA in misbranding actions. This continued rejection by the judicial branch, along with ongoing consumer complaints, eventually pushed Congress to pass the 1967 Fair Packaging and Labeling Act (FPLA).[16] The Interstate Commerce Clause was again used to establish and strengthen federal oversight by requiring standards for appropriate labeling of consumer products in interstate commerce. With this development of standardized information requirements, followed by further legislation and requirements regarding ingredients, nutritional information and allergens, food labeling began to act more concertedly as protection for the consumer. Such authority over labeling and the subsequent standards created a more efficient marketplace for business, and changed our food environment by fostering greater consumer acceptance of purchasing food products previously commonly self-prepared.

The 1967 FPLA Act, although weakened from the version originally presented, did provide improved information for consumers. The regulation promulgated by the FDA required mandatory pieces of information on the label, namely,

- The identity or name of the product and name and place of business of the manufacturer, packer, or distributor.
- An accurate statement of the net quantity of contents at a uniform location on the principal display panel.
- A statement describing the contents of a serving, if a declaration of the number of servings is given.[16]

However, consumers still struggled with assessing nutritional information for packaged foods. Serving size, in particular, remained an issue. By the end of the decade, the critical attention being paid to hunger and malnutrition incited a push for labeling policies to do more than just focus on food composition. President Nixon's 1969 White House Conference on Food, Nutrition, and Health included in its final report a criticism of the manner in which FDA was regulating food labeling and called for improvements that would allow consumers to make informed nutritional dietary choices.[17]

The FDA responded in 1972 with regulations specifying a format for nutrition information on labels. Nutrient reference values called U.S. Recommended Daily Allowances (US RDAs) per standard size serving were used. These values were derived from the highest of the National Research Council's 1968 Recommended Dietary Allowances.[17] Under the FDA's 1972 proposal, providing nutrition information was voluntary—except when nutrients had been added to the food.[18] By 1989, an estimated

60% of processed and packaged foods regulated by the FDA voluntarily included nutrition information on the label.[19]

Food Product Labeling and Public Health

During the same time period that the FDA was pushing industry to provide more information on labels (albeit voluntarily), increased attention was being paid to the connection between health and diet. The Senate Select Committee on Nutrition and Human Needs published the *Dietary Goals for the United States* in 1977 and the first *Dietary Guidelines for Americans* were published in 1980. By the decade's end, *The Surgeon General's Report on Nutrition and Health*[68] and the National Research Council's (NRC's) report *Diet and Health: Implications for Reducing Chronic Disease Risk*[69] were released.

These reports elucidated the connections between diet, morbidity, and mortality among Americans, and suggested that dietary changes could reduce the incidence of some chronic disease. Both provided suggestions for dietary planning, and the Surgeon General's report, in particular, called for the support of the food industry in diet improvement, asking for reformulation of products (namely, lowering fat content) and for increased nutritional information to be provided.

In advocating improvement in the American diet, and pushing for consumers to follow recommendations, it became clear that voluntary industry reporting of nutritional information was at odds with encouraging increased consumer attention to the health importance of one's diet. Standardized and regulated nutrition information that helped support consumer adherence to US dietary recommendations had to be made available.[19]

The FDA proposed regulations in July 1990, while also acknowledging questions about its authority to require nutrition labeling. The FDA made the argument that the nutritional content of a food is a material fact and a food label would be misleading if it lacked the nutrition information as proposed.[17] This argument of authority is based on the 1938 FDCA, which stated a food was misbranded if it "fails to reveal facts material in the light of such representation."[18] The FDA's proposal including a regulation would establish new nutrient reference values for macronutrients, called Daily Reference Values (DRVs), and for vitamins and minerals, called Reference Daily Intakes (RDIs).

In the fall of 1990, Congress passed two Nutrition Acts—the National Nutrition Monitoring and Related Research Act (NNMRRA) and the Nutrition Labeling and Education Act (NLEA). The NNMRRA improved coordination and oversight of food and nutrition data collection efforts. The USDA and HHS were called on to jointly publish the *Dietary Guidelines for Americans* (DGA) every 5 years. The USDA and HHS also developed a 10-year collaborative plan, which included leveraging prior nutrition survey work done by the USDA. NHANES began using USDA food codes and the USDA nutrient database in 2002.[20] The Nutrition Labeling and Education Act (NLEA) implemented policy mechanisms intended to support the goals of the Dietary Guidelines. Generally, the NLEA preempts state requirements about food standards, nutrition labeling, and health claims. It authorized some health claims for foods (health claims will be discussed in a later section of this chapter), defined some important descriptive terms such as "low fat" and "light," authorized claims, and standardized serving sizes and the food ingredient panel.[21]

NLEA and the Nutrition Facts

Although forms of nutrition labels were used prior to the NLEA, the NLEA improved and standardized nutrition information. The Nutrition Facts box on packages is the result of the regulatory effort by the FDA to fulfill the requirements of the 1990 NLEA.

Specific nutrients deemed important for dietary-related public health issues must be included in the Nutrition Facts box. These required nutrients can change over time as evidence and public health issues change, albeit not as quickly as public health leaders might desire, as the regulations must be amended to do so. Following is the list of nutrients in the regulated order of appearance (as of 2014): total calories, calories from fat, saturated fat, trans-fat (added in 2006), cholesterol, sodium, total carbohydrate, dietary fiber, sugars, protein, vitamin A, vitamin C, calcium, iron.[22,23] Consumers are meant to limit these items—fat, saturated fat, trans-fat, cholesterol, and sodium, and ensure enough consumption of the following items: vitamin A, vitamin C, calcium and iron. Figure 4.1 is a sample of the current (as of 2014) Nutrition Facts label.

Nutrition Facts

Serving Size 2/3 cup (55g)
Servings Per Container About 8

Amount Per Serving

Calories 230   Calories from Fat 72

% Daily Value*

Total Fat 8g	**12%**
Saturated Fat 1g	**5%**
Trans Fat 0g	
Cholesterol 0mg	**0%**
Sodium 160mg	**7%**
Total Carbohydrate 37g	**12%**
Dietary Fiber 4g	**16%**
Sugars 1g	
Protein 3g	
Vitamin A	10%
Vitamin C	8%
Calcium	20%
Iron	45%

* Percent Daily Values are based on a 2,000 calorie diet. Your daily value may be higher or lower depending on your calorie needs.

	Calories:	2,000	2,500
Total Fat	Less than	65g	80g
Sat Fat	Less than	20g	25g
Cholesterol	Less than	300mg	300mg
Sodium	Less than	2,400mg	2,400mg
Total Carbohydrate		300g	375g
Dietary Fiber		25g	30g

FIGURE 4.1   2014 Nutrition Facts box. (From U.S. Food and Drug Administration, http://www.fda.gov/Food/GuidanceRegulation/GuidanceDocumentsRegulatoryInformation/LabelingNutrition/ucm385663.htm#images.)

BOX 4.1   WHY 2000 CALORIES?

Why is 2000 calories set as the comparison value to calculate the Daily Reference Value (DRV) for energy-producing macronutrients on the Nutrition Fact box? The 2000-calorie intake standard was chosen, in part, because this is generally the approximate daily calorie requirement for postmenopausal women. Although one might think that the DRV standard on the label would most likely be selected as an average, or as one applicable to most people, or even because the target audience of label-readers might be most likely to be female, one reason behind targeting the caloric needs of postmenopausal women is the determination by the IOM that this group is at the highest risk for excessive intake of calories and fat.

National Research Council, Dietary Reference Intakes: Applications in Dietary Planning, Washington, DC: The National Academies Press, 2003.

For consumers to assess the level of nutrients provided by a serving of the product, the Nutrition Facts label provides a percentage of "Daily Value" (DV). To calculate the nutrient DVs, the FDA established new nutrient reference values called Daily Reference Values (DRVs) for macronutrients and Reference Daily Intakes (RDIs) for vitamins and minerals. Although the FDA had proposed a more progressive scheme, the majority of the RDIs ended up being based on the 1968 Recommended Dietary Allowances, and for nutrients that lacked those, DRVs were created based on scientific evidence as to risk reduction of chronic diseases.[22]

The more progressive approach and updated DRVs the FDA had in mind for nutrient references had to be scuttled after the FDA was hamstrung by the Dietary Supplement Act of 1992, which, due to extensive lobbying by vitamin and supplement manufacturers, contained a moratorium prohibiting until November 1993 any nutrition labeling regulations that used recommended daily allowances or intake values for vitamins and minerals other than those currently in effect. The moratorium took away the FDA's opportunity to publish the new proposal in time to meet the NLEA's required deadlines.[22]

For purposes of the Nutrition Facts label, the DRVs are calculated in reference to a daily diet of 2000 calories. See Box 4.1 for an explanation of why the standard value is 2000 calories.

Proposed Updates to the Nutrition Facts

The FDA has published two rules that propose updates to the Nutrition Facts box, and in fact, the inclusion of "added sugars" is one of the suggested changes. One proposed rule is titled Food Labeling: Revision of the Nutrition and Supplement Facts Labels, and the second proposed rule is titled Food Labeling: Serving Sizes of Foods That Can Reasonably Be Consumed at One-Eating Occasion; Dual-Column Labeling; Updating, Modifying, and Establishing Certain Reference Amounts Customarily Consumed; Serving Size for Breath Mints; and Technical Amendments.

The FDA's proposed changes are summarized as follows, and the proposed label is depicted in Figure 4.2.

1. Greater Understanding of Nutrition Science
   - Require information about "added sugars."
   - Update the daily values (used to calculate the Percent Daily Value) for nutrients like sodium, dietary fiber, and Vitamin D. In addition to the Percent Daily Value, the new label would include the actual amount of mandatory vitamins and minerals and, when declared, voluntary vitamins and minerals.
   - Require the amount of potassium and Vitamin D to be declared on the label, because these are new "nutrients of public health significance," as ascertained through NHANES data. Calcium and iron would continue to be required. Vitamins A and C could be included on a voluntary basis, but are no longer required as the general population is not deficient.

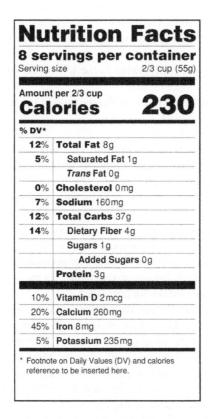

FIGURE 4.2 FDA's proposed Nutrition Facts box. (From U.S. Food and Drug Administration, http://www.fda.gov/Food/GuidanceRegulation/GuidanceDocumentsRegulatoryInformation/LabelingNutrition/ucm385663.htm#images.)

- "Calories from Fat" would be removed because research shows type of fat is more important than the amount; "Total Fat," "Saturated Fat," and "Trans Fat" would be kept.

2. Updated Serving Size Requirements and New Labeling Requirements for Certain Package Sizes

- Change the serving size requirements to reflect how people eat and drink today. By law (specifically the NLEA), the label information on serving sizes must be based on what people actually eat, not on what they "should" be eating.
- Require that packaged foods, including drinks, that are typically eaten in one sitting be labeled as a single serving and that calorie and nutrient information be declared for the entire package. A 20-oz bottle of soda, typically consumed in a single sitting, would be labeled as one serving.
- Dual column labels to indicate both "per serving" and "per package" calories and nutrient information for certain packages that could be consumed in one sitting or multiple sittings, such as 24-oz bottle of soda or a pint of ice cream.

3. Refreshed Design

- Make calories and serving sizes more prominent. The caloric content of foods will be highlighted by bolding and increasing the type size of the font.
- Shift the Percent Daily Value to the left of the label, so it would come first.
- Change the footnote to more clearly explain the meaning of the Percent Daily Value.[24]

Are there other nutrients or information that you think should be mandated to appear in the Nutrition Facts? Should all of the Nutrition Facts items currently listed as of 2014 retain their mandated stature?

Do you think the change in emphasized nutrients will impact characteristics of our food supply? Are we likely to see foods that have greater amounts of potassium? Or simply more advertising and label claims as to potassium-rich foods?

Increased information about the amount of potassium in the foods we are consuming will help people achieve adequate intake. The potassium-specific information will also be helpful for individuals taking certain drugs, as these people may need to consume less than the adequate intake for potassium.[25]

Vitamin A is no longer an issue of deficiency. Sometime ago, one of the authors helped analyze a nutritional data set for children and pointed out that instead of deficiencies, a good number of the children had Vitamin A intake in excess, some dangerously in excess of the IOM's Tolerable Upper Limits. Most likely the excess was due to the consumption of fortified breakfast cereal. *Is there any impact to consumer information from removing Vitamin A from the Nutrient Facts? Should the general population be monitoring dietary intake for excess of vitamin A?*

The FDA states that many of the updates are aimed at addressing current public health problems like obesity. The fact that the NLEA established serving size guidelines was an important step. Previously, manufacturers had complete discretion over

BOX 4.2   SERVINGS: CONNECTION TO USDA DIETARY GUIDANCE

The terms "serving size" and "servings per container" are mandatory components of the Nutrition Facts box. However, as of 2014, the USDA no longer uses the term "servings" for dietary food group recommendations, and has advised manufacturers wishing to promote their food's contribution to the Dietary Guidelines to make statements using cups or ounces, rather than servings.[27] The proposed new Nutrition Facts label keeps the references to serving size and servings per container, but changes the "Amount Per Serving" to "Amount per ___," with the blank filled in with the serving size in common household measures, such as cup.[24]

suggested serving size, which made it difficult for consumers to assess the nutritional and financial value of purchases, especially when comparing competing products. However, the NLEA serving standards set up in the 1990s used historical consumer surveys, which differed even then from current conventional dietary intake (and differ even more so now). As a result, the calorie and nutrient impact of standard serving sizes were still difficult for consumers to assess, particularly for products (often beverages) that appear to be packaged in a single serving container, but the container label indicates the package contains multiple servings. For example, the NLEA 1990 standard serving of juice was considered 4 oz, and a standard serving of a carbonated soft drink or flavored, sweetened coffee 8 oz,[26] but containers that appear to functionally be a single-serving container for these items are often much larger in volume.

The FDA's website contains more details, links to additional information, and a graphic illustrating the serving size changes using soda and ice cream. A pint of ice cream is currently considered to contain four servings, but with the proposed changes, a pint of ice cream would contain two servings.[24] *What data did the FDA rely on to determine what people actually eat? Is the data averaged across age-ranges? One author recalls eating half a pint of ice cream when in college, but now, at a respectable distance, would not.*

Nonetheless, the idea of serving size as more conventionally consumed is a great method for replacing some very mysterious serving sizes, such as one-third of a muffin. A packaged muffin that might now be comprised of two to three servings would instead be described as just one serving. This update in serving size, and the associated update in calories and nutrition facts per serving, will make the caloric and nutritional impact of eating the entire muffin (as one most likely would) much clearer (see Box 4.2).

Nutrient Daily Value Levels: Updates and Nutrient Claims

The public comment period for the proposed changes to the Nutrition Facts ended in the summer of 2014, but the comments are available to read online, and may be illustrative, in particular regarding the values of various nutrients' DVs. One concern raised is that the DVs are outdated. Another concern is that too often the information on a package does not provide the appropriate levels for children, and this is particularly an issue for vitamin A, zinc, and niacin.[28]

The Daily Reference Values also play a role in the nutrient claims that manufacturers can use on packages, and changes to these amounts may impact those claims. Nutrient claims are authorized under the NLEA and provides a descriptive characterization of how much of a specific nutrient is in a food.[29] "Low in fat" or "high in fiber" are nutrient claims, as well as the phrase "good source." These consumer-friendly descriptions can only be used when the product has quantified levels of appropriate nutrients.

For example, for a label to claim that a product is a "good source" of calcium, the food must provide at least 10% of the Daily Reference Value of calcium, and 20% of calcium's DRV to make the claim "high or excellent source." However, companies cannot make these friendly descriptive "good source" claims about just any nutrient. These descriptive characterizations are limited to only those nutrients that the FDA has authorized. Most of the FDA's authorizing regulations apply only to those substances that have an established Daily Value. Additionally, other requirements must also be met to make nutrient claims, including disclosure statements if the food product making the nutrient claim also contains items in excess of certain levels that may increase the risk of disease, such as saturated fat, cholesterol, and sodium.[30] Note the absence of sugar on that list.

Unlike descriptive characterizations, accurate quantitative statements, for example, "2 g of sugar," are always acceptable, with the caveat that quantitative statements must not be qualified. Using the word "only" is a qualifier, because "only" descriptively characterizes the level of a nutrient. In this example, "only" (such as "only 2 g of sugar") characterizes the amount as "low," and therefore would not be allowed.[31]

To keep the public from being misled by claims on food packaging, the FDA also regulates the use of certain statements, such as the phrase "healthy." The FDA considers the use of "healthy" as an implied nutrient content claim in part because a consumer may associate the word "healthy" with current official government dietary advice, and because it characterizes a food item as having "healthy" levels of total fat, saturated fat, cholesterol, and sodium.[31]

Sodium

The possibility of updates to sodium provides an interesting topic, worthy of reviewing the Questions and Answers the FDA has published about the proposed Nutrition Facts update.[24] The proposed updated Daily Value for sodium is 2300 mg, which as an update is quite a minor change from 2400 mg. The proposed value of 2300 mg is based on a 2013 IOM report's conclusion that evidence is inconsistent and insufficient to conclude that lowering sodium intakes below 2300 mg/day will increase or decrease the risk of cardiovascular disease outcomes or mortality in the general U.S. population or in identified subgroups.

However, the 2010 Dietary Guidelines recommend further reduction to 1500 mg/day among groups that are at increased risk of the blood pressure–raising effects of sodium. These groups—individuals ages 51 or older, African Americans, and individuals with high blood pressure, chronic kidney disease, or diabetes—constitute about half of the American population. Thus, FDA is proposing a daily value of 2300 mg but is asking for comment from the public as to if 1500 mg would be more appropriate as well as other alternative approaches for selecting a dietary value for sodium.[24]

"Sodium added" seems evocative of the request to include "added sugars" on the nutrition facts label. *Would it be illustrative or obsfucating to include "added sodium"?*

Furthermore, because much of the sodium in our diet is from sodium added to food during processing, the FDA is separately developing a long-term strategy to reduce the sodium content of the food supply to make it easier for people to consume less sodium. Reducing the sodium content of the food supply is a change to our food environment, not an informational one such as documenting the sodium content on the label. Federal food assistance policies have also already created a market for canned goods lower in sodium through making purchases for commodity distribution, although this is outside the normal retail market. All 12 of the canned vegetables to be offered in 2015 in the Commodity Supplemental Food Program are low-sodium or no salt.[32]

How do you think the FDA can achieve an environmental change in food composition through authority over labeling?

GRAS

The FDA can influence food composition through "GRAS" regulation. "GRAS," or "Generally Recognized as Safe" is discussed in more detail in Chapter 13, and is highlighted in Chapter 9, but we will mention it here as part of the discussion on sodium. Briefly, to be considered GRAS, salt must be "generally recognized" by scientific experts or scientific procedures "to be safe under the conditions of its intended use."

The American Heart Association (AHA) has asked the FDA to review salt's status as GRAS, noting that the issue has been raised several times in the past. In 1978, the Center for Science in the Public Interest asked the FDA to revoke sodium's GRAS status and regulate it as a food additive. However, the FDA wanted to first assess the impact of proposed sodium labeling regulations and a push for voluntary industry participation in lowering sodium, before addressing the GRAS status of salt. The FDA also noted that if the sodium content of processed foods was not "substantially reduced," it would consider taking additional action such as changing salt's GRAS status. At the time (the late 1970s), salt added to processed foods constituted about one-third to one-half of Americans' sodium intake. As Americans now obtain 77% of their sodium from processed foods, "substantial reduction" has not taken place.

The American Heart Association points out that there is no consensus within the scientific community as to salt being harmless, and furthermore the increased amount and ubiquity of salt in the American diet belies the notion of intended use. The AHA provides several carefully drafted options by which the FDA could use GRAS to help lower the sodium content in our food supply.[33] By the time you are reading this chapter, you may know what action the FDA took on salt.

Healthy People and Food Policy

Healthy People acts as the current "operating" policy document of health objectives for the nation. Healthy People began with a 1979 Surgeon General's report, *Healthy People: The Surgeon General's Report on Health Promotion and Disease Prevention*,[10] and subsequently transformed into a HHS initiative developing health objectives for the nation on a 10-year cycle. The current "operating document," Healthy People 2020, has over 1200 objectives organized into 42 topic areas, with baseline measures and specific targets to track for improvements. A subset of 26 objectives, called Leading Health Indicators, has been selected to indicate high-priority health issues.[34] As an excellent historical review comments, Healthy People is notable for longevity as a federal initiative, particularly considering that it has survived through a number of administrations without a compelling legislative mandate or budget.[35]

Healthy People is developed under the auspices of the Department of Health and Human Services. Agencies housed in the HHS contribute most significantly to developing Healthy People, but other departments, such as the United States Department of Education and the USDA also play a role. The U.S. Department of Education is the co-lead of the topic "Disability and Health," and the USDA is the co-lead with the FDA on the topic "Food Safety." Although a federal effort, Healthy People's objectives are national, not just federal, meaning that success in achieving and tracking these objectives is reliant on all levels of government and on nongovernmental organizations.[36] Furthermore, the means for achieving these targets are not for the most part articulated with the Healthy People document—that is, Healthy People does not proscribe how we reach the objectives, although the website contains information about programs and interventions that may contribute to reach the targeted objectives. Programs and plans supporting these objectives, including food-related policy that affects other Healthy People Objectives (such as reducing dental caries), could be developed at any level of government—and by nongovernmental organizations. Public Health Departments in cities, such as Chicago and New York, use Healthy People as an informative planning tool. You can too!

As of 2014, the front page of the HealthyPeople.gov website links to two resources to review performance: the Leading Health Indicators: Progress Update, which reports on the 26 priority health objectives,[37] and DATA2020, an interactive data tool that users can use to explore the data used to track all Healthy People objectives.[38] Three of the Nutrition and Weight Status objectives are Leading Health Indicators: NWS-9 Obesity among adults, NWS-10.4 Obesity among children, and NWS-15.1 mean daily intake of total vegetables. As of October 2014, these three leading health indicators have shown little or no detectable change.[39]

Healthy People 2020's objectives are aimed at four overall goals: attaining high-quality, longer lives free of preventable disease, disability, injury, and premature death; achieving health equity, eliminating disparities, and improving the health of all groups; creating social and physical environments that promote good health for all; and promoting quality of life, healthy development, and healthy behaviors across all life stages.[40]

Healthy People draft objectives are prepared by the lead Federal Agencies, made available to the public for comment, and then reviewed by an interagency workgroup. Although "Nutrition and Weight Status" and "Food Safety" are the most on-point topic groups for food policy, a number of other topic areas may have touch-points with food policy, such as objectives for chronic diseases such as diabetes and heart disease.

Overall, the Nutrition and Weight Status objectives of Healthy People 2020 are grouped into six areas—Healthier Food Access, Healthcare and Worksite Settings, Weight Status, Food Insecurity, Food and Nutrient Consumption, and Iron Deficiency. Many of these objectives incorporate *Dietary Guidelines for Americans* recommendations (see Table 4.1).[41]

Nutrition Facts and Healthy People

Our nutrition and food policy has been promulgated through differing pieces of legislation that have differing intent, funding, and timelines, some being updated from time to time, and others being updated on a scheduled basis. One benefit of this unintentional staggering of guidance, consumer education, and goals is that scientific advances that occur mid-cycle in any one piece of policy may be incorporated in another.

Healthy People is published every 10 years, looking forward (like this text, Healthy People is looking forward from the past). The Dietary Guidelines are published every 5 years, usually released the year after the edition date. NHANES is conducted on a continual basis, and thus the evidence base for our nutritional policy builds in between all these cycles. Policies and programs that are updated occasionally, such as the Nutrition Facts or WIC, may be leveraged toward meeting public health goals of Healthy People, or making it easier for the population to achieve the suggestions of the Dietary Guidelines.

The current (as of 2014) required elements on the Nutrition Facts label have been the same since trans-fat was added in 2006. Speaking only for items that are required in the Nutrition Facts box, sodium, saturated fat, iron, and calcium have population health objectives in Healthy People 2020 (see Box 4.3).
For those nutrients that have specific population health objectives in Healthy People 2020 how can the information on the Nutrition Facts label contribute to promoting dietary objectives?

The other Healthy People objectives for macronutrients (rather than specific foods) focus on solid fats and added sugars. Although a general statement is made that a healthy diet "limit(s) the intake of saturated and trans-fats, cholesterol, added sugars, sodium (salt), and alcohol," the reduction of trans-fat is not an objective.[42]

Public health departments and community organizations have petitioned the FDA to include "added sugars" to the nutrition facts label.[43] *Is there a relevant distinction between sugars and "added sugars"? What do you think might happen (consumer behavior and food manufacturing behavior) if "added sugars" becomes a required component of the Nutrition Facts?*

Progress in three groups in the Healthy People 2020 Nutrition and Weight Status Objectives—Weight Status, Food and Nutrient Consumption, and Iron Deficiency and Anemia—are measured through data collected by NHANES.

TABLE 4.1 Healthy People 2020, nutrition and weight status objectives

OBJECTIVE	DESCRIPTION	BASELINE (YEAR)	2020 TARGET
NWS-1	Increase the number of states with nutrition standards for foods and beverages provided to preschool-aged children in child care	24 states (2006)	34 states and District of Columbia
NWS-2.1	Increase the proportion of schools that do not sell or offer calorically sweetened beverages to students	9.3% schools (2006)	21.3% schools
NWS-2.2	Increase the proportion of school districts that require schools to make fruits or vegetables available whenever other food is offered or sold	6.6% schools (2006)	18.6%
NWS-3	Increase the number of states that have state-level policies that incentivize food retail outlets to provide foods that are encouraged by the Dietary Guidelines	8 states (2009)	18 states and District of Columbia
NWS-4	(Developmental) Increase the proportion of Americans who have access to a food retail outlet that sells a variety of foods that are encouraged by the Dietary Guidelines	N/A	N/A
NWS-5.1	Increase the proportion of primary care physicians who regularly assess body mass index (BMI) of their adult patients	48.7% (2009)	53.6%
NWS-5.2	Increase the proportion of primary care physicians who regularly assess body mass index (BMI) for age and sex of their child or adolescent patients	49.7% (2008)	54.7%
NWS-6.1	Increase the proportion of physician office visits made by patients with a diagnosis of cardiovascular disease, diabetes, or hyperlipidemia that include counseling or education related to diet or nutrition	20.8% (2007)	22.9%
NWS-6.2	Increase the proportion of physician office visits made by adult patients who are obese that include counseling or education related to weight reduction, nutrition, or physical activity	28.9% (2007)	31.8%
NWS-6.3	Increase the proportion of physician visits made by all child or adult patients that include counseling about nutrition or diet	12.2% (2007)	15.2%
NWS-7	(Developmental) Increase the proportion of worksites that offer nutrition or weight management classes or counseling	N/A	N/A

(Continued)

TABLE 4.1 (Continued)   Healthy People 2020, nutrition and weight status objectives

OBJECTIVE	DESCRIPTION	BASELINE (YEAR)	2020 TARGET
NWS-8	Increase the proportion of adults who are at a healthy weight	30.8% (2005–2008)	33.9%
NWS-9	Reduce the proportion of adults who are obese	33.9% (2005–2008)	30.5%
NWS-10.1	Reduce the proportion of children aged 2–5 years who are considered obese	10.7% (2005–2008)	9.6%
NWS-10.2	Reduce the proportion of children aged 6–11 years who are considered obese	17.4% (2005–2008)	15.7%
NWS-10.3	Reduce the proportion of children aged 12–19 years who are considered obese	17.9% (2005–2008)	16.1%
NWS-10.4	Reduce the proportion of children aged 2–19 years who are considered obese	16.1% (2005–2008)	14.5%
NWS-11	(Developmental) Prevent inappropriate weight gain in youth and adults	N/A	N/A
NWS-12	Eliminate very low food security among children	1.3% (2008)	0.2%
NWS-13	Reduce household food insecurity and in doing so reduce hunger	14.6% (2008)	6.0%
NWS-14	Increase the contribution of fruits to the diets of the population aged 2 years and older	0.5 cup (2001–2004)	0.9 cup
NWS-15.1	Increase the contribution of total vegetables to the diets of the population aged 2 years and older	0.8 cup (2001–2004)	1.1 cup
NWS-15.2	Increase the contribution of dark green vegetables, orange vegetables, and legumes to the diets of the population aged 2 years and older	0.1 cup (2001–2004)	0.3 cup
NWS-16	Increase the contribution of whole grains to the diets of the population aged 2 years and older	0.3 ounce (2001–2004)	0.6 ounce
NWS-17.1	Reduce consumption of calories from solid fats	18.9% (2001–2004)	16.7%
NWS-17.2	Reduce consumption of calories from added sugars	15.7% (2001–2004)	10.8%
NWS-17.3	Reduce consumption of calories from solid fats and added sugars in the population aged 2 years and older	34.6% (2001–2004)	29.8%
NWS-18	Reduce consumption of saturated fat in the population aged 2 years and older	11.3% (2003–2006)	9.5%

(Continued)

TABLE 4.1 (Continued)   Healthy People 2020, nutrition and weight status objectives

OBJECTIVE	DESCRIPTION	BASELINE (YEAR)	2020 TARGET
NWS-19	Reduce consumption of sodium in the population aged 2 years and older	3641 mg (2003–2006)	2300 mg
NWS-20	Increase consumption of calcium in the population aged 2 years and older	1118 mg (2003–2006)	1300 mg
NWS-21.1	Reduce iron deficiency among young children aged 1–2 years	15.9% (2005–2008)	14.3%
NWS-21.2	Reduce iron deficiency among young children aged 3–4 years	5.3% (2005–2008)	4.3%
NWS-21.3	Reduce iron deficiency among females aged 12–49 years	10.4% (2005–2008)	9.4%
NWS-22	Reduce iron deficiency among pregnant females	16.1% (2005–2008)	14.5%

**BOX 4.3   NUTRIENT-RELATED HEALTHY
PEOPLE 2020 OBJECTIVES**

NWS-18 Reduce consumption of saturated fat in the population aged
2 years and older
Target: 9.5% (11.3% was the mean percentage of total daily calorie intake
provided by saturated fat for the population aged 2 years and older in
2003–06)
NWS-19 Reduce consumption of sodium in the population aged 2 years
and older
Target: 2300 mg (3640 mg sodium from foods, dietary supplements,
antacids, drinking water, and salt use at the table was the mean total
daily intake by persons aged 2 years and older in 2003–06)
NWS-20 Increase consumption of calcium in the population aged 2 years
and older
Target: 1300 mg (1119 mg calcium from foods, dietary supplements, antacids, and drinking water was the mean total daily intake by persons
aged 2 years and older in 2003–06, age adjusted to the year 2000 standard population)
NWS-21 Reduce iron deficiency among young children and females of
childbearing age
NWS-21.1 Target for children 1–2 years: 14.3% (15.9% of children aged
1–2 years were iron deficient in 2005–08)
NWS-21.2 Target for children 3–4 years: 4.3% (5.3% of children aged 3–4
years were iron deficient in 2005–08)
NWS-21.3 Target for females of childbearing age: 9.4% (10.5% of females
aged 12–49 years old were iron deficient in 2005–09)
NWS-22 Reduce iron deficiency among pregnant females
Target: 14.5% (16.1% of pregnant females were iron deficient in 2003–06)

*Healthy People 2020, Nutrition and Weight Status Objectives. Available
at https://www.healthypeople.gov/2020/topics-objectives/topic/nutrition-
and-weight-status/objectives, accessed November 18, 2014.*

NATIONAL HEALTH AND NUTRITION EXAMINATION SURVEY (NHANES)

NHANES is the principal survey collecting information on the nutrition status of
the U.S. population. It combines household and individual interviews with clinical
tests, measurements, and physical examinations for both child and adult participants,
enabling nutritional data to be directly linked to health status.[44] Data from NHANES

supports the joint Health and Human Services/U.S. Department of Agriculture effort to monitor the diet and nutritional status of Americans and provide information needed for policy and dietary guidelines, such as Healthy People and the *Dietary Guidelines for Americans.*[45]

In 1956, the U.S. Congress passed the National Health Survey Act providing legislative authorization for a continuing survey and special studies to determine the amount, distribution, and effects of illness and disability in the United States. Two surveys—the National Health Interview Survey and the National Health Examination Survey—were developed and deployed by the National Center for Health Statistics within the CDC. Nutritional measures were added in 1970, due to the growing appreciation of the role of diet in chronic disease, and the survey was renamed the National Health and Nutrition Examination Survey (NHANES).[46,47]

Initially, NHANES was conducted episodically: NHANES I, from 1971 through 1974, NHANES II, from 1976 through 1980, and NHANES III, from 1988 through 1994. However, since 1999, the survey has been conducted continuously, with results reported for 2-year cycles, such as NHANES 2009–2010 and NHANES 2011–2012.

In 2002, NHANES, conducted by HHS, was fully integrated with a dietary intake survey conducted by the USDA, the Continuing Survey of Food Intakes by Individuals. The new integrated dietary component is called What We Eat in America and is the food and nutrition component of NHANES.[48] The integration of these two dietary surveys were called for by The National Nutrition Monitoring and Related Research Act of 1990 (NNMRRA), which intended to establish a comprehensive, coordinated program for nutrition monitoring and related population health research. NNMRRA expired in 2000, but the full integration of CSFII (USDA) and NHANES (HHS) was realized. HHS is responsible for sample design and data collection, whereas the USDA is responsible for dietary data collection methodology, the upkeep of the databases, and data review and processing.[49]

NHANES is a continuous, annual survey that collects data from a nationally representative sample of about 5000 people each year drawn from the civilian, non-institutionalized U.S. population aged 2 months and older. The survey is conducted among a nationally representative sample of about 5000 people each year. Certain groups are over-sampled, such as Hispanics, African Americans, persons 60 years and older, and most recently, Asian-Americans, to allow for more precise estimates of their populations.[50]

The survey is conducted through in-home personal interviews, physical examinations in mobile examination centers (MECs), and mail follow-up food frequency questionnaires (FFQs).[47] During the physical health examination at the MEC, anthropometric measurements are taken and blood and urine samples collected to measure items such as iron status, folate, vitamin B-12 status, lipid profile, bone health, and biochemical measures.[51] Thus, dietary data can be linked not only to biomarkers and environmental exposure data but also to numerous other data sets for further analysis.[52]

Here's how the dietary intake data is collected. The NHANES participant is first interviewed at home, during which dietary behavior information is collected. A 24-hour dietary recall is then administered as part of the MEC exam. In about 3–10 days after the MEC exam, the participant will be contacted by telephone to collect a second 24-dietary recall.[53]

The 24-hour dietary recalls are conducted using the USDA's Automated Multiple Pass System, a five-pass recall design used to elicit complete information in a standardized way. The survey participant is provided with visual and measurement aids; a dietary interviewer records detailed information about the reported foods and beverages. During the household and personal interview, survey participants are asked questions related to a number of health and nutrition issues including use of dietary supplements, food security, participation in food assistance programs including food stamps, WIC, school breakfast and lunch programs and food programs for the elderly, frequency of consumption of selected foods, and other behaviors related to diet (see sample questions from 2009–2010 in Table 4.2).

By combining health status questions with laboratory testing, NHANES data can provide information about emerging public health issues such as use of dietary supplements[54] as well as monitor the safety of the food supply. For example, high levels of mercury were found in hair and blood samples collected through NHANES and compared to responses regarding fish consumption. Such analyses led to recommendations to limit the intake of certain mercury-containing fish by young children and pregnant women.[55]

The FDA, among other federal agencies, uses NHANES data to evaluate the safety of the food supply. NHANES food intake data is also routinely used to support proposed and final rules related to nutrition labeling, such as the labeling requirements for trans-fatty acid consumption[52] and establishing serving sizes for the Nutrition Facts panel.[55] NHANES data is also used to evaluate the effects of food fortification regulations. In fact, NHANES data played an essential role in the folate fortification program. NHANES data had linked high prevalence of low serum folate in young women with high prevalence of spina bifida in newborns. The survey also identified foods frequently consumed by the target population, and, after fortification policies were implemented, demonstrated increase in folate intake, increase in serum folate, and decrease in spina bifida rates.[52,55]

TABLE 4.2 Sample NHANES questions

MODULE	SAMPLE QUESTIONS
Consumer Behavior	How often does your family have fruits available at home? During the past 7 days, how many times did you or someone else in your family cook food for dinner at home?
Food Security	In the last 12 months, did you or another adult in your household cut the size of your meals or skip meals because there wasn't enough money for food?
Dietary Screener	During the past month, how often did you drink regular soda or pop that contains sugar? During the past month, how often did you eat a green leafy or lettuce salad with or without other vegetables?
Dietary Behavior	In general, how healthy is your overall diet? Did you receive benefits from WIC, that is, Women, Infants, and Children program, in the last 12 months?

Source: Adapted from Centers for Disease Control and Prevention, National Center for Health Statistics, National Health and Nutrition Examination Survey, Survey Questionnaires, Examination Components and Laboratory Components 2009–2010.

Numerous nutrition and health-related programs in the public, nonprofit, and private sector rely on NHANES data to inform guidelines, eligibility requirements, and program components. The *Dietary Guidelines for Americans* (DGA), updated every 5 years, is the principal federal nutrition policy statement and guides all federal nutrition and food assistance efforts. Dietary assessment data collected through NHANES inform updates to the DGA and help measure interim progress in meeting nutrition goals.

Integrated health programs also rely on NHANES data to develop treatment guidelines and program interventions. NHANES data can identify subpopulations at greater risk for onset of conditions such as high cholesterol and high blood pressure and/or poorer treatment outcomes; such information can be used to develop and target prevention and treatment messaging. These data are invaluable to programs that work to reduce morbidity and mortality of these conditions.

Numerous federal food programs rely on NHANES data to assess program reach and nutritional intake among target populations.[56] Combined with information on household size and income, NHANES data demonstrate if federal programs are meeting enrollment targets. Nutritional intake data combined with health status information for survey participants in subpopulations targeted by federal programs provides information that can be critical to program adjustments. For example, changes to the National School Lunch Program requiring lower fat levels and more fruits and vegetables were made, in part, due to nutrition monitoring data.[57] Similarly, NHANES data was used to analyze nutrient intake in mothers of young children, infants, and children, which, in turn, informed the structuring and contents of new food packages available under the Women, Infants and Children Supplemental Food program. Private sector organizations such as the Dairy Council and National Osteoporosis Foundation also rely upon NHANES for information on current dietary intake levels of certain foods and food groups, as well as changes in consumption patterns over time.[55]

Because NHANES is representative of the noninstitutionalized U.S. population and because the survey is conducted on a continual basis, NHANES data can be used to establish population-based standards, or baselines, for a variety of measures including body mass index, macronutrient intake, and exposure to various substances. For example, laboratory analyses of blood and urine specimens collected as part of NHANES are used to derive population norms for assessing iron, zinc, and folate status.[51,57] Such a baseline enables an objective to be measured and consequently, evaluation of the success of the intervention or policy. See Chapter 11, "Enrichment and Fortification," which highlights the use of population-level NHANES data. In terms of informing policy, it is also important to note that NHANES does not necessarily collect nutrient status for all population groups. In the case of iron, for example, starting in 2003, NHANES limited a certain measurement of iron status reporting to children aged 1–5 years and women aged 12–49.

As noted above with iron status reporting, with each NHANES cycle, test components may be dropped and added. Survey elements are added (or dropped) to reflect emerging interests, but sponsorship and funding also contributes to such decisions. For example, 2013–2014 NHANES is collecting data on taste and smell disorders, as impaired taste and smell can lead to unhealthy dietary changes and exposure to unsafe conditions, or be symptomatic of certain serious health conditions. The National

Institute on Deafness and Other Communication Disorders is sponsoring this survey module.[58]

Depending on what is tested for in the blood and urine sample and what questions are asked in the dietary module, there may be the chance to directly assess associations between dietary intake information (or other environmental exposures) and health. For example, 2003–2004 NHANES found detectable levels of Bisphenol A (BPA) in 93% of 2517 urine samples from people 6 years and older. BPA exposure is primarily food-related, as it can leach into food from the coatings of canned foods and from food storage containers or water bottles.[59] However, the dietary recall portion of NHANES did not inquire about food containers or forms of consumption (canned versus fresh) (see Box 4.4).

BOX 4.4   CONSUMPTION EPIDEMIOLOGY

Because the primary exposure to BPA is believed to be food-related, the BPA level in an individual could possibly act as a biological marker flagging a dietary pattern, such as that of high consumption of ready-to-eat meals. It could also flag health-seeking dietary behavior such as consumption of canned fruits, vegetables, vegetable soups, and juices to meet dietary recommendations, or the consumption of bottled water and vitamin drinks.

Perhaps such a dietary "form of consumption" biomarker could help unravel the tangled associations between diet and disease. Although diet is understood to be a contributory risk factor for cardiovascular disease, certain cancers, and other diseases, the clear successes of identifying micronutrient deficiencies in diseases such as pellagra and rickets have not been similarly replicated. The study of dietary patterns is promising, but literature reviews in the examples of breast cancer[60] and cardiovascular disease[61] still show the need for continued and meaningful dietary assessments. Research methods are continuing to improve, but even studies of dietary patterns may still compare fruit or vegetable consumption as a component of a pattern without making a distinction between the consumption of fresh, frozen, or canned.

The National Health and Nutrition Examination Survey (NHANES), which conducts interviews and physical examinations of a nationally representative sample of about 5000 persons each year, is an important source of information used to help develop public health policy and programs. The detailed information collected helps answer questions about emerging issues such as BPA exposure. A 2008 article analyzing 2003–2004 NHANES data reported that after adjusting for potential confounders, higher urinary BPA concentrations were associated with CVD and diabetes.[62] NHANES has blood, urine, health history, and diet information available. Yet, according to personal communication on January 11, 2010, the authors of this study conducted only a posthoc analysis of the reported diet of these individuals, as NHANES lacked data on food sources or containers.

More recent research may have ameliorated public health concerns about BPA. But the point here is not whether BPA is harmful or harmless. The point is that if form of consumption data was available, we would now be able to take advantage of the potential natural experiment that occurred as some manufacturers and policy makers eliminated or banned BPA, whether rightly or wrongly so.

This example of food consumption–related exposure warrants widening the perspective of *nutritional* epidemiology to include *food* and *form*. The terms "dietary assessment" and "dietary epidemiology" are more inclusive, but perhaps we should push even further to be inclusive not only of food and water, but also of form and conveyance. We might gain clarity on some of the complex information we have gathered on diet/disease association if we were able to analyze such data in connection with the form of food and manner of conveyance: consider this "consumption epidemiology."

Critique of NHANES

Underreporting of dietary intake is a well-known problem in nutritional surveillance, and even though methods used in NHANES have been tested, NHANES is not immune to criticism over dietary intake methods. A 2013 article reviewed NHANES data specifically to study trends in energy intake. The researchers applied a cut-off level of plausibility for energy intake and concluded that over almost 40 years, a majority of respondents had reported implausible amounts of energy intake. The researchers further concluded that the well-publicized trend of increased calorie consumption that had ostensibly been derived from the NHANES data was actually simply due to a methodology change between survey cycles, and, that in general due to the complications of trying to work with historical data and changing methods, reported trends in caloric consumption should be viewed skeptically. In short, the researchers conclude that there is no valid population-level data to support speculations regarding trends in caloric consumption and therefore the argument that increased calorie consumption is a cause of the obesity epidemic is spurious. This study was funded via an unrestricted research grant from The Coca-Cola Company, which the paper notes had no role in the study design, data collection, data analysis, data interpretation, or writing of the report.[63] *Should we take that "no role" statement at face value? Does an unrestricted grant mean the funder hands over money to anyone without knowledge of prior work done by the researchers or inquiry even as to the likely study question?*

Even though the study question concerns caloric intake, the authors' underlying thrust is to call attention to the faults of dietary self-reporting. Under-reporting of food intake is not news to nutrition researchers (the authors of the paper noted above have a different focus of expertise). Although the authors propose the use of a biomarker for energy intake, which is a relevant approach for their topic, such a biomarker would not provide the details of food and diet that are also informative for other areas of public health policy. Reading this article and the published comments will provide insight into some of the difficulties of assessing population-level dietary intake and trends and how those challenges are met. *Does the use of self-reported dietary recall limit the use of NHANES data so severely that we must reject the data and this method?*

However, speaking of the influence of funding, it is also important to examine the source of funding for NHANES. In FY2015, the CDC requested $155,397,000 for health statistics, but this funding is not just for NHANES. The CDC's National Center for Health Statistics collects data through four mechanisms: the National Vital Statistics System, the National Health Interview Survey, National Health Care Surveys and NHANES.[64] It is unclear from this document how much funds will be allocated specifically to NHANES, but the CDC notes elsewhere that roughly one-third of NHANES's funding comes from other agencies who propose and help design survey examinations, questionnaires, and laboratory tests.[58] Proposal solicitation announcements and research proposal guidelines for new survey content are posted on the NHANES website. As of the fall of 2014, the proposal for the 2017–2018 cycle is posted. The memo provides deadlines and describes a two-stage process of sending a letter of intent first, and upon NHANES staff review and approval, a full research proposal. Notably, the letter of intent must include funding source information guaranteed for 2 years (2017–2018), plus any funding required in 2016 for any pilot testing or start up activities. Joint proposals are encouraged from within CDC, other Federal agencies, and nongovernment groups.[65]

Although content may be influenced by funded research proposals, the opportunity for public comment on the information NHANES is collecting (or not collecting) is lacking. A public comment forum such as is provided as part of the development process of the Dietary Guidelines, Healthy People, and proposed regulations by the USDA and FDA, does not appear to exist for NHANES.

Is a forum for public comment relevant for NHANES? In Chapter 5, one criticism is that the committee is composed of "nearly all epidemiologists from elite academic institutions with no direct experience in the practical realities of how food is produced and what average Americans may choose to eat."[66] In turn, we comment that the data epidemiologists rely on includes dietary information reported by individuals' dietary recall as part of the NHANES study. Thus our policy is informed by data on what "average" Americans choose to eat. *Why not inquire what "average" Americans may want to know more about?*

CONCLUSION

Although the USDA may be perceived by consumers as the primary federal food policy player, HHS' agencies FDA and CDC are major contributors to the "with what impact" intersection of food policy with public health. Data from CDC's NHANES helps inform our food and health policy through contributing assessment of nutrient status and dietary intake of the population. In turn, the FDA may use information about critical nutrients related to public health to promulgate food labeling and other regulations related to the nutrient content of packaged foods. The recent proposal to update the Nutrition Facts box included changing some of the emphasized nutrients and the FDA is also considering how it could support environmental level changes in the sodium content of processed foods.

These efforts point at an effort to make nutrition knowledge a means by which Americans can take responsibility over their dietary choices, even if we cannot ensure health and a reduction in illness. In general, food product labeling may have roots in, or be legitimized by, policies instigated by industry or friendly to trade. Nonetheless, nutritional food labeling supports public health objectives and goals, such as those of Healthy People and *The Dietary Guidelines for Americans.*

In addition to supporting public health objectives and goals, both food labeling and NHANES also support food activism. The ability to critically review and reject ingredients used in food product manufacturing would not be as powerful or accessible to as many people without labeling information. As noted above, data analyses on mercury levels and fish consumption collected from NHANES led to recommendations to limit the intake of certain mercury-containing fish by young children and pregnant women.[55] Yet this data was also used to advance the "activist" solution to addressing the problem of high mercury intake—namely, the reduction of mercury in the environment.[67]

NHANES data, including dietary intake data, is also used to inform other food research and food policies besides nutritional labeling, such as obesity prevention initiatives. Although criticism as to the dietary intake methods of NHANES must be considered, in general, the issues with dietary intake methodology are well known. Realizing the extent of how our food and health policies may lean on NHANES data, looking forward it may be desirable to push for increased funding for NHANES, greater transparency as to external funding and the choice of survey data collection, and the opportunity for public comment, as made available by other federal efforts at "with what impact" intersection of food policy and public health.

REFERENCES

1. Select Committee on Nutrition and Human Needs United States Senate. Press Conference, Friday January 14, 1977. Statement of Dr. Philip Lee, Professor of Social Medicine and Director, Health Policy Program, University of California, San Francisco. *Dietary Goals for the United States*, 2nd edition, Washington, DC: U.S. Government Printing Office,1977.
2. Lang T. Food control or food democracy? Re-engaging nutrition with society and the environment, *Pub. Health Nutr.* 2005;8(6a):730–737.
3. United States Department of Agriculture, National Agriculture Library, An Act to Establish a Department of Agriculture. Available at http://www.nal.usda.gov/act-establish-department-agriculture, accessed August 15, 2014.
4. National Archives, Records of the Office of the Secretary of Agriculture 1879–1891, Record 16.3. Available at http://www.archives.gov/research/guide-fed-records/groups/016.html, accessed August 15, 2014.
5. United States Department of Agriculture, Mission Areas. Available at http://www.usda.gov/wps/portal/usda/usdahome?navid=USDA_MISSION_AREAS, accessed August 15, 201.
6. United States Department of Agriculture, Mission Statement. Available at http://www.usda.gov/wps/portal/usda/usdahome?navid=MISSION_STATEMENT, accessed August 15, 2014.

7. United States Department of Agriculture, Office of Budget and Program Analysis, Budget Summary, FY2015 Budget Summary and Annual Performance Plan. Available at http://www.obpa.usda.gov/budsum/budget_summary.html, accessed November 17, 2014.

8. Gosselin M. Beyond the USDA: How other government agencies can support a healthier, more sustainable food system. *Institute for Agriculture and Trade Policy*, February 3, 2010. Available at http://www.iatp.org/documents/beyond-the-usda-how-other-government-agencies-can-support-a-healthier-more-sustainable-foo, accessed August 15, 2014.

9. United States Department of Health and Human Services. About HHS. Available at http://www.hhs.gov/, accessed August 15, 2014.

10. Public Health Service. *Healthy people: The surgeon general's report on health promotion and disease prevention.* Washington, DC: U.S. Department of Health, Education, and Welfare, Public Health Service, 1979; DHEW publication no. PHS 79-55071.

11. U.S. Food and Drug Administration. About FDA. Available at http://www.fda.gov/AboutFDA/WhatWeDo/, accessed October 26, 2014.

12. U.S Food and Drug Administration. Food, Guidance and Regulation, Food Labeling Guide. Available at http://www.fda.gov/Food/GuidanceRegulation/GuidanceDocumentsRegulatoryInformation/LabelingNutrition/ucm2006828.htm, accessed August 15, 2014.

13. Institute of Medicine of the National Academies, About the IOM. Available at http://www.iom.edu/About-IOM.aspx, accessed August 17, 2014.

14. U.S. Food and Drug Administration, Guidance for Industry: A Food Labeling Guide (11. Appendix C: Health Claims), January 2013. Available at http://www.fda.gov/food/guidanceregulation/guidancedocumentsregulatoryinformation/labelingnutrition/ucm064919.htm, accessed August 14, 2014.

15. U.S. Food and Drug Administration, About FDA. Available at http://www.fda.gov/AboutFDA/WhatWeDo/History/Origin/ucm124403.htm, accessed August 15, 2014.

16. Wall EC. The Efficacy of the Fair Labeling and Packaging Act: Then and Now (Harvard Law School, 2002). Available at http://dash.harvard.edu/bitstream/handle/1/8846774/Wall.html?sequence=2, accessed August 17, 2014.

17. Institute of Medicine. Dietary Reference Intakes: Guiding Principles for Nutrition Labeling and Fortification. Washington, DC: The National Academies Press, 2003. Available at http://www.nal.usda.gov/fnic/DRI/DRI_Guiding_Principles_Labeling/guiding_princi ples_labeling_full_report.pdf, accessed October 27, 2014.

18. Wartella EA, Lichtenstein AH, Boon CS (eds.). *Examination of Front-of-Package Nutrition Rating Systems and Symbols: Phase 1 Report.* National Research Council, Committee Report; Institute of Medicine. Washington DC: The National Academies Press, 2010.

19. Gillett TJ. Lessons from Nutritional Labeling on the 20th Anniversary of the NLEA: Applying the History of Food Labeling to the Future of Household Chemical Labeling 37. *Wash. U. J. Law Policy* 2011;267. Available at http://digitalcommons.law.wustl.edu/wujlp/vol37/iss1/12, accessed August 17, 2014.

20. Woteki C. Integrated NHANES: Uses in National Policy. *J. Nutr.* 2003;133:582S–584S.

21. Significant Dates in U.S. Food and Drug Law History. U.S. Food and Drug Administration. Available at http://www.fda.gov/AboutFDA/WhatWeDo/History/Milestones/ucm128305.htm, accessed August 17, 2014.

22. National Research Council. *Dietary Reference Intakes: Applications in Dietary Planning.* Washington, DC: The National Academies Press, 2003.

23. U.S. Food and Drug Administration, Nutrition Facts Label Programs & Materials. Available at http://www.fda.gov/Food/IngredientsPackagingLabeling/LabelingNutrition/ucm20026097.htm, accessed October 22, 2014.

24. U.S. Food and Drug Administration. Proposed Changes to the Nutrition Facts Label. Questions and Answers. Updated August 1, 2014. Available at http://www.fda.gov/Food/

GuidanceRegulation/GuidanceDocumentsRegulatoryInformation/LabelingNutrition/ucm385663.htm#images, accessed October 27, 2014.

25. Institute of Medicine. Dietary Reference Intakes: Electrolytes and Water. Available at http://www.iom.edu/Activities/Nutrition/SummaryDRIs/~/media/Files/Activity%20Files/Nutrition/DRIs/New%20Material/9_Electrolytes_Water%20Summary.pdf, accessed November 14, 2014.

26. CFR101.12. Available at http://www.accessdata.fda.gov/scripts/cdrh/cfdocs/cfcfr/CFRSearch.cfm?fr=101.12, accessed August 17, 2014.

27. United States Department of Agriculture, Center for Nutrition Policy and Promotions. "Guidance on Use of USDA's MyPlate and Statements About Amounts of Food Groups Contributed by Foods on Food Product Labels" May 2013. Available at http://www.choosemyplate.gov/food-groups/downloads/MyPlate/MyPlateOnFoodLabels.pdf, accessed August 17, 2014.

28. Regulations.gov. Environmental Working Group Comment Food Labeling: Revision of the Nutrition and Supplement Facts Labels. Docket FDA-2012-N-1210. Available at http://www.regulations.gov/#!documentDetail;D=FDA-2012-N-1210-0210, accessed October 27, 2014.

29. U.S. Food and Drug Administration. Label Claims for Conventional Foods and Dietary Supplements. December 2013. Available at http://www.fda.gov/food/ingredientspackaginglabeling/labelingnutrition/ucm111447.htm, accessed August 17, 2014.

30. U.S. Food and Drug Administration. Guidance for Industry: A Food Labeling Guide (8. Claims). January 2013. Available at http://www.fda.gov/Food/GuidanceRegulation/GuidanceDocumentsRegulatoryInformation/LabelingNutrition/ucm064908.htm, accessed August 17, 2014.

31. U.S. Food and Drug Administration, Label Claims for Conventional Foods and Dietary Supplements. Available at http://www.fda.gov/Food/IngredientsPackagingLabeling/LabelingNutrition/ucm111447.htm, last updated February 26, 2014.

32. United States Department of Agriculture. Commodity Supplemental Food Program. USDA available foods for 2015. Available at http://www.fns.usda.gov/sites/default/files/csfp/FY2015_CSFP.pdf, accessed November 11, 2014.

33. Regulations.gov. American Heart Association Comment on the Food and Drug Administration (FDA) Notice: Approaches to Reducing Sodium Consumption; Establishment of Dockets: Request for Comments, Data, and Information; Extension of Comment Period. Available at http://www.regulations.gov/#!documentDetail;D=FDA-2011-N-0400-0331, accessed October 22, 2014.

34. HealthyPeople.gov. Leading Health Indicators. Available at http://www.healthypeople.gov/2020/Leading-Health-Indicators, accessed October 23, 2014.

35. Green LW, Fielding J. The US Healthy People Initiative: Its Genesis and Its Sustainability. *Annu. Rev. Public Health.* 2011;32:451–70.

36. Centers for Disease Control and Prevention. *Healthy People.* Available at http://www.cdc.gov/nchs/healthy_people/hp2020.htm, accessed August 17, 2014.

37. HealthyPeople.gov. Healthy People 2020 Leading Health Indicators: Progress Update. Available at http://www.healthypeople.gov/2020/leading-health-indicators/Healthy-People-2020-Leading-Health-Indicators%3A-Progress-Update, accessed October 23, 2014.

38. HealthyPeople.gov. HP2020 Objective Data Search. Available at http://www.healthypeople.gov/2020/data-search/Search-the-Data, accessed October 23, 2014.

39. HealthyPeople.gov. Healthy People 2020 Leading Health Indicators: Update. Available at http://www.healthypeople.gov/sites/default/files/HP2020_LHI_Nut_PhysActiv.pdf, accessed October 23, 2014.

40. HealthyPeople.gov. About Healthy People. Available at http://www.healthypeople.gov/2020/About-Healthy-People, accessed August 17, 2014.

41. HealthyPeople.gov. 2020 Topics and Objectives. Nutrition and Weight Status. Available at http://www.healthypeople.gov/2020/topics-objectives/topic/nutrition-and-weight-status, accessed August 17, 2014.

42. *Healthy People 2020*, 2020 Topics and Objectives, Nutrition and Weight Status. Available at http://www.healthypeople.gov/2020/topicsobjectives2020/objectiveslist.aspx?topicId=29, accessed August 17, 2013.

43. Center for Science in the Public Interest. *Petition to Ensure the Safe Use of "Added Sugars."* Available at https://www.cspinet.org/liquidcandy/sugarpetition.html, accessed August 17, 2014.

44. Madans JH, Sondik EJ, Johnson CLC. Foreword: Future Directions for What We Eat in America-NHANES: The Integrated CSFII-NHANES. *J Nutr.* 2003;133:575S.

45. United States Department of Health and Human Services. Centers for Disease Control and Prevention. National Center for Health Statistics. National Health and Nutrition Examination Survey. Factsheet Updated May 2014. Available http://www.cdc.gov/nchs/data/factsheets/factsheet_nhanes.pdf, accessed September 26, 2014.

46. Centers for Disease Control and Prevention, National Health Interview Survey, About the National Health Interview Survey. Available at http://www.cdc.gov/nchs/nhis/about_nhis.htm, accessed August 17, 2014.

47. Centers for Disease Control and Prevention, National Center for Health Statistics, National Health and Nutrition Examination Survey: History. Available at www.cdc.gov/nchs/nhanes/history.htm, accessed August 17, 2014.

48. United States Department of Agriculture. Agriculture Research Service. What We Eat In America (WWEIA), NHANES: Overview. Last modified November 13, 2013. Available at http://www.ars.usda.gov/News/docs.htm?docid=13793, accessed September 26, 2014.

49. Dwyer J, Ellwood K, Moshfegh AJ, Johnson CL. Integration of the continuing survey of food intakes by individuals and the national health and nutrition examination survey. *J. Am. Diet. Assoc.* 2001;101:1142–1143.

50. Centers for Disease Control and Prevention, National Health and Nutrition Examination Survey, About the National Health and Nutrition Examination Survey. Available at http://www.cdc.gov/nchs/nhanes/about_nhanes.htm, accessed August 17, 2014.

51. Wright JD, Borrud LG, McDowell MA, Wang C-Y, Radimer K, Johnson CL. Nutrition Assessment in the National Health and Nutrition Examination Survey 1999–2002. *J. Am. Diet. Assoc.* 2007;107(5):822–829.

52. Dwyer J, Picciano MF, Raiten DJ, Members of the Steering Committee; National Health and Nutrition Examination Survey. Estimation of Usual Intakes: What We Eat in American-NHANES. *J. Nutr.* 2003;133(2):609S–623S.

53. Centers for Disease Control and Prevention. NHANES Dietary Web Tutorial: Dietary Data Overview. NHANES Dietary Data Collection. Available at http://www.cdc.gov/nchs/tutorials/dietary/SurveyOrientation/DietaryDataOverview/Info2.htm, accessed November 11, 2014.

54. Dwyer J, Picciano MF, Raiten DJ, Continuing Survery of Food Intakes by Individuals; National Health and Nutrition Examination Survery. Future Directions for the Integrated CSFII-NHANES: What We Eat in American-NHANES. *J. Nutr.* 2003;133(2):576S–578S.

55. Chapman N. Securing the future for monitoring the health, nutrition, and physical activity of Americans. *J. Am. Diet. Assoc.* 2005;105:1196–1200.

56. Dwyer J, Picciano MF, Raiten DJ, Members of the Steering Committee; National Health and Nutrition Examination Survery. Collection of Food and Dietary Supplement Intake Data: What We Eat in American-NHANES. *J. Nutr.* 2003;133(2):590S–600S.

57. Woteki C. Integrated NHANES: Uses in National Policy. *J. Nutr.* 2003;133:582S–584S.

58. Centers for Disease Control and Prevention, National Center for Health Statistics. *Inside NCHS.* Newsletter, Issue 2, January 2013. Available at http://www.cdc.gov/nchs/newsletter/2013_January/a2.htm, accessed November 11, 2014.

59. National Institute of Environmental Health Sciences. Bisphenol A (BPA). Available at http://www.niehs.nih.gov/health/topics/agents/sya-bpa/, accessed November 17, 2014.
60. Edefonti V, Randi G, La Vecchia C, Ferraroni M, Decaril A. Dietary patterns and breast cancer. *Nutr. Rev.* 2009;67:297–314.
61. Mente A, de Koning L, Shannon HS, Anand SS. A systematic review of the evidence supporting a causal link between dietary factors and coronary heart disease. *Arch. Intern. Med.* 2009;169:659–669.
62. Lang IA, Galloway T, Scarlett A, Henley W, Depledge M, Wallace R, Melzer D. Association of urinary bisphenol a concentration with medical disorders and laboratory abnormalities in adults. *JAMA* 2008;300:1303–1310.
63. Archer E, Hand GA, Blair SN. Validity of U.S. Nutritional Surveillance: National Health and Nutrition Examination Survey Caloric Energy Intake Data, 1971–2010. *Plos One*. October 9, 2013. Available at http://www.plosone.org/article/info%3Adoi%2F10.1371%2Fjournal.pone.0076632, accessed October 27, 2014.
64. Centers for Disease Control and Prevention: Justification of Estimates for Appropriation Committees. Department of Health and Human Services, Fiscal Year 2015. Available at http://www.cdc.gov/fmo/topic/Budget%20Information/appropriations_budget_form_pdf/FY2015_CJ_CDC_FINAL.pdf, accessed November 11, 2014.
65. Centers for Disease Control and Prevention. National Health and Nutrition Examination Survey. Proposal Guidelines. Last updated May 2, 2014. Available at http://www.cdc.gov/nchs/data/nhanes/NHANES_Content_Deadlines_2017_18.pdf, accessed November 11, 2014.
66. Independent Women's Forum, Two-Dozen Free-market Organizations Voice Concern And Urge Reform of the Dietary Guidelines Advisory Committee, May 28, 2014. Available at http://www.iwf.org/media/2794073/, accessed September 12, 2014.
67. Environmental Protection Agency. Trends in Blood Mercury Concentrations and Fish Consumption Among U.S. Women of Childbearing Age, NHANES 1999–2010. Available at http://water.epa.gov/scitech/swguidance/fishshellfish/fishadvisories/upload/Trends-in-Blood-Mercury-Concentrations-and-Fish-Consumption-Among-U-S-Women-of-Childbearing-Age-NHANES-1999-2010.pdf, accessed November 11, 2014.
68. U.S. Department of Health and Human Services (HHS). Public Health Service Publication 88-50210. Washington, DC: U.S. Government Printing Office, 1988.
69. Committee on Diet and Health Food and Nutrition Board, Commission on Life Sciences. National Research Council. Washington, DC: National Academy Press, 1989.

Activism*

5

The City College of New York (CCNY) has a long history of faculty and student activism, most of it political, concerning freedom of speech. Both Arlene Spark and her father are CCNY alumni, Arthur graduating 34 years earlier than his daughter. Figure 5.1 is a poster from the year of Arthur's graduation.

INTRODUCTION

Food activists care about food and pay attention to how food is produced and the impact that production has on people and the planet we inhabit. For activists, "good food is whole (unprocessed or minimally processed), healthy, produced sustainably, fair and affordable."[1]

Good foods are *traditional* foods. They existed a century ago and, with few exceptions, are not processed outside of the home. Exceptions would be pasteurized vitamin-D fortified milk and iodized salt. A *healthy* food is most likely whole, real, or it has an identifiable source, such as an apple as opposed to a dubious apple filling in a frozen pastry. A food produced *sustainably* is resource-neutral, or close to it. The food comes from an agricultural system that puts as much back into the soil as it extracts, uses water responsibly, and is not a "protein factory in reverse." Food that is *fair and affordable* suggests that people who produce it are treated fairly by being protected from harm and paid reasonable wages. Affordability refers to the doctrine that all people can afford it, and for an activist, affordability is not just related to "food" but to a healthy, nutritious diet of "good food." However, affordability of food of any characterization also likely means that at least some food production is subsidized by the government.

The first part of this chapter examines activism in its several forms, and takes the pulse of the current food movements (the plural is deliberate!). Whatever its form, the aim of food activism is to bring about better health for people and sustainability for the planet through changes in the U.S. food system. The second part of the chapter introduces some of the influential food activists of the past 50 years, all of whom have

---

* Craig Willingham, MPH, contributed primary research and prepared draft versions of portions of this chapter.

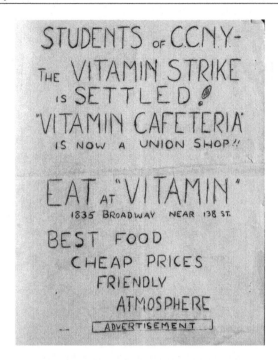

FIGURE 5.1   Example of food activist poster. (With permission from Archives, City College of New York, CUNY.)

written letters to politicians. As all food activists communicate with their elected officials from time to time, a template for writing to members of Congress is provided in the Appendix to this chapter.

Activism, Food Activists, and Their Causes

What are defining characteristics of an activist? If you consider yourself an activist, explain why. If you are a food activist, what are you trying to accomplish?

People who take vigorous action in support of or in opposition to a cause are *activists*. Activists may act alone, but more frequently, they agitate as part of a group of like-minded people. The activist may be self-motivated to take up a cause, or motivated by others. The activist's goal is to persuade people to change their behavior, or to persuade governing bodies to propose new policies and ultimately enact new laws.

Activism is the concerted effort expended by the activist to support or oppose a (controversial) cause. The cause is either a desired behavior change in targeted individuals, or a change in society, politics, the economy, or the environment that will enable the individual to make desirable changes. Alternatively, the cause may be an effort to maintain the *status quo*, that is, to preserve conditions as they currently exist. Thus, activists may focus their efforts (activism) directly on individuals, or indirectly through the world in which the target audience lives. For example, the aim of the activist may be

---

**BOX 5.1 EXAMPLES OF CAUSES CHAMPIONED
BY FOOD ACTIVISTS**

Agriculture and the environment
    Industrialization, fewer farms, older farmers
    Factory farming, animal brutality, antibiotics in food animal production
    Decentralized food supply
    Pesticides, fossil fuel
    Water protection
    Climate change, pollution, dead zones
    Federal support of unhealthy calories (commodity subsidies)
Workforce
    Treatment of seasonal agricultural workers and undocumented workers
    Treatment of food processing workers
    Treatment of food service industry workers
Food marketing
    GM labeling, menu labeling, front-of-package food labeling
    Food advertising to vulnerable populations
Health
    Epidemic type-2 diabetes, especially in low-income communities
    Taxing and regulating sugar-sweetened beverages
    School food and food in other institutional settings (hospitals, government facilities, etc.)
    Incentivizing the purchase of fruits and vegetables
    Regulating portion size
    Institutionalized barriers to healthy food access

---

to persuade people to change their behavior (direct), or to persuade governing bodies to enact new laws, or change or maintain the status quo of existing legislation and policy (indirect).

In any case, the *raison d'être* (cause) of all food activists may be described as the call for a food system that promotes and sustains health for individuals, communities, animals, and the natural world in which we live. *That's a tall order!* No single food activist could possibly expect to work on all the campaigns necessary to bring about the revamping of our food system. Indeed, food activists must focus on the challenges that, for them, have the most meaning. Box 5.1 identifies some causes championed by food activists.

Social media and digital activism

With social media that is now ubiquitous, activists can organize large-scale campaigns that accomplish in hours or days what used to take weeks or months. Whether it is raising awareness or calling citizens to action, technology and social media play a role in connecting people with a common goal of change. Messaging through Twitter, activity on Facebook, email listservs, messaging blasts, and change.org petitions are now inexpensive and a widely available means of communication for many activists.

Food price inflation was one of the grievances that led to the 2011–2012 uprisings in Egypt, popularly known as the Arab Spring. At the height of the rebellion, an activist in Cairo is claimed to have said "We use Facebook to schedule our protests, Twitter to coordinate them, and YouTube to tell the world."[2]

Social media makes it simpler for many people to take action, such as through electronic letter-signing campaigns. Organizations such as Center for Science in the Public Interest (CSPI) typically run several letter-signing campaigns simultaneously. In 2014, CSPI's "take action" campaigns were aimed at school food and food education in schools, food safety, marketing unhealthy foods to children (including not allowing organizations to deduct the expense of such marketing), taxes on sugar-sweetened beverages (SSBs), food additives (particularly dyes in foods consumed by children), the removal of soda from the children's menu board at fast-food restaurants and other improvements to children's restaurant meals and wider spread restaurant calorie labeling.[3] The activist organization posts a form letter (letter template) for constituents to sign electronically. The "signed" letter is then transmitted back to the organization, which forward it to the intended recipient.

What constituted food activism more than two centuries ago?

Almost 250 years of food activism

Before the advent of electronic communication, food activists communicated among themselves by word-of-mouth. Activists joined others with the same agenda to bring about desirable change. The Boston Tea Party, Gandhi's Salt March, the Delano Grape Boycott, and the Nestle Boycott are historic examples of local protests that evolved into seminal national events. These crusades illustrate the power of food-related movements that have a clear focus.

Boston Tea Party

At its core, the Boston Tea Party (1773) was a conflict over "taxation without representation." Because the British colonists living in North America did not have direct representatives in parliament, they had no influence over how they would be taxed. This lack of representation enabled the British government to levy taxes as they saw fit. With no means of fighting taxation, many colonists feared that their property could be taken away through debilitating taxes.

Since the British government had free reign to tax colonists, the years leading up to the American Revolutionary War (1775–1783) saw the introduction of a number of taxation laws, many of which were modified or repealed because of protests from the colonists. But even so, these laws and subsequent protests contributed to growing tensions throughout the colonies, particularly in Boston, where the British government stationed soldiers in 1768.

By law, the British East India Company was the only company allowed to import tea into the colonies. Since the British government had such a close hold on the importation of tea and the ability to tax at will, it instituted high taxes on the product to pay for the costly French and Indian Wars. In 1773, the British government passed the Tea

Act, which authorized the British East India Company to ship tea directly to colonies while the government levied a tax of three pence on each shipment that was unloaded in America. In protest, colonists refused to allow tea to be unloaded from English ships and demanded that the ships return the tea to England so they would not be required to pay the tax. When the British refused to accede to the boycott, the colonists retaliated by dumping the tea into the harbor, one of the most dramatic protests against British colonial rule, cynically referred to as the "Boston Tea Party."

Although not the official beginning of the American Revolution, it was an early, important defining event. The rebellion against colonial rule and subsequent crackdown by the British government galvanized the colonies in their opposition to the Crown, as demonstrated by protests against tea in places outside Massachusetts. Other tea shipments intended for the colonies were forced to return to Britain. The First Continental Congress met 10 months after the Boston Tea Party, and the Declaration of Independence came less than 2 years after that. The colonies were on their way to outright rebellion and eventual independence. The Boston Tea Party remains a seminal moment in the development of the nation.

Salt March[4,5]

Mohandas Karamchand Gandhi (1869–1948), considered the father of Indian independence, worked to both remove British rule from India as well as to improve the lives of India's poorest classes. Gandhi created the concept of *Satyagraha*, a nonviolent way of protesting against injustice. He rejected the term *civil disobedience* because he felt that his actions were just, not disobedient; he also rejected the term *passive resistance* because his actions were not passive. Civil rights leaders Martin Luther King, Jr. and Cesar Chavez used Gandhi's concept of nonviolent protest as a model for their own campaigns.

In the beginning of the twentieth century, India was one of the largest producers of salt in the world with an annual yield of over 1 million metric tons, and throughout the Raj (British Rule), an excellent source of revenue for Britain. The first laws to regulate the salt tax were enacted in 1835, when a salt commission appointed by the British recommended that Indian salt should be taxed to enable the sale of imported English salt. Subsequently, Britain created a monopoly on the manufacture of salt in which production of salt was made an offense punishable by 6-months' imprisonment. People were required to use salt imported from the United Kingdom, a product that was inferior in quality and more expensive than the indigenous Indian salt. In 1878, a uniform salt tax policy was adopted for the whole India. Both production as well as possession of salt were made unlawful by this policy. Salt-revenue officials were authorized to break into places where salt was being illegally manufactured and seize the salt. The India Salt Act of 1882 included regulations enforcing a government monopoly on the collection and manufacture of salt. Salt, which was taxed, could be manufactured and handled only at official government salt depots. When the Salt Tax was doubled in 1923, the resulting high price made it unaffordable, and economic oppression veered into a decline in public health due to consequences arising from deficiency of the iodine, the mineral contained in salt mined from coastal areas where iodine was indigenous.

In response to the oppressive tax on salt, Gandhi began a campaign, the Salt Satyagraha, in which he and his followers embarked on a 200-mile journey on foot from Ashram Ahmedabad to the Arabian Ocean where they planned to gather a few grains of salt. This action formed the symbolic focal point of a campaign of nonviolent protest, targeting the state monopoly on salt. After the long trek to the Indian Ocean, Gandhi picked up a few pieces of salt, which signaled men throughout the rest of the subcontinent to do the same. The raw salt was carried inland to be processed (in pans on the roofs of houses) and then to be sold. Over 50,000 Indians were imprisoned for breaking the salt laws. The protest was only partially successful. The British authorities turned deaf ears to the massive salt tax protests that rocked India during the early 1930s. The Salt Act in its entirety was not formally lifted until 1946, years after the protest marches. Nevertheless, the entire protest was carried out almost without violence, thus demonstrating to the world the strength of peaceful food activism militancy.

Grape boycott [6,7]

In 1965, the Delano Grape Strike started and continued for more than 5 years. The strike began in Delano, California, when the Agricultural Workers Organizing Committee (mostly Filipino farm workers) walked off the farms, demanding wages equal to the federal minimum wage. One week after the strike began, the predominantly Mexican-American National Farmworkers Association led by Cesar Chavez, Dolores Huerta, and Richard Chavez joined the strike. In 1966 the two groups merged, forming the United Farm Workers of America (UFW) in 1966. Leveraging the strengths of both groups into grassroots efforts engendered national attention on how poorly these workers were being treated.

In 1968, Cesar Chavez began his most visible campaign, urging Americans not to buy table grapes produced in the San Joaquin Valley until growers agreed to union contracts. Although there were picket lines in the fields, the real focus moved to the cities where grapes were sold. Hundreds of students, religious workers, and labor activists talked to consumers in front of markets, asking them to do a simple thing: "Help the farmworkers by not buying grapes." Millions of Americans complied with the boycott and stopped buying grapes. In 1970, after losing millions of dollars, growers agreed to sign. The UFW had succeeded in reaching a collective bargaining agreement with the table-grape growers, affecting in excess of 10,000 farm workers.

Nestlé boycott

The relatively recent introduction of multinational infant formula promotion has decreased the global incidence and duration of lactation, most notably in developing (third world) countries. The nature of this influence is an acceleration of long-term trends in existing patterns. For example, mixed feeding patterns found in the Caribbean in the 1970s represented patterns that had been evolving over more than 150 years. By the early 1800s, plantation owners, conscious of the effects of prolonged lactation (14 months or more) on maternal work output and fertility, introduced breast-milk substitutes, such as dilute paps and herbal (bush) teas, as strategies to curb the prolonged lactation period of their slaves. By the 1940s, the urbanization and modernization in third world countries led to an overall decline in breastfeeding, which does seem rationally necessary when viewed in the context of women adapting to these new situations. Nevertheless, the consequences of a decline in breastfeeding have been dire from a nutritional point of view.[8]

The publication of an article in 1971 about the promotion of baby formula in third world countries touched off a media storm of shocking photos of infants dying in the first few months of life. The deaths were attributed to marasmus, which developed in infants whose mothers did not know how to prepare infant formula, could not afford enough of the powdered formula, did not have access to potable water to use in formula preparation, and most significantly, did not understand that using formula would interfere with going back to breastfeeding. The condition was referred to as *commerciogenic malnutrition*.[9]

The infant formula companies used three techniques to create a market that did not previously exist. They convinced new mothers that their products were indispensable for the "good life," linked their products with the most desirable and unattainable concepts, and then provided free samples. Formula marketers systematically undermined its "competition," breastfeeding, at both psychological and physiological levels. Promotional campaigns encouraged the view that breastfeeding is complicated and prone to failure. Breastfeeding is thought of negatively on the basis of beauty (breast sag), work (you have to stay home), classism and racism (women like me do not breastfeed), and fear (it will not work, your baby will starve).

With the knowledge that fear and anxiety can inhibit lactation,[10] companies design marketing strategies that aggravate inherent worries and interfere with the psychophysiology of the human body to sell more of their products. What makes the infant formula case unique is that there are few other products where the marketing can actually undermine a normal bodily function to create the physical need for the commercial product. Formula samples literally hook the mother on formula because lactogenesis decreases once bottle-feeding starts. When the free sample is finished, there is an actual, physical need to buy more formula.[11]

The infant formula industry's inappropriate marketing of infant formula became the rallying point for the largest-ever international campaign to expose and reverse exploitative corporate practices. In particular, Nestlé's corporate accountability was called into question by War on Want, the London-based antipoverty charity organization that highlights the needs of poverty-stricken areas around the world and raises public awareness of the concerns of developing nations. Prompted by concerns about Nestlé's aggressive marketing of infant formula in less economically developed countries, a boycott of Nestlé products was launched in Minneapolis in 1977 by the Infant Formula Action Coalition, (INFACT), now Corporate Accountability International.[12] The boycott was not just for formula; Nestlé owns hundreds of global brands, ones familiar to U.S. consumers are coffee (Nescafé), dairy products (Carnation, Coffee Mate), candy and chocolate (Kit Kat, Nestle Toll House), mineral water (Perrier, San Pellegrino), pet food (Purina, Friskies), meals (Stouffers, Lean Cuisine), and more.[13]

As the boycott spread and increased attention turned on infant formula marketing practices, the World Health Organization and United Nations International Children's Education Fund hosted an international meeting in 1979 that called for the development of an international code of marketing (adopted in 1981), as well as action on other fronts to improve infant and early child feeding practices. The International Baby Food Action Network (IBFAN) was formed by six of the campaigning groups at this meeting.[11] The boycott is on-going.

The Nestlé boycott is coordinated by the International Nestlé Boycott Committee of Baby Milk Action.[14] Company practices are monitored by IBFAN. In a 2010 report,

IBFAN details marketing strategies used by baby food manufacturers, such as unsubstantiated health and nutrition claims, and exposes violations of the *International Code of Marketing of Breastmilk Substitutes*. The report provides a roadmap for needed action.[15]

Arguably, the boycott's most significant achievement is its demonstration that a broad international campaign can raise awareness of unjust practices. People all around the globe are more aware of the value of breastfeeding, while health professionals learned to be skeptical of formula company giveaways. The current international infant formula campaign, initiated by the Nestle Boycott, keeps pressure on formula manufacturers. The campaign has created a coalition of groups in both developing and developed countries and brought the issue to the attention of health officials and government authorities in national and international forums. Awareness has spread about the pervasive influence of multinational corporations.

Despite their differences, the current food movements, the American Revolution, Gandhi's salt march, and the grape and Nestle boycotts have in common the struggle between personal freedom and authority.

Mark Winne asks how long we will allow ourselves to remain dependent on industrial agriculture whose spokespeople insist that it is the *only* way to feed a hungry world. Winne points out that "With the advent of industrialism and its widespread application to our food supply—factory farms, genetic engineering, and agricultural chemicals—the struggle between human freedom and authority has reached a critical juncture. These are ... real-life concerns set squarely on the plate of every eater."[16]

While current food causes have not reached every eater, food activism is responsible, at least in part, for substantial progress, as indicated by these statistics:

- During the 2011–2012 school year, over 40,000 schools participated in farm-to-school activities.[17,18]
- In 2014 there were more than 8200 farmers' markets listed in the USDA's National Farmers Market Directory, more than 200 food hubs supporting local and regional distribution,[19] 200 food policy councils, and branding programs (such as "Jersey Fresh" or "Simply Kansas") in all states.[20]
- According to the National Restaurant Association, the top four trends in restaurant dining expected in 2014 are locally sourced meats and seafood, locally grown produce, environmental sustainability, and healthy kids' meals.[21]

Current food activists rebuke the industrial food chain

The California-based food activist organization, Roots of Change, is less a think tank than a "think and do tank" that calls for citizen advocacy.[22] In 2008, Roots of Change developed a *Declaration for Healthy Food and Agriculture*.[23] The Declaration is modeled on the 1776 U.S. Declaration of Independence in style and in its implication that having access to healthy food is an inalienable right.

The Declaration of Independence declared that the 13 original American colonies regarded themselves as independent states, and no longer a part of the British Empire. The document is considered a major statement on human rights particularly its second sentence, which outlines a general philosophy of government: *We hold these truths to be self-evident, that all men are created equal, that they are endowed by their Creator with certain unalienable Rights, that among these are Life, Liberty and the pursuit of Happiness.* The 56 signers of the Declaration believed that it was necessary to separate from the British Empire because the Empire threatened the colonists' natural rights.

The Healthy Food Declaration is meant to provide policymakers with a set of principles, endorsed by a broad base of organizations and individuals committed to a healthier food and agriculture, to use in creating policy that will lead to a healthier food system. The independence and food declarations differ in that the independence declaration presents principles and asserts that the policy to realize our inalienable rights is that we must be free of British rule, whereas the food declaration presents principles without a specific organizing policy. See Box 5.2 for the Declaration of Healthy Food and Agriculture.

BOX 5.2   DECLARATION OF HEALTHY FOOD AND AGRICULTURE

We, the undersigned, believe that a healthy food system is necessary to meet the urgent challenges of our time. Behind us stands a half-century of industrial food production, underwritten by cheap fossil fuels, abundant land and water resources, and a drive to maximize the global harvest of cheap calories. Ahead lie rising energy and food costs, a changing climate, declining water supplies, a growing population, and the paradox of widespread hunger and obesity.

These realities call for a radically different approach to food and agriculture. We believe that the food system must be reorganized on a foundation of health: for our communities, for people, for animals, and for the natural world. The quality of food, and not just its quantity, ought to guide our agriculture. The ways we grow, distribute, and prepare food should celebrate our various cultures and our shared humanity, providing not only sustenance, but justice, beauty, and pleasure.

Governments have a duty to protect people from malnutrition, unsafe food, and exploitation, and to protect the land and water on which we depend from degradation. Individuals, producers, and organizations have a duty to create regional systems that can provide healthy food for their communities. We all have a duty to respect and honor the laborers of the land without whom we could not survive. The changes we call for here have begun, but the time has come to accelerate the transformation of our food and agriculture and make its benefits available to all.

We believe that the following 12 principles should frame food and agriculture policy, to ensure that it will contribute to the health and wealth of the nation and the world. A healthy food and agriculture policy

1. Forms the foundation of secure and prosperous societies, healthy communities, and healthy people.
2. Provides access to affordable, nutritious food to everyone.

3. Prevents the exploitation of farmers, workers, and natural resources; the domination of genomes and markets; and the cruel treatment of animals, by any nation, corporation or individual.
4. Upholds the dignity, safety, and quality of life for all who work to feed us.
5. Commits resources to teach children the skills and knowledge essential to food production, preparation, nutrition, and enjoyment.
6. Protects the finite resources of productive soils, fresh water, and biological diversity.
7. Strives to remove fossil fuel from every link in the food chain and replace it with renewable resources and energy.
8. Originates from a biological rather than an industrial framework.
9. Fosters diversity in all its relevant forms: diversity of domestic and wild species; diversity of foods, flavors, and traditions; diversity of ownership.
10. Requires a national dialog concerning technologies used in production, and allows regions to adopt their own respective guidelines on such matters.
11. Enforces transparency so that citizens know how their food is produced, where it comes from, and what it contains.
12. Promotes economic structures and supports programs to nurture the development of just and sustainable regional farm and food networks.

Our pursuit of healthy food and agriculture unites us as people and as communities, across geographic boundaries, and social and economic lines. We pledge our votes, our purchases, our creativity, and our energies to this urgent cause.

Roots of Change, Activities, Food Declaration. Available at http://www.rootsofchange.org/content/activities-2/ food-declaration, accessed August 30, 2014. Reprinted with permission of Roots of Change.

Food industry's ability to influence and create food policy: For better and for worse

In 2007, milk suppliers to Safeway's Northwest processing plants stopped using the artificial growth hormone, recombinant bovine growth hormone (rBGH). In 2008, when Walmart, the largest retailer in the United States, stopped selling milk produced from cows treated with rBGH, the move sent a powerful signal to food manufacturers about the growing mainstream demand for healthier food products and almost entirely derailed the practice of using growth hormones in cows. Many grocery store chains have since stopped accepting milk from dairies that use rBGH. The Wegman's grocery store chain began selling only hormone-free milk in 2011. Safeway switched its in-store brands to hormone-free milk (although it also sells other brands produced from cows given the hormone), and Starbucks uses only hormone-free milk in its stores.

Whole Foods' stores in the United Kingdom require labeling for foods that contain genetically modified organism (GMO) ingredients. In 2013, Whole Foods announced a 5-year plan that all products in its U.S. and Canadian stores must be labeled to indicate if they contain GMOs. As such, Whole Foods is the first national grocery chain to set a deadline for full GMO transparency. It is not surprising that the Grocery Manufacturers Association, the trade group that represents major food companies and retailers, opposed the move.[24] Just as food manufacturers developed ways to decrease or eliminate trans-fatty acids in their products once it was required that they include trans-fats on food labels, it is similarly expected that Whole Foods' policy regarding GM food labeling will eventually be adopted by other major food retailers to meet the labeling standard set by Whole Foods.

The ability of industry to influence and create food policy through self-regulation and consumer reaction is examined further in Chapter 12.

Is there a food movement?

During every social movement, there is a moment when its political influence becomes incontrovertible.

- For the civil rights movement, it was the Civil Rights Act of 1964 (P.L. 88–352).
- For the environmental movement, it was the first Earth Day in 1970, celebrated on or around April 22 each year.
- For the contemporary women's movement, it was *Roe v. Wade* (410 U.S. 113 [1973]).
- For creating a disincentive to buy unhealthy food, it was November 4, 2014, when Berkeley passed a law requiring that distributors of sugary drinks pay a one-cent-per-fluid-ounce "soda" tax.

One critic of food activism opined that "there is no 'food movement' outside of Berkeley and the Upper West Side [of Manhattan]."[25] He would have been more accurate locating the food movement in Berkeley and Brooklyn.

Current food activism has reached the point where government recognizes and indeed supports many of the tenets of its adherents, such as strengthening of local and regional food systems, questioning the usefulness of food labels, and, in Berkeley, enacting a tax on SSBs. Nevertheless, to date, there has not been a watershed moment when the influence of food activists has coalesced into an undeniable movement.

The other movements examined in this chapter focused on a single identifiable injustice—taxation without representation, taxation on an essential food, abuse of migrant workers, or commerciogenic malnutrition.

Could GMO labeling be such an identifiable cause? California's Proposition 37 (2012), also known as the "Right to Know" initiative, would have enforced labeling of GM foods, but it failed to pass. Although the votes for GMO labeling were not enough in California in 2012, the votes were enough in Vermont in 2014, when Vermont's Governor signed a GMO labeling bill into law.[26] Maine and Connecticut have also passed GMO labeling laws, but those two states have trigger clauses that require a number of other states to have similar laws before the laws will be in effect. Vermont's law becomes effective in 2016, unless the lawsuit filed against it by the Grocery Manufacturers Association, Snack Food Association, International Dairy Foods Association, and the National Association of Manufacturers prevails.

Does current food activism suffer from being too diffuse and unfocused because so many causes are on the agenda? Can contemporary food activism rally around a single identifiable cause?

In the next half of this chapter we introduce some of the most influential food activists, each of whom supports his or her own cause. When a cause has a critical mass of adherents we refer to it as a movement. Thus far, food activism cannot rally around a single cause because there is so much to be done. What we have are many causes that are bound together by the desire for a food system akin to that outlined in the Healthy Food Declaration.

We close this topic with this observation: food activists have had more success in building an alternative food system then getting political attention from Washington, DC. Although an alternative food economy is a good thing, not everyone has the time, money, or reliable means of transportation to participate. According to author and long-time food activist, Michael Pollan (the Knight Professor of Science and Environmental Journalism at, not surprisingly, UC Berkeley), food activists should focus their efforts on the democratization of the benefits of good food by turning voting with forks into voting with votes.[27]

FIFTY YEARS OF FOOD ACTIVISTS AND THEIR TIMES

Grassroots activism requires the type of commitment to and passion for a cause found only in certain people—those able to take risks and who are good communicators, have initiative and creativity, and are optimistic and good problem-solvers with a positive attitude. People with these traits stand out when we examine the changes in our food system brought about by food activists. We introduce some of those food activists—Waters, Lappé, Jacobson, Gussow, Allen—and the particular movements they engendered.

The Late 60s: Introduction to Food Activism and Its Ideological Roots

To understand the history of food and nutrition activism over the last 50 years, we start by looking at some of the legislative, cultural, and advocacy trends taking hold in the late 1960s. During the time of the Vietnam War (1955–1975), some young adult Americans sought to live a counterculture life, different from that of their parents. As testament to the important role of young adults, in 1967, *Time Magazine* named "Americans Under 25" as their "Man of the Year." One of the iconic publications of that era was the *Whole Earth Catalog* (1968, 64 pages, $5), which provided the rationale and instructions for hitting the restart button on ways of thinking, living, and eating. The Catalog was a response to the big corporate and government one-size-fits-all approach toward agriculture and living that many people felt was hijacking the mid-century. The book's ethos signaled an early nod toward the small-scale regional application of technology to create a more human society, including what the authors believed was a better food system. The catalog was published at a propitious time, when discontent among young adults was at its peak. It introduced ways for young people to live a different kind of life from their parents, one based in part on growing and preserving one's own food. This led some people to become early supporters of the environmental movement and organic gardening. Organic food production of the type promoted by J. I. Rodale, a founder of America's "organic" food movement and editor of the magazine, *Organic Farming and Gardening*, began to be adopted by a new generation that would help to lay the foundation for contemporary interest in "organic agriculture," the modern organic food movement.

The issue of hunger in the United States gained traction soon after the 1967 trip of Senators Robert F. Kennedy (D, NY) and Joseph S. Clark (D, PA) to Mississippi where they saw first-hand children who were emaciated; the television broadcast in 1968 of the CBS documentary *Hunger in America*[28] and the Citizens' Crusade Against Poverty's report of the same year, *Hunger USA*. The CBS documentary put a face on the 10 million Americans who at the time had no idea where their next meal would come from, and the report detailed the scope of hunger in America and the politics behind its persistence. This spurred then Senator George McGovern (D, ND) to form the Senate Select Committee on Nutrition and Human Needs, which began hearings in 1968 in an effort to find solutions to the hunger problem. By 1969 the first round hearings was completed and legislative activity was set in motion.[29] With federal attention to food policy reaching a peak, so too was the nation's, which galvanized future food activists.

Activists in the 1970s and Year Zero

It is difficult *not* to acknowledge 1971 as "year zero" (starting from scratch; a new system that replaces the former culture and traditions) of the several food movements that are still with us. In that year, three events took place in the United States that changed the way we think about food. Alice Waters opened Chez Panisse, Michael Jacobson cofounded the CSPI, and Frances Moore Lappé published *Diet for a Small Planet*. Each event represented an early expression of the political, cultural, and scientific underpinnings of today's focus on food and nutrition.

Frances Moore Lappé

In 1968 Frances Moore Lappé wrote a one-page handout explaining why she believed there was enough food in the world to feed everyone.[30] That handout eventually grew into *Diet for a Small Planet* (the title of Lappe's book is a riff on the title of Gore Vidal's 1958 play, *Visit to a Small Planet*). She was convinced that food scarcity was not inevitable but rather the result of economic policies. In her book, she proposed that people in developed countries should change their eating habits, not just for their health's sake, but as part of creating a more equitably shared global food supply. She stressed the inefficient use of resources as a result of the America's standard meat-centric diet.[31] At a time when many felt that the sentiment of inevitable food scarcity was a foregone conclusion, as predicted by biologist and population expert Paul Erlich in his book, *Population Bomb*, Lappé chose instead to focus on the abundance of waste built into the food systems of the developed world because of our over-reliance on animal- versus plant-protein.[32] She described our food system as "a protein factory in reverse,"[33] wherein it takes 20 pounds of grain to produce one pound of beef. The solution she proposed is to "eat lower on the food chain." A seminal book, *Diet for a Small Planet*, helped lay the groundwork for the rising interest in sustainability (although the word "sustainability" in regards to food systems was not used until more than a decade later).[34]

Alice Waters

That same year, in the same California University city (Berkeley), Alice Waters opened what many consider to be the first establishment that expressly espoused the virtues of locally sourced, organic ingredients.[35] However, the story of Chez Panisse begins a bit earlier. In the mid-1960s, Waters took time off from her studies at the University of California, Berkeley (UC Berkeley or Cal) and spent a semester studying in France, during which she developed a fondness for French rustic cooking.[36] Upon returning to the states she began making small dinners for friends in the style she had come to enjoy during her travels.[37] After graduating and then spending more time abroad, she returned to Berkeley with plans to open a restaurant, which she did in 1971. To her dismay, however, the results of the recipes she brought home were not as good as they were in France because the Bay Area version of the ingredients she used tasted different.[37] This small but important realization inspired her to reform the cuisine in a way that would accent the regional strengths of what was available locally. With Chez Panisse, Alice Waters can rightfully take credit for inspiring a whole generation of chefs and food-minded people to consider the source of ingredients as a fundamental component to any serious cooking.

Michael Jacobson

While the 1970s saw an awakening of interest in how food was grown, it also saw a rise in interest about what is actually in our food, a change attributed, in part, to the CSPI. The organization was started a means to get scientists involved in the social issues of the day, to "stoke the social consciences of scientists and establish the legitimacy of advocacy in the public interest." The organization hoped to provide competent technical input to help Congress craft better legislation for dealing with scientific issues, as well as arm consumers with better information regarding topics they might otherwise be confused about. They also planned to initiate lawsuits and act as coplaintiffs in public interest legal actions.[38] Taken as a whole, their approach would attempt to disengage science from purely empirical and technical concerns and contextualize it from the perspective of what was in the best interest of the public.

While the organization had broad goals, it was not until the late 1970s when CSPI's focus narrowed to health, food safety, nutrition, and preventive medicine that the group began to gain widespread attention.[39] Cofounder Michael Jacobson, PhD, became the face of the CSPI, appearing on television, radio, in print, and before Congress to advocate on behalf of such public health initiatives as SSB taxation, improved food labeling, salt reduction, and stricter guidelines for "organic" labeling of food. The monthly advertisement-free publication of CSPI, "Nutrition Action," reaches 900,000 subscribers. In fact, the organization does not accept funding from corporations and government; their sole source of revenue is donations from the private citizens and subscriptions to their newsletter.[40]

Through the efforts of people like Jacobson, Waters, and Lappé, various food-related causes began to gain momentum and ultimately start reaching a global audience. Add to the mix, Joan Dye Gussow, now emerita professor at Columbia University Teachers College. In her 1978 book, *The Feeding Web*, she laid much of the intellectual foundation for nutritional ecology.[41] Gussow wrote about the connections between food production and the wider ecosystem and how humans in essence are entirely dependent on growing plants for food as opposed to supermarkets and food corporations.[42] She describes the process of crafting the book as having "pulled together articles laying out a picture of how changes in various components of the food system were, or were not, moving in a direction compatible with long term human survival."[43] Her reflections on population growth, organic farming, the use of resources, and the dangers of processed food echoed many of the concerns championed by food and nutrition activists earlier in the decade. In 1986 she was the first nutritionist to use the word *sustainability*.[44]

Activists in the 1980s and Food Policy Councils

Community-based approaches to food and nutrition activism gained a foothold in the 1980s with the formation of food policy councils. Robert Wilson at the University of Tennessee pioneered the food policy council concept. In 1977 he and a group of graduate students conducted a community needs assessment focusing specifically on food availability in Knoxville, Tennessee. The subsequent report highlighted the loss of local farmland and the fragmented state of the food system.[45] The results of their work led Wilson to think about food as a basic human need that should be treated like water, housing, sewage treatment, and other municipal functions, and to then raise the question as to why cities did not have a department of food.[46] A local community action committee, a planning group that included high-level county officials, used the report to successfully apply for a 2-year grant (1979–1981) from the Department of Health and Human Services.[47] Work done under this grant led in 1982 to the passage in the Knoxville city council of a bill establishing the nation's first food policy council. This legislation enshrined in law that food must be a concern of local governments, since food systems play a major role in public health.[48] This move inspired the development of food policy councils in nearly every region of the country under the aegis of the Community Food Security Coalition. Mark Winne, a disciple of Wilson and former executive director of the Hartford Connecticut Food System and cofounder of the Community Food Security Coalition, lectures and writes about food policy councils. Among his publications are *Doing Food Policy Councils Right: A Guide to Development and Action,*[45] and with the Harvard Food Law and Policy Clinic, *Putting State Food Policy to Work for Our Communities.*[49]

Because we all need to eat, consumers play a role in continuing the legacy of food and nutrition activism. One of the most important aspects of this can be seen in the growth of community-supported agriculture (CSAs). These are farmer/consumer arrangements where purchasers invest in a farm for the growing season and receive weekly or monthly shares of the harvest or portions of other items like cheeses, meats, etc.[50] Two of the first such examples of this model came in 1986 with Indian Line Farm in Massachusetts and Temple-Wilton Farm in New Hampshire.[51] At Indian Line Farms Jan Vander Tuin suggested the farm share concept to farm manager Robyn Van En after he had been exposed to it while working in Europe.[52] To test out the idea, Van En, Tuin, and neighboring farmers John Root, Jr., and Charlotte Zanecchia did a season of selling apple and apple cider shares.[53] Encouraged by their initial success, they expanded the shares the following season to a garden patch leased to Van En on Indian Line Farm land and in the process of developing the concept further coined the phrase "community supported agriculture."[54] Temple-Wilton Farm's CSA was started in a similar fashion by Trauger Groh, a farmer who had seen the concept work in Germany through associative economic models. However, instead of selling direct shares, he presented CSA members with a budget and once the costs were covered members were free to take however much they liked.[51] Both examples represented a forward thinking approach toward integrating consumers more intimately within the food system.

In the 1980s the food movement also brought into focus the loss of farmland. Representing the interests of small farmers, the National Family Farms Coalition debuted in 1986, carried along by a wave of activism that emerged in the wake of the decade's farm foreclosure crisis. At this time a loose collective of 37 farmers rights organizations banded together to petition the federal government for higher commodity prices, immediate debt relief, and a freeze on farm foreclosures.[55] All these concerns were bundled together and presented to Congress by Senator Tom Harkin (D, Iowa) as the farm policy reform act of 1986.[56] The bill failed to pass, but the various farmers rights groups galvanized behind the bill coalesced into the National Family Farm Coalition (NFFC, originally the National Save the Family Farm Coalition).[57] Since then, NFFC has become the foremost advocacy/lobbying groups for small farms both in the United States and around the world by working on issues from food security and farm debt refinancing to combating the spread of GM crops.[58]

Tackling the issue of the increasing loss of traditional foodways was the Slow Food movement, which began in 1986 when a group of food-minded activists led by Carlo Petrini, a former leftist journalist, got together to protest the opening of a McDonalds in Rome, Italy.[59] This protest came at a time when the topic of food standardization brought on by the rise in popularity of fast food had become a hotly debated topic.[60] In response to "fast food," Petrini sought to devise a way of thinking about and interacting with food that took into consideration the pleasure of eating (social) and the methods of food production and its social implications (political).[49] With this in mind the group set about crafting what it saw as "an act of rebellion against a civilization based on the sterile concepts of productivity, quantity and mass consumption, destroying habits, traditions and ways of life, and ultimately the environment."[61] After fleshing out its ideas and getting leading public figures of the Italian left on board, Petrini and his compatriots emerged in 1987 with what they called the Slow Food manifesto, taking a snail as its logo.[62] In the decades since, the growing prevalence of issues related to local food production, food security, and food equity has led the Slow Food movement to spread within Italy as well as internationally with more than 10,000 members in 132 countries including the United States.[63]

The 90s: Policy Gains, Biotechnology, and Urban Agriculture

Major legislative change brought on by food and nutrition activists came in the form of the Organic Food Production Act. Through the creation of industry-wide standards, this bill established by law many of the production practices long championed by sustainability proponents. While people like Alice Waters, Frances Moore Lappé, and Joan Dye Gussow had been long time advocates for organic farming, a good part of the momentum that brought this legislation came from the commercial sector.[64] With growth in sales and a maturing of the organic food market, the Organic Trade Association (OTA), together with a number of other commercial groups formed the Organic Food Alliance. Together, they put their collective support behind a broad set of legislative guidelines that would govern the use of the term "organic."[64] The actual legislation, put forth by Senator Patrick Leahy of Vermont, was designed to create a set of universal standards for growing and processing organic food and to regulate enforcement of the rules.[65]

The Organic Foods Production Act (OFPA) was added as part of the 1990 Federal Farm Bill, with its primary purposes being to (1) establish national standards governing the marketing of certain agricultural products as organically produced, (2) assure consumers that organically produced products meet consistent standards, and (3) facilitate interstate commerce in fresh and processed organic foods.[66] With the passage of this version of the Federal Farm Bill we see the first legal definition of the term "organic" with respect to agricultural practices. We also see the first agreed upon definition of "sustainable agriculture."[64]

Ironically, following the passage of the Organic Food Production Act a number of biotech firms began stepping up their efforts to introduce GMO into the food system. As a reaction, in 1992 the Foundation on Economic Trends, headed by Jeremy Rifkin, launched the Pure Food Campaign, a coalition of farm consumers and environmentalists who opposed the spread of GMOs. Pure Food's approach was not to argue that GMOs were dangerous, but rather that they had not been proven safe.[67] They called for a global moratorium on the spread of biotechnology, joining forces with small farmers, consumers, and animal-rights groups in more than 30 countries. Their work in this area is widely seen as having slowed the spread of GMO use in Europe.[68] One of Pure Food's first and most high-profile campaigns was the "Bovine Growth Hormone Boycott," where they developed kits to help community organizers rally against a synthetically produced hormone used to stimulate milk production in cows.[69] This boycott was a proving ground for Pure Food and allowed them to hone their initial tactic of spreading information through video and print press releases. Other strategies have included petitioning the FDA for tighter regulation on the release of genetically modified foods and staging high-profile protests against new efforts to expand the use of biotechnology.

While biotechnology companies pushed for a foothold in the U.S. marketplace, on the local front several cities across the country saw a burst of activity in urban agriculture. One such project began in 1990 when an urban planner named Duane Perry was asked by farmers and merchants to help guide a revitalization project at Philadelphia's Reading Terminal Market, one of the largest and oldest indoor markets in the United States. One of the ideas Perry proposed was the formulation of a new nonprofit called the Farmers' Market Trust, later to become the Food Trust.[70] This organization would act as an extension of the Reading Terminal Market by giving access to the same quality of local fruits and vegetables available there to individuals living in low-income communities.[71] By refocusing the food system in Philadelphia the Food Trust sought to cultivate links between local farmers and inner city communities by giving residents in those neighborhoods access to fresh and nutritious food.[72] Food Trust has gone on to develop projects involving school-based nutrition education programs, community-based fresh food financing initiatives, and sustainable and equitable food systems evaluation.[73]

Will Allen is another proponent of urban agriculture. In 1993, his organization, Growing Power, launched an urban farm in Milwaukee with the goal of bringing the local food concept into the inner city. Allen's inspiration for farming began during his former career as a professional basketball player. While playing for a team in Belgium one of his teammates invited him out to the family farm to lend a hand. The experience profoundly affected Allen and he began doing some small-scale farming on his own. Years later, back in the States, he took down the number on a for-sale sign for a fixer-upper and decided to bid on the place. That property eventually became Growing Power, one of the first urban farms in the Milwaukee area.[74] With Growing Power, Allen has taken the rhetoric of local food and applied it directly to an inner city, underserved setting. The crops grown on his urban farms in Milwaukee, and now in Chicago, provide food for an estimated 10,000 people a year. Through trial-and-error Allen has also developed an innovative approach to farming in urban environments, a skill that he shares with hundreds of volunteers and workshop attendees that flock to his farms every year.[63] In recognition of his work, in 2008, Allen was awarded a John D. and Catherine T. MacArthur Foundation Fellowship "genius" prize.

Urban unity and urban farming were on Ward Cheney's mind when he started the Food Project of Boston in 1991. Cheney, a Massachusetts farmer, had begun to think of ways to transport his skills to an urban environment, and how he could foster change, both in society and the food system.[54] With the Food Project he would help bring young people from vastly different backgrounds together around food and farming, and train them how to radically change the food production and distribution system in America.[75] With its unique focus on young people, the Food Project of Boston was the first urban agricultural program of its kind and stands as model for community-based food activism.[76]

Building urban farms is one of the most direct ways to localize food systems. However, CSAs, since their introduction to the United States in the 1980s, have increasingly acted as a stand in for urban farms with sustainability minded city dwellers. In 1995 the CSA model was on Kathy Lawrence's mind when she started Just Food in New York City. The idea was born out of a series of conversations with Joan Dye Gussow about the impact of food production on the environment.[77] The Just Food solution was to facilitate the creation of new CSAs with two goals: first, make it practical and profitable for farmers to bring their food to New York City neighborhoods, and second, make it cost-effective and appealing for residents to purchase the food.[78] To achieve this, Just Food trained community members to administer CSAs, provided networking opportunities for farmers looking to sell their goods, and supplied technical assistance.[79] They have since expanded into gardening, urban agriculture and culinary classes, supporting access of fresh local produce in food pantries, and policy work to help further the creation of a more equitable food system in New York City.[80]

Putting children in contact with fresh produce has always been a great way to bring children closer to the food system. In 1997 one of the largest programs in the country for connecting children more intimately with their food started at a school in the Santa Monica-Malibu Unified School District in California. Robert Gottlieb, a parent of a student at McKinley Elementary and a professor at Occidental College's Center for Food and Justice, approached Rodney Taylor, the school Food Service Director, about starting a salad bar stocked with produce from local grower. The farm-to-school program, overseen by the Center, would replace the school's underutilized salad bar with fresh seasonal produce.[81] A scaled-down prepilot version of the program was conducted in the summer of 1997 in one of the childcare programs, and a full farmers' market salad bar version was offered in the fall of 1997. Both met with rave reviews.[82] Out of this first project grew the national farm-to-school program that today is administered by the USDA and is in place in dozens of school districts around the country.[83]

Nontraditional Food Activists

Traditionally, we think of a food activist as someone who cares about where our food comes from, supports local and sustainable agriculture, and is involved in some way with dissemination of information about sustainable living. But food activists may also be involved with community outreach programs that promote diet awareness and access to healthy ingredients for safe, healthy food for everyone, and promote actions that can be taken by individuals and communities to make positive changes. All these things fall under the rubric of food activism. We find food activists in houses of worship and in barbershops and hair salons as well as in departments of nutrition at major land grant and private college and universities.

Historically, the clergy have functioned as activists in exhorting their parishioners to follow a particular moral code or system. Religious institutions such as churches, mosques, temples, and synagogues are found in nearly all communities and have significant cultural, political, social, educational, and economic influence. Most have resources, structures, and systems on which to build. They also possess the human, physical, technical, and financial resources needed to support and implement small and large-scale initiatives, and can leverage volunteer resources.

The White House Office of Faith-Based and Community Initiatives enables government to support the work of faith-based and other community organizations.[84] In particular, this Office enables government to use Department of Health and Human Services' funds to encourage public health initiatives that are centered in faith-based organizations, to the extent permitted by law (presumably, the first amendment). Since at least 2001, government-supported faith-based initiatives have enlisted clergy and other church leaders to serve as advocates for various health-promotion and disease prevention initiatives, including programs to prevent diet-related chronic diseases.[85,86]

The practice of using volunteers (natural helpers) is a popular method in advocacy campaigns dating back to the use of networks of informal helpers in public health programs in South African primary healthcare centers in the 1940s and 1950s. Studies of these networks revealed that engaging natural helping networks in healthcare delivery led to a variety of positive changes in health among patients as well as positive changes in health center staff's ability to understand and empathize with the daily lives and challenges of their patient population. This work was later adopted in rural communities in North Carolina, which spawned a number of church-based interventions in the 1980s aimed at serving black populations.[87]

By the 1990s, the practice of natural helpers was being adopted by public health researchers as a way of working effectively with low-income black and Hispanic populations and other specific population groups and segments. Local barbers[88,89] and hairdressers are examples of members of low-income black and Hispanic communities who, as trusted natural helpers, have been enlisted as activists to help remove the obstacles to better health for their customers.

Three programs that train men's barbers and women's hair dressers to advocate on behalf of their client's health are The Black Barbershop Health Outreach Program,[88] Healthy Hair Starts with a Healthy Body,[90] and the Brooklyn-based Arthur Ashe Institute for Urban Health, which has a number of salon-based, barbershop-based, and tattoo salon based programs.[91] The employees in community-trusted barber shops and hair salons serve as health ambassadors who encourage their clients to adopt health-promoting behaviors, such as drinking fewer sugar SSBs, eating more fruits and vegetables, switching from full-fat dairy products to milk and cheeses that are reduced-fat, and removing the salt shaker from the kitchen table.

The 2000s and Beyond: Voices in the Mainstream and the New Activism

The urban agriculture movement that took hold in the 1990s continued to flourish through the next decade, the legacy of which is visible in the work of Added Value, an urban farm in Red Hook, Brooklyn. The idea for the Added Value farm came from a conversation Ian Marvy had with one of the teenagers he encountered while working in the Red Hook Youth Court.[92] When their talk turned to produce, Marvy explained how much money a square foot of vegetables could yield, and the teenager became interested. Marvy and his coworker Michael Hurwitz thought that with the right setup, urban farming might be an attractive option for area teens looking to learn more about food and farming. In 2000 they started a community farm on a disused piece of parkland in Red Hook and called it Added Value, which Marvy describes as making a more sustainable world "through youth empowerment and urban agriculture. "We are taking public space that was programmed for one use, one economy, and one social structure, and transforming it into something else."[93] At 2.75 acres the farm grows everything from Mexican specialty greens to corn and eggplant, while giving teenage workers job skills and providing the neighborhood with fresh food.[94]

Since 2000 various forms of media, including books, film, and the web, started to play an increasing role in shaping food and nutrition activism. Of these, Eric Schlosser's *Fast Food Nation* is considered one of the most detailed and fascinating looks at the ills of the fast-food industry. The inspiration for the book came from an editor from *Rolling Stone*, who asked Schlosser to take a similar investigative approach as he had recently used covering migrant workers for the *Atlantic Monthly*, and apply it to the fast-food industry.[95] The book is a heady mix of statistics, interviews, and observations that casts the standard American diet as a fundamental driver of commerce, environmental degradation, poor health, and a host of other issues.[96] *Fast Food Nation* echoes many of the depictions laid out more than 100 years prior in Upton Sinclair's *The Jungle*, investigation of the meatpacking industry that chronicled the waste and degradation inherent in a highly industrial food systems and outlined the true cost to having cheap, readily available food.[95]

Another book that informed the food and nutrition activism community is *Food Politics* by Marion Nestle. Now the Paulette Goddard Professor in the Department of Nutrition and Food Studies at New York University, in 1986 she worked for the United States Department of Health and Human Services, authoring the Surgeon General's Report on Nutrition, a position that put her at the nexus of foods and politics.[97] With the 2002 publication of her book and now a blog of the same name, Nestle has been a forceful voice on food issues, ethics and policy, making a clear connection between our food choices, their impact on our lives, and our consumption patterns.[98] Her voice and experience have given an academic framework to the various fights undertaken by food and nutrition activists since 2000.

Food advocacy entered the mainstream with the 2004 release of the film *Super-Size Me*. The film ranks as the 15th highest grossing documentary of all-time and has been seen by millions of people around the world. In it, filmmaker Morgan Spurlock takes on a diet of nothing but McDonald's food for 30 days. Over the course of the film the viewer witnesses a marked decline in his health and well-being. His behavior is meant to be over the top and sensational. However, the core message conveying the potentially damaging effects of fast-food heavy diets has led many to question the way they eat.[99] Yet another high-profile development for food activism came in 2006 with Michael Pollan's book, *The Omnivore's Dilemma*, with a central premise of following the stages of four different meals from source to plate.[100] A journalist by trade, Pollan gave an eloquent and detailed account of what happens to the plants and animals we consume and how their methods of cultivation relate to our ethics as a society.[101]

One unique trend in the last decade has been the emergence of several organizations that directly deal with issues related to food access. One such group is the Wholesome Wave Foundation, started in 2007 with seed money from the Newman's Own Foundation and the founders of Priceline.com.[76] Wholesome Wave's approach to food access is to issue grants to farmers' market programs around the country to help build the capacity of farmers' markets to provide access to produce for low-income clients. For example, their Double Value Coupon program provides buying incentives for consumers using federal food assistance benefits at farmers markets, while their innovative VeggieRX program works with doctors to supply produce vouchers to their patients. Founder Michel Nischan, a two-time James Beard Foundation award winner and restaurant owner, launched Wholesome Wave with the aim of creating partnership programs in historically excluded urban and rural communities and to give access to and increase the affordability of fresh, locally grown food to nourish neighborhoods across America.[102] Wholesome Wave embodies the spirit of promoting local food for the benefit of all.

A similar organization, The Fair Food Network, is the brainchild of Oran Hesterman. A former professor at Michigan State University and the former director of the sustainable-agricultural grant-making program at the Kellogg Foundation, Hesterman now brings his years of experience to working toward a more sustainable food system. The Fair Food Network is a nonprofit working throughout Michigan to bring increased access to fresh, sustainably grown food in traditionally underserved areas.[76] The Fair Food Network, in many ways, represents the various streams of the food movement over the last 40 years with its focus on ideologically driven motivations and practical evidence-based solutions.

CONCLUSION

Looking forward from the past, we see how much time, effort, and groundswell of activism is necessary to get political attention from Washington DC. We have explored past activist movements and seen success in those focused on a single identifiable injustice—taxation without representation, taxation on an essential food, abuse of migrant workers, commerciogenic malnutrition. Today we see food activism at work for numerous causes, and perhaps because there are indeed so many food issues on our plate, we have not yet identified a watershed moment when the influence of food activists has coalesced into an undeniable movement.

Yet, now, with the Internet as an organizing and information gathering tool through sites like La Vida Locavore, Grist, Civil Eats, email list-serves like Comfood, and social media interaction through Facebook and Twitter, the ideas and activities that once took years to organize and disseminate can be taken up by a whole cadre of food and nutrition activists in a matter of hours.

You may feel this chapter is incomplete, and it is. Our study of food activism ended with the emergence of work on food access, and has not reached the emerging ideas and technologies of the "sharing economy" and how those may be used to disrupt (or improve) our food system. We have examined the past; we are leaving it to you to look toward the future. If you are interested in food activism, but are not sure where to start, we suggest you check out the sites listed here.

Food Safety News	Daily	http://www.foodsafetynews.com/
Food Politics (Marion Nestle)	Daily	http://www.foodpolitics.com/
Civil Eats	Weekly	http://civileats.com/
U.S. Food Policy (Parke Wilde)	Periodically	http://usfoodpolicy.blogspot.com/
Rudd Center Health Digest	Monthly	http://www.yaleruddcenter.org/newsletter/

This chapter is incomplete for many reasons, not the least of which is that every day something new turns up on the food activist's plate. What impact will the Berkeley soda tax have on other municipalities? How should activists mobilize to protect Vermont's passage of its GMO labeling act? What is to be done about the increased food waste and decreased food intake engendered by school meals that meet nutrition guidelines required by the Healthy, Hunger-Free Kids (HHFK) Act of 2010?

Food waste is discussed in more detail in Chapter 14, so we will not address food waste—and food recovery—in detail here. However, the issue of food waste is one that can be addressed at each level of the food system chain, and, with the tools of an Internet-enabled era that can elicit group responses to timely needs, such as harvesting, serves to illustrate how the purported "sharing economy" can indeed support sharing through both activism and action. If GMO labeling may not be an identifiable cause, food waste may be. The coalescence of the food movement may be upon us, and you can make it happen.

APPENDIX: WRITING LETTERS TO CONGRESS

Because they represent more effort and commitment, personal letters are more effective lobbying efforts than postcards, petitions, and phone calls. When writing letters to members of Congress, the format of the letter is important, regardless of the reason for which the letter is being written. The letter must be courteous, to the point, and when possible, limited to one page.

Two people represent each state in the Senate. To find contact information for a senator, go to http://www.senate.gov/general/contact_information/senators_cfm.cfm.

Members of the House of Representatives proportionately represent the population of the states. The number of voting representatives in the House is fixed by law at no more than 435. Currently, there are five delegates representing the District of Columbia, the Virgin Islands, Guam, American Samoa, and the Commonwealth of the Northern Mariana Islands. A resident commissioner represents Puerto Rico. Each representative is elected to a 2-year term serving the people of a specific congressional district. Representatives can be identified by zip code at http://www.house.gov/representatives/find/. For the district, party, address, phone number, and committee assignments of each representative by state, district, or last name, go to http://www.house.gov/representatives/.

Components of a Letter to a Member of Congress

Date
Your name, address, telephone number, and email
The Honorable _____
U.S. Senate
Washington, D.C. 20510
[or]
The Honorable _____

U.S. House of Representatives
Washington, D.C. 20515
Dear Senator _____: [or] Dear Representative _____:

1. State clearly your purpose for writing the letter. State what it is you want done or recommend a course of action. If your letter pertains to a specific piece of legislation, identify it according to its legislative identifier:
   House Bill: "**H.R.**_____"
   House Resolution: "**H.RES.**_____"
   House Joint Resolution: "**H.J.RES.**_____"
   Senate Bill: "**S.**_____"
   Senate Resolution: "**S.RES.**_____"
   Senate Joint Resolution: "**S.J.RES.** _____
2. Provide information that supports your position. If appropriate, state how the proposed legislation or issue affects you and others. Anecdotal evidence is an effective and persuasive lobbying tool. Provide specific rather than general information about how the topic affects you and others.
3. State any professional credentials you have, especially if your training pertains to the subject of your letter.
4. Use simple language; avoid jargon and abbreviations. Staff workers in Congressional offices are not experts on all issues. "High cholesterol" may be more readily understood than "hypercholesterolemia."
5. Thank the senator or representative for something—their time, their effort, or for their support of current or proposed legislation.
6. Close with "Sincerely" or other appropriate ending. Sign your name and type your name underneath your signature.

REFERENCES

1. Bittman M. Rethinking the word 'foodie.' *The New York Times*, July 24, 2014.
2. Howard PN. The Arab Spring's Cascading Effects. Available at http://www.psmag. com/navigation/politics-and-law/the-cascading-effects-of-the-arab-spring-28575/, accessed September 26, 2014.
3. Center For Science In The Public Interest, Take Action Key Issues. Available at https:// www.cspinet.org/takeaction/#, accessed August 30, 2014.
4. Copley A. *Gandhi: Against the Tide*. Oxford: Basil Blackwell Ltd, 1987.
5. Chadha Y. *Gandhi: A Life*. New York: John Wiley and Sons, 1997.
6. Ferriss S, Sandoval R. *The Fight in the Fields: Cesar Chavez and the Farmworkers Movement*. Orlando, FL: Paradigm, 1997.
7. Ganz N. *Why David Sometimes Wins: Strategy, Leadership, and the California Agricultural Movement*. New York: Oxford, 2009.
8. Marchione TJ. A history of breast-feeding practices in the English-speaking Caribbean in the twentieth century. *Food and Nutrition Bulletin*. 1980;2. Available at http://archive.unu. edu/unupress/food/8F022e/8F022E00.htm#Contents, accessed August 30, 2014.

9. Jelliffe D. Commerciogenic malnutrition? Time for a dialog. *Food Technol.* 1971;25: 55–56.

10. Lau C. Effects of stress on lactation. *Pediatr. Clin. North Am.* 2001;48:221–234.

11. Baer E. Babies mean business. *New Internationalist.* 1982;(110). Available at http://www. newint.org/features/1982/04/01/babies/, accessed August 30, 2014.

12. Corporate Accountability International. Available at http://www.stopcorporateabuse.org/, accessed August 30, 2014.

13. Nestlé, Brands. Available at www.nestle.com/brands, accessed August 30, 2014.

14. Baby Milk Action, Nestlé boycott. Available at www.babymilkaction.org/nestlefree, accessed August 30, 2014.

15. International Baby Food Action Network. *Breaking the Rules, Stretching the Rules 2010.* Available at http://www.ibfan.org/art/BTR_2010-ExecSummary%28final%29.pdf, accessed August 30, 2014.

16. Winne M. *Food Rebels, Guerrilla Gardeners, and Smart-Cookin' Mamas: Fighting Back in an Age of Industrial Agriculture.* Boston: Beacon, 2010.

17. USDA Farm to School Census. Available at http://www.fns.usda.gov/farmtoschool/ census#/, accessed August 30, 2014.

18. USDA. Know Your Farmer, Know Your Food. Available at http://www.usda.gov/wps/ portal/usda/usdahome?navid=KYF_MISSION, accessed August 30, 2014.

19. Cantrell P, Heuer B. Food Hubs: Solving Local. The Wallace Center at Winrock International. Available at http://ngfn.org/solvinglocal, accessed November 9, 2014.

20. USDA. Know Your Farmer, Know Your Food. Available at http://www.usda.gov/wps/ portal/usda/usdahome?navid=KYF_MISSION, accessed November 9, 2014.

21. National Restaurant Association. What's Hot Culinary Forecast 2014. Available at http:// www.restaurant.org/downloads/pdfs/news-research/whatshot/what-s-hot-2014.pdf, accessed November 9, 2014.

22. Roots of Change. Citizen Advocacy. Available at http://www.rootsofchange.org/content/ citizen-advocacy, accessed August 30, 2014.

23. Roots of Change, Activities, Food Declaration. Available at http://www.rootsofchange.org/ content/activities-2/food-declaration, accessed August 30, 2014.

24. Strom S. Major grocer to label foods with gene-modified content. *The New York Times,* March 8, 2013. Available at http://www.nytimes.com/2013/03/09/business/grocery-chain-to-require-labels-for-genetically-modified-food.html?_r=0, accessed September 27, 2014.

25. The Center for Consumer Freedom. Foodie fantasies meet harsh reality. Available at http://www.consumerfreedom.com/2012/11/foodie-fantasies-meet-harsh-reality, accessed August 30, 2014.

26. The Vermont Legislative Bill Tracking System, Current Status, 2013–2014 Legislative Session. Available at http://www.leg.state.vt.us/database/status/summary.cfm?Bill=H.0112, accessed August 30, 2014.

27. Pollan M. Vote for the dinner party. *New York Times Magazine.* October 10, 2012. Available at http://www.nytimes.com/2012/10/14/magazine/why-californias-proposition-37-should-matter-to-anyone-who-cares-about-food.html?pagewanted=all, accessed August 30, 2014.

28. CBS Reports. Hunger in America. May 21, 1968. 60-minute documentary.

29. United States. Advisory commission on intergovernmental relations. *Public Assistance: The Growth of a Federal Function.* Washington, D.C. UNT Digital Library. Available at http://digital.library.unt.edu/ark:/67531/metadc1338/, accessed November 10, 2014.

30. Jensen C. *Stories that Changed America: Muckrakers of the 20th Century.* New York: Seven Stories Press, 2000:270.

31. McFeely MD. *Can she bake a Cherry pie?: American Women and the Kitchen in the Twentieth Century.* Amherst: University of Massachusetts Press, 2000:194.

32. Lappé FM. *Diet for a Small Planet.* New York: Ballantine Books, 1991:479.

33. Gaard GC. Vegetarian ecofeminism: A review essay Frontiers. *J. Women Studies.* 2002;23(3):117.
34. Belasco WJ. *Appetite for Change: How the Counterculture Took on the Food Industry.* Ithaca: Cornell University Press, 2007:327.
35. Lovegren S. *Fashionable Food: Seven Decades of Food Fads.* New York, NY: Macmillan, 1995:455.
36. McNamee T. *Alice Waters and Chez Panisse: The Romantic, Impractical, Often Eccentric, Ultimately Brilliant Making of a Food Revolution.* New York: Penguin Press, 2007.
37. Smith A. *Eating History: 30 Turning Points in the Making of American Cuisine.* New York: Columbia University Press, 2009:376.
38. Holden C. Public interest: New group seeks redefinition of scientists' role. *Science.* 1971;173(3992):131–132.
39. Williams EM, Carter SJ. (Entry for) Jacobson, Michael (1943–) *The A-Z Encyclopedia of Food Controversies and the Law,* 1st edition. Santa Barbara, CA: Greenwood, 2011:249–250.
40. Center for Science in the Public Interest. History of CSPI. Center for Science in the Public Interest Website. Available at http://www.cspinet.org/history/cspihist.htm, accessed November 10, 2014.
41. Cooper A, Holmes LM. *Bitter Harvest: A Chef's Perspective on the Hidden Dangers in the Foods We eat and What You can do about it.* New York: Routledge, 2000:278.
42. Gussow JD. *The Feeding Web: Issues in Nutritional Ecology.* Palo Alto, CA: Bull Publishing Co., 1978:457.
43. Gussow JD. Reflections on nutritional health and the environment: The journey to sustainability. *J. Hunger Environment. Nutr.* 2006;1(1):3–25.
44. Gussow JD, Clancy K. Dietary guidelines for sustainability. *J. Nutr. Educ.* 1986;18(1):1–5.
45. Burgan M, Winne M. *Doing Food Policy Councils Right: A Guide to Development and Action,* 2012. Available at http://www.markwinne.com/wp-content/uploads/2012/09/FPC-manual.pdf, accessed November 10, 2014.
46. Barrantes MR. *Feed Your City: 2004 Berkeley Field Guide to Planning for Food.* Berkeley: City of Berkeley Food Policy Council, 2004.
47. Becker G. Nutrition planning for a city. Community nutritionist. 1982:12–17. Available at http://homepages.wmich.edu/~dahlberg/F6.pdf, accessed November 10, 2014.
48. Haughton B. Developing local food policies: One city's experiences. *J. Public Health Policy.* 1987;8(2):180–191.
49. The Harvard Law School Food Law and Policy Clinic. *Good Laws, Good Food: Putting State Food Policy to Work for Our Communities.* Mark Winne Associates. Available at http://www.markwinne.com/wp-content/uploads/2012/09/food-toolkit-2012.pdf, accessed November 10, 2014.
50. Williamson T, Imbroscio DL, Alperovitz G. *Making a Place for Community: Local Democracy in a Global Era.* New York: Routledge, 2002:412.
51. Duram LA. *Encyclopedia of Organic, Sustainable, and Local Food.* Santa Barbara, CA: Greenwood, 2010.
52. Hayes D, Laudan R. (eds.). *Food and Nutrition,* 1st edition. Tarrytown, NY: Cavendish, Marshall Corporation, 2008.
53. Miles A, Brown M. Teaching direct marketing and small farm viability: Resources for instructors – part 4, community supported agriculture. Available at http://escholarship.org/uc/item/69w2d0nv, accessed October 15, 2014.
54. Henderson EC, Van En R. *Sharing the Harvest: A Citizen's Guide to Community Supported Agriculture, Revised and Expanded.* Vermont: Chelsea Green Publishing, 2007:320.
55. Brown DL, Schafft KA. *Rural People and Communities in the 21st Century: Resilience and Transformation.* Cambridge, UK; Malden, MA: Polity Press, 2011.
56. Ridgway J. New faces, key races and votes to remember. *Mother Jones.* 1986;11(7):40.

57. Browne WP, Lundgren MH. Farmers helping farmers: Constituent services and the development of a grassroots farm lobby. *Agriculture Human Values.* 1987;4(2):11–28.
58. Adams J. *Fighting for the Farm: Rural America Transformed.* Philadelphia, PA: University of Pennsylvania Press, 2003:338.
59. Ducasse A. The slow revolutionary. *Time Magazine Europe.* October 11, 2004;164(14).
60. Miele M, Murdoch J. The practical aesthetics of traditional cuisines: Slow food in Tuscany. *Sociologia Ruralis.* 2002;42(4):312–328.
61. Hodgson P, Toyka R, Petrini C. Slow food. *The Architect, the Cook and Good Taste.* Basel: Birkhäuser Architecture; 2007:138–141. 10.1007/978-3-7643-8483-8_16.
62. Leitch A. Slow food and the politics of pork fat: Italian food and European identity. *Ethnos.* 2003;68(4):437–462.
63. Tracey D. *Urban Agriculture: Ideas and Designs for the New Food Revolution.* Gabriola Island, BC: New Society Publishers, 2011.
64. Meyer DS, Jenness V, Ingram H. *Routing the Opposition: Social Movements, Public Policy, and Democracy,* 1st edition. Minnesota: University of Minnesota Press, 2005:328.
65. Pirello C. *Christina Cooks: Everything you ever Wanted to know About Whole Foods, but were Afraid to ask.* New York: HP, 2004.
66. Gliessman SR, Rosemeyer M. *The Conversion to Sustainable Agriculture: Principles, Processes, and Practices.* Boca Raton, FL: CRC Press, 2010.
67. Andrée P. *Genetically Modified Diplomacy: The Global Politics of Agricultural Biotechnology and the Environment.* Vancouver: UBC Press, 2007.
68. Eichenwald K, Kolata G, Petersen M. Biotechnology food: From the lab to a debacle. *New York Times.* January 25, 2001; Business Day. Available at http://www.nytimes. com/2001/01/25/business/25FOOD.html?pagewanted=all, accessed October 15, 2014.
69. Hertzel LJ. Pus is an ugly world. *North Am. Rev.* 1994;279(2):4–7.
70. Gottlieb R, Joshi A. *Food Justice.* Cambridge, MA: MIT Press, 2010.
71. Rubin S. The grocery gap. *Atlantic.* May 17, 2010. Available at http://www.theatlantic. com/special-report/the-future-of-the-city/archive/2010/05/the-grocery-gap/56677/, accessed November 10, 2014.
72. Frank-Spohrer GC. *Community Nutrition: Applying Epidemiology to Contemporary Practice.* Gaithersburg, MD: Aspen Publishers, 1996.
73. Food Trust. Available at www.thefoodtrust.org, accessed October 24, 2014.
74. Henry Ford Foundation. *Transcript of a Video Oral History Interview with Will Allen [Video].* Dearborn, MI: Henry Ford Foundation, 2011.
75. McCormack KJ, Eisinger S. *Profiting from Purpose: Profiles of Success and Challenge in Eight Social Purpose Businesses.* New York: Seedco, 2005:1. Available at http://community-wealth.org/content/profiting-purpose-profiles-success-and-challenge-eight-social-purpose-businesses.
76. Hesterman OB. *Fair Food: Growing a Healthy, Sustainable Food System for all.* New York: Public Affairs, 2011.
77. Kennedy M, Singh I. A focus on just food. *Inside.* 2010;15(7):16–21.
78. Winne M. *Closing the Food Gap: Resetting the Table in the Land of Plenty.* Boston: Beacon Press, 2008.
79. McFadden S. T*he Call of the Land: An Agrarian Primer for the 21st Century.* Nashville, IN: Norlights Press, 2009.
80. Just Food. Available at www.justfood.org, accessed November 10, 2014.
81. Feenstra G, Kalb M. Farm to school: Institutional marketing. Case Study. Available at http://www.agofthemiddle.org/pubs/farmschool.pdf, accessed November 10, 2014.
82. Mascarenhas M. Farmers' market salad bars and nutrition policy advocacy in LAUSD. Report to the California Nutrition Network, July 2002.

83. United States Department of Agriculture. USDA Farm To School Team. 2010 Summary Report. July 2011. Available at http://www.fns.usda.gov/sites/default/files/2010_summary-report.pdf, accessed November 10, 2014.

84. Presidential Documents. Executive Order 13199 of January 29, 2001. Establishment of White House Office of Faith-Based and Community Initiatives. *Federal Register.* 2001;66(21):8499–8500. Available at http://www.gpo.gov/fdsys/pkg/FR-2001-01-31/pdf/01-2852.pdf, accessed August 30, 2014.

85. Campbell MK, Hudson MA, Resnicow K, Blakeney N, Paxton A, Baskin M. Church-based health promotion interventions: Evidence and lessons learned. *Ann. Rev. Public Health.* 2007;28:213–234.

86. Wilcox S, Laken M, Parrott AW, Condrasky M, Saunders R, Addy CL, Evans R, Baruth M, Samuel M. The Faith, Activity, and Nutrition (FAN) program: Design of a participatory research intervention to increase physical activity and improve dietary habits in African American churches. *Contemp. Clin. Trials.* 2010;31:323–335.

87. Eng E, Rhodes SD, Parker E. Natural helper models to enhance a community's health and competence. Chapter 11. In: DiClemente RJ, Crosby RA, Kegler MC (eds.), *Emerging Theories in Health Promotion Practice and Research*, 2nd edition. San Francisco: Jossey-Bass, 2009:303–330.

88. Releford BJ, Frencher SK, Jr, Yancey AK, Norris K. Cardiovascular disease control through barbershops: Design of a nationwide outreach program. *J. Nat. Med. Assoc.* 2010;102:336–345.

89. Dodge the Punch: Live Right®. Available at http://nkfm.org/communities-families/salonbarber-programs/dodge-punch-live-right, accessed August 30, 2014.

90. Healthy Hair Starts With A Healthy Body®. Available at http://nkfm.org/healthy-hair, accessed August 30, 2014.

91. Salon- and barber shop-based programs of the Arthur Ashe Institute for Urban Health. Available at http://www.arthurasheinstitute.org/arthurashe/programs/salon, accessed August 30, 2014.

92. Dwyer J. Sweat equity put to use in sight of Wall Street. *New York Times.* October 7, 2008. Available at http://www.nytimes.com/2008/10/08/nyregion/08about.html, accessed October 15, 2014.

93. Campbell LK, Wiesen A. United States Forest Service. Northern Research Station, Meristem (Organization). Restorative commons creating health and well-being through urban landscapes. Updated 2009.

94. Johnson L. *City Farmer: Adventures in Urban Food Growing.* Vancouver; Berkeley: Greystone Books, 2011.

95. Boynton RS. *The New Journalism: Conversations with America's Best Nonfiction Writers on Their Craft.* New York: Vintage Books, 2005.

96. Rennison N. *100 Must-Read Life-Changing Books.* London: A & C Black, 2008.

97. Redman N. *Food Safety: A Reference Handbook.* Santa Barbara, CA: ABC-CLIO, 2000.

98. Williams EM. *The A-Z Encyclopedia of Food Controversies and the Law*, 1st edition. Santa Barbara, CA: Greenwood, 2011.

99. Christensen T, Haas PJ, Christensen T. *Projecting Politics: Political Messages in American Film.* Armonk, NY: M.E. Sharpe, 2005.

100. Pollan M. *The Omnivore's Dilemma: A Natural History of Four Meals.* New York: Penguin Press, 2006.

101. Weaver TA. *The Butcher and the Vegetarian: One Woman's Romp Through a World of Men, Meat, and Moral Crisis.* Emmaus, PA: Macmillan, 2010.

102. Schumacher G, Nischan M, Simon DB. Healthy food access and affordability. *Maine Policy Rev.* 2010;(1):124–139.

Federal Food Assistance[*]

6

*I believe the Food Stamp Act weds the best of the
humanitarian instincts of the American people with the
best of the free enterprise system. Instead of establishing
a duplicate public system to distribute food surplus to
the needy, this act permits us to use our highly efficient
commercial food distribution system.*
President Lyndon B. Johnson[1]

INTRODUCTION

Social goals, such as improving nutritional health status and addressing inequities
in commercial distribution of fresh foods can be addressed through food-assistance
programs. Some programs focus on health and nutrition status concerns for specific
populations, such the Special Supplemental Nutrition Program for Women, Infants and
Children (WIC), some programs focus on ensuring foremost that people do not go hungry, such as the Emergency Food Assistance Program (TEFAP), and some programs,
namely the Supplemental Nutrition Assistance Program (SNAP) actually act as a bulwark for income in low-income populations.

Federal food assistance can be provided in various forms. SNAP provides cash
benefits to individuals for purchasing food in normal commercial transactions at retail
channels such as stores and farmer's markets. WIC provides vouchers to individuals
for purchasing specific foods. TEFAP provides commodity (surplus) foods directly to
individuals or to agencies for distribution for at home consumption or use in preparing
congregate dining meals. Programs that provide prepared meals, such as the National
School Lunch Program (NSLP) may receive both cash and commodity food support
from the federal government.

As might be expected, the primary beneficiaries of federal food assistance programs are low-income individuals and families, although some programs are aimed at
assisting communities rather than individuals. There are also collateral beneficiaries

---

[*] Lauren Dinour, DPH, RDN, contributed primary research and prepared draft versions of portions of this
chapter.

141

of food assistance programs—farmers at farmers' markets, mom and pop convenience stores, food corporations, and local and regional communities—that reap an economic benefit from the transfer of money.

This chapter provides a brief introduction to federal food assistance programs generally, including a history of commodity food distribution as food assistance. We illustrate how differing funding and nutritional approaches manifest in differing food policy by comparing highlights of a few programs. An appendix to this chapter provides more details on a number of federal food assistance programs.

FEDERAL FOOD ASSISTANCE PROGRAMS

In 2010, the United States Government Accountability Office (GAO) reported that there were close to 70 programs funded by the federal government in 2008 that provided at least some support for domestic food assistance.[2] The GAO evaluates the effectiveness of public funds expenditures at the request of Congress. In this case, Congress had asked the GAO to examine the prevalence of food insecurity in the United States, spending on food assistance programs, the known effectiveness of the programs, and the implications of providing food assistance through multiple programs and agencies. After identifying close to 70 programs with related food support functions, the GAO highlighted 18 programs that focused primarily on assistance for low-income households and individuals. See Appendix to this chapter.

Although the U.S. Department of Agriculture (USDA) oversees most of the 18 programs, the Department of Homeland Security (DHS) and the Department of Health and Human Services (HHS) also fund food-assistance programs. However, as the USDA is responsible for providing the bulk of food assistance in the United States, this chapter will highlight several USDA nutritional assistance programs.

The Government Performance and Results Act of 1993 (GPRA) and the subsequent GPRA Modernization Act of 2010 requires all federal agencies to publish on the agency's website a multiyear strategic plan containing a comprehensive mission statement, goals and objectives with strategies for attainment, and the means to evaluate progress toward fulfillment of those goals.[3] The USDA has now published the Strategic Plan for FY 2014–2018. Strategic Goal 4 is to ensure that all of American children have access to nutritious and balanced meals. One objective is to either achieve or maintain 79% participation in SNAP by eligible people.[4]

Unfortunately, although this public information on the USDA's goals can be very useful, it is also temporal. The previous strategic plan (FY 2010–2015) is no longer available at the time of this writing, but provided some important details about SNAP participation. In 2007, just over 65% of the eligible people participated in SNAP and the USDA set a goal of reaching 75% participation by 2015. As of 2012, the USDA had exceeded this goal, with 83% of all eligible individuals participating in SNAP. Although these individuals compromised 83% of the eligible population, they received 96% of all possible benefits. (Individuals eligible for larger benefits tend to participate in SNAP at higher rates than those eligible for smaller benefits, so this statistic is telling us that the population that is in most need of the help is accessing SNAP benefits.) Even though participating in SNAP provides economic benefits that extend beyond the individual into the community, there may be a point at which the cost of trying to encourage more eligible people to participate in SNAP is greater than the available remaining benefit.

Food and Nutrition Services Agency

Within the USDA, the Food and Nutrition Services Agency (FNS) runs all of the USDA nutrition programs save for the Community Food Projects Competitive Grant Program under the National Institute of Food and Agriculture.[2] FNS is one of the two agencies in the office of Food, Nutrition, and Consumer Services (FNCS); the other agency is the Center for Nutrition Policy and Promotion (CNPP), which develops and promotes dietary guidance that links scientific research to the nutrition needs of individuals.[5] As of August 2014, the FNS describes its mission as "to end hunger and obesity through the administration of 15 federal nutrition assistance programs," which is perhaps a signal of changing policy emphasis as this is a change from previous statements in 2013 referencing food insecurity and poor nutrition.[6]

About one in every four Americans participates in at least one of USDA's food and nutrition assistance programs.[7] These 15 nutrition assistance programs account for over two-thirds of the USDA's total budget. USDA's food assistance programs are authorized by one of two pieces of legislation—the Farm Bill or the Healthy, Hunger-Free Kids Act of 2010 for programs originally covered in the 1966 Child Nutrition Act.[8] See Table 6.1 USDA's Domestic Nutrition Assistance Programs.

TABLE 6.1 USDA's domestic nutrition assistance programs

Authorized by the 2014 Farm Bill

SNAP	Supplemental Nutrition Assistance Program
TEFAP	Emergency Food Assistance Program
CSFP	Commodity Supplemental Food Program
FFVP	Fresh Fruit and Vegetable Program
SFMNP	Senior Farmers' Market Nutrition

Authorized by the Healthy, Hunger-Free Kids Act of 2010

WIC	Special Supplemental Nutrition Program for Women, Infants, and Children
FMNP	WIC Farmer's Market Nutrition Program
SBP	School Breakfast Program
NSLP	National School Lunch Program
SFSP	Summer Food Service Program
SMP	Special Milk Program
CACFP	Child and Adult Care Food Program
	Nutrition Assistance Block Grants, including Nutrition Assistance for Puerto Rico

Healthy, Hunger-Free Kids Act of 2010

The Farm Bill is discussed in Chapter 2, "Agricultural Policies," so we will focus here on the Healthy, Hunger-Free Kids Act of 2010 (HHFKA). The HHFKA is part of the reauthorization of funding for child nutrition programs originally covered in the 1966 Child Nutrition Act (CNA).[9] HHFKA provided $4.5 billion in new funds for these child nutrition programs for 5 years and is set to expire on September 30, 2015. WIC and other nutrition programs such as the Summer Food Service Program (SFSP), the Child and Adult Care Food Program (CACFP) must be renewed every 5 years because they have actual expiration dates. Although HHFKA also funds the National School Lunch Program and School Breakfast Program, these two programs do not need to be renewed as they are permanently authorized. The offsets used to pay for this legislation included $2.2 billion reaped from ending the increased SNAP benefit funded through the American Recovery and Reinvestment Act 5 months early.[10]

"Old school" food

School lunch is the oldest USDA food-assistance program. School lunch became a national public health initiative in 1946, when the National School Lunch Act (later renamed the Richard B. Russell National School Lunch Act in 2000)[11] authorized the National School Lunch Program (NSLP) while proclaiming that national security depends on encouraging the domestic consumption of nutritious agricultural commodities and safeguarding the health and well-being of the nation's children.[12]

Under the NSLP, nutritionally balanced meals must be made available each day to all children. For those children who meet certain income levels, the meals are provided at no cost or at a reduced price. Children from households with incomes at or below 130% of the poverty level can receive free meals, and those from households with incomes between 130% and 185% of the poverty level can receive meals at a reduced price. The burden of certification is low for the family, which can self-report income. The USDA provides schools with cash reimbursement and donated foods for meals that meet federal nutritional requirements. The USDA purchases food for donations; thus, the NSLP still plays a role in the dual benefit of food assistance to address hunger and support agricultural income. The amount of cash reimbursement depends as to if students pay the full price for lunch or if they are eligible by household income to receive lunch free or for a reduced price, and if more than 60% of students in a school are eligible for free or reduced cost lunches, the reimbursement rate is slightly increased. Rates are also higher in Alaska and Hawaii. As of SY 2014–2015, in the contiguous states, the minimum amount of reimbursement per lunch served is 28 cents and the maximum rate is $3.21.[13]

Over time, the amount of cash reimbursement and the nutritional requirements have been periodically modified. Under President Nixon, program eligibility for free and reduced cost lunched expanded, but subsequent administrations cut subsidies; President Reagan's by more than one-third. Nutritional requirements were modified to help contain costs.

Chapter 10, "Fruits and Vegetables," raises the question if a tomato is a fruit or vegetable. In the 1980s, President Reagan's administration proposed the answer that when it came to school lunch, tomato paste served in the amount of one tablespoon could be credited as a one-fourth cup worth of vegetable. What do many people call a tablespoon of tomato paste? Ketchup.

Although ketchup itself was not actually determined to be a vegetable, the weakening of ideals as to what constitutes food on a child's lunch plate in combination with reduced funding fostered the encroachment of the processed food industry into the school lunchroom. Underfunded school lunch programs needed cheaper food to serve and the processed-food industry was able to oblige with appropriately cheap products fortified with nutrients or turned into low-fat versions to meet nutritional standards.[14]

"New school" food

The NSLP nutritional requirements were most recently updated in 2010 when the Healthy, Hunger-Free Kids Act[15] amended both the Richard B. Russell National School Lunch Act and the Child Nutrition Act of 1966. However, nutritional improvement for school food had started much earlier. In 2004, Congress granted the USDA authority to improve school food nutrition, and the IOM accordingly gave specific science based recommendations to the USDA in 2009.

In addition to funding school meals, HHFKA authorized the USDA to establish new nutritional standards based on the recommendations by the Institute of Medicine and make significant improvements in the food served. For example, the Act requires school food service operations to serve increased portions of fruits, vegetables, and whole grains while limiting sodium, saturated fat, trans fat, and calories in the meals. The vegetable recommendations also provide an emphasis on dark green and orange vegetables and limits starchy vegetables. Additionally, the rules place a calorie cap on lunches: 650 calories for elementary school lunches, 700 for middle schools, and 850 for high schools. Under the guidelines, half of breads and other grain-based foods offered must be whole-grain rich (contain 50% whole grains), until the start of the 2014 school year, when all such foods must be whole-grain rich. (The maximum reimbursement rate noted earlier includes the additional 6 cents incentive that schools receive if they adopt the new regulations.) The bill also supports farm-to-school activities and programs to increase the quantity of organic food served in schools.[16]

The changes in school meals reflect a social determination that the population served by these meals needs improved nutritional offerings. The population-wide increase in childhood obesity and reduction in physical activity is leading to a number of health problems for many children, some of which will follow children into adulthood. Many of these health problems can be prevented or alleviated by improved nutrition. Significantly, HHFKA also authorized the USDA to set nutritional requirements not just for the federal supported meals served at lunch and breakfast, but also for all foods sold on school grounds during the day, and called for a number of initiatives to address childhood obesity prevention. Box 6.1 provides more information on HHFKA.

BOX 6.1   THE HEALTHY, HUNGER-FREE KIDS ACT OF 2010 AND CHILDHOOD OBESITY PREVENTION

Requires USDA to establish updated meal pattern regulations and nutrition standards for the school lunch program.

Provides an additional 6 cents per lunch for schools that are certified to be in compliance with the new meal pattern regulation.

Removes the previous requirement that schools serve milk in a variety of fat contents and instead requires that schools offer a variety of fluid milk consistent with the *Dietary Guidelines for Americans.*

Requires schools to make free potable water available where meals are served.

Requires USDA to establish regulations for local school wellness policies and to provide technical assistance to states and schools in consultation with the U.S. Department of Education and the Centers for Disease Control and Prevention.

Local school wellness policies must include, at a minimum:

1. Goals for nutrition promotion and education, physical activity, and other school-based activities that promote student wellness.
2. Nutrition guidelines for all foods available on each school campus during the school day that are consistent federal regulations and that promote student health and reduce childhood obesity.
3. A requirement allowing parents, students, school food representatives, physical education teachers, school health professionals, the school board, school administrators, and the general public to participate in the development, implementation, and review and update of the local school wellness policy.
4. A requirement that the public be informed about the content, implementation, and periodic assessment of the local school wellness policy.

Requires USDA to establish science-based nutrition standards for all foods sold in schools at any time during the school day outside the school meals programs (allows exemptions for school-sponsored fundraisers if the fundraisers are approved by the school and are infrequent).

Requires local educational agencies to report in clear and accessible manner information about the school nutrition environment to USDA and to the public, including information on food safety inspections, local wellness policies, school meal program participation, nutritional quality of program meals, etc.

Establishes an organic food pilot program that provides competitive grants to schools for programs to increase the quantity of organic foods provided to schoolchildren.

Requires USDA to identify, develop, and disseminate model product specifications and practices for food offered in school programs. USDA must analyze the quantity and quality of nutrition information available to schools about food products and commodities and submit a report to Congress on the results of the study and recommend legislative changes necessary to improve access to information.

Requires USDA to provide technical assistance and competitive grants that do not exceed $100,000 to schools, state and local agencies, Indian tribal organizations, agricultural producers, and nonprofit entities for farm-to-school activities. The federal share of project costs cannot exceed 75% of the total cost of the project.

Requires USDA to establish a program of required education, training, and certification for school food service directors; criteria and standards for selection of School Nutrition State Agency Directors; and required training and certification for local school food service personnel.

> *Healthy, Hunger-Free Kids Act of 2010, Pub. L. no. 111–296, 124 Stat. 3183, 2010. Online. Available at http://www.gpo.gov/fdsys/pkg/PLAW-111publ296/pdf/PLAW-111publ296.pdf. Accessed June 26, 2014.*

Over 5 billion served

In terms of cost, the NSLP is the second largest USDA food assistance program. In terms of potential significant nutritional impact for a discrete population, NSLP is the largest USDA food assistance program. Five billion lunches were served in FY 2013.[17]

In other words, with school lunch, there are 5 billion potential chances over just 1 year to get it right—to not only provide a nutritious, well-balanced meal, but also to set the example for what a healthy meal should look like. The potential impact on population health is tremendous. The modern public health interest in regulating choices of food consumption is predominantly to prevent future danger and cost, as poor health outcomes due to diet tend to develop over time. Improving 5 billion meals now is one way to keep our future health care costs from rising.

Poor nutrition is one of the risk factors for heart disease as well as Type 2 diabetes. Based on NHANES data, less than 1% of adults have an ideal diet. In children, the prevalence of ideal diet is nearly 0%, meaning one in which they meet 4 out of 5 of the following recommendations: (1) consumption of at least 4.5 cups per day of fruits and vegetables, (2) consumption of at least 2 servings a week of fish, (3) consumption of at least 3 serving per day of whole grains, (4) consumption of no more than 36 ounces per week of sugar-sweetened beverages, and (5) consumption of less than 1500 mg per day of sodium.[18]

On a typical day, 21.5 million of NSLP meals were for children from households with income below 185% of the poverty level. In 2010, the National Center for Education Statistics reported that almost half of public school 4th graders were eligible for free or reduced lunches the year before. Among the public school fourth graders that year, 74% of black students, 77% of Latino students, and 68% of American Indian/Alaska Native students were eligible for free or reduced-price lunches. In contrast, only 28% of white students were eligible. Additionally, black and Latino students were more likely to be attending high-poverty public schools in which more than 75% of the students are eligible for free or reduced cost lunch.[19]

In other words, the 21.5 million free or reduced-price meals are not equally distributed among school children, but may instead be concentrated in groups. In turn, this means that nutritional improvements in the NLSP and in other food served at school can contribute not only to the health of individual children, but also to the health of a population cohort. We have at least 21 million chances each day to promote overall help for the children that need it most.

In fact, some of the meals served through the NSLP and SBP go to juveniles and adolescents in detention facilities.[20] Whether a juvenile is in school or a detention facility, the eventual outcome will in most cases be a release back into the larger community, albeit one called graduation for those in school rather than detention. In both cases, the period of incarceration for some and relative containment for others provides a window of opportunity for improved nutrition and dietary behaviors, which should be individually and socially beneficial in the long run.

In 2003, the American Public Health Association pointed out that in the case of incarcerated populations, it "serve(s) the broad public health interest (to provide) testing, treatment, counseling, health education, and health promotion to an otherwise disenfranchised population prior to their ultimate release back into the community."[21] Health promotion includes nutrition improvement, particularly when juveniles in detention and juveniles in high-poverty schools are likely coming from and potentially returning to low-income communities with disproportionate rates of diabetes and obesity.[22]

The challenges of getting nutrition "right"

However, planning a menu to incorporate nutrient standards is challenging, and nutrients can in fact trump what some might consider a healthy meal of food. In particular, the schools often stumble on the required ratio of fat to calories, and thus as noted earlier food manufacturers continue to reformulate products or create new ones that help schools meet guidelines. Whole-wheat spaghetti is one thing; healthier fast food branded pizza is yet another.[23] When the USDA tried to reset the vegetable equivalency standard to as per volume served for tomato paste, which would entail that a slice of pizza could no longer count as including a full-serving of vegetables (but could be served with vegetables), Congress did a end-run via an appropriations act to forbid the USDA from using any of the money to "implement an interim final or final rule regarding nutrition programs under the Richard B. Russell National School Lunch Act (42 U.S.C. 1751 et seq.) and the Child Nutrition Act of 1966 (42 U.S.C. 1771 et seq.) that requires crediting of tomato paste and puree based on volume." The appropriations act also stops the USDA from fully implementing the intended sodium reduction plan until the USDA certifies to having evaluated the relevant scientific data and requires that the USDA cannot establish a whole grain requirement without defining whole grain.[24] As Michele Simon points out, the recommendations the IOM gave to the USDA for school nutrition improvements should already be science based.[25] *Why would Congress overrule a medical expert recommended national nutrition policy?*

The Cost of Federal Food Assistance

The involvement by the USDA in nutrition assistance programs is remarkable, even considering a history of commodities (agricultural surplus) being used for food assistance. Seventy-six percent of the USDA's 2015 budget—over 106 billion—is allocated to nutrition assistance programs.[26]

The federal budget for food assistance programs, although generally increasing over time, has increased significantly since 2008. In FY 2008, the federal government spent more than $62.5 billion on the 18 domestic food nutrition and assistance programs listed in the Appendix to this chapter, 15 of which are covered in the USDA's budget. In 2008, the smallest of the 18 programs spent $4 million and SNAP, the largest, spent $37 billion.[2] Surprisingly, the entire federal food assistance budget for those 18 programs in 2008—62.5 billion—was less than what was spent on SNAP alone in 2011. SNAP cost $75 million in 2011, $78 million in 2012, and almost $80 million in 2013.[27] The USDA has $84 million budgeted for SNAP in 2015.[28]

SNAP increase in spending

What changed between 2008 and 2011 to cause such a huge spike in SNAP spending? The budget for SNAP alone in 2011 was greater than the entire amount spent on federal food assistance programs just 4 years earlier in 2008.

The United States experienced an economic downturn in 2008, which contributed to the particular expansion in SNAP spending. The weakened economy pushed more people down into the eligible income bracket. Additionally, the 2009 Recovery Act temporarily boosted the amount of monthly benefits available and expanded the demographic reach of the program. Loosened eligibility requirements made it easier for childless, jobless adults to qualify for the program. Inputs of more eligible people, as well as more money per person, naturally lead to an increase in program spending.

This loosening of eligibility is an example of food policy used as an economic policy, namely, as a mechanism to help bolster a weak economy. A 2002 study by the Economic Research Service (ERS) of the USDA measured the impact of SNAP (then known as Food Stamps) through a hypothetical model of an increase of $5 billion in funds and concluded that such an increase could act as a stimulus depending on the funding mechanism—emergency financing in a recession stimulates activity but budget neutral funding does not. In the hypothetical model using emergency funding, each dollar spent on SNAP resulted in $1.84 of economic activity.[29] The ERS has further refined this estimate to $1.79; describing it as every $5 generates as much as $9 in economic activity.[30]

The 2009 Recovery Act, which temporarily boosted the amount of SNAP benefits and loosened eligibility, had a planned ending date based on a food inflation cost benchmark. Food price inflation was not as high as anticipated, which pushed the end date out farther in the future then hoped for, and Congress began to look at terminating the increase on a specific calendar date instead in order to allocate that money elsewhere. The benefit increase was rolled back as of October 31, 2013, resulting in a cut of about $10 per person, per month.[31] See Box 6.2 for more details on the SNAP allotment.

BOX 6.2   SNAP BENEFITS TODAY

Administrative rules, including expanding and retracting program eligibility criteria, have continued to be modified over time. The asset limits, income limits, maximum benefits, and calculation of benefits as well as current standards for eligibility to participate in the program are explained in detail on the SNAP website at: http://www.fns.usda.gov/snap/. In general, however, the amount of benefits an eligible household receives, called an *allotment*, depends on the age and number of people in the household and on the net monthly household income. For most households, the gross income (before deductions) must be 130% or less of the federal poverty level and the net income must be at the federal poverty level. As an example, for October 1, 2014 through September 30, 2015, for a household of three to be eligible for benefits, the household cannot have more than $2144 of gross monthly income or $1650 of net monthly income. The maximum benefit allotment is for households that have no income; a household of three people with no income can receive up to $511 per month in benefits.

United States Department of Agriculture, Food and Nutrition Service,
Supplemental Nutrition Assistance Program, updated August 20, 2104,
http://www.fns.usda.gov/snap/eligibility, accessed October 6, 2014.

Furthermore, spending on SNAP should also decline naturally if an improved economy lifts people back above the income eligible criteria. But it is also possible that SNAP is like skirt hemlines—after hovering between the ankle and the floor for many years, once they went up, they weren't coming back down in any meaningful way.

Food assistance program costs

The cost of a food assistance program includes more than just the distributed benefits. Administrative costs of the food assistance programs can be substantial. In 2008, such costs ranged from a tenth to more than a quarter of total costs of food assistance.[2] Administrative costs, however, may also provide an indirect benefit to the local community, and include items, such as nutrition education programs, that may not intuitively be thought of as administrative.

For example, the Food Distribution Program on Indian Reservations (FDPIR) has a relatively high administrative cost. Of $88.5 million by Congress in 2008, about $53.8 million was allocated for food and $34.7 million for local administrative expenses.[32] In 2011, the FDPIR program provided nutrition assistance to an average of 77.8 million persons per month at a cost of $50.71 per food package, with an average monthly per person administrative cost of $48.14.[33]

In other words, on a per-person monthly basis, just over half is spent on food and just under half spent on administration. The administrative cost seems significant, but this amount includes the cost of making food deliveries to remote locations and other local agency tasks which include ordering, storing and distributing program foods, determining applicant eligibility, and providing nutrition education. Nutrition education programs may include individual counseling, cooking demonstrations, nutrition classes, and the dissemination of information on the proper storage of food, thus, providing significant local and community benefits (including economic) under the guise of "administration."[32]

Overlapping federal food assistance programs

Although almost all of the programs are administered through the USDA's FNS agency, the food assistance system remains relatively decentralized. Some programs do provide overlapping benefits, which helps ensure access to those in need but can also lead to inefficient use of funds. The message from agency officials regarding overlap is that no one program alone is intended to meet a household's full nutritional needs, and that overlap is due in part to the fact that the programs were created individually by Congress over the past several decades to address a variety of emerging needs such as specific populations at high risk of malnutrition or hunger.[2]

An example of a smaller program that the USDA considers to be duplicative is the Commodity Supplemental Food Program (CSFP). This program originally targeted low-income pregnant women, children and the elderly, all populations that are covered under other programs. One difference is that the CSFP provided food directly to the population (USDA commodity surplus foods), instead of vouchers or money. The USDA had proposed eliminating this program entirely in 2007–2009, which would have reduced administrative costs of close to 140 million in fiscal year 2008, but Congress continued to fund the program. Some of the administrative cost savings might have been reversed by the potential increased enrollment into WIC,[2] but nonetheless, the over-all cost would have been reduced. In 2014, the overlap with WIC was addressed, and as of February 6, 2014 new enrolment of women and children has ceased. Going forward, CSFP will only serve low-income elderly.[34] *If the USDA proposes a reduction, why would Congress not agree? Are there pros and cons to removing potentially duplicative but population specific nutrition programs?*

"Entitlement": Impact on funding food assistance programs

Certain federal programs, including some food assistance programs, are enacted through "entitlement legislation." Entitlement legislation requires the payment of benefits to all those eligible under the requirements established by law. Therefore, because anyone who is eligible under the program rules has a legally enforceable claim to receive benefits, the funding must essentially be enough to provide benefits to everyone who is eligible.

Thus, for entitlement programs, once a benefit amount per person is allocated, knowing the number of eligible people dictates the funding, effectively preempting any formal process of deliberating funding levels through Congress.[35] In other words, even different types of federal funding processes may cause differences in food policy implementation.

SNAP is an entitlement program but FDPIR and WIC are not. Congress allocates a funding level for WIC as a programmatic whole, meaning that funding is not tied directly to the number of people who are eligible under programs rules. Congress is not required to allocate enough money to go around. If there are more people eligible for WIC than funds allow supporting (the eligibility criteria will be discussed in more detail below), people will be placed on a risk-prioritized waiting list.[36]

WIC's status when it comes to funding was not always clear. The fact that it is not an entitlement program may be a historical artifact of timing. Created as a pilot program in 1972 through an amendment to the Child Nutrition Act of 1966, WIC was made permanent in 1975.[37] Until fiscal year 1978, WIC was funded through a source of discretionary funds (30% of customs receipts) accessible to the Secretary of Agriculture. This funding scheme was a consequence of timing, as the amendment creating WIC was passed after that year's appropriations act was passed.

However, the 1974 Congressional Budget and Impoundment Control Act increased focus by Congress on legislative budgetary control and for FY1978 that source of discretionary funds available to the USDA was eliminated. Congress was now considering how much control they should take over budgetary allocation for WIC. Accordingly, in 1978, the Senate Chairman of the Subcommittee on Agriculture, Rural Development, Committee of Appropriation, asked the Comptroller General of the United States if WIC should be considered an entitlement program.

The Comptroller advised the Senate Committee Chairman that the USDA did not think that WIC was an entitlement program, and furthermore, that GAO historically has taken the position that public interest is better met with congressional control of programs and financing through periodic direct allocations, rather than entitlement programs.[4]

This review of funding philosophy is important. The "nonentitlement" limit on funding impacts WIC's implementation as a nationally consistent food policy. In particular, local agencies may adopt differing means of cost-containment in order to be able to extend benefits to as many as possible.[37] As the approach varies from state to state, the cost-containment and particular state can impact the participants' food access and choices.

"Entitlement": Impact on who eats what, where

WIC authorizes specific foods and standards. Peanut butter is a designated WIC food item. Under federal WIC regulations, peanut butter must conform to a FDA standard of identity, and the following types of peanut butter are acceptable: creamy, chunky, regular, reduced fat, salted, and unsalted.[38] Each state then has the authority to further specify or limit choices within those parameters. The state agency approved peanut butters available to WIC participants in each state may differ widely in subjective characteristics such as brand or other qualities. Some states restrict purchases to some specific brands or to store brands. It is possible that in your state, WIC participants cannot choose the brand or even the type of peanut butter that you would choose for yourself or for your family. In Missouri, for example, only smooth, creamy or regular peanut butter is allowed.[39] No crunchy. Furthermore, only the store brand is authorized, which generally is not only cheaper, but also more likely to have added sweeteners and oils.

Recall the discussion in the introduction chapter about how we characterize food—whether through a focus on nutritional components versus the totality of other characteristics. All peanut butters, since they are made from ground peanuts, have a similar nutritional profile at the micronutrient level, as a source of protein, magnesium, and iron, although other added ingredients such as sugar or oil can vary and create differences in the macronutrient content. *If you eat peanut butter, look at the ingredients in the brand you have chosen. Why do you select that particular brand or type? If the relevant importance of food is simply just the micronutrients noted earlier, why does it matter to you which peanut butter you are eating?*

This is one of the balancing questions of food policy, particularly when it comes to federal assistance. What is the better use of funds? *Providing nutritionally adequate food to more women and children? Providing food optimal for overall public health dietary goals, but to less women and children?*

In the specific case of peanut butter, even though all peanut butter has the same micronutrient profile that meet the specific nutrition program policy goals of WIC, making "natural" peanut butter that contains no added sugar or fats as the only approved type of peanut butter might meet the nutritional goals of WIC as well as promote other food policy dietary goals, such as those of Healthy People 2020.

AMERICAN DIET, HEALTHY EATING INDEX, AND OTHER NUTRITIONAL GOALS OF FEDERAL FOOD ASSISTANCE

Many people in America, regardless of any participation in nutrition assistance programs, do not eat a healthy diet as recommended by the U.S. Dietary Guidelines. The Healthy Eating Index (HEI) is a measure of diet quality as compared to US federal dietary guidance. Originally created in 1995, the HEI was revised in 2006 to meet the 2005 *Dietary Guidelines for Americans*, and updated again in 2012 to meet the 2010 *Dietary Guidelines for Americans*.[40] A 2005 study using the HEI found that the average American scored a 58 out of 100.[41] A perfect score of 100 would have been equivalent to meeting all the (then current) 2005 Food Guide Pyramid guidelines plus recommendations for fat, saturated fat, cholesterol, and sodium.[42] A score of more than 80 implies a "good" diet even though it still allows for failure to meet all of the Food Guide Pyramid recommendations. A more severe measure of a "good" diet would be to require all dietary guidelines to be met.[42] The following section provides a brief discussion of several food assistance programs in relation to the HEI and other nutritional goals.

Food Distribution Program on Indian Reservations

As described in the Appendix to this chapter, FDPIR provides a "food package" consisting of commodity foods. Commodity foods are items purchased by the USDA to remove surpluses from the marketplace, and may include meat, poultry, fish, fruit, vegetables, egg products, dry beans, tree nuts, dairy products, cereals, grains, peanut products, nonfat dry milk and vegetable oils.[43] Since 2002, a work group including Indian Tribe Organization representatives and USDA officials have been continually revising the food package to better meet the nutritional needs and preferences of program participants without significantly increasing food package costs.[44]

Even though the package continues to be updated and is far from perfect, it provides a better "HEI" diet than the average American chooses to consume. As reported to Congress in 2008, the FDPIR food package, if consumed by individuals in the quantities provided, would result in a diet with a HEI-2005 score of 81 out of 100, considerably better than the 58 scored by Americans in general. Even so, the agency has recognized that the FDPIR food packages were still not sufficiently balanced to meet all of the then Dietary Guidelines food group recommendations within appropriate calorie guidelines. Whole grain offerings and other modifications were initiated,[32] and the packages continue to be improved by the working group.

Special Supplemental Nutrition Program for Women, Infants, and Children

Although commonly referred to as WIC, the program is officially titled as Special Supplemental Nutrition Program for Women, Infants, and Children. WIC is targeted toward addressing nutritional deficiencies observed in a specific population.

The program developed out of growing public concern in the late 1960s over hunger in the United States. Action began at the community level—a group of physicians, concerned about seeing pregnant women in their clinics suffering from a lack of food, met with the USDA with the idea of building food commissaries attached to neighborhood clinics. Independently, a Johns Hopkins doctor began a voucher program to distribute food.[37]

These experiences and observations, among others, informed the premise of the WIC program—many low-income individuals are at risk of poor nutrition and health outcomes because of insufficient nutrition during the critical growth and development periods of pregnancy, infancy, and early childhood. The program addresses both hunger and nutrition; and provides education and community resources (WIC clinics) as well as food.

WIC participants receive nutrition education and referrals to health care and social service.[37] Nutrition education is an integral part of WIC and its provision is considered a key WIC benefit. The main assistance, however, is through provision of specific foods aimed at ensuring adequate intake of certain nutrients in order to help prevent future medical and developmental problems. See Box 6.3.

In almost all states, WIC participants are given vouchers to redeem through retail channels. Since the participant is given a voucher for a specific item, cost consciousness is at the state agency level as to the selection of specific food available to be acquired with the voucher. Some states limit the option to the lowest cost brand, or limit the size.

The targeted nutrients from the program's inception: high quality protein, iron, calcium, vitamin A, and vitamin C, are provided through assistance in purchasing specific foods rich in these nutrients—namely, milk, eggs, cheese, dried beans, peanut butter, and certain fortified food products, specifically breakfast cereals high in iron and low in sugar, and juice fortified or naturally high in Vitamin C.[37] Using these nutritional

BOX 6.3   BALANCING HEALTH GOALS: WIC, VITAMIN C, FRUIT JUICE AND DENTAL CARIES

Healthy People objectives are discussed in Chapter 4, "Nutrition Information Policies." Food policies as conducted through federal food assistance, can contribute to achieving the objectives of Healthy People, even through those not specifically documented as a nutrition/food objective. For example, one Healthy People 2020 objective is to reduce dental caries in children. Excessive juice intake has been strongly associated with overweight children and obesity as well as tooth decay in the preschool population. Limiting fruit juice intake in children is known to help reduce dental caries. Yet juice (citrus or vitamin C fortified) is also a way to ensure vitamin C intake in children. These goals can be balanced. The updated WIC package has completely removed juice from the infant package, reduced quantities for all other packages to reflect a recommended intake of about four ounces per day, and counsels participants about limiting juice.

objectives, federal regulations specify the content and amount of food items in WIC "food packages" that are designed for different types of participants (infants, children, breastfeeding women, etc.).

The FNS began updating the WIC food packages in 2007 to align with the 2005 *Dietary Guidelines for Americans* and the infant feeding practice guidelines of the American Academy of Pediatrics. The new packages were implemented under the proposed final guidelines long prior to the final rule being published on March 4, 2014, but the final published rule adds some further modifications such as yogurt as a partial substitute for milk, more whole grain and fish options, and additional fruits and vegetables for children.[45]

As part of the food package update, the FNS sought to eliminate juice for infants as well as to encourage breastfeeding by mothers. The targeted nutrients of concern were also reviewed, albeit still based on dietary intake for a past cohort of women and children. Although NHANES has strong clinical measures, it is cross sectional, and has very limited samples of pregnant women and breastfed children, so the review of the WIC package used data collected by the USDA in the 1994–1996 and 1998 Continuing Survey of Food Intakes by Individuals.[46] *When considering dietary data for a discrete population cohort, is the age of the data irrelevant?*

In any case, the updated analysis determined the priority nutrients for children in the intended population to be iron, vitamin E, potassium, and fiber; for women to be calcium, iron, magnesium, vitamin E, potassium, and fiber. The proposed priority food group for children: vegetables. The proposed priority food group for women: fruit and low/nonfat dairy products.[47]

The updated packages, introduced under an interim rule in 2009, encouraged breastfeeding with incentives that give fully breastfeeding women more food and more money for purchasing fruits and vegetables, and, for the fully breastfeeding infants, baby food meats as well as greater amounts of baby food fruit and vegetables.[48] The final rule, however, commented that the IOM had recommended that all women, not just breastfeeding women, receive the full amount of ten dollars a month for fruits and vegetables, and the difference in the interim rule was only for cost neutrality. The final rule gives the same amount of money for fruits and vegetables to all women participating in the program.[49] Overall, the packages now include a wider variety of foods, increased emphasis on fruits (and a reduction in juice), vegetables and whole grains, and allow the state agencies more flexibility in accommodating cultural food preferences, including nondairy substitutes.

WIC: Nutrition risk criteria

Even though the underlying premise is that a low-income population is more likely to be at nutritional risk, all WIC program participants must still be determined to be at nutritional risk on the basis of a medical or nutritional assessment by a physician, nutritionist, dietitian, nurse, or some other competent professional authority.[50] Nutritional risk categories include dietary risk, anthropometric risk, biochemical risk (e.g., low hematocrit), medical risk (e.g., diabetes mellitus), and other predisposing factors (e.g., homelessness). These criteria are categorized based on severity of potential effect and outcome in order to prioritize participants in case of limited funds, although expansion of WIC program funds during the 1990s allowed a greater number of "lower priority" participants and the importance of the priority system decreased. Anecdotal evidence suggests that in recent years nearly everyone who was eligible and who applied for the program has been able to participate.[37]

If substantial numbers of postpartum women and children currently are found to be eligible on the basis of dietary risk alone versus the more clinically defined risk categories this leads to the question as to if a risk criteria and priority system is truly relevant? Recall that the average American, regardless of income status, has a poor HEI diet. More than 96% of U.S. individuals, regardless of income status, failed to meet U.S. dietary guidance. An even greater percentage of low-income individuals failed to meet those recommendations. In other words, almost all Americans are at dietary risk.

Given a generally recognized society failure to meet the Dietary Guidelines, how meaningful is dietary risk as an eligibility factor among a smaller group? Why not instead presume all women and children ages 2–5 years are at dietary risk, and thus streamline enrollment in WIC to all women and children within the age and income-eligible category?

The highly clinical criteria regarding risk could be retained to prioritize focus and additional services.[51] Conversely, another perspective would be to instead narrowly focus the program to the smaller population of women and children in most critical and clinical need.

Stamps to SNAP: Food Transfer to Income Transfer

SNAP is the rebranded food stamp program. Although the acronym stands for Supplemental Nutrition Assistance Program, the program lacks a specific nutritional goal. Over time, federal food assistance programs like FDPIR and WIC with specific population nutritional health status goals have improved food package offerings or approved items for purchase with an eye toward nutritional status and the Dietary Guidelines. Unlike FDPIR and WIC, SNAP does not have a nutritional objective and makes no reference to the Dietary Guidelines. *Is it significant that SNAP does not have nutritional objective? What does this mean as a food policy?*

Should SNAP have a specific nutritional goal? The lack of a specific nutritional goal makes sense if the underlying intent of this program has changed over the years from a food transfer program to an income transfer program.

SNAP history

The U.S. government had introduced a version of food stamps during 1939–1943 in an effort to deal with agricultural surpluses held back from the market. During this period, the stamps had to be applied specifically to commodities declared to be in surplus. The needs of hungry people were being met, but the driving underlying intent of the policy was economic support of farm income. In order to achieve a larger social goal, individuals' dietary choices were prescribed. Box 6.4 provides a brief history of the domestic use of commodities as food assistance.

BOX 6.4   FEDERAL FOOD ASSISTANCE THROUGH SURPLUS (COMMODITY) FOOD DISTRIBUTION

Select Legislative History of Commodity Distribution as Food Assistance (1930s–1990s)

1933: The Commodity Credit Corporation Charter Act of 1933. The Commodity Credit Corporation (CCC) was established in 1933, primarily to get loans to farmers and help them store nonperishable commodities until prices rose. As farmers were eventually allowed to forfeit their crops to the federal government to repay loans, the government ended up holding commodities.

1935: Section 32 of the Agriculture Act of 1935. Section 32 of this act made available to the Secretary of Agriculture an amount of money equal to 30% of the import duties collected from customs receipts. The sums were to be maintained in a separate fund to be used by the Secretary to remove price-depressing surplus foods from the market through government purchase and dispose of them through exports and domestic donations to consumers in such a way as not to interfere with normal sales. This law provided the basis for donating surplus commodities (and later funding) for federal domestic food programs.

1949: The Agricultural Act of 1949. This Act, and subsequent amendments, granted additional authority for commodity donations to needy people in addition to those made available through the Section 32 of the Agricultural Act of 1935. The Act also authorized the CCC to pay for added processing, packaging and handling costs for foods acquired under price support so that recipient outlets could more fully use them. This allowed for purchasing flour milled from support priced wheat, cornmeal from corn, oils from oil seeds, etc. The theory was that every pound of food purchased in its processed state reduced the amount that would go into CCC stocks. Again, the objective was to increase consumption of these foods that were being acquired by the CCC under price support. During the period from 1935 to 1970, over half of the foods that USDA distributed domestically went to needy families. The expansion of the Food Stamp program, however, meant that school and child feeding programs used an ever-increasing share of the total food distributed.

1961: Executive Order Mandating an Increase in Donated Foods to Needy Households. In January 1961, the first executive order issued by President Kennedy mandated that the Department increase the quantity and variety of foods donated for needy households. This executive order represented a shift in the Commodity Distribution Programs' primary purpose—from surplus disposal to that of providing nutritious foods to needy households.

1960s and 1970s: Commodity Components of New Federal Food Assistance Programs. In the late 1960s and early 1970s, the Nixon Administration proposed phasing out commodity donation programs as part of a broader farm policy effort to end price support programs that required government acquisitions of commodities. This effort was unsuccessful, and in fact may have helped institutionalize commodity donations to domestic food programs. The Congress voted to continue farm price supports that generated government stocks, and mandated commodity assistance for school lunch programs. This time period also saw the enactment of new programs with a commodity component targeting specific populations; such as the School Breakfast Program (SBP), the Summer Food Service Program (SFSP), the Child and Adult Care Food Program (CACFP), the Commodity Supplemental Food Program (CSFP), and the Food Distribution Program on Indian Reservations (FDPIR).

1973: The Agriculture and Consumer Protection Act of 1973. This Act gave the authority to the USDA to make open-market purchases of foods when similar foods were not available under the authorized surplus regulations, and phased out direct commodity distribution. This authority responded to a temporary decline of surpluses in the early 1970s.

1980s and 1990s: Commodity Initiatives. Legislation enacted in the 1980s addressed unemployment and homelessness. In the 1980s when the weak farm economy brought huge government-held stocks of commodities, pressure to increase donations to all outlets increased. For the first time, emergency feeding organizations were guaranteed some form of commodity assistance.

United States Department of Agriculture, Food Distribution Programs, Legislative History Food Distribution Programs, http://www.fns.usda. gov/fdd/fdd-history-and-background, accessed October 15, 2014.

The modern food stamp program (now SNAP) was established in 1964 after a 1961 pilot initiated by President Kennedy.

In the 1961 food stamp pilot program, participants purchased coupons of a higher value than their cash contribution and used the coupons to purchase food at retail stores. Unlike the "New Deal" era food stamp program, food purchases were not limited to surplus commodities. By 1964, as part of the "War on Poverty," pilot programs were operating in 40 counties and 3 cities with 380,000 participants. Congress established a permanent program in August of 1964 by passing The Food Stamp Act of 1964.

The 1964 program was already different in underlying intent from the Depression-era program as there was no longer a requirement for the specific purchase of agricultural surplus (commodity foods) by the consumer, although the counties that were implementing the program could still choose a commodity distribution version of the program. The underlying policy goal of supporting farming income that was important in the "New Deal" era was no longer an emphasis. The new food stamp program was based on the idea that participants simply didn't have enough money to buy an adequate diet. Participants would purchase stamps in the amount they would normally spend on food, and in return, would receive stamps worth a greater amount (an amount that was judged to be more likely to be capable of purchasing a nutritionally adequate diet). However, even though the cost to purchase food stamps was on a sliding scale, such an upfront purchase was not always feasible for those in greatest need.[52]

The failure of efforts by federal food assistance to address hunger in the United States in a meaningful way during this time period was highlighted by media and social-political events such the airing of the CBS television documentary "Hunger in America" and the 1968 encampment in Washington DC by the Poor People's Campaign and their demands of free food stamps to the poorest and reduced price stamps to all. In late 1969, in response to the increased social pressure, the Secretary of Agriculture set a national standard for food stamp benefit levels, decreased the purchase price of stamps and made the stamps free for those of the lowest income group. Legislation in 1970 then turned these new rules and a national standard for eligibility into law. The result of these program criteria, particularly the national standard for eligibility, created essentially a minimum income available to any U.S citizen.[52]

The program expanded rapidly—in the 5 years from 1969 to 1974, participation rose from 2.9 million to 12.9 million and expenditures from a modest 240 million to almost 3 billion. Food inflation and the 1974–1975 recession further contributed to program expansion, with over 20 million people participating by the end of the 1970s.[52]

The shift from food to income transfer is exemplified by the removal of the requirement to purchase the stamps in the 1977 Food Stamp Act,[53] as well as by the replacement of the physical stamps with an electronic benefits transfer card (EBT) for all participants by 2004. The EBT card functions in the same "swipe" manner that a debit card does and was introduced in part to help minimize the potential social stigma of obvious food stamp usage.

Following this technology improvement, in 2008, instead of just amending the 1977 Food Stamp Act, the name of the legislative act itself was changed to the Food and Nutrition Act of 2008 (FNA) in a further attempt to remove the stigma of "food stamps." Additionally, the USDA stresses that SNAP is not welfare, but is instead a "nutrition assistance program designed to help low-income individuals and families buy and consume nutritious foods."[54]

Food or money?

The expansion of food assistance in the 1960s and 1970s was rooted in a widely perceived problem of hunger. The recent expansion in SNAP between 2008 and 2011 indicates that SNAP is again filling a need in our society, but the need for income support may not be as clearly perceived to be a social need in the same manner as hunger.

Discerning whether the underlying goal of SNAP is income support or hunger alleviation is significant. This distinction plays into the social tension over how the money is spent—and what kind of food is purchased. *Does it matter what kind of calories people acquire using SNAP? Why?*

The USDA Thrifty Food Plan: The Basis for SNAP Allocations

The intent of the 1964 act establishing the food stamp program was to enable recipients "to more readily obtain a low-cost nutritionally adequate diet." The sliding scale concept of eligibility for purchase of stamps, however, meant that even though lower income participants paid less for the stamps they in turn received less benefits. The opportunity of obtaining a low-cost nutritionally adequate diet was thus hamstrung by the program design. In 1971, the program's intent was modified to "an opportunity to obtain a nutritionally adequate diet" and the sliding scale of what participants normally spent on food purchases changed into an allotment basis derived from the cost of an "economy food plan" model. The "economy food plan" was a dietary plan created by the USDA as a nutritionally adequate model suitable for short-term, emergency use. The USDA felt that that on average, low-income people spending the amount of money it would cost to eat on the "economy food plan" could achieve a nutritionally adequate diet.[52]

No guarantees

However, in June of 1975, the U.S. Court of Appeals handed down a legal decision that pointed out the 1971 amendment was a shift in policy from supplementing diets toward one of actually guaranteeing an adequate diet. To get out of hot water caused by an implicit guarantee, the USDA created a new food plan for a cost-basis comparison (the Thrifty Food Plan) and Congress revised the wording of the preamble of the 1971 Act into "to permit low-income households to obtain a more nutritious diet." By replacing the word opportunity with permit, this statement backed away from evoking guarantee and retreated to the concept of "supplementation."[52]

Impact of nutrient fortification: More bread, less vegetables

In general, the USDA food plans are a means of estimating food needs and costs of families and population groups. They are updated periodically to incorporate current nutrition and economic research, as well as changing marketplace conditions. For example, between 1964 and 1974 there was an increased availability of fortified breakfast cereals and ready-made baked products, which meant greater general access to certain nutrients, such as iron, was available through grains and breads.

Because the USDA food plan was focusing on discrete necessary nutrient needs as stated in the 1974 Recommended Dietary Allowances, this increased availability of fortified products meant that the new "Thrifty" food plan could increase the usage of lower-cost bread and grain products to meet nutrient needs and reduce the amount of vegetables.

This reductionist diet based on nutrients was also justified under the idea that this consumption pattern (eating bread and grains more so than vegetables) resembled already existing food consumption patterns among low-income people.[55] Fortification thus became a double-edged sword. Meant to improve our diet nutritionally it also allows the diminishment of our diet in totality.

With the new 1974 RDAs in hand, the USDA performed a sophisticated computer analysis to find out if it was feasible to meet these new nutrient recommendations without spending additional money. The "Thrifty Food Plan" was found feasible through this analysis, and is now used as the baseline for allocation of food stamp funding to participants. Although the amount of funds necessary to purchase the "Thrifty Food Plan" diet is technically enough to purchase adequate nutrition, it is a challenge for any shopper without the same nutrition training, information and access to computer software to in fact, do so.

Thrifty food plan updates

The Thrifty Food Plan (TFP) has continued to be periodically modified to meet updated nutrition research, dietary guidance and consumption patterns (the last being available through NHANES data analysis) and is scheduled to be updated in conjunction with each new release of Dietary Guidelines,[56] although as of the writing of this book, the most recent update had been done in 2006.[57] The 2006 TFP provides an allotment of funds, (adjusted for inflation, but with no increase in purchasing power) based on what a family of four would need to purchase to meet the then current United States dietary guidance (2005 Dietary Guidelines), or a "market basket." Compared to the 1999 market basket, the 2006 market basket allocates funds to more vegetables (51%), more milk products (47%), and more fruits (21%), but less to grains (–18%), meat and beans (–29%), and other foods (–45%). The changes in 2006 were based on the encouragement of the 2005 Dietary Guidelines for increased intakes of fruit, vegetables, and fat-free or low-fat milk and milk products and the then My Pyramid recommendations of more of these food groups. It is important to note that the basket calculates a benefit allotment for what people should buy according to U.S. dietary guidance, but does not require they purchase these items.

The TFP is the basis for the maximum food stamp allotment. The TFP "market baskets" specify the types and quantities of foods that people could purchase to be prepared and consumed at home to obtain a nutritious diet at a minimal cost. The TFP is revised from time to time so that the "market baskets" contain foods relevant to the current US dietary guidance and consumption, the most recent revision being in 2006. There are 15 market baskets, one for each of 15 specific age-gender groups, and the cost is calculated each month, thereby providing the basis for inflation adjustments. Box 6.5 shows the TFP for one week based on April 2014 prices for 2- and 4-member families.

> ### BOX 6.5   THE MAXIMUM SNAP ALLOTMENT FOR
> ### ONE WEEK, BASED ON THE TFP IN APRIL 2014
>
> - $89.20 Family of two adults, 19–50 years
> - $84.60 Family of two older adults, 51–70 years
> - $129.90 Family of four, two adults 19–50 years + two children, 2–3 and 4–5 years
> - $148.90 Family of two adults 19–50 years + 2 older children, 6–8 and 11 years
>
> *Official USDA Food Plans: Cost of Food at Home at Four Levels,*
> *U.S. Average, April 2014, http://www.cnpp.usda.gov/Publications/*
> *FoodPlans/2012/CostofFoodApr2014.pdf, accessed September 20, 2014.*

Close, but not quite

The 2006 TFP update was done to show that it was possible to eat a nutritious diet (that of the then current Dietary Guidelines) at the maximum SNAP allotment. However, the USDA does have caveats as to the nutritional content, namely that the 2006 TFP basket allocation does not in fact actually meet the vitamin E and potassium recommendations for some age-gender groups and does not meet the sodium recommendation for many age-gender groups. Accommodating these specific nutrition guidelines would have resulted in market baskets very different from typical consumption habits (in the case of vitamin E and potassium) or would require changes in food manufacturing practices (in the case of sodium).[58]

In other words, even though the American diet typically consumed does not provide sufficient vitamin E and potassium, at least within the assumed socioeconomic range, SNAP is not purporting to financially support a change in consumption habits to improve the participant's dietary nutritional profile, even to meet Dietary Guidelines.

The 2010 Dietary Guidelines recommend that Americans consume more potassium, dietary fiber, vitamin D, and calcium, potassium being the most expensive nutrient. One study found that increasing potassium would add about $380 per year (just over a dollar a day) to the average consumer's food costs.[59]

There are numerous other issues with using the TFP as a basis for allotment. For example, the cost of food does vary across the country, but a price differential is only set for Hawaii and Alaska. And, since the adjustment rate is annual, if inflation rises quickly after adjustment date a loss of spending power will occur.[60] Such variations in cost and access mean that not everyone is able to make the assumed purchases at the assumed prices.

Additionally, a substantial investment of time is required to turn basic unprocessed foodstuffs into consumable food. And although the diet may be (mostly) nutritionally adequate, it is still inconsistent with general USDA dietary guidance as to seeking out a variety of fruits and vegetables. An argument can be made that the next USDA plan up the ladder, the Low-Cost Food Plan, more accurately reflects what foods actually costs and what low-income families actually normally spend on food.[61] As of July 2014, the average monthly food cost allocated for a family of four with two children between the ages of 6 and 11 years is $650.50 under the Thrifty Plan and $854.60 under the Low-Cost Plan.[62]

Perhaps the most telling point that the underlying goal of SNAP is income support is the fact that funds allocated for food are based on how much food items cost that low-income people were reporting they were buying—not necessarily the cost of the most nutritious, most healthy, most dietary-balanced type of food they should be aiming to buy. *Given that there is a actual food plan deemed nutritious developed as an allocation basis of how much one would need to spend, why doesn't SNAP just give vouchers like WIC for the food items comprising the plan?*

SNAP Food Purchases

Under SNAP, the participants select their own foods through normal retail channels of trade with very few restrictions on purchases, the major restrictions being that benefits cannot be used to buy alcohol or prepared hot foods ready for consumption. (States and counties can elect to provide access to restaurant food to elderly, disabled, and homeless SNAP recipients.) The program does contain a nutrition education component, designed to promote healthy food choices consistent with the *Dietary Guidelines for Americans*, but individuals are not required to participate in order to receive benefits. In FY2011, $375 million was spent on nutrition education via SNAP-Ed activities implemented by 52 agencies.[63]

However, even though the program provides nutrition education and incentives are being explored to encourage the purchase of healthy foods; there are no restrictions on "junk" food purchases. Items of limited nutritional value such as soda can be purchased with SNAP benefits.

This is not to say that all SNAP participants just buy junk food. Many participants do select healthy foods, including using EBT benefits at farmers' markets. Also, the foods selected by the participants may have greater specific nutritional benefits or be part of a more overall healthful diet pattern than foods offered by WIC or FDPIR. Unlike WIC, as there are no restrictions on characteristics or brands, if SNAP participants feel that organic milk is healthier, they can choose to buy organic milk with SNAP.

However, we do not actually know the proportions of "junk food" being purchased using SNAP. Even though current technology of bar codes and EBT cards could allow data mining of aggregate purchase data, Congress has not given the USDA the authority to collect this data.[64]

In any case, as the average American diet is relatively poor (recall the HEI dietary index discussion given earlier), even without data on items purchased under SNAP, it makes sense that SNAP recipients may not be purchasing more junk food than any other Americans.

On the other hand, the USDA does know how much funds the retailers are getting, not just the grocers and the farm market, but also the gas stations, convenience markets, and liquor stores. (Although liquor cannot be bought with SNAP benefits, such stores usually have a few purchasable items—chips, sodas, etc.) However, the USDA will not reveal this information and a legal case challenging the USDA on this is pending in the U.S. 8th Circuit of Appeals.[64] As of August 2014, the USDA is soliciting public input about how to provide transparency in retailer data while remaining within legal bounds.[65]

SNAP and "Healthy Food"

In September of 2012, SNAP's website touted: *We put healthy food on the table for more than 46 million people each month.* In the same month, an article in the *Los Angeles Times* cited a study using supermarket loyalty card data that asserted that SNAP benefits might be used to buy up to $2.1 billion a year in sugar-sweetened beverages.[66] The article stirred up a controversy as to the extent of "junk" food being purchased using SNAP benefits. The tagline "healthy food on the table" disappeared from SNAP's website sometime thereafter.

SNAP purchases have no nutrition or health criteria attached. SNAP purchases can, and clearly do, include soda, candy and chips, with an estimated annual nationwide purchase of sugar-sweetened beverages in the amount of 1.7–2.1 billion.[67] *Would it make sense to at least limit the purchase of sugar-sweetened beverages?* The 2010 Dietary Guidelines specifically recommend reducing the intake of sugar-sweetened beverages.

SNAP is regulated under The Food and Nutrition Act of 2008 (FNA). The FNA provides a definition of "food" in the self-referential "we know it when we see it" manner. The FNA defines "food" as "any food or food product for home consumption"— this includes seeds and plants for use in gardens for personal consumption—"except alcoholic beverages, tobacco, and hot foods or hot food products ready for immediate consumption." Exclusions are made for hot meals for congregate feeding programs, and for special populations, such as the homeless, disabled, and elderly.

Beyond these exclusions, food is not further defined. Any item not specifically otherwise excluded therefore falls under the definition of "food." Thus, soda, candy and chips can be bought with SNAP benefits, and the USDA points out that since the current definition of food is a specific part of the Act, any change to this definition would require action by Congress.

Nutritious but "luxury" foods such as seafood and steak can also be purchased. Previously, Congress had considered placing limits on the types of food that could be purchased, but determined at the time that "designating foods as luxury or nonnutritious would be administratively costly and burdensome."[68]

However, with current technology it seems that placing some kind of limitations on food purchase would be possible. In fact, it seems much more feasible when compared to previous limitations, such as the ban on purchasing imported foods written into the Food Stamp Act of 1964.[69] Of course, excluding a type of food, such as soda, from being purchased through SNAP benefits doesn't mean that the participant will stop soda consumption, as they could still spend out-of-pocket for the beverage. In 2004, Minnesota proposed a restriction on junk food, which was rejected by the USDA as it could perpetuate the stereotype that food stamp users made poor food purchasing decisions. In 2011, New York City proposed a 2-year pilot program to prohibit buying soda with SNAP benefits, but the USDA rejected this as well on similar grounds. Some argue that we should not restrict dietary choices, but instead promote the purchase of healthier items.[70] *Why is this argument that we should not restrict dietary choices articulated for SNAP but not for WIC?*

Soda is not the only controversial food item. Energy drinks that have a nutrition facts label are also SNAP eligible food items. (Energy drinks that have a supplement label are considered supplements by the FDA and are not eligible to be purchased with SNAP benefits.) In fact, Monster energy drinks did indeed change their label in 2013 to include nutrition facts, in part to qualify the product for SNAP purchases.[71]

Coupled with the attention in the media on the social cost of long-term dietary patterns that include high levels of junk-food and soda consumption, tax payers may feel like they are paying on both ends. Socially, the consumption of junk food and soda isn't equated with hunger.

SNAP: Hunger and Community Benefits

As of September 2014, the USDA describes SNAP as follows: "SNAP offers nutrition assistance to millions of eligible, low-income individuals and families and provides economic benefits to communities. SNAP is the largest program in the domestic hunger safety net,"[72] generally promoting SNAP's ability to address hunger rather than nutrition.

More information is available in the introduction to the 2008 Food and Nutrition Act. The introductory purpose of the act is as follows: "To strengthen the agricultural economy; to help to achieve a fuller and more effective use of food abundances; to provide for improved levels of nutrition among low-income households through a cooperative Federal-State program of food assistance to be operated through normal channels of trade; and for other purposes."[73]

Even though we may think of the SNAP program simply as a means of food-purchasing assistance for individuals, the underlying purpose of the 2008 Act incorporates additional elements of food policy. Specifically, this act also emphasizes the support of agriculture (a logical approach for a USDA program, given the mission of the USDA to support agriculture) and the support of a normal trade channel which means that retailers, local or otherwise, healthy-food oriented or otherwise, reap an economic benefit from food being purchased through a retail channel, rather than distributed from the government.

In other words, even though SNAP does not impact "what" people eat in the specific nutritional manner that WIC and FDPIR can, SNAP as food policy does influence "who eats and to what impact"—and the beneficiaries of the impact are not solely the consumer. Convenience stores see enough of a benefit to lobby the USDA not to reveal how much each store authorized to accept benefits receives, with one trade association alone spending $1 million on lobbying in the first quarter of 2012.[74] As of September 2013, over 250,000 firms (not individual retail outlets) were authorized by the USDA to participate in SNAP, an increase of over 31% since FY 2009. The USDA is willing to admit that in FY2013, 82% of all SNAP benefits were redeemed at supermarkets or supercenters.[75]

The federal government is able to leverage the increased participation among retailers, however. The 2014 Farm bill now requires SNAP retailers to stock more fresh foods and to pay for their electronic benefit transfer (EBT) machines. The 2014 Farm bill also includes $100 million in funding (over ten years) for Food Insecurity Nutrition Incentive grants, to support programs offering bonus incentives for SNAP purchases of fruits and vegetables.[76]

SNAP and WIC Population Overlap and Participation

The goals of SNAP and WIC programs are different, but there is potential overlap in the low-income population served. If all criteria of WIC are also met, SNAP participants can participate in both programs at the same time. There is also overlap between FDPIR and SNAP populations; and although one of the original goals of FDPIR was to make food assistance available as an alternative to the food stamp program, SNAP has become easier to access given expansion in numbers and type of participating retailers. Although the participants cannot be on both FDPIR and SNAP programs at the same, they can move back and forth with relative fluidity. For example, FDPIR participants may choose to use instead SNAP to stock up 1 month using food stamps on staples, such as coffee, that are not included in the FDPIR package.[77]

Just as the implementation of FDPIR is impacted by the structural environment participants inhabit (limitations of rural access, etc.), the benefits provided by WIC and SNAP can also be influenced by where a participant lives. As WIC is federally funded but implemented through state agencies, the implementation and cost containment strategies can vary widely, thus the states' choice of available foods for the participants can vary—not at the specific nutritional level but in other areas, such as brands, ingredients, and whether conventional or organic. Even though the food may be equivalent in nutrient quality the other factors can contribute to perceived quality, healthfulness and taste preferences, thereby influencing participation by potential beneficiaries. SNAP, on the other hand, is abstractly egalitarian in benefits as the decision of what to purchase is left to the participant, but the participants' choices of what to purchase can also be influenced by which SNAP retailers are accessible to the participant and what those retailers choose to carry in stock.

SNAP or WIC?

SNAP participation is anticipated to increase through 2014 and start to decline in 2015,[78] yet WIC enrollment peaked at 9.3 million in August 2009. As of June 2014, 8.1 million mothers and children under 5 are participating in WIC. Although this decrease in participation is also reflective of a decrease in U.S. birthrate since 2007,[79] some question if the ease of acquiring and using SNAP benefits over WIC benefits is enticing women who are eligible for both to not enroll in WIC. To receive WIC benefits, participants are required to see a health professional to discuss nutrition, and in most states must reenroll every 6 months once the child is more than a year old. SNAP benefits can be considerably more generous and much easier to use with less stigma due to EBT transactions versus presenting vouchers that must match to specific foods. (The 2010 HHFKA is promoting universal adoption of EBT for WIC by mandating that all WIC state agencies implement EBT systems by October 1, 2020.)[80]

Because WIC has been shown to be an effective cost-saver in the long run due to the beneficial outcome of the prenatal services,[81] the GAO has recommended that all pregnant women under 185% of the Federal poverty level be eligible. Because of the perceived long-term social good benefits resulting from the form of food policy implemented through WIC, a transformation of WIC to an entitlement-like funded program could be a beneficial and effective use of funds. This social benefit also makes it important to understand why some eligible women might choose not to participate in WIC, and calls into question why the hurdle to acquire benefits may be more difficult for WIC than for NLSP or for SNAP. *Why not streamline enrollment in WIC to all women and children within the age and income-eligible category?*

CONCLUSION

Federal food assistance programs arose at different times, for different social and political needs, and the programs continue to be modified as health concerns, dietary guidance, and economic and political conditions change. In other words, we do not have a unified thematic policy for food assistance in the United States. The "who eats what" details of each program, including funding mechanisms, and nutritional objectives (or lack thereof) reveal differing underlying policies.

The NSLP and WIC have nutritional objectives. These programs are interested in promoting future health and preventing future cost, as poor health outcomes due to diet tend to develop over time. Nutritional assistance programs and policies should be expected to change over time, as even a demographically targeted population will change over time. The changes in school meals reflect a social determination to improve children's diets, as does the updates to the WIC food packages.

SNAP lacks nutritional objectives; and thus can really be characterized as income assistance rather than nutrition assistance. The ability to choose soda under SNAP is controversial, because the consumption of soda is seen as leading to future poor health outcomes and increased cost; but the program is not intended to promote future health. Is it now time to update SNAP; at least to increase the benefits to the level of the Low-Cost Food Plan?

When it comes to nutrition, we all need assistance. No matter how we score our diet, whether through HEI or NHANES, the average American diet is nutritionally poor. Looking forward, how can we leverage federal food assistance to improve everyone's diet?

APPENDIX: LIST AND DESCRIPTION OF THE 18 FOOD AND NUTRITION ASSISTANCE PROGRAMS IDENTIFIED BY THE GAO, GROUPED BY AGENCY WITH COSTS

United States Department of Agriculture

1. Child and Adult Care Food Program (CACFP)
   CACFP supports the provision of nutritious meals and snacks to children and adults who attend eligible day care or outside-school-hours-care programs. 3.3 million children and 120,000 adults are served daily. The current form of CACFP was authorized in the 1989 National School Lunch Act. An earlier version was piloted in 1968 and made permanent in 1978. Adult day care centers were added in 1987 and emergency shelters in 1999. In addition to cash reimbursement, USDA makes donated agricultural foods or cash-in-lieu of donated foods available to institutions participating in CACFP.[82]
   2013 cost: $2.9 billion.[83]

2. Commodity Supplemental Food Program (CSFP)
   CFSP provides nutritious USDA (commodity) foods to low-income elderly persons at least 60 years of age. CSFP food packages do not provide a complete diet, but rather are good sources of some nutrients typically lacking in the diets of the target population. Among the foods provided are canned fruits and vegetables, canned meats, bottled unsweetened fruit, and dried beans. Until February 6, 2014, CSFP also provided commodity foods to low-income pregnant and breastfeeding women, other new mothers up to 1 year postpartum, infants, children up to age six. The Agricultural Act of 2014 (Farm Bill) amended the eligibility requirements of the CSFP. New women and children enrollees will not be accepted, but those that were already participating in the program may receive benefits until their eligibility ends. The program is authorized under Section 4(a) of the 1973 Farm Bill (Agriculture and Consumer Protection Act of 1973).[84]
   2013 cost: $202 million.[85]

3. Community Food Projects Competitive Grant Program
   This program was initially authorized in the 1996 Farm Bill legislation. It provides support to communities, rather than individuals, to promote self sufficiency and food security in low-income communities. Three types of grants will be funded in FY2014: (1) Community Food Projects (CFP), (2) Planning Projects (PP) and (3) Training and Technical Assistance (T & TA) Projects. Grants can be up to $250,000 and may require matching dollars.[86]
   2014 cost: $5 million available in 2014, prior funding was authorized through 2012.[87]

4. Food Distribution Program on Indian Reservations (FDPIR)

FDPIR provides commodity foods to low-income households living on Indian reservations, and to Native American families residing in designated areas near reservations. This program grew out of the 1936 Needy Family Program, a state-administered program distributing surplus agriculture commodities. By the 1950s, five commodity items were provided: rice, cornmeal, flour, dry beans, and nonfat dry milk. Rolled oats, canned luncheon meat, fruits, vegetables, and juices were added in the 1960s. The Needy Family Program became less relevant when the Food Stamp program started in 1964, and the Food Stamp Act of 1977 created the FDPIR to replace the Needy Family Program specifically on reservations and to serve as an alternative for food stamps as there was concern about the distances people would have to travel to acquire and use food stamps. The package has been modified periodically to meet nutritional standards and incorporate culturally preferred foods. The modern package has a higher nutritional rating than the standard diet of average Americans.[32]

2013 cost: $102 million.[85]

5. Fresh Fruit and Vegetable Program

Encourages consumption of fruits and vegetables by making free fruit and vegetables available at no cost to all children in participating elementary schools (the schools that have the highest free and reduced price enrollment for the National School Lunch Program). The fruit and vegetables are distributed as snacks outside of meal time. Piloted in 2002, it was established as permanent program in the 2004 Child Nutrition and WIC Reauthorization Act, and expanded to selected schools in all 50 states in the 2008 Farm Bill.[88] This is a different program from the Department of Defense (DOD) Fruit and Vegetable Program; in which the USDA allocates moneys to schools for their use to buy additional fruit and vegetables than those available from the USDA by purchasing through DOD procurement and delivery channels.

2013–2014 school year cost: $165.5 million.[88]

6. National School Lunch Program

The National School Lunch Program (NSLP) is a federally assisted meal program operating in public and nonprofit private schools and residential child care institutions. All meals served under the program receive federal subsidies. It provides nutritionally balanced, low-cost or free lunches to children each school day. Most of the support USDA provides to schools comes in the form of a cash reimbursement for each meal served. As of 2013, the basic cash reimbursement rates, if school food authorities served less than 60% free and reduced-price lunches during the second preceding school year are as follows:

- Free lunches (and snacks): $2.86 ($0.78)
- Reduced-price lunches (and snacks): $2.46 ($0.39)
- Paid lunches (and snacks): $0.27 ($0.07)

School food authorities that are certified to be in compliance with the most recent meal requirements receive an additional 6 cents of reimbursement for each meal served.

Higher reimbursement rates are also in effect for Alaska and Hawaii, as well as for schools with high percentages of low-income students. The program was established under the National School Lunch Act, signed by President Harry Truman in 1946.[89]

2013 cost: $11.1 billion.[90]

7. Nutrition Assistance Block Grants (Puerto Rico, American Samoa, Northern Mariana Islands)

   Provides food assistance to low-income households in the U.S. territories of the Commonwealth of Puerto Rico, American Samoa, and the Northern Mariana Islands, in lieu of the Food Stamp Program as of 1980 (1982 for Puerto Rico). The 2008 Farm bill funded a study of the feasibility of including Puerto Rico in SNAP. Currently Puerto Rico distributes the benefit through an electronic debit card, like SNAP, but up to 25% may be redeemed for cash. The 2014 Farm Bill is phasing out the provision of cash, but with exceptions for vulnerable populations and is funding a feasibility study on SNAP participation for the Northern Mariana Islands.[76]

   2014 cost: $2 billion for Puerto Rico, $8 million for American Samoa, $12 million for Northern Mariana Islands.[91]

8. School Breakfast Program

   Provides cash assistance to States to operate nonprofit breakfast programs in schools and residential childcare institutions. Established in 1966 as a 2-year pilot project, it was permanently authorized in 1975. As part of the legislation making the SBP permanent, Congress declared its intent that the program "be made available in all schools where it is needed to provide adequate nutrition for children in attendance."[92]

   2013 cost: $3.5 billion.[93]

9. Senior Farmers' Market Nutrition Program

   The Senior Farmers' Market Nutrition Program (SFMNP) awards grants to States, United States Territories, and federally-recognized Indian tribal governments to provide low-income seniors with coupons that can be exchanged for eligible foods (fruits, vegetables, honey, and fresh-cut herbs) at farmers' markets, roadside stands, and community supported agriculture programs. This program aims to improve nutrition through increased consumption of fresh produce while also supporting farmers' markets. Piloted in 2001, it received funding from the 2002 Farm Bill, the 2008 Farm Bill, and again in the 2014 Farm Bill.[94]

   2013 cost: $21.12 million.[95]

10. Special Milk Program

    Provides milk to children in schools and childcare institutions who do not participate in other Federal meal service programs. The program reimburses schools for the milk they serve. It was established in 1966 under the Child Nutrition Act.[96]

    2013 cost: $10.7 million.[97]

11. Summer Food Service Program

    Federal resource available for local sponsors who want to provide meals and snacks with a summer activity program. Piloted in 1968, it was authorized in

1975. Like the National School Lunch Program, it has particular nutritional requirements that must be met in order for the program to be reimbursed for the meal.[98]

2013 cost: $373 million.[99]

12. Supplemental Nutrition Assistance Program (SNAP)

Supplemental Nutrition Assistance Program (SNAP), formerly known as "food stamps," provides financial assistance specifically for purchasing food. Food stamps were first introduced during 1939–1943 to address the problem of unmarketable food agriculture surpluses during wide spread unemployment. A version of a food stamp program was piloted again from 1961–1964 and then permanently established with the Food Stamp Act of 1964.[100] The program now provides an EBT (debit card) card instead of actual "stamps," and an increasing number of retailers have signed on to become authorized retailers. Unlike other federal food assistance programs, SNAP does not currently have specific nutritional standards for participant individuals and households. The 2014 Farm Bill added a number of provisions regarding EBT equipment that would allow SNAP to be used in more direct-to-consumer marketing channels; such as being able use the benefits to participate in Community Supported Agriculture (CSAs) ventures, as well as a provision authorizing pilot projects to test online and mobile technologies for purchases made with EBT.[101]

2013 cost: $79.9 billion.[102]

13. The Emergency Food Assistance Program (TEFAP)

The TEFAP program makes commodity foods available to the States. USDA buys the food, including processing and packaging, and ships it to the States. The amount received by each State depends on its low-income and unemployed population. States provide the food to local agencies that they have selected, usually food banks, which in turn, distribute the food to soup kitchens and food pantries that directly serve the public. TEFAP was first authorized in 1981 to distribute surplus food, the goal being to reduce huge federal food inventories and storage costs while at the same time helping low-income persons. The pressure to distribute the foods meant that for the first time, emergency feeding programs were guaranteed some form of commodity assistance. However, by 1988, stocks of some surplus food had been depleted. With the Hunger Prevention Act of 1988, the focus shifted to providing funds to the USDA to purchase commodities specifically for TEFAP. Foods acquired with appropriated funds are in addition to any surplus foods donated to TEFAP by USDA.[103] Although this program supports feeding the hungry, it also acts as a market stabilizer for over-production of foods. If the market can absorb produced food, less food will be pulled off to allocate for domestic hunger.

2013 cost: $691 million.[85]

14. WIC (Supplemental Nutrition Program for Women, Infants and Children)

WIC was piloted in 1972, and made permanent in 1974. WIC serves to safeguard the health of low-income women, infants, and children up to age 5 who are at nutritional risk. A nutritional risk assessment must be made by a health professional. WIC provides vouchers for purchase of specific food items intended to

support the nutritional needs of the target population. WIC provides grants to States to support distribution of supplemental foods, health care referrals, and nutrition education. Most State WIC programs provide vouchers that participants use to acquire supplemental food packages at authorized food stores. WIC is not an entitlement program, as Congress does not set aside funds to allow every eligible individual to participate in the program. Instead, Congress authorizes a specific amount of funds each year, which could entail a waiting list if more eligible people were seeking assistance than funds could support.[104]
2013 cost: $6.45 billion.[105]

15. WIC Farmers' Market Nutrition Program
Congress established the WIC Farmers' Market Nutrition Program (FMNP) in 1992. This program makes fresh, unprepared, locally grown fruits and vegetables from local farmers markets available to WIC recipients.[106]
2014 funding: 16.5 million.[107]

DHS Federal Emergency Management Agency (FEMA)

16. Emergency Food and Shelter National Board Program
The Emergency Food and Shelter National Board Program (EFSP) is a Federal program administered by the U.S. Department of Homeland Security's Federal Emergency Management Agency (FEMA) and has been entrusted through the McKinney-Vento Homeless Assistance Act of 1987 (PL 100–77) "to supplement and expand ongoing efforts to provide shelter, food and supportive services" for hungry and homeless people across the nation. A National Board, chaired by FEMA, governs the EFSP representatives from American Red Cross; Catholic Charities USA; National Council of the Churches of Christ in the U.S.A.; The Salvation Army; United Jewish Communities; and, United Way of America.
2013 cost: estimated assistance provided in the amount of $113,804,918.[108]

HHS Administration on Aging

17. Elderly Nutrition Program: Home-Delivered and Congregate Nutrition Services
This is the largest Older Americans Act program; other OAA programs include Elder Rights Protection, Home and Community Based Long-Term Care, and Health and Wellness programs. Elderly nutrition is supported in two areas, Congregate Nutrition Services, established in 1972 and Home-Delivered Nutrition Services, established in 1978. Services are targeted to those in greatest social and economic need with particular attention to low income individuals, minority individuals, those in rural communities, those with limited English proficiency and those at risk of institutional care.[109]
2013 cost: $621.59 million.[110]

18. Grants to American Indian, Alaskan Native, and Native Hawaiian Organizations for Nutrition and Supportive Services
Nutrition and Supportive Services for Native Americans programs were first established in 1978. Grants are provided to eligible Tribal organizations to promote the delivery of home and community-based supportive services, including nutrition services and support for family and informal caregivers, to Native American, Alaskan Native, and Native Hawaiian elders.[111]
2013 cost: $26.16 million[110]

REFERENCES

1. Johnson LB. Remarks upon signing the food stamp act. August 31, 1964. Available at http://www.presidency.ucsb.edu/ws/?pid=26472, accessed November 18, 2014.
2. GAO, *Domestic Food Assistance: Complex System Benefits Millions, but Additional Efforts Could Address Potential Inefficiency and Overlap among Smaller Programs,* GAO-10-346 Washington, DC: U.S. Government Printing Office, 2010
3. H.R. 2122. The GPRA Modernization Act of 2010. January 5, 2010, http://www.gpo.gov/fdsys/pkg/BILLS-111hr2142enr/pdf/BILLS-111hr2142enr.pdf, accessed October 6, 2014.
4. United States Department of Agriculture, Strategic Plan for FY 2010–2015. Available at http://www.ocfo.usda.gov/usdasp/sp2014/usda-strategic-plan-fy-2014-2018.pdf, accessed November 19, 2014.
5. United States Department of Agriculture, USDA Mission Areas, Food and Nutrition Service, last updated March 23, 2012. Available at http://www.usda.gov/wps/portal/usda/usdahome?contentidonly=true&contentid=missionarea_FNC.xml, accessed October 6, 2014.
6. United States Department of Agriculture, Food and Nutrition Service, About FNS. Available at http://www.fns.usda.gov/about-fns, accessed September 9, 2014.
7. Oliveira V. The Food Assistance Landscape: FY 2011 Annual Report, EIB-93, USDA, Economic Research Service. Available at: http://www.ers.usda.gov/media/376910/eib93_1_.pdf, accessed October 6, 2014.
8. The Healthy, Hunger-Free Kids Act of 2010. P.L. 111-296. Available at http://www.gpo.gov/fdsys/pkg/PLAW-111publ296/pdf/PLAW-111publ296.pdf, accessed June 26, 2014.
9. Gunderson G. The National School Lunch Program Background and Development. Available at hthttp://www.fns.usda.gov/nslp/history_6, accessed August 8, 2014.
10. National Conference of State Legislatures, Healthy, Hunger-Free Kids Act of 2010 Summary. March 24, 2011. Available at http://www.ncsl.org/research/human-services/healthy-hunger-free-kids-act-of-2010-summary.aspx, accessed October 7, 2014.
11. Richard B. Russell National School Lunch Act, Public Law 79-396, 60 Stat. (1946): 239. Available at http://www.ssa.gov/OP_Home/comp2/F079-396.html, accessed June 24, 2014.
12. Fox MK, Hamilton W, Lin B-H. *Effects of Food Assistance and Nutrition Programs On Nutrition and Health: Volume 3, Literature Review.* FANRR19-3, U.S. Department of Agriculture, Economic Research Service, 2004. Available at http://www.ers.usda.gov/media/872926/fanrr19-3fm_002.pdf, accessed May 10, 2015.
13. United States Department of Agriculture. FNS. School Meals, Rates of Reimbursement. Last modified July 16, 2014, http://www.fns.usda.gov/school-meals/rates-reimbursement, accessed October 7, 2014.
14. Levine S. *School Lunch Politics: The Surprising History of America's Favorite Welfare Program.* Princeton: Princeton University Press, 2008.

15. Healthy, Hunger-Free Kids Act of 2010, Public Law 111-296, 124 Stat. (2010): 3183–3266. Available at http://www.gpo.gov/fdsys/pkg/PLAW-111publ296/pdf/PLAW-111publ296. pdf, accessed June 26, 2014.

16. United States Department of Agriculture, FNS, Clarification of the Policy on Food Consumption outside of the Foodservice Area, and the Whole Grain Rich Requirement, April 23, 2014. Available at http://www.fns.usda.gov/sites/default/files/SP41-2014os.pdf, accessed September 15, 2014.

17. U.S. Department of Agriculture. Food and Nutrition Service. Child Nutrition Tables. National School Lunch Program, Participation and Lunches Served. Data as of November 07, 2014. Available at http://www.fns.usda.gov/pd/child-nutrition-tables, accessed November 19, 2014.

18. Go AS, Mozaffarian D, Roger VL, Benjamin EJ, Berry JD, Borden WB, Bravata DM et al. American Heart Association Statistics Committee and Stroke Statistics Subcommittee. Heart disease and stroke statistics—2013 update: A report from the American Heart Association. *Circulation.* 127, e6–e245, 2013. Available at: http://circ.ahajournals.org/content/127/1/e6.full.pdf, accessed January 7, 2014.

19. National Center for Education Statistics. Status and Trends in the Education of Racial and Ethnic Minorities. Available at http://nces.ed.gov/pubs2010/2010015/indicator2_7.asp#5, accessed November 19, 2014.

20. 7 CFR 210, Part A. National School Lunch Program, http://www.fns.usda.gov/sites/default/files/7cfr210_13_1.pdf, accessed October 7, 2014.

21. American Public Health Association. Correctional HealthCare Standards and Accreditation. 2004. Available at http://www.apha.org/legislative/policy/policysearch/index.cfm?fuseaction=view&id=1291, accessed October 7, 2014.

22. Greifinger R (ed.). *Public Health Behind Bars: From Prisons to Communities*, New York: Springer, 2007. Ramaswamy M. Freundenberg N. Health Promotion in Jails and Prisons: An Alternative Paradigm for Correctional Health Services Chapter 13, 229–248.

23. Poppendieck J. *Free for All: Fixing School Food in America.* Berkeley, CA: University of California Press, 2010.

24. Public Law 112-55. Consolidated and Further Continuing Appropriations Act, 2012.http://www.gpo.gov/fdsys/pkg/PLAW-112publ55/html/PLAW-112publ55.htm, accessed October 7, 2014.

25. Simon M. School Food politics: What's missing from the pizza-as-vegetable reporting. November 17, 2011. *Eat Drink Politics*, http://www.eatdrinkpolitics.com/2011/11/17/whats-missing-from-the-pizza-as-vegetable-reporting/, accessed October 7, 2014.

26. United States Department of Agriculture, FY2012 Budget Summary and Annual Performance Plan, http://www.obpa.usda.gov/budsum/FY12budsum.pdf, accessed September 9, 2014.

27. United States Department of Agriculture, Food and Nutrition Service, SNAP Annual Summary, Supplemental Nutrition Assistance Program Participation and Costs, Data as of September 5, 2014, http://www.fns.usda.gov/sites/default/files/pd/SNAPsummary.pdf, accessed September 9, 2014.

28. United States Department of Agriculture, FY 2015 Budget Summary and Annual Performance Plan, http://www.obpa.usda.gov/budsum/FY15budsum.pdf, accessed September 9, 2014.

29. Hanson K, Golan E. Issues in Food Assistance: Effects of Changes in Food Stamp Expenditures Across the U.S. Economy, 2002, United States Department of Agriculture, Economic Research Service, http://www.ers.usda.gov/publications/fanrr-food-assistance-nutrition-research-program/fanrr26-6.aspx#.VA9JzSiRndk, accessed September 9, 2014.

30. Hanson K. *The Food Assistance National Input-Output (FANIOM) Model and Stimulus Effects of SNAP* ERR-103. Washington, DC: Economic Research Service, United States

Department of Agriculture, October 2010, http://www.ers.usda.gov/publications/err-economic-research-report/err103.aspx#.VA9LTCiRndk, accessed September 9, 2014.

31. Dean S. Rosenbaum D. SNAP Benefits Will be Cut for All Participants in November 2013, Center on Budget and Policy Priorities, updated January 9, 2014, http://www.cbpp.org/cms/?fa=view&id=3899, accessed September 9, 2014.

32. Harper E. et al. FDPIR food package nutritional quality: Report to Congress, US Department of Agriculture, Food and Nutrition Service, Office of Research and Analysis, Special Nutrition Programs Report FD-08-FDPIR. Alexandria, VA: USDA November, 2008.

33. United States Department of Agriculture, Office of Budget and Program Analysis, FY 2013 FNCS Explanatory Notes, http://www.obpa.usda.gov/30fns2013notes.pdf, accessed September 9, 2014.

34. United States Department of Agriculture, FNS, Commodity Supplemental Food Program, last updated September 4, 2014, http://www.fns.usda.gov/csfp/commodity-supplemental-food-program-csfp, accessed September 9, 2014.

35. Staats E.E. Comptroller General, Entitlement Funding and Its Appropriateness for the WIC Program, Report to Sen, Thomas F. Eagleton, Chairman Senate Committee on Appropriations, April 13, 1978.

36. United States Department of Agriculture, Food and Nutrition Service, WIC Eligibility Priority System, last modified November 20, 2013, http://www.fns.usda.gov/wic/wic-eligibility-priority-system, accessed September 9, 2014.

37. Oliveira V et al. *The WIC Program: Background,Trends, and Issues,* Food and Rural Economics Division, Economic Research Service, U.S. Department of Agriculture. Food Assistance and Nutrition Research Report No. 27. Washington DC: USDA, 2002.

38. United States Agricultural Department, Food and Nutrition Services, WIC Food Packages, updated September 5, 2014, http://www.fns.usda.gov/wic/wic-food-packages-regulatory-requirements-wic-eligible-foods#PEANUT%20BUTTER, accessed September 9, 2014.

39. United States Agricultural Department, Food and Nutrition Services, WIC State Agency Food Lists, http://health.mo.gov/living/families/wic/wicfoods/pdf/MissouriWICApproved FoodList2014-2015.pdf, accessed September 9, 2014.

40. United States Agricultural Department, Center for Nutrition Policy and Promotion, Healthy Eating Index, http://www.cnpp.usda.gov/healthyeatingindex.htm, accessed September 9, 2014.

41. Cole N, Fox MK. Diet Quality of Americans by Food Stamp Participation Status: Data from the National Health and Nutrition Examination Survey, 1999–2004 Report submitted to the U.S. Department of Agriculture, Food and Nutrition Service. Cambridge, MA: Abt Associates, Inc., 2008.

42. Committee on Dietary Risk Assessment in the WIC Program,5 Food-Based Assessment of Dietary Intake. *Dietary Risk Assessment in the WIC Program.* Washington, DC: The National Academies Press, 2002.

43. United States Department of Agriculture, Food and Nutrition Service, Food Distribution Programs, updated February 25, 2014, http://www.fns.usda.gov/fdpir/about-fdpir, accessed September 9, 2014.

44. United States Department of Agriculture, Food and Nutrition Service, Food Distribution Programs, http://www.fns.usda.gov/fdpir/fdpir-food-package-review-work-group, accessed September 9, 2014.

45. United States Department of Agriculture, Food and Nutrition Service, WIC, updated September 8, 2014, http://www.fns.usda.gov/wic/final-rule-revisions-wic-food-packages, accessed September 9, 2014.

46. Institute of Medicine, *Planning a WIC Research Agenda: Workshop Summary.* Washington DC: The National Academies Press, 2011.

47. National Research Council, *Proposed Criteria for Selecting the WIC Food Packages: A Preliminary Report of the Committee to Review the WIC Food Package.* Washington DC: The National Academies Press, 2004.
48. United States Department of Agriculture, Food and Nutrition Service, WIC Food Packages Maximum Monthly Allowances, updated December 2, 2013, http://www.fns.usda.gov/wic/wic-food-packages-maximum-monthly-allowances, accessed September 9, 2014.
49. United States Department of Agriculture. Food and Nutrition Service. Women, Infants and Children. Final Rule: Revisions in the WIC package. http://www.fns.usda.gov/sites/default/files/03-04-14_WIC-Food-Packages-Final-Rule.pdf, accessed October 20, 2014.
50. National Research Council, *WIC Nutrition Risk Criteria: A Scientific Assessment.* Washington, DC: The National Academies Press, 1996.
51. National Research Council, *Dietary Risk Assessment in the WIC Program.* Washington, DC: The National Academies Press, 2002.
52. Andrews MS, Clancy KL. The Political Economy of the Food Stamp Program in the United States, In *Food Subsidies in Developing Countries: Costs, Benefits, and Policy Options*, Chapter 5. International Food Policy Research Institute. Baltimore: Johns Hopkins University Press, 1993, 61–78.
53. United States Department of Agriculture, Food and Nutrition Service, Supplemental Nutrition Assistance Program, Short History, updated November 20, 2013, http://www.fns.usda.gov/snap/short-history-snap, accessed September 9, 2014.
54. United States Department of Agriculture, SNAP, 10 Facts You Should Know about SNAP, updated July 25, 2013, http://origin.www.fns.usda.gov/snap/outreach/Translations/English/10facts.htm, accessed September 9, 2014.
55. Peterkin B. United States Department of Agriculture, Consumer and Food Economics Institute, Agriculture Research Service, *Food Plans for Poverty Measurement*, Technical Paper XII, 1976.
56. Supplemental Nutrition Assistance Program Benefits and the Thrifty Food Plan, Office of Inspector General, Audit Report 27703-1-KC December 2009.
57. United States Department of Agriculture, Center for Nutrition Policy and Promotion, USDA Food Plans: Cost of Food, http://www.cnpp.usda.gov/USDAFoodPlansCostofFood, accessed September 9, 2014.
58. Carlson A et al. *Thrifty Food Plan, 2006,* (CNPP-19) United States Department of Agriculture, Center for Nutrition Policy and Promotion, 2007.
59. Monsivais P, Aggarwal A, Drewnowski A. Food Policy: Following Federal Guidelines To Increase Nutrient Consumption May Lead To Higher Food Costs For Consumers, *Health Affairs.* 201130:81471–1477; published ahead of print August 3, 2011, doi:10.1377/hlthaff.2010.1273.
60. National Research Council, *Supplemental Nutrition Assistance Program: Examining the Evidence to Define Benefit Adequacy.* Washington, DC: The National Academies Press, 2013.
61. Food Research and Action Center, *Replacing the Thrifty Food Plan in Order to Provide Adequate Allotments for SNAP Beneficiaries,* December, 2012. http://frac.org/pdf/thrifty_food_plan_2012.pdf, accessed May 10, 2015.
62. United States Department of Agriculture, Official USDA Food Plans: Cost of Food at Home at Four Levels, U.S. Average, July 2014, http://www.cnpp.usda.gov/sites/default/files/usda_food_plans_cost_of_food/CostofFoodJul2014.pdf, accessed September 9, 2014.
63. Ben-Shalom Y, Fox MK, Newby PK. Characteristics and Dietary Patterns of Healthy and Less-Healthy Eaters in the Low-Income Population, U.S. Department of Agriculture, Food and Nutrition Service, Office of Research and Analysis. Alexandria, VA: 2012, http://www.fns.usda.gov/sites/default/files/HEI.pdf, accessed September 9, 2014.
64. Freyer FJ, Wielawksi IM, What Do Food Stamps Buy, *Los Angeles Times,* August 2, 2013, http://articles.latimes.com/2013/aug/02/opinion/la-oe-wielawski-food-stamps-20130802, accessed September 9, 2014.

65. United States Department of Agriculture, USDA seeks public input to increase transparency of SNAP retailer data, August 4, 2012, http://www.fns.usda.gov/pressrelease/2014/016814, accessed September 9, 2014.

66. Mestal R. Food Stamps buy up to $2.1 billion a year in sugary drinks, study says,*Los Angeles Times,* September 18, 2012, http://articles.latimes.com/2012/sep/18/news/la-heb-snap-food-program-billions-sugary-soft-drinks-20120918, accessed September 9, 2014.

67. Andreyeva T, Luedicke J, Kathryn E, Henderson K, Amanda S, Tripp A. Grocery Store Beverage Choices by Participants in Federal Food Assistance and Nutrition Programs. *Am. J. Prev. Med.* 2012;43:411–418.

68. United States Department of Agriculture, SNAP Eligible Food Items, last updated July 8, 2014, http://www.fns.usda.gov/snap/retailers/eligible.htm, accessed September 9, 2014.

69. United States Department of Agriculture, SNAP Legislative History, http://www.fns.usda.gov/sites/default/files/History_of_SNAP.pdf, accessed September 9, 2014.

70. McGeehan P. U.S. Rejects Mayor's Plan to Ban Use of Food Stamps to Buy Soda, *New York Times,* August 19, 2011, http://www.nytimes.com/2011/08/20/nyregion/ban-on-using-food-stamps-to-buy-soda-rejected-by-usda.html, accessed September 9, 2014.

71. Meier B. In a New Aisle, Energy Drinks Sidestep Some Rules, *New York Times,* March 19, 2013, http://www.nytimes.com/2013/03/20/business/in-a-new-aisle-energy-drinks-sidestep-rules.html?_r=0, accessed September 9, 2014.

72. United States Department of Agriculture, SNAP, http://www.fns.usda.gov/snap/supplemental-nutrition-assistance-program-snap, accessed September 9, 2014.

73. United States Department of Agriculture, Food and Nutrition Act of 2008; http://www.fns.usda.gov/sites/default/files/PL_110-246.pdf, accessed September 9, 2014.

74. Rosiak L. Top secret: $80B a year for food stamps, but feds won't reveal what's purchased, *The Washington Times,* June 24, 2012, http://www.washingtontimes.com/news/2012/jun/24/top-secret-what-food-stamps-buy/#ixzz2A4kVty1z, accessed September 9, 2014.

75. United States Department of Agriculture, SNAP Retailer Management 2013 Annual Report, http://www.fns.usda.gov/sites/default/files/snap/2013-annual-report.pdf, accessed September 9, 2014.

76. Aussenberg RA. *SNAP and Related Nutrition Provisions of the 2014 Farm Bill (P.L. 113-79),* April 24, 2014, Congressional Research Service, http://nationalaglawcenter.org/wp-content/uploads/assets/crs/R43332.pdf, accessed September 9, 2014.

77. Finegold K et al. *Tribal Food Assistance: A Comparison of the Food Distribution Program on Indian Reservations (FDPIR) and the Supplemental Nutrition Assistance Program,* 2009. The Urban Institute. Available at http://www.urban.org/research/publication/tribal-food-assistance, accessed May 10, 2015.

78. Congressional Budget Office, An Overview of the Supplemental Nutrition Program, April 19, 2012, http://www.cbo.gov/publication/43175, accessed September 9, 2014.

79. United States Department of Agriculture, WIC program data, last modified September 5, 2014, http://www.fns.usda.gov/pd/wicmain.htm, accessed September 9, 2014.

80. Special Supplemental Nutrition Program for Women, Infants and Children (WIC): Implementation of the Electronic Benefit Transfer-Related Provisions of Public Law 111-296, Proposed Rule. *Federal Register.* 2013;78(40):13549-13563. Available at: http://www.fns.usda.gov/wic/regspublished/EBT-proposedrule.pdf, accessed July 9, 2013.

81. Avruch S, Cackley AP. Savings achieved by giving WIC benefits to women prenatally. *Public Health Rep.*1995;110: 27–34, http://www.ncbi.nlm.nih.gov/pmc/articles/PMC1382070/.

82. United States Department of Agriculture, Child and Adult Care Food Program, last updated July 28, 2013, http://www.fns.usda.gov/cacfp/why-cacfp-important, accessed September 9, 2014.

83. United States Department of Agriculture, FNS, Child Nutrition Tables, last updated June 10, 2014, http://www.fns.usda.gov/sites/default/files/pd/ccsummar.pdf, accessed September 9, 2014.

84. United States Department of Agriculture, Commodity Supplemental Food Program, last updated September 4, 2014, http://www.fns.usda.gov/csfp/commodity-supplemental-food-program-csfp, accessed September 9, 2014.

85. United States Department of Agriculture, FNS Food Distribution tables, last updated September 5, 2014, http://www.fns.usda.gov/sites/default/files/pd/fd$sum.pdf, accessed September 9, 2014.

86. United States Department of Agriculture, National Institute of Food and Agriculture, last updated February 27, 2014, http://www.nifa.usda.gov/fo/communityfoodprojects.cfm, accessed September 9, 2014.

87. United States Department of Agriculture, National Institute of Food and Agriculture, last updated October 31, 2011, http://www.nifa.usda.gov/funding/cfp/cfp_synopsis.html, accessed September 9, 2014.

88. United States Department of Agriculture, FNS Fresh Fruit and Vegetable Program Fact Sheet, last updated August 20, 2014, http://www.fns.usda.gov/ffvp/fresh-fruit-and-vegetable-program, accessed September 9, 2014.

89. United States Department of Agriculture, FNS National School Lunch Program, last updated August 20, 2014, http://www.fns.usda.gov/nslp/national-school-lunch-program-nslp, accessed September 9, 2014.

90. United States Department of Agriculture, FNS Child Nutrition Tables, National School Lunch Program, last updated June 10, 2014, http://www.fns.usda.gov/sites/default/files/pd/cncost.pdf, accessed September 9, 2014.

91. United States Department of Agriculture, FNS budget 2014 explanatory notes, http://www.obpa.usda.gov/exnotes/FY2014/30fns2014notes.pdf, accessed September 9, 2014.

92. United States Department of Agriculture, School Breakfast Program, last updated August 20, 2014, http://www.fns.usda.gov/sbp, accessed September 9, 2014.

93. United States Department of Agriculture, FNS Child Nutrition Tables, School Breakfast Program, last updated June 10, 2014, http://www.fns.usda.gov/sites/default/files/pd/cncost.pdf, accessed September 9, 2014.

94. United States Department of Agriculture, Seniors Farmers' Market Nutrition Program, last updated August 28, 2014, http://www.fns.usda.gov/sfmnp/overview, accessed September 9, 2014.

95. United States Department of Agriculture, FNS, Seniors Farmers' Market Nutrition Program, Grant and Program Data, last updated May 9, 2014, http://www.fns.usda.gov/sfmnp/sfmnp-profiles-grants-and-participation, accessed September 9, 2014.

96. United States Department of Agriculture, Special Milk Program, last updated June 11, 2014, http://www.fns.usda.gov/smp/special-milk-program, accessed September 9, 2014.

97. United States Department of Agriculture, FNS Child Nutrition Tables, Special Milk Program, last updated June 10, 2014, http://www.fns.usda.gov/sites/default/files/pd/cncost.pdf, accessed September 9, 2014.

98. United States Department of Agriculture, FNS Summer Food Service Program, last updated July 19, 2013, http://www.fns.usda.gov/sfsp/program-history, accessed September 9, 2014.

99. United States Department of Agriculture, FNS Summer Food Service Participation and Cost, last updated September 5, 2014, http://www.fns.usda.gov/sites/default/files/pd/sfsummar.pdf, accessed September 9, 2014.

100. United States Department of Agriculture, FNS, Supplemental Nutrition Assistance Program, A Short History of Snap, last updated November 20, 2013, http://www.fns.usda.gov/snap/short-history-snap, accessed September 9, 2014.

101. National Sustainable Agriculture Coalition, 2014, Farm Bill Drilldown: Local and Regional Food Systems, Healthy Food Access, and Rural Development. February 11, 2014, http://sustainableagriculture.net/blog/2014-farmbill-local-rd-organic/, accessed September 18, 2014.

102. United States Department of Agriculture, FNS, Supplemental Nutrition Assistance Program, Participation and Costs, updated September 5, 2014, http://www.fns.usda.gov/sites/default/files/pd/SNAPsummary.pdf, accessed September 9, 2014.

103. United States Department of Agriculture, FNS, About TEFAP, last updated November 9, 2013, http://www.fns.usda.gov/tefap/about-tefap, accessed September 9, 2014.

104. United States Department of Agriculture, FNS, About WIC—WIC at a Glance, updated November 19, 2013, http://www.fns.usda.gov/wic/aboutwic/wicataglance.htm, accessed September 9, 2014.

105. United States Department of Agriculture, FNS, WIC Program Participation, last updated September 5, 2014, http://www.fns.usda.gov/sites/default/files//pd/wisummary.pdf, accessed September 9, 2014.

106. United States Department of Agriculture, FNS, WIC Farmers' Market Nutrition Program, last updated August 28, 2014, http://www.fns.usda.gov/fmnp/wic-farmers-market-nutrition-program-fmnp, accessed September 9, 2014.

107. United States Department of Agriculture, FNS, WIC Farmers' Market Nutrition Program Overview, last updated August 28, 2014, http://www.fns.usda.gov/fmnp/overview, accessed September 9, 2014.

108. General Services Administration, Catalog of Federal Domestic Assistance, EmergencyFood and Shelter National Board Program, https://www.cfda.gov/?s=program&mode=form&tab=step1&id=da6ee76bb87d6f1b6084c30354cf0e32, accessed September 9, 2014.

109. Department of Health and Human Services, Administration of Aging, Nutrition Services, http://www.aoa.gov/AoA_programs/HCLTC/Nutrition_Services/index.aspx, accessed September 9, 2014.

110. Administration for Community Living, Budget, last updated March 5, 2014, http://www.acl.gov/About_ACL/Budget/ACL-FY2014-funding-information.aspx, accessed September 9, 2014.

111. Department of Health and Human Services, Administration of Aging, Services for Native Americans, http://www.aoa.gov/aoa_programs/hcltc/native_americans/index.aspx, accessed September 9, 2014.

Milk*

7

*Remember that a quart of whole milk a day
for each child, to be used as a beverage
and in cookery, is not too much.*
C.L. Hunt and H.W. Atwater (1917)[1]

INTRODUCTION

This chapter takes a critical look at milk and dairy products. We examine the factors that have shaped the role of fluid milk (referred to simply as "milk") and other dairy products (evaporated milk, condensed milk, fat-free and low-fat milk, buttermilk, cheese, and yogurt, sour cream) in the American diet. We provide some historical context to dairy food production, processing, and consumption, the nutrient composition of dairy products, and the "Milk Reformers" agitation for "nature's perfect food." We review the evolution of dietary recommendations for milk and dairy products (or how milk became one of the basic food groups), we discuss challenges to the belief in milk as the perfect food, and we finish with policies for milk in federal food assistance programs. The history of dairy production and promotion in the United States provides context for making an informed decision regarding milk consumption and for developing related nutrition and health policies.

NATURE'S PERFECT FOOD

There has been a long-standing cultural belief in the United States that milk is "nature's perfect food." Yet, dairy products in the nineteenth century and in the beginning of the twentieth century were often produced under unsanitary conditions and carried harmful bacteria that wreaked havoc on child mortality rates across the country. With such a rocky start, how did milk become the "perfect" staple of the country's diet? The distance that dairy products have traveled since their introduction into the human diet

---

* Matthew Nulty, MPH, RDN and Lia Wallon, MPH contributed primary research and prepared draft versions of portions of this chapter.

is long, and in the United States, the dairy products we consume most often do not resemble their ancient or even colonial raw and sour predecessors. Examining modern milk consumption through a historic lens reveals why milk and dairy products came to occupy such a large and seemingly important part of the average American's diet and can inform future food and nutritional policies.[2,3]

Modern milk has traveled a long way from the colonial farmhouse, and even farther from the animals we keep for milk production. Dairy cows have been an important part of life in America since the first English settlers arrived in Jamestown in the early 1600s. Families that could keep two or more cows staggered the "dry" times to have milk year round. Cattle continued to move west with the settlers, and as towns grew, farmers kept more animals and sold any surplus milk they had. In the mid-1800s, as cities expanded, farms became further removed from consumers, and transportation of milk before it spoiled became a problem. The advances of the railroad, and the consequential "milk train" that stopped at each small town meant that milk could be transported as far as 50 miles by train. Sanitation and refrigeration remained problematic, however, and further technical solutions were required before dairy farming could become a major industry with national brands.[4,5]

For over a century, nutritionists, government agencies, and organizations like the American Dairy Council have promoted milk as "nature's perfect food." When we think about milk today, there is a general, albeit erroneous, assumption that it was always consumed cold and "sweet." But before the days of refrigeration and before science had discovered vitamins and minerals, dairy was more often consumed in a variety of fermented products such as buttermilk, yogurt, and curds, and used as feed for livestock. The cold and "sweet" modern version of milk and other dairy products actually gathered prominence in the 1850s when urban American mothers were encouraged to use it as a substitute for breast milk. The newly expanding number of city dwellers, perhaps themselves nostalgic for the perfection of nature, created the demand for fresh milk. Subsequently, a prominent dairy industry was created and along with it with a growing list of consequences: surpluses, shortages, sanitation regulations, waste disposal, and environmental contamination. The history culminates in a myth of the naturally occurring perfection of sweet, pasteurized, and homogenized milk.[4-6]

The Nutrient Composition of Milk

Milk and dairy products have been heralded as near perfect foods in part because they contain substantial quantities of many nutrients, such as protein and calcium, and in the case of fortified milk, vitamins A and D (see Table 7.1).

About Cows

A basic element of a cow's physiology is important to consider when examining modern milk in the United States. *Cows do not naturally produce milk year-round.* A cow produces milk during spring and summer when she is in calving season. The milk is produced to feed her calf. A cow, without artificial help, will produce very little, if

TABLE 7.1 Selected nutrients in 1 cup of vitamins A and D fortified milk

	TYPE OF MILK			
	FAT-FREE	*LOW-FAT 1%*	*REDUCED 2%*	*WHOLE*
Nutrients and % Daily Values				
Protein (g)	8	8	8	8
Lactose (milk sugar) (g)	12	13	12	12
Total Fat (g)	0	2.4	4.8	7.9
Saturated Fat (g)	0	1.5	3.1	4.6
Potassium (mg)	382	366	342	322
Potassium (%DV)	11%	11%	10%	9%
Vitamin A (IU)	500	478	464	395
Vitamin A (%DV)	10%	10%	13%	11%
Vitamin B_{12} (µg)	1.2	1	1.3	1.1
Vitamin B_{12} (DV)	20%	17%	22%	18%
Vitamin C (mg)	0	0	1	0
Vitamin C (%DV)	0%	0%	2%	0%
Vitamin D (IU)	115	117	120	124
Vitamin D (DV)	29%	30%	30%	31%
Calcium (mg)	299	305	293	276
Calcium (%DV)	30%	31%	29%	28%
Folate (µg)	12	12	12	12
Folate (%DV)	3%	3%	3%	3%
Iron (mg)	0.1	0.1	0.1	0.1
Iron (%DV)	1%	1%	1%	1%
Magnesium (mg)	27	27	27	24
Magnesium (% DV)	7%	7%	7%	6%
Phosphorus (mg)	247	232	224	205
Phosphorus (%DV)	25%	23%	22%	21%
Thiamin (Vitamin B_1) (mg)	0.1	0.1	0.1	0.1
Thiamin (%DV)	7%	7%	7%	7%
Riboflavin (Vitamin B_2) (mg)	0.45	0.45	0.45	0.41
Riboflavin (%DV)	26%	26%	26%	24%
Niacin (Vitamin B_3) (mg)	0.2	0.2	0.2	0.2
Niacin (%DV)	1%	1%	1%	1%

Source: For all nutrient values: U.S. Department of Agriculture, Agricultural Research Service. 2012. USDA *National Nutrient Database for Standard Reference, Release 25.* Nutrient Data Laboratory Home Page, http://www.ars.usda.gov/ba/bhnrc/ndl.

Notes: % DV = Daily Value, based on energy and nutrient recommendations for a general 2000-calorie diet. If a food has 5% or less of a nutrient, it is considered to be low in that nutrient; if it has 20% or more, it is considered to be a good source of that nutrient. There are no DVs for the energy nutrients.

Units: g = grams; mg = milligrams; µg = micrograms; IU = International Units.

any, milk during fall and winter months. As demand for milk has grown, so has the manipulation of cow's birthing cycles. It becomes more difficult to think of milk as "nature's perfect product" when artificial insemination, selective breeding, growth hormones, and grain-based feed are used to increase milk production year-round and milk is homogenized and ultra-pasteurized.

In 1950, there were approximately 22 million cows in the United States and the average yearly milk production per cow was 5314 pounds. In 2000, the number of cows had dropped by more than half to approximately 9 million cows, but yearly production had more than tripled to 18,204 pounds per cow.[7] In other words, 60% fewer cows produced 3½ times the milk. Given the astounding number of gallons produced, it is obvious that we have come a long way from days when families milked their own cows. *Why do we need this much milk? Who benefits from this increased milk production?*

"Nature's perfect food" has been unnaturally manipulated to fit modern nutritional, aesthetic, and mythical values. *Do nutritional guidelines concerning dairy products as recommended by the government represent actual nutritional interests of our population?*

The "Perfect Food": Nineteenth Century

For generations, the United States has regarded milk as the "perfect food." Where and why did this concept originate? Fluid milk's prominent role in our modern day consciousness emerged from an idealized concept of perfection that arose during the birth of the American republic at the end of the eighteenth century, from which the American people aimed to create a model society based upon Enlightenment Ideals for the world to praise for its perfection, and was then supported by the Second Great Awakening, a nineteenth century Christian revival movement. The Second Great Awakening rejected earlier held beliefs of predestination, and instead called for individual responsibility over moral behavior and intolerance of wickedness. A wide-cast net striving for moral reform to attain perfection transformed into social reform movements, often using the Bible for guidance on everyday life, including that of diet.[4]

Robert Hartley (1796–1881), America's first major pure milk agitator and likely the country's first consumer advocate, was the first American to make a sustained argument that milk was the perfect food. His lifelong activism, dedicated to social reform, emerged from temperance and mission societies in New York City in the early 1800s. Growing up in rural upstate New York, Hartley believed that the city was an epicenter of human degeneracy and that the city dwellers could improve their lives by living a pure life like those in the countryside. His keen interest in perfecting human society led him to investigate New York City's milk supply, which at the time was commonly contaminated and the cause of death in many infants. Hartley's 1842 *Essay on Milk* was a treatise not only on milk, but also on the ideal of perfection. His defense of milk's perfection was based on two Biblical claims: (1) consumption of milk is universal through time and space and (2) milk is the most complete source of nutrition available. His goal was to reconnect New York City, increasingly urbanized and depraved, with a food carrying the more virtuous ideals and values of country living.[4]

Milk for Infants and Children: The Decline of Breastfeeding and the Rise of Social Welfare

Ensuring a clean milk supply became critically necessary for infant survival in the growing urban centers in the mid-nineteenth century, as women of all classes were breastfeeding infants with greater infrequency and turning to milk instead for infant nourishment. Physicians, pediatricians, and psychologists promoted the idea that civilized women living in the industrialized world were increasingly physically incapable of breastfeeding; claiming that the animal instincts necessary to breastfeed went into decline as women adapted to life in the cities and became farther removed from their primitive ancestors, and artificial feeding was necessary to save their babies.

There was much debate over the origins of the reducing the ability to breastfeed. Some argued that it was physiological. John Spargo's 1908 book *The Common Sense of the Milk Question* even speculated that it might have been the tight corsets and other similar clothing that caused a decline in women's ability to produce milk. Others argued that it was simply a belief purported by those who thought breastfeeding was uncouth.[3,4]

Some reasons for the decline in breastfeeding included: a desire by men to keep their wives fertile, the middle classes wishing to emulate the upper classes found artificial feeding to be cheaper than employing a wet nurse, women who had to work outside of the home (sometimes themselves as wet nurses) being unable to breastfeed throughout the day, poor women could not afford to acquire enough food to assure a consistent supply of breast milk while conversely, upper class women were purposefully restricting their food intake to remain thin and therefore compromising their ability to produce breast milk. Cow's milk was the recommended replacement, as it was thought to be sufficiently similar to human breast milk.[2–4]

Cow's milk does share some similarities with human milk as the water, fat, and solids content are similar, but there are important differences that make it a less than "perfect food" for infants and children. Cow's milk has approximately three times as much protein as human milk, and a much greater percentage of casein protein, which may exacerbate allergic reactions. Fluid cow's milk has less total fat, but a higher percentage of short-chain fatty acids and fewer polyunsaturated fatty acids. There is less lactose in cow's milk compared to human breast milk and about double the potassium. For the human baby fed cow's milk, these differences affect the bioavailability, digestibility, and absorption of nutrients, as well as any potential benefits that might be reaped from nonnutritional factors.[8]

The nutritional differences between human and cow's milk were somewhat understood by scientists and physicians by the late 1800s but were not generally appreciated by the public. Diluting cow's milk with water and adding sugar and other substances before feeding to infants was common practice, but there were no standards or guidelines for preparing cow's milk as a substitute for breast milk, which meant that the amounts of water (often contaminated) and sugar added varied from mother to physician. Preparations of pasteurized milk diluted with lime-water and fortified with sugar of milk meant to be nutritionally equivalent to breast milk were created by doctors and served at Nathan Straus' milk depots in New York City.[9] (Straus was an early milk reformer concerned with children's welfare and is discussed later in the chapter.)

The newly burgeoning children welfare movement in the United States and some European countries helped to lead to the mega-promotion of milk. Infant and child mortality rates were rising (often as a result of tainted milk). Simultaneously, birth rates in industrialized nations such as Britain, France, the United States, and Australia were declining among the middle and upper class, while continuing steadily among lower classes. Many members of well-to-do society started to worry about a loss of the "civilized" white race.[3] Philanthropists, physicians, public health officials, social workers, and politicians used the child mortality statistics as a rallying cry to push whatever agenda was perceived superior, whether it be promoting artificial feeding or sanitizing the often lethal fluid milk supply.[2,4,6,10]

Making Milk Perfect: Science and Technology

The second half of the nineteenth century brought government involvement and technological advances, both of which supported increased milk production. In 1895, the Division of Agrostology and the Dairy Division were created under the USDA. These two divisions studied how different types of grasses affected milk color and odor, inspected dairy farms for cleanliness, and experimented with dairy product manufacturing. Later, as a result of dairy surpluses left over after World War I, the divisions organized educational campaigns to increase consumption of dairy products. This was just the beginning of the government's involvement in promoting dairy products to eliminate surpluses, which will be discussed in more detail later in the chapter.[11]

Dairy farming technology innovation began as early as the 1850s with rudimentary bucket and pitcher pumps, leading to increasingly sophisticated attempts of suction and vacuum pumps and milking tubes.[12] The "Pneumatic Foot Power Milker," patented in 1892, decreased the time it took to milk a cow by facilitating a farmer to milk two cows at once, thus enabling farmers to increase their herd size without increasing the number and cost of laborers needed to milk the cows.[13] The development of refrigerated rail cars, tuberculin testing for cows, pasteurization, homogenization, glass bottles, and automatic bottling equipment, all contributed to safe milk as well as the increasing ability to commercialize milk.[4,5,6,11]

The *germ theory of disease*, probably the single most important contribution of microbiology to medicine and public health, was developed in the middle of the nineteenth century. The discoveries of Snow, Pasteur, and Koch contributed to the understanding that microorganisms, too small to be seen without the aid of a microscope, can invade the body and cause certain diseases. Until the acceptance of the germ theory, diseases were often perceived as punishment for a person's evil behavior. When entire populations fell ill, the disease was often blamed on miasmas, swamp vapors, or foul odors from sewage. The work of these three men, along with others, led to a firm establishment of germ theory by the turn of the twentieth century.

- In the 1850s, English physician John Snow's (1813–1858) work connecting contaminated water to cholera outbreaks helped in bringing an end to the belief that cholera was caused by miasma (pollution or a noxious form of "bad air"), although his hypothesis that cholera was caused by contaminated water was not fully accepted until 1866, after Snow's death.[14]

- In the 1860s, French chemist and microbiologist Louis Pasteur (1822–1895) described the process of fermentation for the first time and invented the process of pasteurization, namely the method of using heat to prevent the spoilage of liquid foodstuffs, such as milk, wine, and beer. Pasteur found that through briefly heating the liquids, he could kill most bacteria and molds already present in the foods. Pasteurization provides the trade-off between applying enough heat for enough time to kill most microorganisms present to delay or prevent spoilage and applying so much heat that the foodstuff is negatively altered. Along with Robert Koch, Pasteur is often credited as the father of germ theory and bacteriology.[15]
- In the 1870s, German physician bacteriologist Robert Koch (1843–1910) proposed a sequence of experimental steps (Koch's Postulates) for directly associating a particular microbe with a specific disease. Using this technique, Koch discovered the bacteria that cause anthrax (*Bacillus anthracis*), tuberculosis (*Mycobacterium tuberculosis*), and cholera (*Vibrio cholera*).[16]

Lacking the knowledge of germ theory meant a lack of understanding of hygienic procedures to assure a safe milk supply or even to pinpoint the source of illness as contaminated milk. Urban dairies were often colocated with breweries feeding the fermented spent-grain ("swill") leftover from brewing beer to the cows. This milk was often whitened with chalk or other material to make up for the yellowish tint caused by the grain mix.[2] However, milk from outside city limits was not necessarily safer, as it traveled longer distances without refrigeration, passed through multiple contamination points, and the farm may not have been any cleaner than the city "brewery" dairies. Farmers would unknowingly spread disease by failing to clean a cow's udder or their own hands before milking. Cows were kept with other animals and often milked in the same area where they all slept, ate, and defecated. A farmer might store milk in unclean containers, letting it sit uncovered and collecting airborne debris and contaminants.[3–6]

During the mid to the end of the nineteenth century, high infant and children mortality rates were a problem across class lines, and mortality increased significantly in the summer when heat fostered the contamination of milk by lethal bacteria. Contaminated milk led to "summer complaint," an illness of diarrhea, dehydration, and often death. Milk was a particularly good host for such contamination because its high protein content created a helping hand in bacterial growth, allowing bacteria not only to live, but also to thrive.[3–6]

Although the developing understanding of the germ theory meant that cows could be tested for tuberculosis, and milk could be heat-treated to kill most, not all, disease-causing bacteria, large gaps in nutrition and sanitation knowledge still lingered. Two other important obstacles to a consistently safe milk supply for everyone, rich or poor, remained; namely, an effective means of educating the public and creating enforceable health and production regulations.[5]Along with the growing child welfare movement, several prominent milk reformers began championing for safe milk, and paved the way for twentieth century milk regulations.

Twentieth Century Milk Reformers: The Regulation of the "The Milk Question"

"The Milk Question" occupied the minds of politicians and concerned citizens at the beginning of the twentieth century. The contradiction between the necessity of milk and the potential danger of milk was a significant problem. *How could a safe supply of a necessary food be created when that food could also be harmful to health?* This question provoked the publication of reports, scholarly papers, and books, such as 1912s *The Milk Question*, authored by Dr. Milton Rosenau director of the U.S. Public Health Service's Hygiene Laboratory, which followed John Spargo's 1908 *The Common Sense of the Milk Question*. Unlike many milk reformers and promoters, Rosenau did not promote the ideal of milk as a universal food throughout history. Rosenau recognized that other countries flourished without milk as a staple in their diet. However, as our society had created a dependence on milk as an essential food, it was now faced with a public health crisis, as milk was apt to be dangerous to our health.[17]

Several citizens played pivotal roles in cleaning up milk and cementing it further in the national psyche as the ideal food. Robert Hartley, the early milk reformer mentioned earlier in this chapter, led the New York Association for Improving the Condition of the Poor on a crusade against infant mortality. His original interest in promoting a clean milk supply was not in the product itself (although this was very much a prime belief of his) but in the promotion of perfecting humanity. His 1842 essay, *An Historical, Scientific and Practical Essay on Milk as an Article of Human Sustenance*, continued to promote the new idea of milk as "nature's perfect food." He believed that it was a complete form of nutrition and therefore must be consumed to be healthy. Armed with the Bible as an evidence that humans had always consumed milk, Hartley exposed the filthy conditions of swill dairies in the cities and espoused the need to ensure safe and clean milk so humans, especially children, could partake of this perfect beverage.[4,5]

Nathan Straus, Macy's co-owner and prominent philanthropist, established milk depots throughout New York City in 1893 (other U.S. cities also established milk depots) where poor mothers could obtain pasteurized milk for their infants. Three years later, his stations were supplying 600,000 bottles of milk a year, contributing to the drop in infant mortality rate in New York City from 248 per 1000 infants in 1885 to 80 per 1000 in 1919.[18]

Hartley was not the only prominent milk reformer to use history as the argument for milk consumption. However, what these early reformers failed to mention was that although dairy products had been consumed throughout human history, dairy in the form of unfermented liquid milk was very rarely consumed. Instead, fermented and cultured dairy products in varying forms of thick yogurts, brined, salted, or sun-dried cheeses and fermented beverages (even some that were alcoholic) were most commonly consumed historically. These forms did not require refrigeration and could last into the colder months when animals typically lacked fresh pasture and did not produce much, if any, milk.[4–6]

Fermentation produces lactic acid, an important characteristic of fermented dairy products. Because fermentation breaks down lactose, the sugar that is abundant in milk, into lactic acid, fermentation significantly lowers the levels of lactose in dairy products.

Furthermore, the lactic acid prevents the growth of harmful bacteria, acts as a natural preservative, creates a tart flavor and causes the casein proteins to thicken.[19] Fresh, "sweet" (nonfermented) milk does not keep fresh, sweet, and safe to consume for very long; which became a significant problem for the milk supply to urban cities as the distance increased between farmer and consumer.

Pasteurization, developed in 1860 in France, kept milk safe longer and enabled milk to be transported for farther distances, but this process was not adopted in the United States for milk until the late 1890s, and was not widely adopted until the turn of the century. Pasteurization was one of the most important and controversial changes made in the nascent United States dairy industry and the clean milk supply movement had several iterations before it ended in universal pasteurization. One of the alternative proposals at the time to help in keeping milk safe was an inspection process that resulted in certified, branded milk production operations, and this solution held prominence until the early 1900s when pasteurization promoters gained ground.[4]

The social discussion and resolution between the two choices show the same tensions that exist today in other food safety matters. See Chapter 13, "Food Protection." *Who bears the burden of responsibility, risk and cost? Does the producer or the consumer? Is a certified sanitary process of production good enough or does the food need to be fundamentally made safe through pasteurization? Does inspection by the state entail a guarantee of safety by the state?*

For example, home sterilization of milk is an option to keep milk safe; and sterilizing milk by boiling it at home before feeding it to infants had been prescribed before pasteurization regulations for milk producers were passed. However, home sterilization was often done haphazardly without proper instruction on length of time or temperature.[20]

Proponents of milk certification believed that pasteurization killed beneficial bacteria and altered the taste of milk. They argued that it was possible to create safe distribution of raw milk through education and careful inspection of farms and milk bottling centers. Moreover, some people believed that with a requirement to pasteurize, dairy farmers would be lazy, keeping unsanitary milking facilities and producing contaminated milk, in turn relying on pasteurization to then make the milk clean. And in fact, the public had the right to be wary, as some companies did at first pasteurize—and repasteurize—old milk to make it saleable. An inspection process could provide a backbone to pasteurization regulation. Some pediatricians also believed that pasteurized milk was less desirable nutritionally and possibly even detrimental to infant health. In fact, even many supporters of pasteurization believed that lactic acid did have some benefits and would have preferred to forgo heat-treating milk. As Spargo points out, "It is the familiar case of enduring the lesser evil to be rid of the greater."[2–4]

Proponents of pasteurization argued that it was not possible to create a safe distribution method for raw milk, because they believed that raw milk was inherently a health hazard, while others believed the logistics of such a system of inspection were insurmountable. As an example, in 1906, New York City mandated the inspection of milk farms serving the city and assigned 33 inspectors to the task—of inspecting 44,000 farms in six states, some as far as 400 miles away. Certification worked as an expensive product sold to the wealthy who could afford to buy not just milk, but also the assurance of the farmer as to the desired qualities beyond the actual milk;

much like small-farmed, sustainable, humane sources of meat sought by the wealthy today. However, it seemed unlikely that the state could duplicate the rigorous and labor-intensive certification process to meet the public health demand of creating a safe milk supply for all. In short, pasteurization regulations were not simply a *de facto* adoption of technology to solve a public health problem, but instead a deliberate choice made from available options.[4] You will see a similar choice being made in Chapter 11, "Enrichment and Fortification," by the decision to add vitamins to milled refined flour instead of changing the milling process to make the resulting flour less refined.

Pasteurization made milk safer. Yet, it also contributed enormously to consolidation of the milk market; and to further increase the distance between farmer and consumer. Smaller fluid milk producers and dairies that could not afford pasteurizing equipment went out of business. The number of milk dealers in Detroit dropped by more than 50% within just three months after the passing of the city's pasteurization law in 1915, and similar results followed in Milwaukee, Chicago, and Boston after pasteurization laws were passed.[2,5]

The Developing Dairy Industry: The Creation of a Market

Advances in packaging followed by other processing techniques, namely mechanical cream separators and homogenization, also highly enhanced the burgeoning dairy industry by giving farmers and processors the ability to produce a wider variety and increasing number of dairy products in less time. The development of mechanized glass bottling in the late 1800s made direct marketing to consumers easier by providing the consumer with a means of visibly assessing quality. The glass bottle revealed the line of cream formed on top and gave customers a visible characteristic by which to judge the quality of the milk: more cream meant better milk. Mechanical cream separation in the 1880s helped to spur increased butter production to such a degree that specialized facilities that only made butter, called creameries, sprang into business. Finally, the turn of the century brought homogenization, another drastic change that transformed the fluid milk supply.[6]

A milk homogenizer was patented in 1899 by August Gaulin of France. Homogenized milk was first sold in the United States in 1919, although it did not become widely marketed and accepted until the 1930s. Fluid milk is an oil-in-water emulsion in which the fat globules are less dense than water and rise to the top, thus producing the visible cream line noted above. A homogenizer passes the milk through a tiny orifice under high pressure, breaking up the fat globules into much more numerous and smaller globules. With the milk fat evenly dispersed throughout the fluid, there was no cream line for the customer to judge quality by, and the milk consequently looked different and had a richer mouth-feel.[21]

Dairy farmers and milk bottling plants favored homogenization for several reasons: it allowed milk from different farms, herds, and type of cow to be mixed together; and, combined with mechanical cream separation, producers could remove all the fat and mix it back in, in whatever quantity they desired.[6] Homogenized cream would look like higher quality cream due to the thickening effect. Ice cream producers, in particular, found homogenization useful as it allowed a smooth ice cream mixture to be made out of any milk, including powdered skim milk and water.[22] Consumers may have enjoyed the resulting convenience and taste of homogenized milk, but it is important to note that homogenization has no food safety benefit, and the adoption of homogenization was for the benefit of the industry, not for the consumer.

Just as pasteurization was taking over the fluid milk market and larger dairies were pushing out smaller producers, new scientific research was revealing the nutritional components of food and the role of food in preventing certain diseases caused by nutritional deficiencies. The year 1910 marked a significant change not only in the knowledge of nutrition with the discovery of vitamins but also the advent of intense nutrition promotion by food companies. Early nutrition advice from government and public health officials focused on how to prepare healthy, sustaining meals on a budget, but now with the discovery of vitamins, the consumption of specific foods could be promoted. Vitamin A, which is found in milk, had been shown to promote growth in mice in laboratory studies. This discovery had a profound effect on mothers, who had already been receiving information from pediatricians, public health officials, and others on the importance of milk in a growing child's diet, and lent even more credibility to the movement to protect child welfare. The consumption of milk, a perfection of nature, should surely lead to the perfection of children. Food manufacturers used the newly understood food components to promote their products as "protective food," echoing the language from USDA brochures. A whole new push on reaching not just good health, but optimum health, was used in promoting increased food consumption.[2,4,5,23]

The dairy industry was one of the most effective food industries in using newly understood nutrition science to promote its prized product. The promotional and educational campaigns were effective in changing public perception that milk was not only beneficial and necessary for infants and children but for adults as well. Several large dairy companies, such as Borden's and National Dairy Products, increasingly gained greater control over the dairy distribution market and leveraged the most influence over the consumer market. They eventually formed a national organization (the National Dairy Council (NDC)) to promote milk's health benefits in schools and to this end, provided lesson plans. The information disseminated in school was reinforced by advertisements in newspapers and magazines across the country of children drinking the recommended daily quart of milk. Malt and chocolate flavored milks were introduced to increase children's milk consumption. In the end, schools proved to be the place where the dairy products and promotions were to have the longest-lasting effect on increasing milk consumption.[2]

Increased milk consumption and the specialized production of cream, butter, and cheese resulted in increased by-products and waste. The market for skim milk had not yet been developed in the early twentieth century, which meant that there were substantial amounts of the fat-free milk product left over from butter and cream production and cheese makers were left holding on to ample amounts of whey. Unlike historic family farm dairy and cheese production, the waste could not be fed to the pigs. Thus, like any other industrial waste product in the early twentieth century, these byproducts were being dumped into streams or otherwise improperly disposed. The increasing amount of waste demanded new solutions, which includes creating new products and adding the byproducts directly into existing food products.[2,4,5,24]

The case of surplus skim milk and its transformation into skim milk powder is an important turning point in the history of milk consumption in the United States. Americans tended to shun skim milk, thinking it inferior to whole milk. But what to do with all the extra skim milk produced by the manufacturing of butter and cream? Even worse, when the Great Depression slowed Americans' demand for fluid milk, the dairy producers were left with even greater surpluses that could only spoil if a solution was not found. The solution turned out to be drying.[25,26]

Technological advances allowed creameries to dry their surplus skim milk and create a shelf-stable product nutritionally comparable to fluid milk. The government used skim milk powder in relief food programs and recommended people to use the surplus byproducts for animal feed. Although this was a sensible use of surplus, when nutritionists promoted skim milk they had to battle the stigma that skim milk was a food fit only for poor people and hungry pigs.[23]

The battle was won in the schools. The USDA joined the NDC in promoting "milk-for-health" campaigns in schools across the country and promotion of skim milk continued through the 20s and 30s. In 1936, a *New York Times* article highlighting New Yorkers' increased milk consumption mentions the beneficial vitamins and nutrients found in milk surplus products: buttermilk, skim milk curds, and whey, and recommends skim milk powder as a good substitute for fresh milk. The article also mentions the historical backing authorities relied on for making the claim that milk is "a 'nearly perfect food.'"[27]

The beginning of World War II changed milk consumption practices with food rationing and diversion of industrial resources to the War effort. Soldiers and American allies were provided with generous amounts of skim milk powder. Skim milk powder proved to be a lightweight and nutritious staple that could easily be carried by soldiers and mixed into many of their rations. Back home, Americans were encouraged to consume skim milk powder as well. Nutritionists capitalized on the circumstances to promote skim milk's high nutritional quality and economic value, but as the war ended, so did American's willingness to drink skim milk. But without the soldiers to absorb the surplus, the surplus of skim milk and skim milk powder increased to the point of demanding significant involvement by the government, and the government turned to school-aged children. Reimbursement programs for milk provided in schools had already been conducted in various cities during the years 1940–1946, and in 1946, the school milk programs were incorporated into the National School Lunch Act. Schools across the country were also the recipients of nonfat dried milk purchased by the USDA. The government also purchased nonfat dried milk for relief efforts in rebuilding Europe.[4,5,28]

But even government support could not absorb all surplus milk byproducts. The industry had to find additional solutions, and so it did, by touting cottage cheese and skim milk as the perfect diet food and by incorporating nonfat milk and milk proteins into virtually any part of the food supply, including hot dogs, frozen desserts and as a flavor enhancer. The revolution of using milk byproducts in nondairy and even nonfood forms is one of the most important scientific transformations of dairy products, further cementing their importance in not only the American diet, but the economy as well. Highly processed products extracted from dairy such as casein, whey, and skim milk powder can be used in everything from hot dogs to paint. However, advertising was necessary to convince consumers that these solutions were desirable, especially the promotion of non-fat milk as a diet food. The promotion of milk as a perfection of nature had made it through high infant mortality rates, polluting byproducts and postwar surpluses. In order to sustain this success, the dairy industry needed further help from the government.[4,23]

Questionable relationships between government and industry developed as the government stepped in to help in addressing the growing environmental and economic problems presented by surpluses and byproducts of dairy production facilities.[4,5] With the government's support in marketing and dietary guidance, milk solidified its place in the American diet.

The USDA and the Marketing of Dairy

After so many years of promotion, the push for skim milk has finally succeeded. Americans are now consuming more skim milk products than ever before. Ironically, this increased consumption of skim milk products has now resulted in a surplus of whole milk and milk fat products. To help in reducing the surplus in whole milk and milk fat, USDA's Dairy Production Stabilization Act of 1983 established the National Dairy Promotion and Research Board (NDB) to carry out promotion and research programs that help to build demand and expand domestic and international markets for dairy products.[29] To further this end, Dairy Management Inc.™ (DMI) was created in 1995 to increase demand for and drive sales of U.S. dairy products and ingredients on behalf of America's dairy producers. This agency sponsors milk programs such as the Real® Seal program for cheese, the "Fuel Up to Play 60" program, and the "Got Milk?" campaign.

DMI is funded solely by contributions from dairy farmers that are collected through a mandatory system known as the "check-off" program. The Dairy Stabilization Act requires that dairy farmers participate in this program and that the revenues must be paid to DMI.[30] The current rate of assessment is 15 cents per 100 pounds of milk (about 12.5 gallons).[29]

Dairy Management works with state and regional dairy promotion organizations to ensure the future success of dairy by integrating marketing, promotion, advertising, public relations, nutrition education, and nutrition, product and technology research programs. Along with international, state and regional organizations, DMI manages the National Dairy Council® (NDC). Founded in 1915, NDC provides nutrition information through national, state and regional dairy council organizations.[31] The NDC is the producer-funded nutrition education and research arm of DMI, and for most of the twentieth century served as the *de facto* source of nutrition information for school teachers. Box 7.1 discusses the potential conflict of dairy advertising and U.S. dietary guidance.

The National Dairy Council: The Nutrition Education People

Anyone who has been practicing in the nutrition and dietetics field for the last 30–50 years will tell you that the NDC was light-years ahead of other food industries in publication, distribution, and delivery of nutrition education materials in the United States,

BOX 7.1   DAIRY ADVERTISING AND HEALTH CLAIMS

The case of Michael Zemel, PhD, now professor emeritus of nutrition at the University of Tennessee, provides an example of inherent conflicts in USDA's historical roles as both marketer of agriculture products and the architect of U.S. dietary guidance.

Research conducted by Dr. Zemel on calcium and weight loss became the backbone of a multimillion advertising campaign by the dairy industry in October of 2003. Dairy Management, Inc. (DMI) conceived of the advertising, and as noted in the text of this chapter, the USDA oversees work done by DMI. Two slogans used by the campaign were "Body by Milk," which targeted adolescents, and "Milk your diet. Lose weight!."

The April 2004, issue of the peer-reviewed journal, *Obesity Research*, contains a report of the Dairy Council-supported research conducted by Zemel et al. The authors conclude that calcium and dairy accelerate weight and fat loss during energy restriction in obese adults.[59] In small print, the journal editors state that the article must be regarded as an advertisement since the costs of its publication were defrayed, in part, by the payment of page charges, and although the support of the Daily Council was noted, Zemel failed to disclose that he had filed patents on calcium weight loss methods. In a later issue, the Editor of the journal confirmed that although patents are public records, the policy is for authors to declare any potential conflicts of interest, including patents, and requested that Dr. Zemel do so in the future.[60]

The marketing campaign of dairy consumption for weight loss went beyond the findings in Zemel's study. The Dairy Council's advertisements credited the inclusion of three to four servings of milk, yogurt, and cheese with producing 24 pounds of weight loss, but failed to disclose that the participants in Zemel's study were obese and were asked to restrict calories. Some advertisements also promoted 24 ounces in 24 h claim, a claim not made by Zemel, and the industry could not point to any study with this finding. The milk-weight loss claim was never replicated. In fact, another researcher paid by Dairy Management tried to replicate the study but found no weight-loss benefits of drinking milk. Zemel also stated later that his work is only relevant to people not consuming enough calcium; in other words, if we have sufficient calcium, just adding more would not be effective.[61]

In April of 2005, Physicians Committee for Responsible Medicine (PCRM), a nonprofit health advocacy association, brought suit with the FTC against the advertising campaign. The FTC regulates commercial advertising; and it is unlawful to deceive consumers through an advertisement that omits material information likely to affect a consumers' choice to purchase a product. Health claims are presumed to be material. Generally, the FTC follows FDA policies, such that the FTC is likely to reach the same conclusion as the FDA when it comes to supporting scientific basis for claims, and if a health claim has not been approved by the FDA, the FTC requires that claims clearly convey to what extent the claim is supported by science. A claim that is qualified by evidence at odds with a larger body of scientific knowledge is potentially misleading.

The PCRM suit argued that DMI's billing of dairy foods as health claims would not pass the scrutiny of the FDA. In May 2007, the FTC, having met with USDA and campaign staff, reported the agreement by the Fluid Milk Board, the Dairy Board and parties affiliated with their marketing campaigns to retract the ads and marketing materials "until further research provides stronger, more conclusive evidence of an association between dairy consumption and weight loss."[62]

You may find it instructive to read PCRM's suit, available online at http://www.fda.gov/ohrms/dockets/dockets/05p0224/05p-0224-cp00001-Exhibit-03-PCRM-vol1.pdf/. It contains, among other topics, a critical analysis of Zemel's studies, a compendium of studies that attempted—but failed—to replicate Zemel's findings, and a discussion of how dairy advertisements distorted the results and significance of his research.

particularly to school teachers and school children. For many years, the NDC used the self-proclaimed title "the Nutrition Education People" on many of their education materials, and rightfully so. They published curricula for all age levels, such as a popular science curriculum for students called "Food...Your Choice." They even produced a teaching guide for sport coaches entitled "Food Power: A Coach's Guide to Improving Performance." These historic materials (pre1980) were comprehensive, informative, and colorful (Personal communication, Matthew Nulty, MPH, RDN).

No longer self-titled "the Nutrition Education People," the NDC now fits into the DMI mission by bringing together nutrition science researchers, registered dietitians, and communications experts who are committed to educate the public about the health benefits of milk and milk products. A variety of stakeholders such as health professional, educators, school nutrition directors, academics, industry leaders, consumers, and the media are engaged with nutrition programs, materials, and research to support government nutrition recommendation. The NDC website provides extensive health education material for consumers and for professionals, handouts, meal plans, infographics, webinars, as well as numerous research reports, science summaries, and the third edition of the *Handbook of Dairy Foods and Nutrition*, "a complete reference to dairy nutrition research that includes chapters on dairy's role in heart health, dental health, bone health."[32]

The NDC and its affiliates have also been a prominent voice in lobbying nationwide for the dairy industry. In 2013, the dairy industry spent over 8 million in lobbying.[33] With this much money aimed at lobbying, the dairy industry is able to conduct a streamlined and organized approach; much of which is aimed at dietary guidance and the implementation of federal food assistance programs and policies. For example, in 1983, when the USDA was proposing changes for the Thrifty Food Plan (see Chapter 6, "Federal Food Assistance"), the NDC submitted a well-written seven-page letter, making a strong case for the USDA not to set "arbitrary standards for dietary intake of fat, cholesterol, sodium, and caloric sweeteners" and not to allow "reductions in the amounts of cheese and other nutrient-dense foods in the family food plans...in an effort to reduce the amount of fat, cholesterol, sodium, or caloric sweeteners in those food plans" (Personal communication from Matthew Nulty, MPH, RDN, to Dr. Spark).

Even today, the NDC and its affiliates regularly lobby both state and federal governments on a regular basis concerning policies affecting the interests of the dairy industry. Their comments about proposed policies and programs are public information and are available online during specified open comment periods. For example, the NDC's comments for the 2015 Dietary Guidelines are available when we search the public comments by organization, published at the Health.gov 2015 Dietary Guidelines website.[34]

At the time of writing this chapter, NDC has submitted 15 comments, many of which provide updated research to support the continued promotion of health benefits associated with milk. Many of the NDC's comments discuss the desirable health behavior of building a habit of drinking milk in children, and one comment essentially asks the 2015 Dietary Guidelines Advisory Committee (DGAC) not to lump flavored, sweetened milk in with other sweetened beverages, as flavored milk still provides the same nutrient profile of regular milk and is not a significant contributor of added sugar. However, others might view consumption of flavored milk as building a habit of drinking sweetened beverages that is not quite such a desired behavior.

Other comments emphasize milk as a source of calcium. After attending a meeting in which the 2015 DGAC discussed nondairy sources of calcium, the NDC submitted a comment that milk provides many nutrients, not just calcium, advised that recommendations of nondairy alternatives would need to be evaluated through dietary patterns, and pointed the committee in the direction of work previously done for the 2010 report.

The NDC also provided a fascinating comment as to sustainability, which includes research reporting that healthy diets are less sustainable and a second comment that provides information about what the dairy industry has been doing to support sustainability. Generally, the NDC appears to be suggesting that the DGAC hold back on sustainability as the research is still young, there may be significant unintended nutritional impacts, and there is more research from other government groups that will shortly be available.

Finally, another comment by the NDC provides support for the idea of diet modeling using reduced-fat cheese to help Americans in achieving dietary guidance recommendations; which hints at the idea that dietary guidance could consist of suggestions to "eat more less." This comment also shows how well the dairy industry has been able to expand the amount of milk products consumed by Americans. Cheese has already been an amazing success story. The NDC/DMI has helped U.S. cheese consumption increase from slightly over 14 pounds annually per person in 1975 to 32 pounds per person in 2008 (Figure 7.1).[35]

USDA has oversight responsibility for the dairy and fluid milk promotion programs. All advertising, promotion, research, and education budgets are approved by USDA, which is reimbursed for its oversight activities.[36] In other words, the dairy marketing programs encouraging us to eat more milk, cheese, and other dairy products, are supported in part by USDA.

Among its other initiatives, DMI developed recipes with cheese for home cooks and restaurant chefs, and over time, food companies and restaurants in the U.S. have increased the amount of cheese in the food chain.

Per capita consumption of selected dairy products, 1975–2008

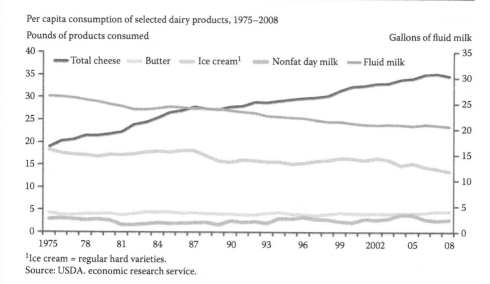

¹Ice cream = regular hard varieties.
Source: USDA. economic research service.

FIGURE 7.1 Annual per cheese consumption has increased from slightly over 14 pounds to 32 pounds. (Adapted from Davis, C.G. et al., Long-Term Growth in U.S. Cheese Consumption May Slow, LDP-M-193-01, USDA, Economic Research Service. Available at http://www.ers.usda.gov/media/146945/ldpm19301_1_.pdf, accessed May 1, 2014.)

Cheese is no longer just an appetizer (cheese and crackers) or a continental-style coda of the meal (fruit and cheese). Dairy Management has successfully promoted increased use of cheese in prepared food products. Many restaurants have adopted cheese-enhanced menu items, such as cheese fries (french fried potatoes with melted cheese), Cheezy Bites (Pizza Hut®), which features string cheese baked into the pizza crust, and steak quesadilla (Taco Bell®), with cheddar, pepper jack, and mozzarella.[37]

For decades, the U.S. government has promoted diets reduced in total fat and saturated fat. The USDA and Health and Human Services (HHS) in their *Dietary Guidelines for Americans* (DGA), the National Heart, Lung, and Blood Institute's (NHLBI) Dietary Approaches to Stop Hypertension (DASH) eating plan, and recommendations from voluntary health organizations, such as the American Heart Association, all promote a diet that is low in total and saturated fat. To achieve this goal, Americans are advised to switch from whole milk dairy products to reduced-fat or fat-free dairy products. See Table 7.2: Fat in various types of milk.

While cheese is perceived as nutritious, it takes 8 ounces (1 cup) of milk to produce just 1½ ounces of cheese. Thus, cheese has all the calcium and protein that milk has, but also all of the fat and calories, and in a much smaller package. Cheese is, therefore, a food we are advised to eat in moderation. Notice the paradox. Through the Dairy Production Stabilization Act, the USDA supports programs to increase cheese consumption while at the same time, the HHS and USDA's Dietary Guidelines advise us to cut back on fatty foods, such as whole milk and cheese. The next section discusses the development of federal dietary guidance concerning dairy products.

TABLE 7.2   Fat in various types of milk

	GRAMS OF FAT IN 1 CUP	FAT AS A % OF WEIGHT	KCALORIES IN 1 CUP	KCALORIES FROM FAT	FAT AS A% OF CALORIES
Type of Milk					
Whole (full-fat)	7.9	At least 3.25	149	71	48
Reduced-fat	4.8	2–3.25	122	43	35
Low-fat	2.4	Less than 2	102	22	21
Skim	0	0–0.10	83	0	0
Summary	0–7.9 g of fat in 1 cup of milk	By weight, milk is 0–4% fat	83–149 kCalories in 1 cup of milk	Fat contributes 0–56 kCalories to 1 cup of milk	Milk derives 0–48% of its kCalories from fat

Source: For all nutrient values: U.S. Department of Agriculture, Agricultural Research Service, 2012. USDA *National Nutrient Database for Standard Reference, Release 25*, Nutrient Data Laboratory Home Page, available at http://www.ars.usda.gov/ba/bhnrc/ndl.

Note: Calculations are based on 3.25% whole milk, 2% reduced-fat milk, 1% low-fat milk, and 0% skim milk.

THE EVOLUTION OF DIETARY RECOMMENDATIONS FOR MILK AND DAIRY PRODUCTS

The history of USDA nutrition guidance includes over 100 years of recommendations regarding the consumption of milk and dairy products.[38,39] During the twentieth century, there were two clear trends regarding milk and dairy products in U.S. food guidance: the development of a specific dairy food group and an increase in the amount of dairy products recommended. Although the earliest food groups clustered both milk and meat together as protein-rich foods, milk and meat were soon separated. During the 1940s, a *food group specifically for milk and dairy products* was created and milk has had its own food group ever since.

The recommendation for "milk and milk products" was 2 cups or more from the 1940s through the 1970s, but the minimum level has crept upwards. As of the 2010 Dietary Guidelines, it is now 3 cups or equivalents, for children over the age of 9, adolescents, and adults.

Earliest Guides

Generally, the first few decades of dietary recommendations focused on receiving adequate nutrition. The USDA's first food guidelines were developed by Dr. Wilbur Olin Atwater. In 1902, *Principles of Nutrition and Nutritive Value of Food*, Atwater advocated variety, proportionality, and moderation; minimizing waste; measuring calories; and an efficient, affordable diet that focused on nutrient-rich foods. Atwater established the groundwork for future food guides by recommending what he regarded as ordinary and necessary food materials—meats, fish, eggs, milk, butter, cheese, sugar, flour, meal, and potatoes and other vegetables.[39] Thus, dairy products are considered necessary food materials from the start of federal guidance.

In the first official USDA-sponsored food guide,[40] *Food for Young Children* (1916), nutritionist Caroline Hunt sorted foods into five groups: milk and meat, cereals, vegetables and fruits, fats and fatty foods, and sugars and sugary foods. Even though milk and meat were combined into one food group, Hunt notes that "milk is such an important food for children that it is desirable to speak about it by itself." Evoking the ideal of nature and perfection, Hunt describes milk as "the natural food of babies and the most important food for young children. A quart of milk a day is a good allowance for a child."[41]

Food for Young Children was followed the next year with the 14-page *How to Select Foods: What the Body Needs*,[1] aimed at the general public, providing "a simple method of selecting and combining food materials to provide an adequate, attractive, and economical diet" and promoted the same five food groups to adults. These same five food groups—with milk and meat in one group—were maintained in subsequent reports through the 1920s.[40]

Milk's partnership with meat as a protein source contributed to the first government messages to consume more milk as part of the conserving food campaigns initiated by America's entry into World War I in April of 1917. The newly formed US Food Administration, led by Herbert Hoover (before he became president), initiated a "Food Will Win the War" campaign. Hoover was convinced that voluntarily conserving resources would be effective enough to avoid food rationing to keep enough food allocated for the war effort.[42] Although the 1918 book *Foods that Will Win the War and How to Cook Them*, was not published by the US government, it educates the population about the need to conserve meat consumption for the war effort and how to do so, including recipes. *Foods that Will Win the War* advised that the solution is to "reduce meat consumption to the amount really needed and then to learn to use other foods that will supply the food element which is found in meat. This element is called protein, and we depend upon it to build and repair body tissues." Milk, skim milk, cheese, and cottage cheese are all cited as protein foods that can replace meat.[43] Correspondingly, the baking recipes in *Foods that Will Win the War* often use milk to bolster the reduced use of eggs; thus also fostering milk's role as necessary staple in our pantry and diet.

The discovery of vitamins led to a greater emphasis on nutrients within dietary guidance messages, and eventually, the establishment of Recommended Dietary Allowances (RDAs) for the intake of vitamins, although this too was a result of wartime concerns. The original impetus for the RDAs was to prevent diseases caused by nutrient deficiencies for national security reasons. Government needed guidelines for feeding American troops and children in school lunch programs (who in the future could become soldiers). The first RDAs provided recommended intakes for calories and nine nutrients, including protein and calcium.[44] As milk is a convenient source for both protein and calcium, the development of the RDAs no doubt also contributed to making milk an "easy sell" nutritionally.

Milk Recommendations Based on Food Groups

Food group recommendations are covered in more detail overall in Chapter 3, "Dietary Guidance." This section focuses on the development of a dairy food group and subsequent recommendations.

As noted above, milk and meat together provided one of the five food groups promoted in USDA guidance published in the first two decades of the twentieth century. As the Great Depression gripped the United States, the USDA developed dietary standards and food plans to provide a buying guide to help people in maintaining a healthy nutritious diet while shopping on a limited budget. These guides increased the number of food groups from five to 12.[45,46] With the expanded number of groups available, milk was given its own food group, further enhancing its nutritional prominence. The recommendation for milk consumption was also increased to 2 cups per day.[46] The spotlight on milk likely helped the USDA market the dairy surplus (particularly skim milk); necessary to keep farming income stable during this time.

When wartime food rationing arose in the United States in 1943 during the Second World War, USDA and the War Food Administration published the National Wartime Nutrition Guide that promoted seven food groups to help in maintaining nutritional standards under rationing conditions which restricted access to food. Milk retained its own food group, even with the reduction of number of food groups from 12 to 7. The model provided a large number of food groups to choose from in case of limited supplies of a specific type of food. Recommendations for servings were not provided because daily consumption from each group was not necessary.[39] This "Basic 7" guidance was plagued by two fundamental shortcomings. No guidance was provided as to the sizes of servings that were recommended, and people found it difficult to remember seven food groups.

A Daily Food Guide ("Basic Four") was introduced in 1956, and corrected the weaknesses of the Basic 7 by providing a simplified guide with fewer food groups and, for the first time, some guidance as to portion numbers and sizes. Milk retains its own food group, and the quantity recommended is two or more servings (1 cup of milk and 1½ ounces of cheese count as a serving). Fluid milk is specifically recommended, with the amount varying by age. Two to three cups are recommended for children under the age of 9, 3 or more for children aged 9–12, 4 or more for teenagers, and 3 or more for adults.[47]

Figure 7.2 shows a USDA poster used to promote the Basic Four. Notice the strategic up front and on the top location of the milk group. The Basic Four seems to have been tailored made for the dairy industry. In fact, the Dairy Council produced its own chart *A Guide to Good Eating* with four rectangular quadrants that read from top to bottom (Figure 7.3). Not surprisingly, the milk group appeared at the top.

In 1979, the USDA released the Hassle-Free foundation diet, which altered the milk recommendations to 2 servings from the "milk-cheese" group (a serving being equivalent to 1 cup of milk or 1½ ounces of cheese). In response to the Surgeon General's 1979 report titled *Healthy People: The Surgeon General's Report on Health Promotion and Disease Prevention*, the USDA and the United States Department of Health and Human Services (DHSS) joined forces to issue authoritative and consistent guidance on diet and health. This, in turn, led to the development of the first DGA in

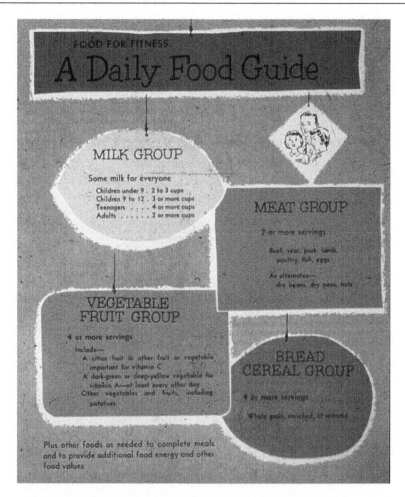

FIGURE 7.2 USDA daily food guide poster promoting the basic four food groups. (From Food and Nutrition Information Center, National Agricultural Library, USDA, 1956–1979.)

1980, a document that would be revised by-law every five years to keep pace with the research in the area of nutrition and health.[40,48] From this document came the Food Guide Pyramid total diet, which recommended two to three servings from the "milk, yogurt, and cheese" group (with the serving sizes being 1 cup of milk, 1½ ounces of cheese, and 1 cup of yogurt).

The Food Guide Pyramid remained the food icon of the DGA up through the 2005 edition at which point the icon was switched to the MyPyramid. This abstract guide used brightly colored bands to represent different food groups in which dairy products, represented by the blue-colored band, were simply called the "Milk" group. This version did away with the concept of servings and instead called for daily consumption of 3 cups of milk or milk products. Material published on the USDA's website provided a detailed list of 1 cup equivalents for milk products.

FIGURE 7.3 Dairy Council poster, "A guide to good eating." (From Food and Nutrition Information Center, National Agricultural Library, USDA, 1956–1979.)

The 2010 DGAs changed the MyPyramid to the consumer-friendly MyPlate image. The major change for milk between the MyPyramid and the MyPlate was the renaming of the group from "Milk and Milk Products" to "Dairy Products," the latter term being one that was widely used until the launch of the MyPyramid.

With the powerful voice of the NDC in education and DMI in lobbying, milk and dairy products have secured a solid footing in national dietary guidance and government-sponsored food assistance programs. Even with publication of MyPlate, which replaced the Pyramid as the newest iteration of USDA graphic dietary guidance, dairy remains predominantly featured. The plate is divided into quadrants of unequal size for

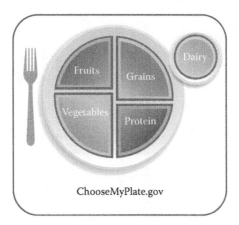

FIGURE 7.4   USDA's MyPlate prominently depicts dairy as a glass. (From United States Department of Agriculture.)

vegetables and grains (the largest sectors) and fruit and protein (the smallest), and to the right of the plate, a tumbler representing dairy (Figure 7.4).

Table 7.3 contains a summary of the recommendations for dairy products in each edition of the Dietary Guidelines since 1980. The *Guidelines* matured from merely mentioning dairy foods (milk, cheese, and yogurt) in 1980 to extolling the virtues of a dairy-free diet 30 years later. Looking forward from the past: skim milk made its first appearance in the DGA in 1985; the number of recommended servings (2–3) of dairy foods was first mentioned in 1995 along with listing some nondairy sources of calcium for the first time; in 2000, the number of servings of dairy was spelled out for each stage of the lifecycle; in 2005, a table was presented devoted exclusively to nondairy sources of calcium; and in 2010, dairy-free vegetarian diets were recognized as healthy food patterns.

Dietary Guidance: Recommended Intake of Calcium

The Food and Nutrition Board (FNB) of the National Academy of Science's Institute of Medicine establishes recommended daily intakes for nutrients, including calcium. Select micronutrients of public health interest are highlighted on the Nutrition Facts label found on most packaged food; calcium is one of the four required nutrients.[49]

However, although the FNB provides guidance on nutrient intakes, it does not recommend in what manner people should consume food to meet those nutrient intakes. In the case of calcium, as for many nutrients, a variety of foods can contribute calcium to the diet. Unfortunately, the plant-based sources, unlike milk, are not always packaged with a Nutrition Facts label, and it may be easy for consumers to overlook such sources of calcium. Undoubtedly, the fact that milk has calcium content readily discernable at a glance helps to support milk's status, as does the current discussion of a population wide deficiency of vitamin D.

TABLE 7.3 Recommendations for dairy products in the Dietary Guidelines, 1980–2010

EDITION AND YEAR	RECOMMENDATIONS FOR DAIRY PRODUCTS
1st 1980	• Avoid too much fat, saturated fat, and cholesterol (front cover and p. 3). • To assure yourself an adequate diet, eat a variety of foods daily, including selections of milk, cheese, and yogurt (p. 5). • To avoid too much fat, saturated fat, and cholesterol, limit your intake or butter and cream (p. 13).
2nd 1985	• Avoid too much fat, saturated fat, and cholesterol (front cover and p. 5). • To assure yourself an adequate diet, eat a variety of foods daily in adequate amounts, including selections of milk, cheese, and yogurt, and other products made from milk (p. 5). • To avoid too much fat, saturated fat, and cholesterol, use skim or low-fat milk and milk products (p. 16).
3rd 1990	• Choose a diet low in fat, saturated fat, and cholesterol (front cover and p. 5). • A daily food guide. Food group: Milk, yogurt, and cheese. Suggested servings: 2–3 (p. 6). • For a diet low in fat, saturated fat, and cholesterol. Milk and milk products: **Have two or three servings daily**. Count as a serving: 1 cup of milk or yogurt or about 1½ ounces of cheese. Choose skim or low-fat milk and fat-free or low-fat yogurt and cheese most of the time. One cup of skim milk has only a trace of fat. 1 cup of 2% fat milk has 5 g of fat, and 1 cup of whole milk has 8 g of fat (p. 17).
4th 1995	• Choose a diet low in fat, saturated fat, and cholesterol (front cover and p. 5). Foods high in fat should be used sparingly: many foods in the milk group. The fats from meat, milk, and milk products are the main sources of saturated fats in most diets. • For a diet low in fat, saturated fat, and cholesterol (p. 32). Choose skim or low-fat milk, fat-free or low-fat yogurt, and low-fat cheese. **Have two to three low-fat servings daily**. Add extra calcium to your diet without added fat by choosing fat-free yogurt and low-fat milk more often. (One cup of skim milk has almost no fat, 1 cup of 1% milk has 2.5 g of fat, 1 cup of 2% milk has 5 g [1 teaspoon] of fat, and 1 cup of whole milk has 8 g of fat.) If you do not consume foods from this group, *eat other calcium-rich foods*. • Most foods in the milk group are good sources of calcium. Some foods in this group are high in fat, cholesterol, or both. Choose lower fat, lower cholesterol foods most often. Read the labels. *Some good sources of calcium are*: milk and dishes made with milk, such as puddings and soups made with milk; cheeses such as Mozzarella, Cheddar, Swiss, and Parmesan; yogurt; *canned fish with soft bones such as sardines, anchovies, and salmon; dark-green leafy vegetables, such as kale, mustard greens, turnip greens, bok choy. Read labels for tofu processed with calcium sulfate and tortillas made from lime-processed corn* (p. 10).

(Continued)

TABLE 7.3 (Continued)   Recommendations for dairy products in the Dietary Guidelines, 1980–2010

EDITION AND YEAR	RECOMMENDATIONS FOR DAIRY PRODUCTS
5th 2000	• Choose sensibly. Choose a diet that is low in saturated fat and cholesterol and moderate in total fat (inside front cover). • Choose a diet that is low in saturated fat and cholesterol and moderate in saturated fat (pp. 14–15). **Two to three servings needed each day from the Milk, Yogurt, and Cheese Group (Milk Group)**—preferably fat-free or low-fat 2 or 3 (3 servings for children and teenagers 9–18 years and adults over 50 year; others need 2 servings; during pregnancy and lactation, the same as for nonpregnant women). This includes lactose-free and lactose-reduced milk products. One cup of *calcium-fortified soy-based beverage* is an option for those who prefer a nondairy source of calcium. Choose fat-free or reduced-fat dairy products most often: 1 cup of milk or yogurt, 1½ ounces of natural cheese such as Cheddar, and 2 ounces of processed cheese such as American. • Use plant foods as the foundation of meals (p. 16): *Enjoy meals that have tortillas at the center of the plate, accompanied by* a moderate amount of low-fat foods from the milk group; go easy on foods high in fat. • *If you usually avoid all foods from the dairy group, be sure to choose foods from the other groups that are good sources of calcium and vitamin D* (p. 16). Some sources of calcium[a] (p. 17) are yogurt[b]; milk[c,b]; natural cheeses such as Mozzarella, Cheddar, Swiss, and Parmesan[b]; *soy-based beverage with added calcium; tofu, if made with calcium sulfate (read the ingredient list); breakfast cereal with added calcium[c]; canned fish with soft bones such as salmon, sardines[d]; fruit juice with added calcium; pudding made with milk[b]; soup made with milk[b]; and dark-green leafy vegetables such as collards and turnip greens.* • Know the different types of fat. Keep intake low of saturated fat-rich dairy products such as cheese, whole milk, cream, butter, and regular ice cream; keep intake low in dairy fats, which are rich sources of cholesterol (p. 28). Choose fat-free or low-fat milk, fat-free or low-fat yogurt, and low-fat cheese most often. Try switching from whole to fat-free or low-fat milk. This decreases the saturated fat and calories but keeps all other nutrients the same (p. 29). Grams of saturated fat in 1 cup whole milk (5.1) versus low-fat 1% milk (1.6) and 1 ounce regular Cheddar cheese (6.0) versus low-fat Cheddar cheese (1.2) (p. 31).
6th 2005	• **Chapter 2: Adequate nutrients within calorie needs. Substitutions for milk and milk products (p. 9). Those who avoid all milk products should eat nondairy calcium-rich food listed in Appendix B-4. The bioavailability of these foods varies.** • **Chapter 5: Food groups to encourage (p. 24).** *Consume 3 cups per day of fat-free or low-fat milk or equivalent milk products.* **(According to MyPlate, 1 cup of calcium-fortified soymilk [soy beverage] counts as an equivalent to 1 cup of milk.)** *(Continued)*

TABLE 7.3 (Continued)   Recommendations for dairy products in the Dietary Guidelines, 1980–2010

EDITION AND YEAR	RECOMMENDATIONS FOR DAIRY PRODUCTS
	• Appendix A-2: USDA Food Guide (p. 54). *Calcium-fortified soy beverages are an option for those who want a nondairy calcium source.* • *Appendix B-4: Nondairy sources of calcium (p. 59):* Footnote states that both calcium content and bioavailability should be considered when selecting dietary sources of calcium. Some plant foods have calcium that is well absorbed, but the large quantity of plant foods that would be needed to provide as much calcium as in a glass of milk may be unachievable for many. Many other calcium-fortified foods are available, but the percentage of calcium that can be absorbed is unavailable for many of them. • Appendix B-5: Dairy food sources of calcium.
7th 2010	• Chapter 3: Foods and food components to reduce (pp. 20–32). • Chapter 4: *Foods and nutrients to increase* (pp. 33–40). Choose foods that provide more potassium, dietary fiber, calcium, and vitamin D, which are nutrients of concern in American diets. These foods include vegetables, fruits, whole grains, and milk and milk products (fat-free or low-fat milk and milk products, such as milk, yogurt, cheese, or *fortified soy beverages*). Those who do not consume milk or milk products should consume foods that provide the range of nutrients generally obtained from the milk group, including protein, calcium, potassium, magnesium, vitamin D, and vitamin A. Soy beverages fortified with calcium and vitamins A and D are considered part of the milk and milk products group because they are similar to milk both nutritionally and in their use in meals (p. 38). The *intake* of milk and milk products, *including fortified soy beverages*, is less than recommended for most adults, children, and adolescents aged 4–18 years, and many children ages 2–3 years. **Recommended amounts are 3 cups per day of fat-free or low-fat milk and milk products for adults and children and adolescents ages 9–18 years, 2½ cups per day for children ages 4–8 years, and 2 cups for children ages 2–3 years. Those who do not consume milk or milk products should consume foods that provide the range of nutrients generally obtained from the milk group,** including protein, calcium, potassium, magnesium, vitamin D, and vitamin A. *Soy beverages fortified with calcium and vitamins A and D are considered part of the milk and milk products.*

(Continued)

TABLE 7.3 (Continued)   Recommendations for dairy products in the Dietary Guidelines, 1980–2010

EDITION AND YEAR	RECOMMENDATIONS FOR DAIRY PRODUCTS
	Calcium and vitamin D (p. 41). Calcium recommendations may be achieved by consuming recommended levels of fat-free or low-fat milk and milk products and/or consuming *alternative calcium sources (Appendix 14).Removing milk and milk products from the diet requires careful replacement with other food sources of calcium, including fortified foods. Calcium in some plant foods is well absorbed, but consuming enough plant foods to achieve the RDA may be unrealistic for many* (p. 41). In the United States, *most dietary vitamin D is obtained from fortified foods, especially fluid milk and some yogurts. Some other foods and beverages, such as breakfast cereals, margarine, orange juice, and soy beverages, also are commonly fortified with this nutrient.* (p. 41). For many years, most fluid milk has been fortified with vitamin D to increase calcium absorption and prevent rickets. Fortified foods. Vitamin D-fortified milk is now the major dietary source of vitamin D for many Americans. Other beverages and foods that are often *fortified with vitamin D include orange juice, soy beverages,* and yogurt. *Natural sources of vitamin D include some kinds of fish, for example, salmon, herring, mackerel, and tuna, and egg yolks, which have smaller amounts* (p. 49). Appendices 9, 14, and 15. Vegan adaptation of the USDA food patterns and food sources ranked by amounts of calcium and vitamin D.
	Chapter 5: *Benefits of vegetarianism* (p. 45). Building healthy eating patterns. Supplements and *fortified foods* (p. 49). USDA food patterns (p. 50). **Recommended dairy servings for a 2000 kcalorie diet: Mediterranean (Greece: 1, Spain: 2.1); DASH (2.6); and USDA Food Pattern, which *includes fortified soy beverages (3.0)*.** The vegan adaptation of the USDA pattern includes **2–3 cups from the vegan "dairy group:" calcium-fortified beverages and foods from plant sources. Sample products are: calcium-fortified soy beverage, calcium-fortified rice milk, tofu made with calcium-sulfate, and calcium-fortified soy yogurt** (p. 81).
8th est. Fall 2015	The public comments on the Report from the *Dietary Guidelines for Americans* Advisory Council (DGAC) (fall 2014); after considering DGAC's scientific recommendations and the public's comments, the *Dietary Guidelines for Americans* policy document is prepared (spring 2015); the 8th edition of the *Dietary Guidelines for Americans* is released (fall 2015).

Source: HHS and USDA, 2015 *Dietary Guidelines for Americans,* March 2013, available at http://health.gov/dietaryguidelines/2015-dga-timeline.pdf. Accessed April 13, 2013.

Note: Nondairy sources of calcium are *italicized* and servings of dairy products per day are **bolded.**
a Read food labels for brand-specific information.
b This includes lactose-free and lactose-reduced milk.
c Choose low-fat or fat-free milk products most often.
d High in salt.

Armed with information from the FNB, it is the role of the Dietary Guidelines to make the recommendations on how Americans can meet appropriate nutrient intakes. Chapter 4 of the 2010 DGA is titled "Food and Nutrients to Increase," and states that intakes of vegetables, fruits, whole grains, milk, and milk products are lower than recommended; and as a result, dietary intakes of several nutrients, including potassium, calcium, and vitamin D are low enough to be of public health concern. (Although milk is a convenient source of vitamin D, this is due to fortification, and other dietary sources can provide potassium and calcium.) Chapter 4 of the 2010 Dietary Guidelines does note that calcium recommendations may be achieved by consuming milk and/or alternative calcium sources. Further details on alternative sources are relegated to an appendix, but the appendix lists only select foods that provide calcium; the only nondairy products listed are: fortified cereal, fortified orange juice, tofu, sardines, and fortified soymilk. The 2010 Dietary Guidelines also states that calcium in some plant foods is well absorbed, but that consuming enough plant foods to achieve the RDA may be unrealistic for many.[50]

At this point, it is helpful to recall the history of the RDAs noted in Chapter 3, "Dietary Guidance." The original RDAs were set at 50% more than the average requirement to account for variability in the population, and thus cover most normal individuals. The Institute of Medicine (IOM) currently describes the RDAs as corresponding to "2 or so above the median needs," or the level of intake that would "cover" (meet) the requirements of at least 97.5% of the population.[51] In other words, many people can obtain sufficient calcium without having to achieve the level of the RDA. According to the IOM, in the United States, an estimated 72% of calcium from food intake comes from milk, cheese and yogurt, and from foods to which dairy products have been added (e.g., pizza, lasagna, and dairy desserts); leaving the remaining approximate 28% to come from all other sources (vegetables, grains, legumes, fruit, meat, poultry and fish, egg, and miscellaneous).[52] For an adult aged 19–50, the RDA for calcium is 1000 mg, thus, using this estimate, the average daily American diet would contain 280 mg of nondairy calcium, without making even any effort to seek out nondairy calcium sources.

The FNB established RDAs for the amounts of calcium required for bone health and to maintain adequate rates of calcium retention in healthy people. For the 2010 RDAs for calcium, see Table 7.4: Calcium RDA throughout the lifecycle.

The case for dairy as the go-to calcium source is attributed to its high bioavailability, meaning the ability for a high percentage of the calcium to be absorbed and used by the body. Although dark, leafy vegetables such as arugula, kale, and spinach contain an amount of calcium comparable to that of dairy products; these plant foods also contain compounds called oxalates that chelate (bind) to calcium and inhibit its absorption in the gastrointestinal tract.[53] Comparing the relative merit of sources of calcium requires not only a comparison of amount but also the extent of absorbability. To absorb the same amount of calcium as 240 mL (about 1 cup) of milk, we would have to eat 1½ cups of white beans, or 2 cups of broccoli, or 1½ cups of kale.[54]

Thus, good sources of calcium do include some beans and dark green leafy vegetables, the caveat being that we would need to consume a greater amount of these foods than dairy. Other nondairy sources of calcium include calcium-set tofu, Chinese cabbage, calcium-fortified orange juice, and calcium fortified plant-based beverages such as soy milk, almond milk, and rice milk. Although grains are not particularly rich in calcium, the use of calcium-containing additives in these foods accounts for a

TABLE 7.4   Calcium RDA throughout the lifecycle, starting at 1 year of age

AGE GROUP	RECOMMENDED DIETARY ALLOWANCE (RDA) PER DAY[a]
Children 1–3 years	700 mg
Children 4–8 years	1000 mg
Children 9–18 years	1300 mg
Adults 19–50 years	1000 mg
Adults 51–70 years	
Men	1000 mg
Women	1200 mg
Adults >70 years	1200 mg
Pregnancy and Lactation	
14–18 years	1300 mg
19–50 years	1000 mg

[a] Food and Nutrition Board, Consensus Study, Dietary Reference Intakes for Calcium and Vitamin D. Released 11/30/2010. Available at http://www.iom.edu/Reports/2010/Dietary-Reference-Intakes-for-Calcium-and-Vitamin-D.aspx, accessed April 10, 2013.

substantial proportion of the calcium ingested by people who consume a large amount of grains.

Finally, water can be a source of calcium as well. It has been reported that the median calcium content of municipally treated water may range from 20 to 105 mg per liter.[55]

The National Dairy Council's Role in Dietary Recommendations

Since its founding, NDC has emphasized the importance of calcium through all stages of the lifecycle, particularly during the growing years to promote the development of strong bones and to amass sufficient bone mass to forestall the development of osteoporosis later in life. In the name of education, the NDC has been in the forefront of convincing health professionals and educators that because milk and most dairy products are so rich in calcium, it would be difficult to meet the recommended intake of calcium without them.

In 2003, borrowing from the successful 5 A Day produce campaign (see Chapter 10, "Fruits and Vegetables"), Dairy Marketing Inc. and the NDC launched 3-A-Day® of Dairy for Stronger Bones," a marketing and nutrition education campaign which continued through 2010. The 3-A-Day program objectives were to increase total consumption of dairy products and reinforce dairy as the leading source of calcium by providing simple guidance about dairy food selections. Health professional outreach remained a critical component of the 3-A-Day program. The American Academy of Family Physicians, the American Academy of Pediatrics, the American Dietetic Association (now The Academy of Nutrition and Dietetics), the National Medical Association, the School Nutrition Association, and the National Hispanic Medical Association all supported and partnered with 3-A-Day™. Research conducted by DMI. indicated that Hispanic immigrants to the United States are often great consumers of dairy when they first arrive, but dairy consumption decreases with integration into U.S. culture, and the partnership was cited a means to "better educate Hispanics, both those who've recently arrived and those who've lived in the United States for generations, on the need for three daily servings of low-fat and fat-free dairy."[56]

Not surprisingly, the NDC operationalizes the RDA for calcium by expressing it in terms of the *number of servings of dairy products* necessary to obtain the recommended amount of calcium. In February 2011, the NDC asserted that the new 2010 *Guidelines* "maintains the 2005 Guidelines' recommendation that Americans aged 9 years and older consume three servings of low-fat or fat-free dairy everyday."[57] Despite the fact that the Dietary Guidelines provides recommendations for nondairy as well as dairy sources of calcium and vitamin D, nondairy sources of these nutrients are (naturally) rarely mentioned in nutrition education materials produced by the NDC.

As a consumer, you may be thinking—of course the NDC is going to promote dairy. *Why is this significant?* Because educators of all types—including registered dietitians— may overly rely on NDC's educational material due to prominence and saturation.

While the USDA has taken small steps to recognize the value of nondairy sources of calcium and vitamin D, the dairy industry is apparently threatened by the government's progress. In 2013, the Dairy Council of California distributed a white paper to registered dietitians criticizing the USDA's MyPlate guidance suggesting that calcium fortified soymilk is a reliable means for someone who is lactose intolerant to get the health benefits of dairy. The Dairy Council of California provides the following warning:

> Many people are choosing milk "substitutes"—soy, almond and rice beverages, among others—for a variety of reasons. Few realize, however, that the nutritional package these alternative beverages deliver is simply not the same as real milk.[58]

In fact, fortified soy beverages and reduced fat milk have similar nutrient profiles. The fact that soy beverages have almost double the amount of calcium as milk is an artifact of bioavailability. The calcium in the soy beverage is absorbed at only three-quarters the rate of the calcium in cow milk. See Table 7.5 for a nutrient comparison.

The next section discusses other challenges to the belief in milk as the perfect food.

TABLE 7.5 Nutrient comparison of fortified soy beverage and types of milk

	FORTIFIED SOY BEVERAGE	WHOLE COW'S MILK	REDUCED-FAT (2%) MILK	LOW-FAT(1%) MILK	SKIM(FAT FREE) MILK
Calories	130	150	120	100	85
Fat (g)	3.6	8	5	3	0.4
% Fat	30	48	38	27	4
Calcium (mg)	585[a]	290	297	300	302
Vitamin B$_{12}$ (μg)	1.2	0.87	0.89	0.90	0.93

[a] Calcium in fortified soy milk is absorbed at the rate of 75% of calcium in cow's milk.

CHALLENGES TO THE BELIEF IN MILK AS THE PERFECT FOOD

For almost a century, the NDC has trumpeted the value of calcium in the diet, and the value of milk and dairy products as the richest source of calcium. With the powerful voice of the NDC both in education and lobbying, milk and dairy products have secured a permanent spot in the national dietary guidance and government-sponsored nutrition programs.

The potential risks of milk consumption are ignored in NDC educational and reference material. Yet epidemiological evidence has reported conflicting information concerning dairy consumption and risk or benefits related to chronic diseases.

Can We Be Confident of the Safety of High Intakes of Milk or Calcium?

Unlike plant-based foods (fruits, vegetables, beans, and legumes), the consumption of which is almost universally recommended for health maintenance and disease prevention (see Chapter 10, "Fruits and Vegetables," and Chapter 3, "Dietary Guidance"), the case for milk and dairy products is weaker. Although consumption of large amounts of milk and/or calcium is associated with the decreased risk of developing some noncommunicable diseases, milk and/or calcium consumption is associated with the increased risk of developing other such diseases.

In terms of our overall dietary choices, while many dairy products are available in reduced-fat and nonfat forms, the butterfat removed from some products is inevitably restored to our food environment to be consumed as premium ice cream, butter, or baked goods. The use of surplus milk fat is a process analogous to surplus corn being made into high fructose corn syrup, which becomes an ingredient in many processed foods.[63]

- A diet high in calcium has been implicated as a probable risk factor for prostate cancer.[64–66] Several cohort studies found suggestions that the risk for prostate cancer may vary by the fat content of milk, with a higher risk associated with low-fat and nonfat milk.[67,68] A 2014 literature review concludes that milk intake appears to be associated with increased risk of prostate cancer.[69]
- Butter fat (the fat in milk), is a rich source of saturated fats. A high saturated fat intake (greater than 10% of total calories) is a recognized risk factor for heart disease.[70,71]

Even osteoporosis is not a clear win for the dairy industry. Despite the strong case for milk as the best source of calcium, osteoporotic fractures occur at higher rates in milk-consuming Western countries than countries in Asia and sub-Saharan Africa with less dairy consumption.[72] The connection is unclear, but one reason for this difference may be that high intakes of protein lead to decreased calcium absorption.[73]

Infectious Diseases Attributed to Milk

Today, we tend to overlook the risk of foodborne illness and premature death that was once associated with milk consumption. Milk has the potential to serve as a vehicle of disease transmission and has, in the past, been associated with disease outbreaks of major proportions. Unpasteurized milk is likely to harbor foodborne pathogens such as *E.coli*, *Campylobacter*, and *Salmonella*, which are shed in animal feces and can contaminate milk during milking.

With the promotion for milk as the "perfect food" in the mid-nineteenth century, many infants who lived in cities were fed raw cow's milk as a replacement for breast milk and in an effort to provide the "perfect food." The assumption by city dwellers of the healthfulness of milk backfired tremendously as contaminated milk contributed to an increase of infant mortality. In fact, one historian referred to the nineteenth century city milk supply as "white poison."[4] Despite increasing evidence that milk was potentially deadly to drink, especially for infants, milk was still fed to babies and young children.

Even as late as 1938, milkborne outbreaks constituted one-quarter of all disease outbreaks due to infected foods and contaminated water. Currently, milk and dairy products are associated with less than 1% of reported foodborne outbreaks.[74] In fact, since 1993, most milkborne infections have been attributed to raw milk. During the 13-year period (1993–2006), there were 121 dairy-related outbreaks reported to CDC, 60% of which were linked to raw milk products. Three-quarters of the outbreaks occurred in states where the sale of raw milk was legal at the time. Moreover, the outbreaks associated with raw milk more often led to illness of greater severity and hospitalization than outbreaks associated with pasteurized milk.[75]

The use of the commerce clause by the federal government to regulate our food is discussed in more detail elsewhere in this book. The FDA has prohibited distribution of nonpasteurized dairy products in interstate trade since 1987, with an exception allowing for sale of cheese that has been aged for 60 days under legally regulated conditions. Illnesses and outbreaks associated with consumption of nonpasteurized products continue to occur, despite the federal ban on the sale of these products in interstate commerce, the broad use of pasteurization by the dairy industry, and the infrequency with which nonpasteurized dairy products are consumed.[75]

Lactose Intolerance: Lactase Persistence

Low-fat and fat-free dairy products are ubiquitous and a relatively inexpensive, versatile and tasty source of nutrients. Nevertheless, *dairy is not for everyone*. People with milk allergies should not consume dairy and those with lactose intolerance may prefer not to consume dairy.

It has been estimated that less than 1% of the population is allergic to cow's milk, although the prevalence of allergy is higher when self-reported. In general, studies that used self-report found a higher prevalence of food allergies than studies that used clinically observed measures such as skin prick testing, food-specific IgE determinations, or double-blind, placebo-controlled food challenges. For cow's milk, the pooled estimate of allergy prevalence was 3.5% (95% CI, 2.9–4.1%) when self-reported and 0.6–0.9% as assessed by the other three methods.[76]

Lactose intolerance, however, is much more prevalent than cow's milk allergy. The consumption of cow's milk in adulthood by humans defies the evolutionary role of milk in all mammalian species, which is to *nourish infants*. Mother's milk is the first *and only* food of an infant mammal. Lactase, an enzyme in the newborn and infant's digestive tract, acts to prepare lactose, the carbohydrate in milk, for digestion. After infancy, as mammals start to consume a variety of foods, their lactase production decreases and in most cases disappears altogether. For most humans, there is a gradually decreasing gene expression for the lactase enzyme, which leads to lactose intolerance in adulthood.[79]

See Box 7.2: Lactose digestion, absorption, and lactase insufficiency.

Biblical arguments by Robert Hartley and the other milk agitators aside, keeping animals for milk production was not universally practiced. Descendants of populations that did depend on various animals' milk production are more likely to be able to digest lactose as adults. People with Northern European heritage have a much greater chance of being able to continue to digest milk as adults, with a 5% risk of lactose intolerance. Conversely, lactose intolerance prevalence can be higher than 90% among adults in some East Asian communities, and is also very common in people of West African, Middle Eastern, and Southern European descent.[77]

Most adults are lactose intolerant, due to the gradually decreasing activity (expression) of the *LCT* gene after infancy. *LCT* gene expression is controlled by a DNA sequence called a regulatory element. The regulatory element is actually located within a nearby gene called *MCM6*, and the ability to digest lactose into adulthood depends on which variations individuals have inherited from their parents.[77] At least five human populations around the world have independently evolved mutations regulating the expression of the lactase gene, which results in continued lactose tolerance

---

BOX 7.2   LACTOSE DIGESTION, ABSORPTION, AND LACTASE INSUFFICIENCY

The carbohydrate in milk is lactose, a disaccharides (2-sugar unit) composed of a molecule of the monosaccharide (1-sugar unit) glucose to which is attached a molecule of the monosaccharide galactose. In order to be absorbed through the intestinal wall into the body, lactose must be broken apart (digested) into its monosaccharide constituents. Digestion of lactose is catalyzed by the enzyme, lactase. When there is enough lactase present in the small intestine, the milk sugar from the milk we drink is digested and absorbed. However, if there is not enough lactase present in the small intestine to digest the lactose, the lactose molecules, which are too large to be absorbed, continue down the intestinal tract to the large intestine, where the molecules ferment. The fermentation results in gas, which causes cramps. The fermented sugar attracts water, which results in diarrhea.

in adulthood.[78] Adults who can still digest lactose are considered lactase persistent. Anthropology professor Andrea S. Wiley describes the significance of using the term "lactase persistent" rather than lactose intolerance; stating "*lactose* implies that lactose is part of the diet; *lactose maldigestion* or *malabsorption* would never manifest itself if lactose was not being consumed."[79]

Lactose Intolerance: Ailment or a Normal State?

Given the high prevalence of lactose intolerance, and the fact that lactose persistence has been attributed to relatively recent gene mutations, at least in the Northern European lactose-persistent population,[78] it is necessary to ask if lactose intolerance deserves to be perceived and treated not as an ailment but as a naturally occurring phenomenon. In fact, lactase nonpersistence is the most common phenotype. *If a population majority is likely to be lactose intolerant, should government dietary guidance first promote nondairy sources of calcium?*

The first three editions of the DGAs include the recommendation for 2–4 servings of milk or dairy products every day. The earliest Guidelines do not mention nondairy sources of cow's milk that deliver comparable amounts of milk's leader nutrients—calcium, vitamin D, and protein. At the time, some people claimed that the failure to suggest alternatives to milk and dairy foods indicated ethnocentrism (bordering on unintentional racial bias), in part because it was clear by the late 1980s that reduced lactase activity was the genetic norm. They suggested that a positive public health measure would be to include nondairy sources of milk's leader nutrients as part of the "Milk Group."[80] Indeed, current national nutrition guidance does include nondairy sources of calcium in the Milk Group, and federal food assistance programs such as program for Women, Infants, and Children (WIC) and the school meals programs must now assure that nondairy sources of calcium are available as appropriate.

The Dairy Industry's Public Response to Lactose Intolerance

When the near perfection of milk is challenged, the NDC has a history of responding with damage control through counter arguments that are characterized as "research-based." Sometimes, in anticipation of a challenge, the NDC volleys a preemptive strike. Lactose intolerance is a perennial topic addressed by the Dairy Council.

In *Lactose Intolerance & Minorities: The Real Story* (2008), the NDC presented strategies for minority populations to enjoy milk and dairy products even if they had a history of gastrointestinal distress after consuming milk. Three studies were cited (all sponsored by the NDC and all with the same lead author), which demonstrate that mostly White and Asian self-diagnosed lactose intolerant men and women were able to consume dairy products under various circumstances. In particular, the subjects tolerated 1 cup of milk with breakfast,[81] 2 cups of milk divided between breakfast and dinner,[82] and 1 cup of milk for breakfast, 1 ounce of cheese and 8 ounces of yogurt for lunch, and 1 cup of milk and 1 ounce of cheese at dinner.[83]

The Dairy Council thus advised individuals with symptoms of lactose intolerance to

1. *Adjust the amount of lactose consumed* by having a small amount of milk with food and gradually increase the serving size until symptoms just begin to develop.
2. *Train for tolerance* by gradually increasing the intake of milk to improve tolerance to lactose.
3. *Drink milk with a meal or snack* to slow gastric emptying and/or delivery of lactose to the colon.
4. *Choose wisely,* as some dairy foods are better tolerated than others.
5. *Try lactose-free or lactose-reduced milk products,* which contain all the same nutrients as milk with lactose that has not been hydrolyzed.

Not surprisingly, the "Real Story" is no longer available online. The newest version published by the NDC, finalized in 2011, is titled "Lactose Intolerance Among Different Ethnic Groups," with updated research. The studies have been replaced with newer ones, along with information from the 2010 National Institutes of Health (NIH) Consensus Development Conference on Lactose Intolerance and Health. The Consensus Conference expert panel concluded that the majority of people with lactose malabsorption do not have clinical lactose intolerance"; and those diagnosed with "lactose malabsorption can ingest 12 g of lactose (the equivalent of 1 cup of milk) without significant symptoms, particularly if ingested with other foods." The tips are updated as well:

1. Sip it (start with a small amount and build tolerance slowly).
2. Try it (seek out low-lactose or lactose-free products).
3. Stir it (mix with other foods to help in slow digestion).
4. Slice it (use cheese which is naturally low in lactose: cheddar, colby, Monterey jack, or swiss).
5. Spoon it (the active cultures in yogurt may help digestion).

Overall, the document retains the sense that self-diagnosis must not be relied upon, and that people with lactose intolerance can learn new strategies to consume dairy products.[84]

While the preponderance of information promulgated by the Dairy Council may be accurate; occasionally, they can be accused of not telling the whole truth, or of providing irrelevant information that obfuscates a truth they would rather not be told. For example, in the following 3-sentence "myth buster" that appeared online in April of 2013 (and is still online at the time of this writing), the Dairy Council of California© claims that soy beverages are not appropriate alternatives to cow's milk.

Although the protein and fat content are similar, soy beverages are naturally low in calcium, containing only about 10 mg per serving [emphasis theirs]. Manufacturers often fortify soy beverages with calcium, but the amount added is not regulated and can vary from 80 to 500 mg per serving. Moreover, the absorption of the calcium from soy beverages may not consistently be as high as from cow's milk.[85]

The first sentence is true, but irrelevant. Since the 2000 Dietary Guidelines, official dietary advice for Americans has included *fortified* soy beverages: "one cup of soy-based beverage *with added calcium* is an option for those who prefer a nondairy source of calcium," and soy-based beverage with added calcium is included in a list of calcium sources.[86]

The second sentence is also true, but irrelevant. In federal nutrition guidance, it is assumed that the fortified soy beverages in the Milk Group will contain the same amount of absorbable calcium as in cow's milk. Implicit in federal nutrition guidance is that soymilk is fortified to 500 mg of calcium per 1 cup serving. One cup of soymilk fortified to 500 mg of calcium and the 285 mg of calcium naturally occurring in one cup of cow's milk yield approximately the same amount of absorbable calcium.[87]

The last sentence is true, but not the whole truth. It is true that the rate of calcium absorption from fortified soy beverages is lower than the rate of calcium absorption from cow's milk. The 3 cups (710 mL) of cow's milk recommended daily by the 2010 Dietary Guidelines provides 855 mg total calcium and 186 mg absorbable calcium, while the same amount of calcium-fortified soymilk provides 1104 mg total calcium with 200–233 mg absorbable calcium, depending on the type of calcium used to fortify the beverage. Thus, the extra calcium used to fortify soymilk makes up for the reduced bioavailability. Both beverages are good sources of absorbable calcium.[87]

How can we assure consumers that one cup of calcium-fortified plant-based beverages delivers the same amount of absorbable calcium as an equivalent serving of cow's milk? We suggest that in the future:

- In order to be included in the Milk Group, soy beverages must be specified as fortified with 500 mg of calcium per cup, and
- Manufacturers should fortify other plant-based beverages with the amount of calcium necessary in order to absorb the same amount of calcium as in a cup of cow's milk, or
- Dietary guides should specify the amount of fortified plant-based beverage that must be consumed in order to provide the same amount of absorbable calcium as a cup of cow's milk.

The national recommendation to consume low-fat (1%) and fat-free (0%) milk has recently been questioned by Harvard physician-nutritionists who assert that few randomized clinical trials have examined the effects of 0–2% milk compared to whole (4%) milk on weight gain and other health outcomes. Milk guidance is based on the assumption that replacing the full-fat milk with lower calorie milk will automatically reduce calorie intake and thus prevent unintentional weight gain. However, it is also plausible that for people consuming a low-quality diet, replacing full-fat milk with an equal quantity of a lower fat variety will lead to compensation for the decrease in calories by consuming other low quality foods that are consistent with the rest of the diet. Pending clinical trials that indicate otherwise, among their suggestions for future guidance is to recommend a broader acceptable range of intake (0–3 cups per day) and focus on limiting sugar-sweetened milk.[88]

While all recent federal food guidance has presented milk and dairy products in its own food group, nonfederal organizations do not have the same agenda as the U.S. government. Food guidance from nongovernment organizations does not always elevate dairy products to the pantheon of dietary essentials. Two examples are the New American Plate (American Institute for Cancer Research, 1999) and the Healthy Eating

Plate (Harvard School of Public Health and Medical School, 2011). See Chapter 3, "Dietary Guidance," for figures of these plates.

When the 2015 Dietary Guidelines are published, you will be able to ascertain if the new guidelines provide adequate nutrition information about other sources of calcium, and perhaps more importantly, address lactose intolerance as a naturally occurring, common condition. Our Dietary Guidelines help to drive the milk policies manifested in our federal food assistance programs; which are discussed in the next section.

POLICIES FOR MILK IN FEDERAL FOOD ASSISTANCE PROGRAMS

Through the educational and lobbying efforts of dairy interest groups, such as the NDC, milk, and other dairy products are at the core of most federally funded food assistance programs. Substantial amounts of dairy products are produced, distributed, and consumed for government programs. These programs include the National School Lunch and Breakfast Programs, the Special Milk Program, and the Supplemental Nutrition Program for Women, Infants, and Children (WIC).

The early child food assistance programs involving school food programs and milk were justified not only on their nutritional grounds, but also for promoting national defense and absorbing surplus commodities. As observed in a 1961 report, "The encouragement of milk consumption in the school market has been an effective means of moving large quantities of milk to a constructive yet noncompetitive use. In 1960, 2.5 billion pounds were consumed in schools under the National School Lunch and Special Milk Program."[89]

Public provision of lunch to children precedes all other government food assistance programs. Philadelphia began offering school lunches during the first decade of the twentieth century and New York City started serving lunches to children in the Bronx and Manhattan in 1919. Federal involvement began during the Depression when the Federal Surplus Commodity Corporation distributed surpluses to schools, a program that continued through World War II. During those years, feeding children was incidental to fostering employment by hiring school food service workers and disposing of surplus foods. After the war, concern over the high number of military draftees who failed their physical exam because of nutritional deficiencies led Congress to establish a permanent national lunch program for school children "as a measure of national security, to safe-guard the health and well-being of the nation's children and to encourage the domestic consumption of nutritious agricultural commodities and other food..."[90]

National School Lunch and Breakfast Programs

The National School Lunch and Breakfast Programs have been mainstays of American public schools since the mid-twentieth century. Schools can be reimbursed for these meals, but in both programs, milk is a required component of a reimbursable meal.[91]

From the 1940s through the 1970s, whole milk was served as part of the NSLP. In the mid 1970s, the USDA required that schools offer low-fat or skim milk as an option.[92] By 2004, lactose-free milk was specifically allowed, but alternative beverages only grudgingly allowed for students whose disability restricts their diet, and only on receipt of a written statement from a licensed physician that identifies the disability and that specifies the substitute for fluid milk.[93]

The school year 2012–2013 brought a significant change. Now, the milk options provided must be consistent with the 2009 IOM report on school meals[94] and the 2010 Dietary Guidelines. Thus, whole milk and even 2% milk is no longer allowed. The standards allow only for low-fat (1%) or fat-free plain milk or fat-free flavored milk in the school meal programs. The regulations do not set any specific calorie or sugar limits for fat-free flavored milk sold on the meal line, but with the corresponding new calorie restrictions for meals, which include milk, schools will be seeking flavored milk with the least possible calories or possibly may limit flavored milk altogether.[95]

The Special Milk Program

The Special Milk Program was inaugurated in 1954, when the 83rd Congress authorized the use of Commodity Credit Corporate funds to reimburse schools for milk served above regularly served amounts. In other words, Congress was explicitly providing an incentive to stimulate greater milk consumption among school children.[96]

This program currently provides milk to children in schools and childcare institutions who do not participate in other Federal meal service programs. The program reimburses schools for the milk they serve. Schools or institutions must offer only pasteurized types of unflavored or flavored fat-free or low-fat (1%) milk. These milks have to meet all state and local standards. Much like the NSLP and the SBPs, the Special Milk Program has eligibility criteria based upon household income. For school year 2013–2014, the federal reimbursement rate is 20.25 cents for each half-pint of milk sold to children, and the full purchase price of the milk is reimbursed for milk received free. The Special Milk Program peaked in the 1960s; in 1969 3 billion half-pints of milk were served. Due to the expansion of the school breakfast and lunch programs, the amount of milk provided through the SMP has declined significantly, to 61 million half pints in 2012 at a cost of $12.3 million.[97]

The Supplemental Nutrition Program for Women, Infants, and Children (WIC)

The dairy industry also benefits from the "food packages" offered through the WIC Program. Dairy products provided key components in the WIC food packages; which were developed in 1972 when the program was launched. In 2009, USDA introduced a new set of food packages based on recommendations from the Institute of Medicine that aligned with the 2005 DGA and infant feeding practice guidelines of the American Academy of Pediatrics.[98] The proposed food package reduced the amount of milk provided to women and children at least 2 years of age, requires whole milk for children until the age of two, eliminates whole milk for almost all women and for children over the age of 2, adds tofu and soy as milk alternatives, and provides an extra pound of cheese for fully breastfeeding mothers. The Final Rule was published on March 4, 2014, with some changes relevant to this chapter's focus on milk.[99]

- Milk alternatives: The Final Rule removed the requirement that a health care professional licensed to write medical prescriptions must provide documentation for children to receive soy-based beverages and tofu as milk alternatives. A WIC nutritionist can authorize this substitution through individual nutritional assessment, including but not limited to, milk allergy, lactose intolerance, and vegan diets; but the rule indicates that lactose-free dairy products should be offered first. This reflects the program's goal of providing more culturally acceptable foods while softening the potential of critical disbelief over individual's requests.
- *Less whole milk*: Food and Nutrition Service (FNS) also received numerous comments about letting women have whole milk, but most of these comments were from local WIC agency staff. Although the substance of the comments is not discussed in the final rule, these comments may also have been focused on cultural acceptability, but in this case the FNS declined, commenting that "Whole milk adds unnecessary saturated fat and cholesterol to the diets of participants," and that nutritional needs can be met through fat-reduced milks and other foods.
- *Addressing obesity*: Comments also pointed out that the American Academy of Pediatrics recommends fat-reduced milk for children over the age of 1 if overweight or obesity is of concern, and accordingly, in the final rule, the FNS implemented this option at the level of the State agency.
- *Dietary guidelines*: The interim rule allowed 2% milk for women and children over the age of 2 years. To further follow the Dietary Guidelines, the Final Rule is proscribing 1% milk as the standard for some of the food packages.
- *Lactose intolerance*: Under the interim rule, FNS allowed, with medical documentation, additional amounts of cheese to be issued beyond the substitution rate for milk so that State agencies could accommodate participants with lactose intolerance. However, this was in part due to the fact that at the time of the interim rule, not enough soy products were available in the marketplace that met WIC nutrition requirements and the interim rule did not address yogurt (which had been recommended by the IOM). Since yogurt is included in the final rule, and as there are more appropriate alternatives available, the packages no longer allow for additional cheese beyond the substitution rate.
- *Tofu*: The final rule clarifies that tofu must be calcium set, but may also contain other coagulants.
- *Yogurt*: was not included in the interim rule due to the questions of cost containment. Comments received pointed out that yogurt provides priority nutrients and is convenient, popular, and culturally acceptable. Also noted was a pilot study, conducted by the California WIC Program in conjunction with

TABLE 7.6   The dairy components of WIC food packages, 1972 and 2009

WOMEN AND CHILDREN (FOOD PACKAGES IV–VII) WIC MONTHLY MILK AND CHEESE ALLOWANCES	
OLD RULES (1972)	FINAL NEW RULE (2014)
• No restrictions on milk fat content. • Maximum milk prescriptions ranged rom 24 to 28 quarts of milk per month. • Cheese could be substituted for milk at a rate of one pound per three quarts; cheese could replace a total of 12 quarts of milk.	• Women and children age two and older receive fat reduced milk; whole milk is provided to children with one year of age. • Maximum milk prescriptions reduced to provide 16 quarts for children and up to 24 quarts for women (16 quarts for postpartum women, 22 quarts for partially breastfeeding women, and 24 quarts for fully breastfeeding women). • Added new milk substitution options (tofu, cheese, and soy beverage), removed extra substitution for cheese in the case of lactose intolerance. • Authorized an additional pound of cheese for fully breastfeeding women.

the NDC, which demonstrated the feasibility of providing yogurt in WIC food packages. With the Final Rule, FNS allows plain or flavored yogurt that has less than ≤40 g of total sugar per 1 cup of yogurt, but only low fat or nonfat for women and children over the age of 2. Note: under WIC rules, some items are allowed to have sugar, but some must not have any sugar, such as frozen fruit, dried fruit, salsa, spaghetti sauce, canned tomato sauce and paste, and canned vegetables. *Why not allow only plain, unsweetened yogurt? What nutritional benefit comes from the added sugar?*

See Table 7.6 for comparison of WIC dairy packages.

The Child and Adult Care Food Programs (CACFP)

Participating programs for children and the elderly are required to provide specific food groups in specific quantities in order to receive reimbursement for the meals served. The required CACFP food groups are: milk, bread/bread alternate, fruit, vegetable, and meat/meat alternate (includes yogurt and cheese). Some soy milks serve as creditable substitutions for cow's milk for participants whose parents or guardians submit a written request for the substitution.

Children through 18 years of age:

- Breakfast requires *milk*, fruit or vegetable, and bread or grain product.
- Lunch and dinner consist of *milk*, bread or grain product, and two different fruits and/or vegetables. Meat or *meat alternate (includes yogurt and cheese)*.

- Snacks include two of the following components: *milk*, fruits or vegetables, bread or grain product, meat or *meat alternate (includes yogurt and cheese)*.

Adults at least 60 years of age:

- Breakfast requires *milk,* fruit or vegetable, and bread or grain product
- Lunch requires *milk,* fruit or vegetable, bread or grain, meat and *meat alternate (includes yogurt and cheese)*
- Supper requires fruit or vegetable, bread or grain, meat and meat alternate (includes yogurt and cheese)
- Snack consists of any two of the following components: *milk,* fruit or vegetable, bread or grain, and meat and *meat alternate (includes yogurt and cheese)*

A milk or dairy product must be served at every meal (except for snacks). Note that as yogurt and cheese count as meat alternatives, each lunch and supper meal could possibly have two forms of dairy.[100]

Supplemental Nutrition Assistance Program (SNAP)

The dairy industry has *not* been able to penetrate the Supplemental Nutrition and Food Program (SNAP, formerly known as the Food Stamp Program). SNAP is the only major food assistance program that does not require that milk is a part of the program. As discussed in Chapter 6, "Federal Food Assistance," SNAP does not require the purchase of any particular food items, even if the items represent food groups deemed essential to the health of people enrolled in the other federal food assistance programs.

It is a source of considerable consternation to many nutrition advocates that the SNAP program does not do a better job of encouraging its participants to make wise food choices. Conversely, others argue that it is paternalistic to implicitly assert that the government knows what is best for SNAP participants. *Why isn't this argument made for WIC?*

Two basic types of approaches to remedy the situation have been proposed: (1) impose specific restrictions on what types of foods may be purchased through SNAP beyond the current bans on alcoholic beverages or (2) provide additional benefits to encourage the purchase of healthier foods, such as fresh fruits and vegetables, and low-fat and fat-free dairy products.[101]

CONCLUSION

Overall, the inclusion of milk in federal assistance programs emphasizes milk as a necessary food for children and disposes of surplus milk, such as via the three dairy products available in the Commodity Supplemental Food Program: American cheese blend (skim), instant milk (most likely nonfat), and 1% milk.[102]

Dairy products and their purported essential nutritional properties are taken for granted in contemporary United States society as having always been a staple of the country's diet. The perceived necessity of milk and dairy products in our diet is well ensconced in our culture and dietary guidance. And this (condensed) history of milk raises many questions about U.S. nutrition policy. If the promotion of milk in schools and nonfat milk as a diet food was primarily for disposing of an agricultural nuisance, what can we infer as to the validity of current nutritional advice?[23]

Since the early twentieth century, the government has promoted milk as a good source of protein, vitamins, and minerals. With the powerful voice of the NDC both in education and lobbying, milk and dairy products have secured a permanent spot in the national dietary guidance and government-sponsored nutrition programs. Milk may have been perceived in the beginning of the twentieth century as "nature's perfect food," but this viewpoint failed to consider that milk was most often historically consumed as a fermented product, which, by drastically reducing the levels of lactose, would have been more easily digestible for adults who were not lactose-persistent.[79]

Is it appropriate for milk is touted as necessary, by the government, in schools and popular culture, when significant parts of the population experience physical discomfort from eating dairy foods?[5]

There are a wide variety of vegetables and fortified food products that provide calcium with at least, if not more, bioavailability than milk.[103] Whether or not modern dairy products are an essential part of the human diet is gaining vocal debate in many parts of society; including through continued and conflicting research examining the relationship between dairy consumption and risks or benefits for chronic disease.

The "Milk Question" of the twentieth century was "How could a safe supply of a necessary food be created when that food could also be harmful to health?" Looking forward, the "Milk Question" of the twenty-first century may likely be "Is milk a necessary food for everyone?"

REFERENCES

1. Hunt CL, Atwater HW. How to Select Foods: What the Body Needs. 1917. USDA. Farmers' Bulletin 808. Available at http://archive.org/details/howtoselectfoods00hunt, accessed September 12, 2014.
2. Levenstein H. *Revolution at the Table: The Transformation of the American Diet.* New York: Oxford University Press, 1988.
3. Spargo J. *The Common Sense of the Milk Question.* New York: The Macmillan Company, 1908.
4. DuPuis M. *Nature's Perfect Food: How Milk Became America's Drink.* 1st edition. New York: New York University Press, 2002.
5. Smith-Howard K. Perfecting Nature's Food: A Cultural and Environmental History of Milk in the United States, 1900–1970 [dissertation]. University of Wisconsin Madison, 2008.
6. Mendelson A. *Milk: The Surprising Story of Milk through the Ages.* 1st edition. New York: Random House, 2008.

7. Don PB. *The Changing Landscape of U.S. Milk Production.* United Economic Research Service, Statistical Bulletin Number 978. June 2002. United States Department of Agriculture. Available at http://www.ers.usda.gov/publications/sb-statistical-bulletin/sb978.aspx#.VEPjf764l0s, accessed October 19, 2014.

8. Prentice A. Constituents of human milk. *Food Nutr. Bull.* 1996;17(4):126–141. United Nations University. http://archive.unu.edu/unupress/food/8F174e/8F174E04.htm, accessed October 18, 2014.

9. Pure Milk for the Masses: Success of Mr. Straus's East Third Street Depot, *The New York Times,* July 30, 1893.

10. Melvin P. Milk to motherhood: the New York milk committee and the beginning of well-child programs. *Mid-America* 1983;65(3):111–133.

11. USDA National Agricultural Library. Early Developments in the American Dairy Industry. Available at http://www.nal.usda.gov/speccoll/images1/dairy.htm, accessed October 21, 2014.

12. Van Vleck R, American Artifacts: Scientific Medical & Mechanical Antiques. Early Cow Milking Machines. Available at http://www.americanartifacts.com/smma/milker/milker.htm, accessed October 21, 2014.

13. Graybeal JA.Mehrings Pneumatic Foot Power Milker, *Carroll County Times,* October 21, 2001. Available at http://www.hsccmd.org/Documents/Carroll%20County%20Times%20Yesteryears/2001/10-21-2001.pdf, accessed October 21, 2014.

14. Halliday S. Death and Miasma in Victorian London: an obstinate belief. *BMJ* 2001;323(7327):1469–1471.http://www.ncbi.nlm.nih.gov/pmc/articles/PMC1121911/, accessed October 19, 2014.

15. Debré P, Forster E. *Louis Pasteur. Trans.* Forster E. Baltimore: The Johns Hopkins University Press, 1998.

16. Koch R. Biogrpahy. The Nobel Prize in Physiology or Medicine, 1905. Nobelprize.org. The Official website of the Nobel Prize. Available at http://www.nobelprize.org/nobel_prizes/medicine/laureates/1905/koch-bio.html/, accessed October 21, 2014.

17. Rosenau MJ. *The Milk Question.* Cambridge, MA: Houghton Mifflin The University Press, 1912.

18. Wilcox L. Worms and germs, drink and dementia: US Health, Society, and Policy in the Early 20th Century. *CDC. Preventing Chronic Disease, Public Health Research. Practice Policy* 5(4); October 2008. Available at http://www.cdc.gov/pcd/issues/2008/oct/08_0033.htm, accessed October 7, 2014.

19. Fankhauser D. Making Buttermilk. Available at http://biology.clc.uc.edu/fankhauser/cheese/buttermilk.htm, accessed October 19, 2014.

20. Apple R. *Mothers and Medicine: A Social History of Infant Feeding, 1890–1950.* Wisconsin: The University of Wisconsin Press, 1987.

21. Tunick M. Dairy Innovations over the Past 100 years. *J. Agric. Food Chem.* 2009;57:8093–8097.

22. Chicago Produce Co. Inc. *Chicago Dairy Produce* 1915;22:25.

23. Bammi V. Nutrition, the historian, and public policy: A case study of U.S. nutrition policy in the 20th century. *J. Soc. Hist.* 1981;14(4):627–648.

24. Froker RK. The problem of dairy surpluses. *J. Dairy Sci.* 1954;37(1):113–116.

25. Abrahams P. Agricultural adjustment during the new deal period the New York milk industry: A case study. *Agric. Hist.* 1965;39:92–101.

26. Ippolito R, Masson R. The social cost of government regulation of milk. *J. Law Econ.* 1978;21(1):33–65.

27. Mackenzie C. More Milk on New York's Tables. *New York Times* January 19, 1936.

28. Gunderson GW, USDA Food and Nutrition Service. The National School Lunch Program Background and Development. Available at http://www.fns.usda.gov/cnd/lunch/AboutLunch/ProgramHistory.htm,accessed October 21, 2014.

29. USDA. Agricultural Marketing Service. Dairy Production Stabilization Act of 1983 (7 U.S.C. 4501–4514). (As Amended through May 7, 2010.) Available at http://www.ams. usda.gov/AMSv1.0/getfile?dDocName=STELDEV3021892, accessed October 21, 2014.

30. Moss M. *Salt Sugar Fat How the Food Giants Hooked Us.* New York: Random House, 2013.

31. DMI. History. Available at http://www.dairy.org/about-dmi/history, accessed October 21, 2014.

32. National Dairy Council. Available at http://www.nationaldairycouncil.org/Pages/Home. aspx, accessed October 19, 2014.

33. Center for Responsive Politics. Annual Lobbying on Dairy. 2013. Available at http://www. opensecrets.org/lobby/indusclient.php?id=A04&year=2013, accessed October 19, 2014.

34. Health.gov. Office of Disease Prevention and Health Promotion. Dietary Guidelines for Americans, 2015. http://www.health.gov/dietaryguidelines/dga2015/comments/readCom ments.aspx, accessed October 21, 2014.

35. Favis CG, Blayney DP, Dong D, Stefanova S, Johnson A. *Long-Term Growth in U.S. Cheese Consumption May Slow/LDP-M-193-01.* Economic Research Service/USDA. Available at http://www.ers.usda.gov/media/146945/ldpm19301_1_.pdf, accessed October 21, 2014.

36. USDA. Report to Congress on the National Dairy Promotion and Research Program and the National Fluid Milk Processor Promotion Program. 2010 Program Activities. Available at http://www.ams.usda.gov/AMSv1.0/getfile?dDocName=STELPRDC5100700, accessed October 20, 2014.

37. Moss M. While Warning About Fat, U.S. Pushes Cheese Sales. *New York Times*, November 6, 2010. Available at http://documents.nytimes.com/documents-on-marketing-cheese#document/p4, accessed October 21, 2014.

38. USDA. Center for Nutrition Policy & Promotion. *A Brief History of USDA Food Guides.* Available at http://www.choosemyplate.gov/food-groups/downloads/MyPlate/ ABriefHistoryOfUSDAFoodGuides.pdf, accessed October 21, 2014.

39. Welsh SO, Davis C, Shaw A. *USDA's Food Guide: Background and Development.* USDA, Human Nutrition Information Services, Hyattsville, MD. Miscellaneous Publication Number 1414. September 1993.

40. Davis C, Saltos E. *Dietary Recommendations and How They Have Changed Over Time.* Chapter 2. America's Eating Habits: Changes and Consequences 1999 May 1999;Agriculture Information Bulletin No. (AIB750):33–50.

41. Hunt CL. Food for Young Children. 1916. USDA. Farmers' bulletin no. 717. Available at http://openlibrary.org/books/OL24364099M/Food_for_young_children/, accessed September 12, 2014.

42. Trenholm S. Food Conservation during WWI: "Food Will Win the War". October 16, 2012. *The Gilder Lehrman Institute of American History.* Available at http://www.gild-erlehrman.org/collections/treasures-from-the-collection/food-conservation-during-wwi-%E2%80%9Cfood-will-win-war%E2%80%9D, accessed October 21, 2014.

43. Moorhouse A, Goudiss CH. *Foods that Will Win the War and How to Cook Them.* New York: The Forecast Publishing Company, 1918.

44. National Research Council. *Recommended Dietary Allowances.* Washington, DC: The National Academies Press, 1941, http://www.nap.edu/catalog.php?record_id=13286, accessed June 8, 2014.

45. Stiebeling H. *Food Budgets for Nutrition and Production Programs.* Washington, DC: United States Department of Agriculture, Miscellaneous Publication No 183, 1933.

46. Stiebeling H, Ward M. *Diets at Four Levels of Nutritive Content.* Washington DC: United States Department of Agriculture, Circular No. 296, 1933.

47. U.S. Department of Agriculture. *Historical Dietary Guidance.* Available at http://fnic. nal.usda.gov/dietary-guidance/dietary-guidelines/historical-dietary-guidance, accessed October 20, 2014.

48. Dietary Guidelines Advisory Committee. *Report of the Dietary Guidelines Advisory Committee on the Dietary Guidelines for Americans, 2010, Appendix E-4, History of the Dietary Guidelines for Americans*, 2010. Available at http://cnpp.usda.gov/Publications/DietaryGuidelines/2010/DGAC/Report/E-Appendix-E-4-History.pdf, accessed September 12, 2014.

49. U.S. Food and Drug Administration, Nutrition Facts Label Programs & Materials. Available at http://www.fda.gov/Food/IngredientsPackagingLabeling/LabelingNutrition/ucm20026097.htm, accessed October 22, 2014.

50. U.S. Department of Agriculture and U.S. Department of Health and Human Services. *Dietary Guidelines for Americans*, 7th edition. Washington, DC: U.S. Government Printing Office, December 2010.

51. Ross AC, Manson JE, Abrams SA, Aloia JF, Brannon PM, Clinton SK, Ramon AD et al. The 2011 report on dietary reference intakes for calcium and vitamin D from the Institute of Medicine: What clinicians need to know. *J. Clin. Endocrinol. Metab.* 2011;96:53–58.

52. Institute of Medicine. *DRI Dietary Reference Intakes, Calcium Vitamin D*. Washington, DC: The National Academies Press, 2011. Available at http://www.iom.edu/Reports/2010/Dietary-Reference-Intakes-for-Calcium-and-Vitamin-D/DRI-Values.aspx, accessed October 16, 2014.

53. McGuire M, Beerman KA. *Nutritional Sciences: From Fundamentals to Food*. United States: Thomson Wadsworth, 2007.

54. Weaver CM, Proulx WR, Heaney R. Choices for achieving adequate dietary calcium with a vegetarian diet. *Am. J. Clin. Nutr.* 1999;70(Suppl): 543S–548S.

55. Park YK, Yetley EA, Calvo MS. *Calcium Intake Levels in the United States: Issues and Considerations*. FAO Corporate Document Repository. Available at http://www.fao.org/docrep/w7336t/w7336t06.htm, accessed November 11, 2014.

56. Checkoff Targets Hispanic Consumers. *Daily Herd*. December 10, 2007. Available at http://www.dairyherd.com/dairy-news/latest/checkoff-effort-targets-hispanic-consumers-113937834.html, accessed October 20, 2014.

57. National Dairy Council. *The Dairy Download*. February, 2011. Available at http://www.nationaldairycouncil.org/PressandMedia/DairyDownloadNewsletter/Pages/DairyDownload2011_02.aspx, accessed October 16, 2014.

58. Dairy Council of California. *Milk and Dairy: The Forgotten Food Group?* 2013. Available at http://www.healthyeating.org/Portals/0/Documents/Health%20Wellness/White%20Papers/Dairy_Contributors.pdf, accessed October 21, 2014.

59. Zemel MB, Thompson W, Milstead A, Morris K, Campbell P. Calcium and dairy acceleration of weight and fat loss during energy restriction in obese adults. *Obes. Res.* 2004;12(4):582–590.

60. Atkinson RL. Editor's Note. *Int. J. Obesity.* 2005;29:1392.

61. Schardt D. Milking the Data. Does dairy burn more fat? Don't bet your bottom on it. *Nutrition Action Health Letter* September 2005. Available at http://www.cspinet.org/nah/09_05/milking.pdf, accessed November 3, 2014.

62. Federal Trade Commission May 3, 2007 letter to Physicians Committee for Responsible Medicine. Available at http://milk.procon.org/sourcefiles/FTCResponsetoPCRM.pdf, accessed November 3, 2014.

63. Pollan M. *Omnivore's Dilemma: A Natural History of Four Meals*. New York: The Penguin Press, 2006.

64. World Cancer Research Fund, American Institute for Cancer Research. *Food, Nutrition, Physical Activity, and the Prevention of Cancer: A Global Perspective*. Washington DC: AICR, 2007.

65. Giovannucci E, Liu Y, Platz EA, Stampfer MJ, Willett WC. Risk factors for prostate cancer incidence and progression in the Health Professionals Follow-up Study. *Int. J. Cancer* 2007;121:1571–1578.

66. Giovannucci E, Rimm EB, Wolk A, Ascherio A, Stampfer MJ, Colditz GA, Willett WC. Calcium and fructose intake in relation to risk of prostate cancer. *Cancer Res.* 1998;58:442–447.
67. Park S, Murphy SP, Wilkens LR, Stram DO, Henderson BE, Kolonel LN. Calcium, vitamin D, and dairy product intake and prostate cancer risk: the Multiethnic Cohort Study. *Am. J. Epidemiol.* 2007;166(11):1259–1269.
68. Rohrmann S, Platz EA, Kavanaugh CJ, Thuita L, Hoffman SC, Helzlsouer KJ. Meat and dairy consumption and subsequent risk of prostate cancer in a US cohort study. *Cancer Causes Control* 2007 02;18(1):41–50.
69. Mandair D, Rossi RE, Pericleous M, Whyand T, Caplin ME. Prostate cancer and the influence of dietary factors and supplements: A systemic review. *Nutr. Metab.*, 2014;11:30. Available at http://www.nutritionandmetabolism.com/content/11/1/30, accessed October 21, 2014.
70. USDHHS. National Heart, Lung, and Blood Institute (NHLBI). *Expert Panel on Integrated Guidelines for Cardiovascular Health and Risk Reduction in Children and Adolescents: Summary Report.* NIH Publication No. 12-7486A. October 2012. Available athttp://www.nhlbi.nih.gov/guidelines/cvd_ped/peds_guidelines_sum.pdf, accessed October 21, 2014.
71. National Institutes of Health, National Heart, Lung, and Blood Institute, National Cholesterol Education Program. *Third Report of the Expert Panel on Detection, Evaluation, and Treatment of High Blood Cholesterol in Adults (Adult Treatment Panel III).* Bethesda, MD: National Institutes of Health, 2002. Available at http://www.nhlbi.nih.gov/guidelines/cholesterol/index.htm, accessed October 21, 2014.
72. Prentice A. Diet, nutrition and the prevention of osteoporosis. *Pub. Health Nutr.* 2004;7:227–243.
73. Committee to Review Dietary Reference Intakes for Vitamin D and Calcium, Food and Nutrition Board, Institute of Medicine. *Dietary Reference Intakes for Calcium and Vitamin D.* Washington, DC: National Academy Press, 2010.
74. U.S. Department of Health and Human Services. Public Health Service. *Food and Drug Administration. Grade "A" Pasteurized Milk Ordinance (Including Provisions from the Grade "A" Condensed and Dry Milk Products and Condensed and Dry Whey— Supplement I to the Grade "A" Pasteurized Milk Ordinance).* Public Health Service/Food and Drug Administration, 2009 Revision. Available at http://www.fda.gov/downloads/Food/GuidanceRegulation/UCM209789.pdf, accessed October 21, 2014.
75. Langer AJ, Ayers T, Grass J, Lynch M, Angulo FJ, Mahon BE. Nonpasteurized dairy products, disease outbreaks, and state laws—United States, 1993–2006. *Emerg. Infect. Dis.* 2012;18(3):385–391.
76. Schneider Chafen JJ, Newberry SJ, Riedl MA, Bravata DM, Maglione M, Suttorp MJ, Sundaram V et al. Diagnosing and managing common food allergies: a systematic review: Systematic review. *JAMA.* 2010;303:1848–1856.
77. U.S. National Library of Medicine, Genetics Home Reference. Available at http://ghr.nlm.nih.gov/condition/lactose-intolerance, accessed October 20, 2014.
78. Krüttli A, Bouwman A, Akgül G, Della Casa P, Rühli F, Warinner C. Ancient DNA analysis reveals high frequency of European Lactase Persistence Allele (T-13910) in Medieval Central Europe. *PLoS ONE* 2014 January 23;9(1):e86251. doi:10.1371/journal.pone.0086251.
79. Wiley AS. "Drink milk for fitness": The cultural politics of human biological variation and milk consumption in the United States. *Am. Anthropol.* 2004;106(3):506–517.
80. Bertron P, Barnard N, Mills M. Racial bias in federal nutrition policy, Part II: Weak guidelines take a disproportionate toll. *J. Natl. Med. Assoc.* 1999;91:121–157.
81. Suarez FL, Savaiano DA, Levitt MD. A comparison of symptoms after the consumption of milk or lactose-hydrolyzed milk by people with self-reported severe lactose intolerance. *N. Engl. J. Med.* 1995;333(1):1–4.

82. Suarez FL, Savaiano D, Arbisi P, Levitt MD. Tolerance to the daily ingestion of two cups of milk by individuals claiming lactose intolerance. *Am. J. Clin. Nutr.* 1997;65(5):1502–1506.

83. Suarez FL, Adshead J, Furne JK, Levitt MD. Lactose maldigestion is not an impediment to the intake of 1500 mg calcium daily as dairy products. *Am. J. Clin. Nutr.* 1998;68(5): 1118–1122.

84. National Dairy Council. *Lactose Intolerance Among Different Ethnic Groups.* December 2011. Available at http://www.nationaldairycouncil.org/SiteCollectionDocuments/LI%20 and%20Minorites_FINALIZED.pdf, accessed October 20, 2014.

85. Dairy Council of California©. Soy Milk Is A Good Alternative to Milk. Milk Myth #7. Available at http://www.healthyeating.org/Milk-Dairy/Milk-Myth-Busters/Article-Viewer/Article/35/milk-myth-7-soy-milk-is-a-good-alternative-to-cows-milk.aspx, accessed October 21, 2014.

86. U.S. Department of Agriculture and U.S. Department of Health and Human Services. *Dietary Guidelines for Americans, 2000.*5th edition. Washington, DC: U.S. Government Printing Office, December 2000.

87. Zhao Y, Martin BR, Weaver CM. Calcium bioavailability of calcium carbonate fortified soymilk is equivalent to cow's milk in young women. *J. Nutr.* 2005;135:2379–2382.

88. Ludwig DS, Willett W. Three daily servings of reduced-fat milk: An evidence-based recommendation? *JAMA Pediatr.* 2013;167(9):788–789.

89. Parry SP, Downen ML. *The Federal School Lunch and Special Milk Program in Tennessee with Implications for the Dairy Industry.* The University of Tennessee Agricultural Experiment Station. 1961. Available at http://trace.tennessee.edu/cgi/viewcontent.cgi? article=1264&context=utk_agbulletin, accessed November 3, 2014.

90. Eisinger PK. *Toward an End to Hunger in America.* Washington, DC: The Brookings Institution, 1998.

91. USDA Food and Nutrition Service. *Final Rule Nutrition Standards in the National School Lunch and School Breakfast Programs,* January 2012. Available at http://www.fns.usda. gov/sites/default/files/dietaryspecs.pdf, accessed October 21, 2014.

92. Levine S. *School Lunch Politics: The Surprising History of America's Favorite Welfare Program.* Princeton: Princeton University Press, 2008.

93. The Richard B. Russell National School Lunch Act (As amended through P.L. 108–269, July 2, 2004), Sec.9 (2) FLUID MILK. Available at http://milk.procon.org/sourcefiles/ nationalschoollunchact.pdf, accessed October 21, 2014.

94. National Academy of Sciences. *School Meals: Building Blocks For Healthy Children.* Washington, DC: The National Academies Press, 2009. Available at http://www.nap.edu/ catalog.php?record_id=12751, accessed October 21, 2014.

95. Federal Register. *Nutrition Standards in the National School Lunch and School Breakfast Programs.* January 26, 2012.Available at https://www.federalregister.gov/ articles/2012/01/26/2012-1010/nutrition-standards-in-the-national-school-lunch-and-school-breakfast-programs, accessed October 21, 2014.

96. Gunderson GW. *School Milk Programs.* Last modified June 17, 2014. Available at http:// www.fns.usda.gov/nslp/history_11#Footnotes, accessed November 3, 2014.

97. USDA. *Food and Nutrition Service. Special Milk Program.* Available at http://www.fns. usda.gov/smp/special-milk-program, accessed October 20, 2014.

98. Institute of Medicine. Food and Nutrition Board.*WIC Food Packages: Time for a Change.* Washington DC: National Academies Press, 2005. Available at http://www.iom.edu/ Reports/2005/WIC-Food-Packages-Time-for-a-Change.aspx, accessed October 21, 2014.

99. United States Department of Agriculture. Food and Nutrition Service. *Women, Infants and Children.* Final Rule: Revisions in the WIC package. Available at http://www.fns.usda.gov/ sites/default/files/03-04-14_WIC-Food-Packages-Final-Rule.pdf, accessed October 20, 2014.

100. USDA. Food & Nutrition Service. CACFP. *Meals and Snacks.* Available at http://www.fns. usda.gov/sites/default/files/Child_Meals.pdf, accessed October 21, 2014.

101. Mercier S. *Review of U.S. Nutrition Assistance Policy: Programs and Issues.* Washington DC: Agree Transforming Food & Ag Policy, June 2012. Available at http://foodandagpolicy. org/sites/default/files/AGree%20Review%20of%20US%20Nutrition%20Assistance%20 Policy_1.pdf, accessed October 21, 2014.

102. United States Department of Agriculture. Commodity Supplemental Food Program. USDA available foods for 2015. Available at http://www.fns.usda.gov/sites/default/files/ csfp/FY2015_CSFP.pdf, accessed November 11, 2014.

103. Bertron P, Barnard ND, Mills M. Racial bias in federal nutrition policy, Part I: The public health implications of variations in lactase persistence. *J. Natl. Med. Assoc.* 1999;91(3):151–157.

Meat*

8

The poor man must walk to get meat for his stomach, the
rich man to get a stomach to his meat.
Benjamin Franklin[1]

INTRODUCTION

The United States is intensely carnivorous. In 2012, we processed 8.6 billion chickens, 250 million turkeys, 32.1 million cattle, 113.2 million hogs, and 2.2 million sheep and lambs.[2] That is more than a chicken per person living on planet Earth as of September 2014,[3] which should give pause to thought on the efficiency of our meat production, distribution, and processing system.

The first section of this chapter reviews U.S. dietary guidance, the role of the USDA, and meat consumption and trends. The second section of the chapter provides a discussion of the cost of meat consumption as it relates to human health, discussing the factors enabling industrial meat production and meat consumption (such as subsidized feed, antibiotics, CAFOs, politics, and taxes), and the associated health and environmental costs, with a brief comment on global trade. The final section of this chapter discusses some connotations and values we place in information on labels relevant to meat.

MEAT IN OUR DIET

United States Dietary Guidance: Meat and Protein Are Not Synonymous

One of the underlying complexities of dietary guidance generally is the need to accommodate the concept of nutrients as components of food (see Chapter 3 for the full story).

---

* Andrea Wilcox, MPH, RDN, contributed primary research and prepared draft versions of portions of this chapter.

Our current dietary suggestion graphic, MyPlate, has a segment titled "Protein Foods,"[4] which was formerly known as the "Meat, Poultry, Fish, Dry Beans, Eggs and Nut" food group. In addition to protein, "Protein Foods" also supply vitamin B (niacin, thiamin, riboflavin, and B6), vitamin E, iron, zinc, and magnesium.[5]

Although "protein" is a much snappier and space-saving title, protein is a nutrient, not a food, as Marion Nestle points out.[6] We eat foods, not nutrients. Foods happen to be sources of nutrients. Meat is a source of protein, but so are nuts, seeds, tofu, beans, grains, and, yes, even vegetables and fruits. The necessary constituent amino acids that make up protein can be acquired through a varied diet without eating meat. However, protein that conveniently contains all necessary amino acids in an ideal ratio for our metabolic use is often characterized as "high-quality" protein. This characterization supports the idea of equating meat with protein.

The first food-group dietary guidance (Caroline Hunt's 1916 buying guide, see Chapter 3) placed milk and meat together as protein-rich foods, but thereafter dairy became a separate food group and has retained this singular status to the present day. The exclusion of dairy from the "Protein Foods" group becomes relevant when assessing overall dietary protein and nutrient consumption.

Considering the impact of the MyPlate graphic as a visual reminder, does it reinforce consuming protein-rich foods and milk at every meal?

In 1916, the recommendation for meat and other protein-rich foods was two to three servings, based on a 3-oz serving size. The suggestion that protein-rich foods should comprise 4–6 oz of food consumed each day is a message that has remained relatively constant over time.

However, over time, dietary guidance from the USDA regarding the specifics of meat consumption has gone through the wringer with the food industry. For example, in 1979, the first "Healthy People" report from the Surgeon General made the recommendation that Americans eat less red meat. That recommendation has never been made again in such words. Instead, the guidance has been softened to "choose lean meat," which does not make the same implication that it would be desirable to eat less meat.[7] In 1991, the introduction of the Food Pyramid was delayed with the explanation that further research was needed, but it was felt at the time that the withdrawal was actually due to dismay from the meat and dairy industry.[8]

Today, industry continues to exert pressure on the development of dietary guidelines, even if simply through lobbying for the appointment of desirable nutrition officials to the Dietary Guideline committee. Keep in mind that the U.S. Dietary Guidelines set standards for what food the federal programs buy and provide. As the purchasing power of the federal government is quite large, it creates a market that influences product availability. Changes or suggestions in the Dietary Guidelines can impact the food industry by eliminating or introducing new business prospects (make all the things fat-free!) as well as the health of the public.

The current protein intake recommendation from the USDA for adults who are not active is 5–6.5 oz per day.[9] Because protein is a nutrient, not a food, the USDA translates this into food suggestions. To achieve 1 oz of dietary protein, the USDA recommends consuming 1 oz of meat, poultry, or fish, or one-fourth cup cooked beans, or one egg, or one tablespoon of peanut butter, or 0.5 oz of nuts or seeds.[10]

Meat as a Component of Our Diet

Meat is a convenient source of quality protein. Protein quality references both the ability of our body to digest the protein as well as the amino acid profile of the protein. Proteins are made up of various amino acids, in differing ratios. This means that protein from an animal source will have a different amino acid profile than a protein from a plant source.

"Protein" describes groupings of amino acids, which vary in underlying characteristics that are relevant to human nutrition. "Essential" amino acids are those that we need to consume in our diet. "Nonessential" amino acids are ones that our body can synthesize in sufficient amounts from other amino acids. A "high-quality" or "complete" protein source contains all of the essential amino acids at the ideal ratios for synthesizing the nonessential ones.[5]

Animal-based foods are complete protein sources. However, those consuming a plant-based diet can still get all the essential amino acids and live healthfully without eating meat. Many plant-based foods contain most, if not all, of the essential amino acids. The issue is the quantity needed to consume to achieve the relevant levels of each essential amino acid. However, consuming a variety of plant-based foods, including traditional combinations such as beans with rice, can reduce the overall gross amount necessary to consume to achieve "complete" protein.[11]

Quinoa is a popular topic at the time of writing—both for its high nutritional content and due to debates over sustainability. Quinoa is a seed that, when prepared, is eaten in preparation like a grain. Depending on the variety, it generally has more protein than grains. Even more importantly, the quality of the protein is high. Under FAO's essential amino acid scoring pattern for 3–10-year old children, quinoa exceeds the recommendation for all eight essential amino acids.[12]

Quinoa is technically a complete protein. However, animal-based complete proteins are packaged in the amounts that we have determined are appropriate to consume in relation to overall dietary nutritional guidance, which makes promoting meat consumption significantly simpler for dietary guidance. Note that the dietary guidance for protein uses a baseline of one-to-one ounce of protein to ounce of meat.

How Much Protein?

2010 U.S. Dietary Guidelines recommend a daily amount of protein as generally 0.80 g of good-quality protein per kilogram of body weight, with greater needs at certain lifespan times.[5] (Keep in mind that U.S. dietary advice is not necessarily a globally universally accepted standard. A WHO report from 2007 recommends a bit less at an average of 0.66 g.)[13] Under the U.S. standard, a 150-lb adult requires 54 g of protein daily. The 2010 Dietary Guidelines note that 3 oz of lean meat or poultry contains approximately 25 g of protein, whereas "cereals, grains, nuts, and vegetables contain about 2 g of protein per serving."

Remember that the Dietary Guidelines are not meant to be a consumer-facing document. For consumers, the USDA references servings in relation to ounces. The current

USDA protein intake recommendation for adults who are not active is 5–6.5 oz of protein per day.[9] One ounce of meat, poultry, or fish constitutes 1 oz of protein for purposes of the USDA, and so does one-fourth cup cooked beans, or one egg, or one tablespoon of peanut butter, or 0.5 oz of seeds.[10] Protein-wise, consuming 1 oz of nuts is equivalent of consuming 2 oz of meat.

Information about protein content of foods can be found in the Nutrition Facts box on packages. For example, one sample can of mixed nuts stated a serving size of 1 oz with 6 g of protein, and another sample can stated a serving size of one-fourth cup (30 g) with 5 g of protein.

Protein and other nutrient information for a variety of foods can also be reviewed by looking up items in the USDA's Agricultural Research Service Nutrient Data Laboratory's National Nutrient Database for Standard Reference. According to this database, one-half cup of uncooked quinoa has 12 g of protein, and one cup of cooked protein has 8 g of protein.[14] However, the 2010 Dietary Guidelines state "cereals, grains, nuts, and vegetables contain about 2 grams of protein per serving," which makes it sound quite difficult to achieve 54 g of protein without eating meat. *How can you reconcile this statement with the information noted above?*

USDA and Meat: Mission Conflicted?

President Abraham Lincoln signed the USDA into existence in May of 1862. It was charged with the task "to acquire and to diffuse among the people of the United States useful information on subjects connected with agriculture in the most general and comprehensive sense of that word, and to procure, propagate, and distribute among the people new and valuable seeds and plants."[15]

With the use of the phrase "seeds and plants," this original mission seemed to emphasize crop-based agriculture. The USDA subsequently became significantly involved with meat through the 1906 Meat Inspection Act (addressed in Chapter 13), as well as through actions by the meat industry fostering the inclusion of meat research in the agricultural experiment stations attached to land grant education institutions.[16]

Today, the USDA's mission of disseminating useful information connected with agriculture (and meat) includes dietary guidance as well as the promotion of agriculture foods. As of 2014, the USDA's mission is "We provide leadership on food, agriculture, natural resources, rural development, nutrition, and related issues based on sound public policy, the best available science, and efficient management."[17]

This leadership includes research and promotion programs, such as "check-off" marketing programs that are established at the request of the specific commodity industries and implemented through federal law at various times. The beef check-off program was created through the 1985 Farm Bill legislation.[18] The "check-off" moniker harkens back to a "check to opt in" structure to the marketing program, but now the programs are funded through a mandatory assessment on the producers. The producers of the relevant commodity foods pay a small percentage of the wholesale price received upon the sale of crop or animals produced. Box 8.1 lists the various commodity food Research and Promotion Programs.

**BOX 8.1 COMMODITIES WITH RESEARCH
AND PROMOTION PROGRAMS**

- Beef
- Blueberries
- Dairy Products
- Eggs
- Fluid Milk
- Haas Avocados
- Honey Packers and Importers
- Lamb
- Mangos
- Mushrooms
- Peanuts
- Popcorn
- Pork
- Potatoes
- Processed Raspberries
- Sorghum
- Soybeans
- Watermelons

*United States Department of Agriculture, Agricultural Marketing
Service, Research and Promotion Programs. Available at http://www
.ams.usda.gov/AMSv1.0/ams.fetchTemplateData.do?template=
TemplateB&leftNav=ResearchandPromotion&page=
ResearchandPromotion, last modified May 1, 2014,
accessed September 7, 2014.*

The USDA is heavily involved in the administration of the check-off programs, appointing board members and approving promotion campaigns through the Agricultural Marketing Service, among other things. The pooled funds are used for research and promotion, aimed at developing and maintaining consumer markets. For example, the dairy program clearly states that the program is meant to "drive increased sales of and demand for dairy products and ingredients."[19] Not surprisingly, a dairy check-off funded study found that whey protein was more effective than soy protein for developing lean muscle.[20] However, the USDA also points out that research funded through the check-off programs can provide ancillary side-benefits that include public health.[21]

These check-off programs are quite effective in increasing our purchase (and consequently, consumption) of the intended commodities. A 2012 analysis of the effective return of the pork check-off program for 2006–2010 reported that every check-off dollar invested during that time returned an estimated $17.40.[22] The dairy check-off home page states that partnership efforts over the past 5 years have increased sales by more than 12 billion pounds.[23]

However, although all producers are contributing through the mandatory assess-ments, not all producers may see such a return. A pork marketing campaign touting lean-ness may not benefit heritage pork producers who are counting on pork fat to carry the day.[24] Unhappy producers have challenged the mandatory nature of the assess-ment program, equating the mandatory participation to forced speech. This challenge reached the Supreme Court in 2005, and the Court held that the beef check-off program does not violate the First Amendment, finding that the check-off programs are federal government programs, and thus the messages of the program are official "government speech," which citizens do not have a First Amendment right not to fund.[25]

Why might we care about these programs? The Supreme Court has determined that these assessment programs are federal government programs. Marketing is intended to increase consumption, and marketing is generally effective. If the federal government is involved, then this marketing becomes food policy that impacts what we eat, and effectively increases our consumption, often of foods we already enjoy to an unhealth-ful proportion in our diet.

We all eat—and in fact, we all need to eat. We do not need to be sold on the need to eat, generally. Although advertising will trigger our interest in specific food items, we will remember to eat without being encouraged to do so through advertising. Why is the federal government supporting advertisements reminding us that we could eat beef for dinner, and that pork is white meat, and that eggs are incredibly edible?

If these programs are essentially federal programs, as the Supreme Court deter-mined, does that in turn imply that the government is telling us what to eat? Even if the government is not "deciding for us," is it promulgating a dietary norm through such promotion and advertising?

We briefly discussed Mill's Harm Principle in the introduction chapter and the con-cept that "persuading and entreating" is acceptable but compulsion is not. (The Harm Principle is discussed in more detail in Chapter 12.) Although the check-off programs may appear to be an example of the government just helping to "persuade" us to eat more beef, dairy, and eggs, the programs do more than advertise. The dairy check-off program contributed market research analysis to Pizza Hut as to the preferred blend of cheeses by "millennials" (people born between the early 1980s and mid 1990s), which resulted in the creation of the 3-Cheese Stuffed Crust Pizza.[26] One slice of the 3-Cheese Stuffed Crust Pizza has between 7 and 11 g saturated fat.[27] Yet, ChooseMyPlateDairy promotes a "Key Consumer Message" to "switch to fat-free or low-fat milk," because many cheeses, whole milk, and dairy products are high in saturated fat.[28]

The current (2010) U.S. Dietary Guidelines recommend that American reduce con-sumption of saturated fat, limiting saturated fat consumption to 10% of total daily calo-ries, and even better, to 7% of total daily calories. If consuming 2000 calories daily, this means not exceeding 22 g at a maximum and ideally, not exceeding 15 g.[5] Two pieces of 3-Cheese Stuffed Crust Pizza and you are done for the day. In other words, the U.S. dietary guidance recommendation of eating less saturated fat is not exactly being sup-ported by the message of the dairy check-off program.

We are getting mixed messages from our government! But viewing check-off pro-grams as "the government" may not be completely fair, as these programs are coor-dinated under the mission of a particular department of the government, namely the USDA. In this case, we should consider "eat more cheese" or "eat more beef" not as a message from the government generally, but instead a message specifically from the USDA.

So, is "eat more" an appropriate message from the USDA? It fits with the agriculture promotion mission. But how does the agriculture promotion mission of the USDA fit with the dietary guidance mission of the USDA? Would the "eat more beef" message from the USDA be more appropriate if the USDA was not involved with dietary advice that arguably should include "eat less meat"?

This is just one example of how food policy can be completely at odds with health policy.

Additionally, although the USDA's mission technically includes promoting all agriculture, the meat industry is powerfully connected. A 2012 USDA internal newsletter, focusing on "green" improvements to the Washington DC headquarters, included a suggestion that employees could reduce their personal environmental impact by choosing to eat a vegetarian dish at the cafeteria.[29] This suggestion referred to the "Meatless Monday" campaign, a public health awareness campaign reborn in 2003 at Johns Hopkins drawing on historical U.S. government war-effort campaigns asking the public to do their part by eating less of certain items. Participating in a "meatless day" is one way to eat less meat.[30] See Box 8.2 for more on the "Meatless Monday" campaign.

That USDA internal newsletter, in the passage mentioning the "Meatless Monday" campaign, noted the environmental benefits of reducing meat consumption, and that one could choose to participate in the campaign if so desired as the cafeteria did offer vegetarian entrees. No suggestion was made that eating a vegetable dish was mandatory. The meat industry in turn however had, well, a cow.

BOX 8.2   MEATLESS MONDAY

The rationale for the reborn "Meatless Monday" campaign is health-centric but touches on sustainability and environmental issues. Many people making a small change can do big things. Mitigating small health risks among large number of people can result in health improvements for the individuals making the small change, but also result in larger, structural environmental and social changes that can contribute to everyone's health. Although many people may not wish to go completely vegetarian, most can commit to a small change by eating less meat; and there is social incentive around participating in a common activity 1 day a week.

If many people make a small change in their own meat reduction, especially if conducted through institutions that reach many people and have buying power, there can be a magnification of all the small changes into a larger change. The power of the simple message "Meatless Monday" has taken hold domestically and globally, being taken up by communities, numerous institutions of higher education, K-12 schools, including the Los Angeles Public school system which serves around 650,000 meals, and even the Norwegian army.

Meatless Monday. Available at http://www.meatlessmonday.com,
accessed September 16, 2014.

The president of the National Cattlemen's Beef Association stated "This move by USDA should be condemned by anyone who believes agriculture is fundamental to sustaining life on this planet,"[29] and a spokesman said "it was slap in the face of the people who every day are working to make sure we have food on the table to say 'Don't eat their product once a week.'" By the next day, the newsletter had been removed, with the official statement from the USDA that it had been "posted without clearance" and that the USDA does not endorse Meatless Monday.[31]

Instead of being concerned about the check-off program, which fits the mission of the USDA, would it make more sense to remove the dietary guidance component from the USDA?

Trends in Meat Consumption in the United States

The amount of meat we have consumed, and the type, has changed over time. Chapter 9 discusses the advances in grain yield starting in the late 1800s. The impressive advancements in grain production also provided the tools and means for increasing our meat consumption. Higher yields meant that grain did not need to be rationed for the livestock working in the fields; more of it could be used to feed animals raised solely for our later consumption, and of course as farming equipment mechanized, there were less working animals to feed. In 1900 just over 10% of the world's grain harvest was fed to animals, which were mostly working for their supper. By the late 1990s over 40% of the global grain harvest was going to feed animals intended to become our supper.[32]

The USDA has a Food Availability Data series, with data extending back to 1909 that can be studied as a proxy for consumption and used to review trends over time. The USDA estimates that overall meat available for consumption in the United States increased from 51 kg per capita in 1901 to over 80 kg (170 lb) in the 1990s. The high as of 2004 was 184 lb, declining to 171 lb by 2011. The proportion of beef (as total meat consumption) was at about 50% in 1910, but had declined to 35% in 2000.[32] Is this surprising for a nation that is associated with the rise of the fast food hamburger? Perhaps not, since we also produced the chicken nugget.

Although this long cycle describes an overall increase in meat consumption, meat consumption was actually trending downward from 1909 through World War I and the Great Depression until the mid-1930s.[33] The higher volume of grain being produced had to be consumed. Cold cereals were being marketed as convenient "health foods." The "Meatless Monday" campaigns of World War I incited patriots to do their part by leaving meat for the war effort. And, the emerging science of nutrition began questioning a primary emphasis on meat, in part due to a shift in attention to the exciting discovery of vitamins. The old idea of organizing a diet solely around principles of protein, fats, and carbohydrates had to be modified to incorporate the new idea of health—vitamins and vegetables. New York City, long before Michael Bloomberg's efforts to ban trans-fats and soda, banned hot dogs from school lunches.[16]

In response, the meat industry took action by forging a stronger relationship with the USDA to promote meat consumption, and, with gifts of funding fellowships, persuaded many of the agricultural experiment stations attached to land grant education institutions to fund meat research coordinated by the industry and the USDA.[16] Meat consumption rose again.

Currently, meat consumption is trending downward, having possibly peaked in 2004.[34] Meat and dairy trends between 1970 and 2012 show both increases and decreases. In 2012, each American had available to consume, on average, 35 lb more poultry; 2 lb more fish and shellfish (boneless, trimmed equivalent), and 22 lb more cheese (!) than in 1970. Conversely, each American, on average, had 34 lb less red meat available, 59 fewer eggs, and 12 gallons less milk in 2012 than in 1970.[35]

The Change to Chicken

Like other components of our modern diet, there has been a general trend toward the homogenization of meat consumption. We are increasingly eating beef, pork, and poultry and decreasingly other meats. The British eat just half the mutton they did in 1960, and even the more culinary expansive French eat less mutton, goat, and horse now than in the past.[32] Chicken is ubiquitous today in the United States, whether dining out or in. But it has not always been so. In 1928, Americans were (per capita) eating a half a pound of chicken per year. By 2005 we were eating half a bird a week.[36] Figure 8.1 illustrates the increase in chicken consumption over time.

Although Hoover himself did not actually promise a chicken in every pot (that famous slogan came from the Republican Party, not Hoover specifically), chicken was

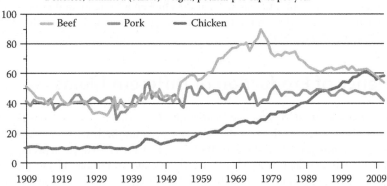

FIGURE 8.1 The amount of chicken available for consumption has steadily increased since the 1950s. The amount of beef available for consumption has declined from the mid-1970s to 2011. Calculated by ERS/USDA based on data from various sources (see food availability documentation). (From United States Department of Agriculture, Economic Research Service, Food Availability (per capita) Data System (FADS), summary findings, http://www.ers.usda.gov/data-products/food-availability-(per-capita)-data-system/summary-findings.aspx, accessed May 10, 2015.) Note: Figures are calculated on the basis of raw and edible meat. Excludes edible offals, bones, viscera, and game consumption for red meat. Includes skin, neck, and giblets for poultry (chicken and turkey). Excludes use of chicken for commercially prepared pet food. Food availability data serve as proxies for food consumption.

not then commonly regularly eaten, but instead primarily kept for eggs. Due to containing a lower amount of saturated fat, chicken meat was more perishable than beef and had to be consumed quickly if slaughtered. Chicken was a seasonal dish, as farmers brought young hatched males (nonegg layers) to the market in the spring or old roosters and spent layers to the markets in the fall.[16]

The 1906 Meat Inspection law did not apply to poultry because of a described lack of "consumer demand" for such inspection, as chickens tended to be purchased less frequently, and when purchased, more often more directly from the farmer or local butcher.[37] The demand for chicken, as well as the expectation of cheap chicken, developed post–World War II.

National grocers contributed to the development of the "broiler boom" of the 1950s by using chicken as "loss-leaders" to get shoppers in the door. The pressure of low poultry prices pushing down profits meant that producers raised even more chickens, which, in turn depressed prices even further. One large retailer even sponsored chicken breeding contests, which further encouraged the development of the fatter and faster (to grow) the chickens of our future.[16]

Poultry production, just as wheat had been in the 1880s, was an exciting new way to make money. The "gold rush" winners were the integrators who invented the broiler industry by controlling the process from egg through to chicken sold at the grocery store. Integration made chicken the cheapest meat product on the market; and integration allowed the development of the value-added products that could combat the loss-leading practices of the retailers.[16]

Poultry factory farms made integration possible, and were perceived then as marvels of the new age of technology rather than a threat to traditional livelihoods and communities. Furthermore, the creation of the chicken as a national diet mainstay was subsidized through government food policies, namely the research and poultry improvements programs of the USDA.[16]

Efficiency of Production of Different Meats

Although somewhat of a what-came-first question, the increased consumption of chicken is also related to the fact that chicken became the fastest and most efficient meat we can commercially produce. In 1960 it took just over 70 days on average to raise a chicken to slaughter, but by 1995 it took only 50 days. Additionally, even though the chickens were being fed more efficiently (i.e., less food), we were getting more food out of the chickens, as remarkably, chickens had become bigger.[32]

Pigs are more efficient than cattle as meat-makers in numerous ways. Pigs put on weight faster and more efficiently than do cows, they produce more offspring (a pig litter numbers from 8 to 18 compared to a cow who usually has one calf), and a short gestation time. Chickens need the least amount of grain and are the best at converting plants to animal protein. Cows (and other ruminants) are the best at converting plants that other domestic species cannot use. To this extent, using grasslands that are not arable as crop fields to produce beef could be an appropriate, efficient way to raise meat for human consumption, albeit not enough to match the amount we currently enjoy. Domestically and globally, there may be land able to sustain foraging for cattle that cannot support intensive crop farming.[38]

Homogenization of Breeds

Homogenization of breeds, as well as our taste, stems from the need for efficiency. The separation of the breeds into egg-layer and broiler came with industrial production specialization. Now, in industrial egg-layer production, the culling of the male chicks happens once sex is determined shortly after hatching and the male chicks (hundreds of thousands) are killed rather than being raised for food.

The efficiencies required for factory meat production also resulted in "value-added" hogs or chickens as animals. In other words, the processors pushed producers to raise streamlined hogs or chickens of relatively homogenous size and weight. Indoor confinement, especially for hogs, contributed to homogenous production, as farmers were able to replicate absolutely ideal temperature conditions and exercise complete control over diet.[16] And naturally, farmers begin selecting for qualities and characteristics in breeds that allowed for homogenous production and the traits that the processors and integrators desired, thus resulting in a homogenous livestock population in the United States. In 1998, 99% of all turkeys raised in the United States were one breed—the broad-breasted whites.[39]

This means that not only have trends in types of meat consumed changed over time, so has the meat itself. Alligators and rattlesnakes used to "taste like chicken"—do they still? The loss of biodiversity in farm animals may lead to increased risk of disease wiping out a population, which is a problem for our own food security at a population level. It also has created a homogenous experience in taste for us. With exposure early on in life to a common taste, we surely develop an expectation as to how things taste.

The discussion over corn-fed and grass-fed cows (in more detail later in this chapter) reveals how the same type of meat can differ in taste and nutrition. An additional impact of the need to make efficiencies in national food product production may have inherently changed our tastes and expectations about foods. Consider how much even hot dogs have changed. Children fed hotdogs in the late 1960s were eating fatty hot dogs (almost a third fat) containing less than 12% protein, but the typical hot dog from the late 1930s was about 19% fat and 20% protein.[16]

Dietary Guidelines Influence: Boneless Chickens, Tailless Turkeys, and Fishless Oceans

Dietary guidance influences trends in our consumption. Federal dietary guidance can be powerful enough even to cause the creation of new foods, such as chicken nuggets.

Boneless chicken

The first (and only) federal government suggestion to actually eat less red meat contributed to a decline in fast-food sales of burgers and McDonald's wanted an alternative to keep customers in the restaurant. Lo, as described in *Fast Food Nation,* the chicken nugget.[40]

Chicken began to be butchered for direct prepackaged consumer sale by the 1950–1960s,[16] but even as late as the 1980s most chicken (over 80%) consumed in the United

States was still relatively unprocessed. By the 2000s, this percentage had reversed to most chicken being processed. A generation gap manifested between those who recalled when chicken usually looked like chicken, and those that, for the most part, knew chicken as a boneless or processed food item.[41]

Additionally, the chicken producers had found that chickens just were not that profitable. The profits that came from sheer scale, consolidation and efficiency in producing chickens were not sustainable due to low margins and unpredictable commodity cost. Until, that is, the producers came up with processed versions of chicken that could be sold as "value-added." The value-added items not only cost consumers more money, but also ironically were less likely to be the healthy alternative that consumers were seeking.[41]

Even those consumers that avoid chicken "products" and shop for cuts of chicken meat are often purchasing boneless, skinless chicken breasts, and may not think to ask what happens to the rest of the bird. That is, until we go out to eat fried chicken wings for a happy hour bucket special. Those wings have to go somewhere, and so does the rest of the bird. Our efforts to have a convenient, healthful diet that includes lower fat meat options may in turn compromise our own health domestically and others globally.

Turkey tails

Because food is inextricably linked to trade, the impacts of our food policies can impact not just our own food and health, but that of others. If the poultry processors are meeting our needs for convenience and their needs for profit by selling us just the best parts of the birds, where is the rest of the bird going?

Turkey tails were heading south, to the Pacific Island of the Republic of Samoa, in the amount of more than 20 kg per person. Turkey tails are not the most healthful cut of the bird. Interestingly, they are not even listed in the USDA nutrient database. Foster Farms, however, has published nutritional information, along with the suggestion that they are great in soup or grilled. A 4-oz serving of a raw turkey tail contains 380 calories, almost all of which (330 calories) are from fat. One serving contains about 36 g of fat, which is more than half of the daily recommendation for a 2000 daily calorie diet.[42]

The Samoan population is struggling with very high rates of obesity, high blood pressure and diabetes, and as a public health measure, banned the import of turkey tails in August of 2007. In 2008, at a WHO meeting in Manila, Samoa asked the other delegates for help in counteracting food marketing from multinational food companies, stating that traditional diets were protective against diabetes and high blood pressure, but those diets were threatened by globalization and free trade. The response of the U.S. delegates was a recommendation that prevention programs be focused at the community or individual level, and a request that the meeting's final resolution contain a statement "to encourage citizens to take responsibility for their own health."[43]

Samoa had some time previously (in 1998) begun the application process to join the World Trade Organization. Their application was accepted in 2012, albeit with the requirement that the turkey tail ban be lifted by May 2013. A 2-year transition period banning domestic sales and imposing a high import duty would allow the country time to implement public health programming concerning healthy diets. After 2 years, the duty would be reduced and turkey tail sales could continue, with the import duty reduced or replaced by another tax regulation or by recommendations from the health program. The World Trade Organization's mission is naturally the promotion of trade, not health.[44]

Conversely, here in the United States, we can consume turkey tails as a treat for those seeking new culinary experiences at $12 a serving, even when the suppliers are desperately selling at 0.70 a pound.[45] Perhaps there will be enough of such a market here in the United States to even out the distribution and consequential health impact of turkey tail consumption.

In sum, our dietary guidance can have a global impact through market creation. If we are not eating the fat, someone else is. If we are not buying all the chicken parts, someone else is. A researcher studying these escalating health problems in Samoa pointed out that imported turkey tails were replacing the healthier local diet staples in part simply because they were cheaper.[46]

Fish

Our latest dietary guidance recommends the consumption of seafood, which may be influencing our appetite for fish. Our 2010 Dietary Guidelines recommend consuming 8 oz per week of seafood as a source of omega-3 fatty acids and it may help reduce the risk for heart disease.[47]

Of course, not all fish or seafood provides the same nutrient profile. Fish can provide a low-fat source of protein, but can be breaded, fried, and served with butter or creamy sauces. The advice to eat seafood for health can easily be turned into a hall-pass to eat deep fried shrimp.

An increased global appetite for seafood (also fostered by distribution and refrigeration technologies) has resulted in massive overfishing. Not only have we overfished the species we consume, we are devastating the species we do not consume. Extensive by-catch is being discarded.[48] Management of fishing has been shown to work in New Zealand and the United States but it is unclear if this can work in all areas.[49]

The collapse of world fisheries also provides another example of how policies regarding other issues can become inherently policies about food. Government polices established post World War II may have encouraged a global fishing industry. U.S. policies in particular opposed any restriction on American boats of any kind, even though overfishing was already occurring.[50]

Farming fish might be absolutely necessary in our future if we want to keep eating fish while allowing fishery stock to rebound. Fish-farming has been successfully done in some areas for centuries, such as in the Trebon region of the Czech Republic, which has been farming fish since the sixteenth century.[51] Fish farms can be conducted in differing ways, such as suspension cages, closed containment, and others, and some methods may be superior in environmental protection and conditions to other methods. But fish-farming could also be just moving all of our factory farm problems from land to sea. Farm-raised fish, just like their land-counterparts, can end up in crammed conditions necessitating antibiotics and pesticides. The discharge from a fish farm can be polluting to water, just as discharge from an animal farm can pollute, and the development of a method for certifying an organic standard for farmed fish that would meet the standards of OFPA has accordingly been controversial, and to date unsuccessful.[52]

Furthermore, some argue that farm-raised fish may not have the nutritional profile for which we are advised to consume seafood to begin with. Just like their land-based industrially raised animal brethren, farm-fish may be given a diet of corn, soy, or other food that does not contain the same nutritional omega-3s that wild fish garner from aquatic plants and other fish.[53]

And of course, we can make farmed fish bigger and fatter, just like we have done with animals. Conventional breeding for prized characteristics can of course occur, but a genetically modified salmon has also been created through the insertion of an ocean pout gene and a Chinook salmon gene into Atlantic salmon, with the result of a salmon that grows quickly to "market size." Interestingly, the FDA has decided to regulate genetically engineered animals under the "new animal drug provisions" under the Food, Drug, and Cosmetic Act, because the process of genetic engineering meets the definition of a "new animal drug," namely that it is intended to affect the structure or function of the body of the animal.[54]

Programs such as the Monterey Bay Aquarium's Seafood Watch provide important education to consumers on sustainable fish consumption, including a review of some eco-certifications.[55]

Efforts at stewardship are continuing on a global level; however, to help save the fish, we may also need to focus our attention, public education, and dietary guidance on other nutrient sources of omega-3 fatty acids (EPA and DHA). Krill are tiny crustaceans with similar fatty acids as those found in fish, and a potentially sustainably harvested source. We can also gain the health benefits of fish consumption from emphasizing other plant-based sources of omega-3 fatty acids, such as walnuts and flaxseeds. *Could walnuts or flaxseed be emphasized as directly as fish in our dietary guidance?*

Costs to the Consumer

In other chapters of this book, we touch on the idea of who bears the risk, such as in food safety, and who bears the burden of cost or harm when considering the social cost of health outcomes that are associated with individual dietary choices. This idea of social cost is also known as external costs. Production of any kind creates external costs that are borne by society instead of solely by the producer. This happens across industries, including food production.

The nature of external costs is contentiously argued, both as to what is an external cost and as to the extent of the cost burden. Examples of external costs could be taxpayer subsidies, healthcare costs, environmental costs of cleaning up damage, just cleaning up after animals or ourselves, or having to switch methods of production. Some argue that government subsidies are also an external cost, even though not normally thought of such.[56]

Another economic factor to consider is that big food producers are big employers, and in this case, the interest of the producer becomes the interest of the state that is home to the operations. If we could leave the equation there it might be simple, but these jobs do come at a cost. Meatpacking is a dangerous occupation. Speeding up processing for efficiency harms the workers. Our cheap food comes at a potentially high cost to a worker employed for low wages. In 2013, meatpackers worked for a mean hourly wage of $11.99, annually $24,930.[57]

Meat, particularly ground beef, is in fact inexpensive. In 2008 it was cheaper for us to eat a hamburger at home than it had been in 1980; well, at least the beef patty was cheaper. Over the same time period in which the cost of beef decreased, the cost of buying the vegetable toppings rose. By 2008, you could buy three times amount of beef as in 1980 for the same price. This cheap meat was brought to the table through increased efficiency in raising animals as well as the slaughtering and processing.[53] Similar efficiencies have lowered the per unit production cost of chicken, including doubling the average live weight of a broiler chicken from 1934 to 2004, which leads to the thought that perhaps we were not eating much chicken early in the twentieth century because the eggs provided more nutrition benefits than the scrawny bird itself. In any case, as a result of lower production costs and bigger chickens, chicken was more affordable in 2004 than in 1980, at $1.74 per pound in 2004 versus $2.22 in 1980.[58]

Did the efficiencies that created such low-cost simply shift the burden to consumers in other areas? The next section discusses the factors enabling industrial meat production and meat consumption (such as subsidized feed, antibiotics, CAFOs, politics, and taxes) and the associated health and environmental costs, with a brief comment on global trade.

THE COST OF MEAT

The idea of factory farming was not always considered in a negative light. Around the turn of the century through to the 1920s, the idea of applying the processes of the factory to the farm was an embrace of technology and efficiency that felt appropriately modern.[16] Industrial or "factory" farms focus on efficient mass production with the use of modern day technology and machinery to maximize profit; larger, consolidated, operations are more efficient than a number of smaller farms. In 1900, the average farm size was 147 acres as compared to an average farm size of 441 acres today.[59]

Due to this shift in farming practice, American's increasing appetite for meat products comes with often overlooked environmental and human health costs. The factory farming practices can be harmful to workers, nearby communities, and others at a greater distance through air pollution, water pollution, and the spread of disease.

Contributing Factors Enabling Industrial Farming and the Meat Industry

The railroad's associated impact on food production and delivery, discussed in Chapter 9, also played a similar role in the development and eventual consolidation of the beef-packing industry, and, as the packers found distribution of other food stuffs profitable, encouraged the burgeoning development of a national food system. Like the later chicken producers, the packers also found that diversification into food products and byproducts helped maintain profits.[16]

Although today we see antitrust laws in the United States referenced in corporate mergers in industries such as transportation, the Sherman Antitrust law's roots are essentially food policy, stemming from the rise and fall of a "cattle bubble" in the late 1880s. The Sherman Antitrust law was meant to protect trade from being restrained. However, it did not apparently do much to actually keep the beef packers from expansion and consolidation.[16]

Today, consolidation of the players in the chain between the farmer and the consumer (packers, soybean crushers, seeds, grocery retail) affects the prices for farmers. The entrance of nontraditional retailers, such as Walmart and Costco, into food retailing has contributed to further consolidation among retailers. In 2010, the USDA and the U.S. Department of Justice held joint workshops regarding antitrust issues in the agricultural sector and published a report in 2012. As the workshops provided an open forum, many issues were raised by attendees that fall outside the scope of antitrust law, which makes the report well worth reading as an introduction to modern farming challenges. Unfortunately, as the report states, "the antitrust laws focus on competition and the competitive process, and do not serve directly other policy goals like fairness," and in turn, this means that "a merger or practice cannot be challenged solely on the grounds of endangerment to food safety, the environment, or the health of a rural community."[60]

Land use policies also contributed to trends of consolidation in the meat industry. Between 1935 and 1950 the federal government diverted 9 million acres of western federal grazing lands to other uses, while other federal projects encouraged crop farming and housing. Overall, the west lost 45 million acres of grazing range. In other word, federal policies unrelated to food and agriculture also encouraged the eventual development of the concentrated animal feeding operation (CAFO).[16]

Concentrated animal feeding operations

Industrial farming started booming in the 1920s when it was discovered that adding vitamins A and D to animal feed permitted farmers to keep animals indoors year round.[61] However, after World War II, the development of industrial farming ran into a roadblock when the increase in the confinement of livestock led to a higher incidence of disease in livestock. Changing farming methods and the introduction of different feeds, as well as antibiotics, helped solve this problem.

Concentrated animal feeding operations consist of animals confined in abnormally high stocking rates. CAFOs began appearing in California as early as the 1940s, with paved lots to make manure removal easier. Contributing factors to the development of CAFOs included the increase in population (in turn, increasing demand for beef), less labor available to farm, and less land available for traditional grazing methods due to increasing costs.[16]

Farmers became able to keep thousands of animals in confinement while simultaneously promoting faster growth and preventing disease, which in turn maximized meat production.

Today's factory farms are much larger than historical farms and can house thousands or tens of thousands of animals. However, this has not led to an increase in the farming workforce, because advanced machinery can take up the work. A CAFO can operate with only a handful of farmers. Fewer farmers also mean less business at the local bakery, store, or diner. Furthermore, the size of the operation and the required efficiencies of production and supply mean that the CAFOs may not contribute to the local economy because they have to get their supplies from elsewhere. Like Walmart, when a CAFO enters a region, declines in local economy, population, and infrastructure can occur.[62]

Buoyed by industrialization and mass production by CAFOs, the majority of meat in the U.S. is produced by a handful of large corporations. As of 2000, four companies (Tyson, Cargill, JBS Swift, and National Beef) were responsible for 80% of the country's beef production. The fifth on the list at the time was Smithfield Foods. Smithfield was then also the top pork producer in the country, and with Tyson, ConAgra and Cargill, together produced 60% of the pork in the United States at the time.[63]

Concentration of livestock industry is a global trend; as of 2013, Smithfield Foods is owned by the WH Group, a Chinese company formally known as Shuanghui International. The WH Group is the largest pork company in the world, as it owns Smithfield Foods, is a majority shareholder in China's largest meat processing business (the Henan Shuanghui Investment and Development Corporation), and has a 37% shareholding in Campofrio Food Group, the largest pan-European packaged meat products companies.[64]

An investor group led by Goldman Sachs bought Shuanghui International in April 2006, and Goldman Sachs was still a 5.2% shareholder in 2013.[65] The concern is not one of nation-borders, but instead over the increasing power of market concentration. Large farms have an advantage in most commodities because as the size of an operation grows, the average cost of production per unit declines, resulting in higher average returns. These competitive forces will likely continue to diminish the number of small farms and continue to shift production to larger farms.[66] Industrial farming has caused many changes in farming communities, such as increasing farm size and gross farm sales, while conversely lowering retail sales, housing quality, and wages for farm workers.[67] Without CAFOs meat production would decrease. This is not an option for the agricultural industry. Some might argue it is not an option for the American consumer, and that is a question that as consumers, we may need to decide.

The Farm Bill

The Farm Bill was initiated in 1933 when the United States passed the Agricultural Adjustment Act and is revised every 5–7 years. See Chapter 2 for more details on the Farm Bill. Corn subsidies were introduced after World War II when technology advancements and the use of new petrochemical fertilizers resulted in bumper crops and a surplus of corn.[61] The USDA implemented policies to help farmers utilize the corn surplus, one of which was through feeding it to livestock. From 1995 to 2012, corn subsidies in the United States totaled $84.4 billion.[68] By 2008, the Farm Bill had become the largest source of public subsidies to animal factories. It provides income support to growers of certain commodities, including feed grains and dairy, and this subsidization of animal feed crop production contributes to the profitability of CAFOs. Tufts University's Global Development and Environmental Institute estimates that between 1997 and 2005, the industrial animal sector saved over $35 billion due to subsidies that lowered the price of purchased feed.[61]

Most subsidies go to the largest meat and poultry companies that dominate the industry. It is estimated that between 1997 and 2005, Tyson Foods, which controls over 20% of the chicken market, received subsidies of some $2.6 billion, while Smithfield Foods received $2.54 billion as a result of using low-priced animal feed.[61] When the industry is dominated by a small handful of large factory farms that receive substantial amounts of money from the government, smaller farms have difficulty competing.

The Environmental Quality Incentives Program (EQIP) has become a large source of funding for CAFOs. The EQIP conservation program initially focused on providing funds to help small livestock operations, create safer ways to handle waste, and comply with environmental regulations. With the 2002 Farm Bill reauthorization, however, EQIP turned its focus to CAFO infrastructure funding and away from environmental compliance. Previously EQIP funds had been restricted from being used for large-scale waste lagoons, animal waste spraying systems, and other waste facilities. With the 2002 bill, these restrictions were lifted, indicating that efforts were shifted from cost-efficient applications to those applicants with the greatest pollutant potential, resulting in an increase in CAFO waste throughout the country. CAFOs received an estimated $100 million per year in EQIP funding from 2002 to 2006, and the 2008 Farm Bill provides even more EQIP subsidies for animal factories.[61] EQIP was reauthorized in the 2014 Farm Bill with increased payment limit for general EQIP from $300,000 to $450,000 per contract, allowing large CAFOs the chance to get even more of these funds. The EQIP Organic Initiative payment limit is $80,000 per contract.[69]

Animal physiology and subsidized feed

Corn is compact and convenient, making it feasible to feed tens of thousands of animals in confined areas. Subsidized corn and soy in the United States play a significant role in mass meat production, as livestock now consumes 60% of the corn and 47% of the soy produced in the United States.[70] A high-calorie grain-based diet maximizes growth and promotes weight gain in the least amount of time for cheap and efficient production, made even cheaper by access to subsidized feed.

However, a diet of grain can be detrimental to the health of livestock not physiologically designed to eat grain. Cows, sheep, and other grass grazers have a rumen, a section of the digestive tract that produces gas and enables them to convert grass into high-quality protein. When these animals are fed too much grain and too little roughage, rumination stops and a layer of slime can trap the gas. This causes the rumen to inflate and press against the cow's lungs. Unless treated immediately, the cow can die of suffocation. To avoid dealing with the negative health consequences from grain-feeding, producers attempt to grow cows quickly and reduce length of time to slaughter, essentially aiming to kill the cow before it falls ill.[61]

Cows raised on grass reach slaughter weight too slowly to satisfy the demands of industrial farming. In the 1950s, farmers commonly slaughtered a cow when it reached 2 or 3 years old; now in 14 months, a cow can reach a satisfactory slaughter weight of 1200 lb thanks to massive quantities of corn, protein supplements, drugs, and growth hormones.[61] The rapid weight gain in grain-fed cows results in meat containing more saturated fat than their grass-fed counterparts. With little concern regarding whether livestock consumes calories healthful to it or us, but simply that they become food fast and cheaply, the fast growth of animals is truly the "fast food" of the industrial farming system. As cattle are much less efficient at turning grain into meat than grass, even some proponents of cattle husbandry argue that grain feeding makes little sense and is a misuse of resources.[32]

Antibiotic and growth hormones

Animal feed, close confinement, exposure to waste, and a stressful environment all contribute to increased susceptibility for disease among livestock. These conditions have resulted in the common use of nontherapeutic antibiotics, such as through inclusion in animal feed for all animals. Such nontherapeutic use started on industrial farms when it was discovered that antibiotics resulted in rapid weight gain, up to 3% more weight than the animals normally would gain. The mechanism that causes the weight gain is still unclear, but there is some evidence that the death of the flora in the animals' intestines makes more nutrients available to the animal.[71]

Like other technological advances in the middle of the twentieth century, antibiotics improved farming production, which improved the ability of farmers to make a living while decreasing consumer cost.[72] However, today antibiotics are also used to compensate for the unsanitary conditions of overcrowded animal production, some to inhibit gas production in the rumen of cows, and some to reduce the incidence of liver infection. An industrial farm staff veterinarian commented that if antibiotics were not used in beef production it would not be feasible to feed so much grain. Using fewer antibiotics, without change in feeding practices, would result in more cattle deaths, higher priced beef, and a less profitable industry.[61]

Public health activists do not disagree with the use of antibiotics for animals that are sick. The issues with antibiotic treatment are threefold—first, the animals would not have a greater risk for disease if they were not confined by the thousands and were fed more appropriate diets, and thus under "normal" conditions would not require antibiotics. Secondly, treating a sick animal is not controversial; however, the nature of commercial farming, particularly with birds, can prohibit individual treatment. If one bird is diagnosed, the whole flock gets treated with the addition of medication to the drinking water, which inherently results in antibiotic overuse. Thirdly, animals are treated with antibiotics for prophylactic purposes and to promote faster growth, and there is concern that the overuse of antibiotics for nontherapeutic reasons is leading to antibiotic resistance in humans.

Another reason CAFOs are successful in mass meat production is through the use of growth hormones. Synthetic estrogen, a hormone, is commonly implanted in cattle to promote faster growth. An implant of synthetic estrogen costs $1.50 and adds between 40 and 50 lb to the cow, for a return of at least $25.[61] Dairy cows may be given a genetically engineered hormone, rBGH, to increase milk production and ultimately increase profits. The use of hormones to promote growth and milk production has been banned in Europe, but not in the United States because no risk to human health has been proven. However, measurable hormone residues are found in the meat products and this is of concern in public health.[73]

Although the FDA must approve the use of drugs in industrial animal farming, the studies presented to the FDA are often primarily from the manufacturer of the drug, such as is the case with ractopamine, a growth accelerator used in pork, cattle, and turkeys. In today's global market, the approval of such drugs can impact trade as many countries have banned ractopamine.[74]

Politics and money

In 2012, the year Congress was meant to reauthorize the Farm Bill, agribusiness spent close to $138 million on lobbying.[75] In addition to gifting politicians with money and investing in lobbying, the agribusiness also pushes for laws that protect the industry. For example, as of 2010, 13 states have "veggie libel laws" that forbid people from speaking judgmentally about agriculture. In 1996, an episode of Oprah Winfrey's talk show discussed mad cow disease in the livestock industry and the possibility of mad cow disease spreading from cows to humans. Oprah exclaimed, "It has just stopped me from eating another burger," and was sued by the Texas Beef Producers for making a disparaging statement about the industry. Although Oprah won the case, this example demonstrates the ability of "Big Ag" to influence our legal system.[61]

In 2008, the EPA amended the Clean Water Act permit regulations for CAFOs to require applications for National Pollutant Discharge Elimination Systems (NPDES) permits for CAFOs that discharged or proposed to discharge waste into water. The National Pork Producers Council sued the EPA, and on March 15, 2011 the U.S. Court of Appeals for the Fifth Circuit affirmed the NPDES permit requirement for CAFOs that actually discharge, but rejected the requirement for those proposing to discharge. Conversely, the EPA also faces lawsuits from environmental groups after withdrawing a 2011 proposed NPDES CAFO reporting rule.[76] According to the Union of Concerned Scientists, the federal government uses at least $7 billion tax to subsidize and cleanup after CAFOs.[61] If factory farms had to pay the full costs of their pollution there would be a greater incentive to invest in more environmentally friendly technology.

Health Costs of Meat Consumption and Production

Health risks due to diet: Heart disease, obesity, and cancer

Research studies indicate that meat consumption, and in particular, red meat consumption, is associated with an increased risk of heart disease. The Nurses' Health Study provides evidence that diets low in red meat, containing nuts, low-fat dairy, poultry or fish were associated with a 13–30% lower risk of coronary heart disease compared with diets high in meat. On the contrary, low-carbohydrate diets high in animal protein were associated with a 23% higher total mortality rate, whereas low-carbohydrate diets high in vegetable protein were associated with a 20% lower total mortality rate.[77]

The saturated fats and cholesterol in meats are considered the driving agents of the association to increased risk of heart disease, but research is continuing to seek additional risk factors. For example, high levels of plasma carnitine (abundant in red meat) accompanying high levels of trimethylamine-N-oxide has been associated with atherosclerosis.[78] Dietary cholesterol is only found in animal products and it is known that high plasma levels of cholesterol can be a risk factor for heart disease. Additionally, meats higher in saturated fats seem to be the most detrimental to health.

The 2010 Dietary Guidelines recommend limiting our consumption of saturated fat, with suggestions to purchase low-fat or fat-free milk and to trim fat from meat. The Guidelines also provide a chart showing the major sources of saturated fat in the American diet as percentages: full-fat cheese (8.5%), pizza (5.9%), grain-based desserts (5.8%), dairy-desserts (5.6%), chicken including chicken mixed dishes (5.5%) and sausage, franks, bacon, and ribs (4.9%).[81] However, the suggestions from the Dietary Guidelines do not seem to follow logically from this chart. For example, it is pretty difficult to trim fat from sausage, franks, bacon, and ribs. Also, the suggestions do not mention cutting back on cheese and pizza, instead focusing on replacing one category of milk with another.[79] Furthermore, beef and burgers are two different categories, with beef at 4.1% and burgers at 4.4%.[79] If those two categories were combined, beef in any form would be right after cheese as a major source of saturated fat.

Likewise, the Dietary Guidelines note moderate evidence of a relationship between higher cholesterol intake and higher cardiovascular disease risk, and comments that cholesterol intake can be reduced through limiting consumption of the specific foods high in cholesterol.[79] Chicken can contain more cholesterol per gram than low-fat beef. For example, 100 g of skinless, boneless, raw chicken breast contains 73 mg of cholesterol and 100 g of 97% lean ground beef contains 63 mg of cholesterol.[80] And of course, although skinless, boneless, grilled chicken breast sandwiches are available at fast-food restaurants, so are ones that are breaded, deep-fried, with bacon and cheese.

When it comes to beef, studies have been conducted on the impact of the cow's diet on the nutrient profile of the resulting meat. Studies have demonstrated that grain-fed meat has substantially more fat than grass-fed, and the fats that are found in grass-fed meat tend to be healthier. Grass-fed meat has more omega-3 fatty acids, beta-carotene, and conjugated linoleic acids (healthy fats), all of which are beneficial for heart health.[61] Is it possible that consuming grass-fed lean meats can contribute a compromise between our desires and a healthier lifestyle?

Studies also link meat intake to weight gain and obesity. Recent data demonstrates that consumption of meat protein appeared to be associated with weight gain over 6.5 years, with 1 kg (2.2 lb) of weight gain per 125 g (4.4 oz.) of meat consumption. Additional research done by the Center for Human Nutrition at the Johns Hopkins School of Public Health supports the link between meat consumption and increased body weight. Researchers tested the associations between total meat consumption and adiposity measures including body mass index (BMI), waste circumference, obesity, and central obesity using data conducted for the National Health and Nutrition Examination Survey (NHANES). They found a statistically significant association between meat consumption and BMI. Higher meat consumption was associated with a higher BMI and waist circumference, whereas consumption of fruits and vegetables was inversely associated with BMI. Those with high meat consumption were about 27% more likely to be obese, and 33% more likely to have central obesity when compared to those with low meat consumption.[81] It should be noted that those who had higher meat consumption had higher energy intakes in general, but findings based on national data emphasize the importance of examining the effect of meat consumption on health and obesity.

Another study examined the association of the consumption of food groups and specific food items to the prospective annual changes in waist circumference for a given

BMI. Researchers found that among European men and women, higher fruit and dairy consumption was associated with a lower gain in waist circumference whereas the consumption of white bread, processed meat, margarine, and soft drinks were positively associated with a change in waist circumference.[82] These studies support the conclusion that higher intakes of meats and processed meats can lead to both increased energy intake as well as increases in BMI and waist circumference.

Meat consumption, red meat in particular, has also been linked to cancer. Recent meta-analyses show that people who have a high intake of cured meats and red meat tend to have a modest increased risk of colorectal cancer. Researchers suggest that red meats are worse than poultry or fish because red meats contain more iron that is bound to the hemoglobin in blood, which can increase the formation of N-nitroso compounds. These compounds have been shown to cause tumors in animal models. The World Cancer Research Fund recommends limiting the consumption of red meat and avoiding the consumption of processed meats altogether.[83]

Environment-based health risks: Cancer

Drinking water contaminated by factory farms is linked to cancer as well. Delmarva, Delaware is a large poultry producing area, and has exceedingly high rates of cancer, with Sussex County having the highest in the nation. One concern is the higher levels of arsenic that are found in the drinking water in areas where chicken manure is spread. Some of these chicken farms have eight barns each housing up to 40,000 birds. These chicken CAFOs contaminate the water source with arsenic. The most common forms of cancer associated with arsenic exposure are skin, lung, liver, prostate, and bladder.[84]

Prairie Grove, which is located in Washington County, Arkansas, also has abnormally high cancer rates. Dozens of cancer cases and other serious disorders were reported among residents of Prairie Grove, many of who were children and adolescents.[84] Washington County, the home of Tyson Foods, is one of the largest poultry-producing places in the country, and Prairie Grove was in the midst of such production. Until 2004, all children in Prairie Grove attended one K-12 school located in front of a chicken CAFO. Arsenic-containing soil was spread on fields surrounding the school and town. Court records demonstrate that since 1990, when chicken litter began to be spread on the field in substantial amounts, an abnormally high number of young people in Prairie Grove had been diagnosed with some type of cancer. Attorneys for these ill-stricken residents claimed that cancer rates in Prairie Grove were 50 times higher than the national average. In 2001 alone, three 14-year-old boys were diagnosed with a very rare form of testicular cancer, one example of many incidents of rare cancers in young residents of Prairie Grove.[83] Unfortunately, the young residents of Prairie Grove have provided us concrete data linking exposure to arsenic from factory farms to an increased risk of cancer.

Food safety and health risks

Each year about one in six Americans (48 million) suffers from a foodborne illness, with 128,000 people being hospitalized and 3000 dying.[85] Meat is historically and inextricably linked with food safety in the United States (see Chapter 13 on food protection). Pathogens that are commonly associated with meat include *Salmonella enteritidis* and *Escherichia coli*. Some strains of *E. coli*, such as *E. coli* O157, are of particular risk to humans due to the production of shiga toxin. Ingesting as few as 10 microbes can cause a fatal infection in humans. *E. coli* O157 is common in feedlot cattle, and because the digestive tracts of grain-fed cows have an acidity level closer to humans, acid-resistant strains are more likely to develop that have a greater chance of surviving in human stomachs and become potentially fatal.[61]

An estimated 265,000 *E. coli* infections occur annually in the United States, 36% of which are likely the shiga toxin-producing strain O157.[86] Ironically, in the FDA's role of "Drug Administration" the FDA has also approved the use of a new drug used to treat the devastating condition of hemolytic uremic syndrome sometimes occurring from shiga-toxin *E. coli* infections. This new drug promises good results and the avoidance of lifetime dialysis for stricken children.[87] Drug companies do not make money when the condition requiring treatment is eliminated, however. The fact that such drug approval comes from the same agency that is working to limit such infections may lead one to consider if pulling food safety out into a separate and unified agency would be more appropriate.

Salmonella enteritidis is a bacteria found in poultry that can cause foodborne illness in humans. Annually, the CDC estimates that *Salmonella* causes 1.2 million illnesses and 450 deaths in the United States.[88] Poultry farms that fail to comply with food safety regulations and that have unsanitary conditions increase the risk of *Salmonella* infection. In 2011, Sparboe Farms, the nation's then fifth largest shell egg producer, and then supplying major retailers such as Target and McDonalds, had an alleged 13 food safety violations that included dead rats, insects, and dead, decaying chickens, as well a lack of compliance with required environmental testing for *S. Enteritidis*.[89] With 300 million dozen eggs being produced annually by such a factory, the risk of widely spreading *Salmonella* is high.

Consider the numbers of animals processed in 2012 as cited in the first paragraph of this chapter: 8.6 billion chickens, 250 million turkeys, 32.1 million cattle, 113.2 million hogs, and 2.2 million sheep and lambs.[2] Between 2008 and 2011, federal inspectors found almost 45,000 violations of health standards at the 616 hog processing plants. One such violation is the observance of fecal matter on the carcass, which, through processing and mixing of one carcass with many others through processing, could spread bacteria very widely, risking the health of numerous people. During the same time period, however, overworked and undertrained inspectors suspended operations at the plants only 28 times.[90]

Also, no doubt given the efficiencies in time required for profit-maximization, the inspections are under pressure by industry to limit processing downtime. Our historical process relying on visual and manual inspection is being modified to include more sampling for pathogens, which cannot be observed visually. A new computer system creates a schedule and automatically selects samples to send to the testing labs. The results are received electronically, and faster than in the past, which allows the opportunity to track down pathogen contamination before it leads to outbreaks, unless the system fails as it did at 18 plants in 2012, leaving at least 100 million lb of beef not properly sampled.[91]

The industry-developed product of mechanically tenderized beef highlights the scenario in which label information regarding processing could be important. Information about how the processing method could impact a consumers' ability to discern food safety risk. Although consumers have become familiar with the industrial processing of hamburger and associated risk of pathogen contamination, the foodborne disease outbreaks associated with cuts such as whole steak is another paradigm shift into the new industrial realities of meat processing.

Mechanical tenderizing is done at the processor, with a machine that presses blades or needles (sometime also with marinade) into the meat. This increases the risk of cross-contamination of fecal matter entering below the surface of the meat, and thus any introduced pathogens are less likely to be destroyed in the cooking process. The 2009 FDA model Food Code made the new recommendation of ensuring mechanically tenderized meat reaches 155 degrees while cooking,[92] not 145 as normally advised for steak. The 2013 model Food Code has retained this provision.

However, although additional steps regarding cooking times and temperatures can be taken to improve the safety of mechanically tenderized meat, most restaurants, grocery stores, and consumers do not know if the purchased meat is mechanically tenderized. The USDA has proposed a rule labeling mechanically tenderized beef as such and with cooking temperature instructions which will go into effect May 2016.[93]

Health risks from animal husbandry: BSE and antibiotic resistance

Bovine spongiform encephalopathy (BSE), more commonly known as Mad Cow Disease, is a progressive neurological disorder of cattle that results from infection by a transmissible agent called a prion. Researchers conclude that BSE possibly originated from feeding infected sheep meat-and-bone meal to cattle, or from feeding meat-and-bone meal from BSE-infected cattle to other cattle.[94]

In 2003, Sunny Dene Ranch tested positive for BSE, the first case ever reported in the United States.[84] BSE occurred in the United States again in 2004, 2006 and, most recently in 2012 in a California dairy cow.[94] BSE is a public health concern as it can be transmitted from cows to humans through consumption of contaminated beef. The human form, variant Creutzfeldt–Jacob Disease (vCJD), is a degenerative, fatal brain disorder.[95] There have been 153 cases of vCJD worldwide, mostly in the UK; however one can live with vCJD for some time before becoming symptomatic, and more people could be unknowingly carrying the disease. The FDA began regulating the contents of animal feed to prevent BSE in 1997, and in 2008 prohibited the use of the highest risk cattle tissues in all animal feed.[96]

The use of antibiotics in livestock production has led to the development of antibiotic resistance in farm environments and concern over the risk of transfer of resistant agents from animals to humans.[97] NARMS, the National Antimicrobial Resistance Monitoring System, a joint effort of federal agencies including the FDA, CDC, and USDA, began monitoring antibiotic-resistant pathogens in our food supply in 1996, and has reported an increase in resistant pathogens. The 2011 annual retail meat report from NARMS, released in early 2013, indicates that 51 percent of bacteria found on ground turkey are resistant to ampicillin.[98]

As exposure to antimicrobials increases, so does the chance of the development of resistance among bacteria. Resistant bacteria and antibiotic compounds accumulate in manure when livestock are given feed that contains antibiotics. When such manure is applied to agricultural soils, there is concern that this significantly increases antibiotic-resistant bacteria in the soil, which can then spread from soil to urban environments through wind or water and further contaminate the environment, including through attachment to crops.[99]

The issue of antimicrobial resistance has become more apparent as the incidence of drug-resistant infections has increased in humans.[67] Many research studies conclude that antibiotic-resistant bacteria in humans are becoming more prevalent. Now understanding that overprescribing antibiotics can result in antibiotic-resistant bacteria, we have begun to use antibiotics more carefully in human treatment, but conversely, antibiotic usage has increased in treating animals over the time period 2001–2011,[99] in part through practices such as described above of treating an entire flock of chickens by putting an antibiotic in the drinking water.

However, one critical point in arguments over antibiotic use is that not all antibiotics given to animals are of significance to human medicine. In 2009, the FDA reported that 45% of antibiotics sold for animal use were ones not used for disease in humans, and tetracyclines made up another 42%. Tetracyclines are now in limited use for human treatment, with other antibiotics being the first choice of treatment.[100]

Conversely, fluoroquinolones are a critical drug for human treatment. Fluoroquinolone-resistant bacteria were rare until 1995, when the FDA approved the use of baytril in drinking water for poultry.[71] In 2005, the FDA issued a final rule that baytril was no longer approved for treating bacterial infections in poultry.[101] However, a 2012 study of feather meal, made from poultry feathers and commonly added to feed for swine, cattle, fish, and even chicken, found residues of fluoroquinolone in 8 of 12 of the samples, along with other pharmaceutical products.[102]

For humans, antibiotics are available for use only with a prescription, but historically antibiotics were easily made available over-the-counter for use in farm animals. As part of the 2009 reauthorization of the Animal Drug User Fee Act (ADUFA), the FDA was authorized to collect some information on antibiotics sold for use on industrial farms. As of late 2013 the FDA has proposed that licensed veterinarians supervise the use of antibiotics for farm animals, which entails getting a prescription. Furthermore the FDA is asking for voluntary compliance from drug companies to change labels that detail how drugs can be used, so that they cannot be used for growth. By June 2014, the FDA has secured voluntary engagement from all 26 affected drug manufacturers to promote the judicious use of antimicrobials in livestock; however, prophylactic use of antibiotics for disease prevention is not being prohibited at this point.[103]

Environmental Impacts of Meat Production

Industrial farming can lead to large amounts of waste, air and water pollution, and greenhouse gas emissions, and is also a factor in resource depletion.

Livestock and waste production

Factory farming and CAFOs produce massive amounts of waste from thousands of animals. A 100-acre CAFO generates the same amount of sewage as a city of 100,000 inhabitants. A new hog farm in Utah will produce more animal waste than all the sewage created by humans in Los Angeles, California. The USDA estimates that more than 335 million tons of dry matter waste is produced each year on farms in the United States.[61] The specialization of industrial farming has separated meat production from agricultural production, and thus, the animal manure, which could have been used for fertilizer, literally goes to waste.

Not only do CAFOs produce enormous amounts of waste, but also regulations for the management and disposal of animal waste have historically been more lenient than regulations for human waste. It is nearly impossible to keep up with the waste that is produced on farms, and much of it is placed into lagoons. These lagoons are essentially outdoor holding tanks, which can be hazardous to the environment. Large amounts of the waste from the lagoons often seep into groundwater, get released into the atmosphere, and can mix with rainwater. Some farms have an automated clean up system for animal waste products, which may use up to 150 gallons of water per cow per day. Farms may also spray liquid manure onto the soil for fertilizer. This application can spread viruses, bacteria, antibiotics, metals, nitrogen, and phosphorus, resulting in air and water pollution.[61]

Air and water pollution

Factory farming contributes to air pollution through the production of animal feed and waste, burning of fossil fuels, and the shipping of meat products to distributors. Chronic inhalation of pollutants can be hazardous to people who work at and live nearby factory farms. It can result in problems such as respiratory irritation, asthma, and neurological disorders.[104] Hydrogen sulfide, one of the gases produced by decomposing manure, is of particular concern. When liquid manure slurry is agitated as part of the pit-emptying process, hydrogen sulfide can be rapidly released, reaching lethal levels in seconds and causing animal and worker death in buildings above the manure pits. In 2008, residents living near a dairy farm in Minnesota were advised to flee their homes due to toxic levels of hydrogen sulfide rising from the dairy's waste lagoon, which exceeded by 200 times the state air quality standard. Eventually, residents were forced to leave and the dairy farm was allowed to continue operations.[61]

Runoff from farms can contaminate groundwater and nearby waterways with fertilizers, nuclear waste, animal waste, heavy metals, and residues from antibiotics, hormones, and pesticides. The EPA has reported that hog, chicken, and cattle waste has polluted 35,000 miles of rivers in 22 states and has significantly contaminated groundwater in 17 states. These contaminates in waterways can lead to outbreaks of harmful algae and bacteria and consequently, massive fish kills. Each year, nutrient runoff from feed grain fields flows down the Mississippi River and into the Gulf of Mexico, contributing to an 8000 square-mile oxygen depleted dead zone harming aquatic life.[61] A recent United Nations sponsored study found that the natural nitrogen and phosphorous cycles are being seriously imbalanced by fertilizers, 80% of which is attributable to meat production, causing extensive problems.[105]

Another example of water pollution from factory farms occurred in 2005, when a lagoon collapsed at a dairy farm in New York; 3 million gallons of waste spilled into the Black River, resulting in a fish kill of 250,000, as well as residents no longer being able to use this river for water supply and recreation.[61] Not only can polluted water harm fish, but it can also be harmful to humans. Arsenic and residues from antibiotics can be found in water near factory farms. Consumption of arsenic-contaminated water can lead to disease and antibiotic residues can foster the growth of antibiotic-resistant bacteria, both of which are of public health concern.

Agriculture is the largest user of water resources in the United States and CAFOs are the largest per capita consumer of water. CAFOs use enormous amounts of water to irrigate feed crops, hydrate livestock, move waste, and service slaughterhouse disassembly lines. The dewatering of rivers for hay and other crop production, along with the loss of groundwater supplies by pumping, and having processing regions located in dry regions, all result in water degradation.[61] Not only does this significantly decrease the fish supply due to habitat change and water loss, it also makes CAFOs responsible for extensive external costs related to depleting the Earth's natural resources.

Dependency on fossil fuels and greenhouse gas production

Industrial agriculture is widely dependent on fossil fuels for machine use in each step of the production process and for the production of fertilizer and agricultural chemicals. The use of grain-based feeding itself has a large environmental impact because industrial agriculture is powered by fossil fuel. Growing the amount of corn required to produce meat through factory farming requires substantial oil usage. If a cow eats 25 lb of corn a day and reaches a goal weight of 1250 lb, it will have consumed an estimated 284 gallons of oil.[61] Some countries' total energy input into food production exceeds energy output.[106] For this reason, energy expenditure in factory farming is of significant environmental concern.

The burning of fossil fuels produces greenhouse gas emissions. Greenhouse gases trap heat in the atmosphere and contribute to climate change. Some greenhouse gases occur naturally, but substantial amounts are from human processes, including livestock production. According to a 2006 United Nations Food and Agriculture Organization report, the livestock sector accounts for 18% of global greenhouse gas emissions, which is more than the entire world's transportation emissions. Another study indicates that global livestock production is responsible for 32 billion tons of carbon dioxide per year, or 51% of all greenhouse gas emissions.[61]

These greenhouse gases can be harmful to the environment and to humans. Methane and nitrous oxide are two intoxicating greenhouse gases associated with industrial animal production. These gas emissions occur from the decay of organic waste and from agricultural and industrial activities.[107] The livestock sector produces 37% of the methane and 65% of the nitrous oxide emissions.[108] Environmentally sustainable agriculture is now a question at hand. In the event of inaction, climate change will affect food yields and human health.[106]

Water policy may by default become food policy. Adequate water supply is a concern for our future. Meat production is generally water-intensive, taking 20–50 times as much water kilogram by kilogram, to produce meat as compared to vegetables, but it does vary by type of animal. Cattle feeding is the most water-intensive. Broilers have the lowest water-drinking needs, less than one-tenth that required for cattle,[32] but water is also needed to produce the animal feed and to clean off animal waste, etc.

Global Hunger, Meat, and Trade Policy

The "turkey tail" discussion earlier in the chapter highlights the issue that food policy is also trade policy, which may be conducted through multinational corporations with their own interests as well as through national government interests and policies.

Because meat is highly resource-intensive to produce, arguments for reducing meat production often include discussions on global resources and global hunger. Although critical, it is beyond the scope of this chapter to delve into the relationship between meat and world hunger, but it is important to keep in mind that food policy, even when nominally related to hunger, may be bound to global trade and political issues. As an example, in the late 1960s and early 1970s, the world seemed at risk of famine. When Russia's grain crop failed, they bought ours. Although likely very much in reality a corporate deal, the public perceived this as a nationalism-charged issue. However, if the world was in fact likely to suffer starvation, our government could not appear indifferent. The governmental decision to promote more food production in the 1970s may not have been alone an agricultural decision, but also as an economic and political decision.[16]

Similarly, although we produce an enormous quantity of feed for animals, the comparison of relevant human food production from the same acreage is not a "one-to-one," as naturally we are producing types of grains for our animals that we ourselves may not wish to consume (corn, sorghum, soybeans). A realistic transition would account for our current dietary preferences in grains (wheat, rice) with leftover materials from processing used as animal feed.[32]

As noted above, some countries have refused to purchase meat from the United States due to concerns over drugs, such as the β-agonist ractopamine used to promote growth. In 2012, the United States exported approximately $300 million of beef alone to Russia. In January of 2013, Russia prohibited the importation of meat containing ractopamine, and the exports dropped precipitously. Perhaps not surprisingly, in late 2013, the USDA introduced a certification program that will allow the use of a "Never Fed Beta Agonists" label.[109]

More knowledge about how our meat has been raised, slaughtered, and processed has become more important for many consumers. The final section of this chapter discusses some connotations and values we place in information on labels relevant to meat.

CONSUMER VALUES AND MEAT

It is important to understand the values people consider when comparing and selecting foods. People may use the information provided by labels as the means to make a decision while shopping. Sustainable, local, grass-fed, organic, humanely raised—these are all value choices that some consumers may balance against cost.

The Importance of Labeling

Producers have, even historically, sought to use everything they could to make meat processing profitable. "Pink slime," a topic of controversy during the writing of this book, is an example of an industry solution for increasing profitability that also highlights how scientifically "safe" can be at odds with the values people place in food.

"Pink slime" is a nickname. The manufacturers call it "lean finely textured beef." It is produced from the pieces trimmed off of a beef carcass when carved for cuts of meat. The chopped up pieces are spun in a centrifuge to remove the fat and then treated with ammonia or citric acid to kill pathogens. As filler, the "pink slime" can replace some of the actual ground beef, and thereby lowers the overall fat content of the product. The USDA approved the product for use in preparing hamburger patties as "fat-reduced beef" in 1991, but at the time would not allow it to be included in packages of ground beef, under the distinction that premade hamburgers could have fat added, but ground beef could not. But just a few years later, in 1993, the original manufacturer convinced the USDA that the same product, termed "lean finely textured beef," could be used more widely, including in packages of ground beef.[110]

The USDA considers "pink slime" to be ground beef. And, because USDA considers this product to be ground beef, the manufacturer is not required to list "lean finely textured beef" as an ingredient in packages of ground beef. A consumer buying a package of ground beef would have no reason to believe that it contained filler. The nickname "pink slime" was coined by a former USDA scientist who did not consider "lean finely textured beef" to be actual ground beef, and felt that including it in ground beef without disclosure was simply fraudulent labeling.[111]

The USDA scientist was not complaining about safety or nutrition. "Pink slime" might be safe enough considering the necessary ammonia treatment, and it may contribute to the apparent overall nutritional profile of the meat, but does it meet consumer's expectations and values as "ground beef"? Conversely, consider the "eating the whole animal" trend. *Diners may now pay a premium for animal parts formerly discarded. How does this differ from "pink slime"?* Consider the value we place in knowing what we are eating and making deliberate choices.

The Importance of Being Organic

Recall from other chapters that the USDA and FDA share food regulation roles. Although the FDA has the lion's share of responsibility over packaging information, the USDA is responsible for enforcing meat regulations. Title XXI of the 1990 Farm Bill, the Organic Food Production Act (OFPA), authorized the USDA to establish the National Organic Program (NOP) to administer national standards covering organic food production.[112]

USDA organic standards, in addition to prohibiting synthetic pesticides in the animal feed, does provide a standard of minimums as a reference point for the consumer, such as pasture requirements. Ruminant animals must graze pasture during the grazing season for their region, which must be at least 120 days per year, and must obtain a minimum of 30% dry matter intake from grazing pasture during the grazing season. The animals must have year round access to the outdoors. Producers must have a pasture management plan and manage pasture as a crop to meet the feed requirements for the grazing animals and to protect soil and water quality.[113] These pasture requirements for ruminant animals became law in June of 2010.

Prior to the 2010 rule, the requirement had been simply "access to pasture," which, as it was vague, allowed a range of variation in application. This came to light in January 2005, when a newspaper article described an organic dairy farm in Colorado that consisted of over 5000 cows who were fed almost exclusively grain (albeit organic grain) and given access to an outdoor feedlot for air and exercise. Long-standing smaller organic dairy producers argued that this was not a fair reading of legislative intent, as "access to pasture" meant that the cows must get the benefit of fresh air, exercise, and nutrition of grazing on pasture. The producer argued it was not possible to have the cows graze in arid Colorado. Other organic stakeholders argued that organic farming *practices*, particularly the treatment and feeding of livestock, was an integral part of the meaning of the term "organic," and that if the area could not appropriately support a large organic dairy herd, then only a small one, or none at all, should exist in Colorado.[114]

The 2010 regulation further defining organic in relation to dairy production as a result of this kerfuffle is an example of tension in food policy outcome—what impact is there to defining organic? Have we created an underlying philosophy that reflects ideals valued by consumers, such as sustainability? Does "organic" support small and medium farms or has it become a marketing tool that large agriculture can more easily leverage?[115] Is organic agriculture just a marginally better version of large agriculture? Although continuous total confinement of ruminants inside or in a yard, feeding pad, or feedlot is prohibited under the organic standards, if one were to visit a large-scale organic dairy would we be able to discern the difference?

The USDA has regulated the use of the term "organic," "free-range," "cage-free," and "grass-fed." Grass-fed animals receive a majority of their nutrients from grass throughout their life, while organic animals' pasture diet may be supplemented with grain. The grass-fed label does not limit the use of antibiotics, hormones, or pesticides, unless also labeled as organic. The terms "humane" and "pasture-raised" are not USDA regulated, and such claims can vary widely as to the actual treatment of the animal, and there are various third-party certifiers of "humane."[116]

The Importance of Location

Country of origin labeling (COOL) for certain foods, including meats, was approved by Congress in 2002, in part due to consumer demand, and was implemented by the USDA in 2009. In May of 2013, the USDA revised the rules to give consumers specific information on where animals were born, raised, and slaughtered. More significantly, each package of meat, except for ground meat, must contain meat only from the same source.[117]

Not surprisingly, the trade association American Meat Institute (AMI) has filed a lawsuit claiming that this revision is in violation of the statute as drafted by Congress and furthermore violates the First Amendment by requiring it to disclose country-of-origin information to retailers, who will ultimately provide the information to consumers. AMI also argued that government does not have a substantially enough interest in COOL to sustain the rules. In July 2014, the U.S. Court of Appeals of the District of Columbia ruled against AMI. The court disagreed with all arguments. Regarding the First Amendment claim, the court stated that arguments were not persuasive to find COOL neither factual nor controversial, and accordingly does not trigger an application of precedent case law. The court also found that the government in fact has substantial interest, supported by a history of such disclosures in other industries, a demonstrated consumer interest, and the public health concerns of food-borne illness.[118]

CONCLUSION

The mission of the USDA includes providing leadership on food, agriculture, and nutrition; these objectives can manifest in conflicting policies and dietary guidance. The impact of dietary guidance on the types of foods and food products made available in our food system is tremendous: a recommendation to choose leaner meats lower in saturated fat instead of a recommendation to eat less meat has helped secure the omnipresence of skinless, boneless, chicken. What happens to the rest of the chicken? In the final chapter of this book, "Looking Forward," we discuss waste; in this chapter we see how the producers can create a market elsewhere—if we are not buying the skin, bone, or tails, someone else is.

In general, industrial animal farming is a profit-driven industry that creates external costs to the environment and human health. Our food policies must balance "cheap food" against these external costs. Some of these external costs may be modifiable into more sustainable processes, as new innovations such as placing anaerobic digesters on dairy farms may not only be able to process the manure but also take on food waste to generate renewable energy. Other external costs, such as the use of water, may simply not be sustainable, and our expectations about meat in our diet and our food costs may have to change. Food products using pea-protein to simulate the texture of meat are one example of innovative solutions that may address our taste for meat.

Looking forward, food policy conducted at a local level, such as ordinances that allow for backyard chickens, can foster changes that can influence our expectations about animal husbandry, the meat industry, and our dietary choices. Following the small changes illustrated in other areas, such as the "Meatless Monday" campaign, people can "vote with their fork." By adopting "flexitarianism" or even "demitarianism," eating meat can be made special by cutting consumption and also thoughtful by purchasing meat and dairy products that support a healthy, humane food system with the expectation of eventually moving toward a more sustainable agricultural food system.

REFERENCES

1. Benjamin Franklin. *Poor Richard's Almanac.*
2. The American Meat Institute, The United States Meat Industry at a Glance. Available at http://www.meatami.com/ht/d/sp/i/47465/pid/47465, accessed September 6, 2014.
3. U.S. Department of Commerce, United States Census Bureau, U.S. and World Population Clock. Available at http://www.census.gov/popclock/, accessed September 7, 2014.
4. United States Department of Agriculture, What's in the Protein Foods Group? Available at http://www.choosemyplate.gov/food-groups/protein-foods.html, accessed September 7, 2014.
5. U.S. Department of Agriculture and U.S. Department of Health and Human Services. *Dietary Guidelines for Americans,* 2010, 7th edition, Washington, DC: U.S. Government Printing Office, December 2010.
6. Nestle M. *Food Politics,* January 4, 2012. Available at http://www.foodpolitics.com/2012/01/peevish-about-protein/, accessed September 6, 2014.
7. Nestle M. *Food Politics: How the Industry Influences Nutrition, and Health.* California: University of California, 2007.
8. Burros M. Eating well; Testing of food pyramid comes full circle. *The New York Times,* March 25, 1992. Available at http://www.nytimes.com/1992/03/25/garden/eating-well-testing-of-food-pyramid-comes-full-circle.html?scp=1&sq=eating%20well:%20testing%20of%20food%20pyramid&st=cse, accessed September 6, 2014.
9. United States Department of Agriculture, How much food from the Protein Foods Group is needed daily? Available at http://www.choosemyplate.gov/food-groups/proteinfoods_amount_table.html, last modified June 4, 2011, accessed September 6, 2014.
10. United States Department of Agriculture, ChooseMyPlate, What Counts as an Ounce Equivalent in the Protein Foods Group? Available at http://www.choosemyplate.gov/food-groups/protein-foods-counts.html, accessed September 6, 2014.
11. Centers for Disease Control and Prevention, Nutrition Basics, Protein, updated October 4, 2012. Available at http://www.cdc.gov/nutrition/everyone/basics/protein.html, accessed September 6, 2014.
12. Food and Agriculture Organization of the United Nations, International year of quinoa 2013. Available at http://www.fao.org/quinoa-2013/what-is-quinoa/nutritional-value/en/, accessed September 6, 2014.
13. WHO Technical Report Series 935, Protein and Amino Acid Requirements in Human Nutrition, 2007, Geneva, Switzerland.
14. United States Department of Agriculture, Agriculture Research Service, National Nutrient Database for Standard Reference, Release 27. Available at http://ndb.nal.usda.gov/ndb/, accessed September 6, 2014.
15. The Library of Congress. A Century of Lawmaking for a New Nation: U.S. Congressional Documents and Debates, 1774–1875, *Statutes at Large, 37th Congress,* 2nd Session, An Act to establish a Department of Agriculture, May 15, 1862. Available at http://memory.loc.gov/cgi-bin/ampage?collId=llsl&fileName=012/llsl012.db&recNum=418, accessed September 6, 2014.
16. Ogle M. *Meat We Trust: An Unexpected History of Carnivore America.* Boston: Houghton Mifflin Harcourt, 2013.
17. United States Department of Agriculture. About USDA, Mission Statement. Available at http://www.usda.gov/wps/portal/usda/usdahome?navid=MISSION_STATEMENT, accessed October 28, 2014.
18. Neuman W. Audit finds problems in Cattlemen's spending. *The New York Times,* August 3, 2010, p. B1.

19. Dairy Management and the Dairy Checkoff. Available at http://www.dairy.org/about-dmi, accessed September 6, 2014.

20. Dairy Check-off Program. Available at http://www.dairy.org/news/2013/july/checkoff-funded-study-reaffirms-wheys-role-in-developing-muscle, accessed September 16, 2014.

21. United States Department of Agriculture, Agricultural Marketing Service, Research and Promotion Programs, last modified August 27, 2013. Available at http://www.ams.usda.gov/AMSv1.0/ams.fetchTemplateData.do?template=TemplateB&leftNav=ResearchandPromotion&page=ResearchandPromotion.

22. United States Department of Agriculture, Agricultural Marketing Service, Research and Promotion Programs, Pork Promotion and Research Program Background Information. Available at http://www.ams.usda.gov/AMSv1.0/ams.fetchTemplateData.do?template=TemplateN&navID=PorkProgramBackgroundInformation&rightNav1=PorkProgramBackgroundInformation&topNav=&leftNav=&page=PorkBackgroundofPorkCheckoffProgram&resultType=&acct=lspromores, accessed December 16, 2013.

23. Dairy Management Inc, News, Dairy Checkoff Partnerships Grow 2013 Sales, December 12, 2013. Available at http://www.dairy.org/news/2013/december/dairy-checkoff-partnerships-grow-2013-sales, accessed September 7, 2014.

24. Wilde P. Sorry, McWilliams, the New York Times got the USDA cheese story right. *Grist.* November 20, 2010. Available at http://grist.org/article/food-2010-11-18-the-new-york-times-got-the-usda-cheese-story-right/, accessed September 7, 2014.

25. *Johanns v. Livestock Marketing Association*, 544 U.S. 550,2005, 335 F. 3d 711. Available at http://www.law.cornell.edu/supct/html/03-1164.ZS.html, accessed September 7, 2014.

26. Dairy Check-off Program. Available at http://www.dairy.org/news/2013/october/checkoff-helps-pizza-hut-introduce-new-3-cheese-pizza, accessed September 7, 2014.

27. Pizza Hut. Available at https://order.pizzahut.com/nutrition-information, accessed September 7, 2014.

28. United States Department of Agriculture, ChooseMyPlate, Health and Nutrition Benefits of Dairy. Available at http://www.choosemyplate.gov/food-groups/dairy-why.html, accessed September 7, 2014.

29. Walsh B. Why it was silly for the Beef Industry to Freak out over USDA's "Meatless Monday" Newsletter. *Time.* July 27, 2012. Available at http://science.time.com/2012/07/27/why-it-was-silly-for-the-beef-industry-to-freak-out-over-usdas-meatless-monday-newsletter/, accessed September 7, 2014.

30. Meatless Monday. History. Available at http://www.meatlessmonday.com/about-us/history/, accessed September 16, 2014.

31. Harmon A. Retracting a plug for meatless Mondays. *The New York Times,* July 25, 2012. Available at http://www.nytimes.com/2012/07/26/us/usda-newsletter-retracts-a-meatless-mondays-plug.html?_r=0, accessed September 7, 2014

32. Smil V. Eating meat. *Population Develop. Rev.* 2002;28(4):599–639.

33. Carrie RD, Cross AJ, Koebnick C, Sinha R. Trends in meat consumption in the United States. *Public Health Nutr.* 2011;14(4):575–583.

34. Larsen J. Peak meat: U.S. meat consumption falling. *Earth Policy Institute,* March 7, 2012.

35. United States Department of Agriculture, Economic Research Service, Food Availability (per capita) data system. Available at http://www.ers.usda.gov/data-products/food-availability-%28per-capita%29-data-system/summary-findings.aspx, updated April 7, 2014, accessed September 7, 2014.

36. Boyd W, Watts M. Agro-industrial just-in-time: The chicken industry and postwar American capitalism. In: Goodman D, Watts M (eds.). The legal basis for food safety regulation in the USA and EU. *Globalising Food: Agrarian Questions and Global Restructuring.* London, UK: Routledge, 1997,192–224.

37. Morris JG, Jr., Potter ME (eds.). The legal basis for food safety regulation in the USA and EU. *Foodborne Infections and Intoxications,* 4th edition, London: Academic Press, 2013, 514.

38. United States Environmental Protection Agency, Agriculture, Beef Production, last updated June, 27, 2012. Available at http://www.epa.gov/oecaagct/ag101/printbeef.html, accessed September 7, 2014.
39. Food and Agriculture Organization of the United Nations. Biodiversity for food and agriculture: Farm animal genetic resources, 1998.
40. Schlosser E. *Fast Food Nation: The Dark Side of the All-American Meal.* Boston: Houghton Mifflin Harcourt, 2001.
41. Striffler S. *Chicken: The Dangerous Transformation of America's Favorite Food.* New Haven, CT: Yale University Press, 2005.
42. Foster Farms, Fresh Turkey Tails. Available at http://www.fosterfarms.com/products/product.asp?productcode=7140, accessed September 7, 2014.
43. Gale J. Stopping Turkey Tails at the Border Pits Trade against Health. *Bloomberg News,* November 28, 2011. Available at http://www.bloomberg.com/news/2011-11-29/stopping-turkey-tails-at-the-border-pits-trade-against-health.html, accessed September 7, 2014.
44. World Trade Organization, WTO news, October 28, 2011. Available at http://www.wto.org/english/news_e/news11_e/acc_wsm_28oct11_e.htm, accessed September 7, 2014.
45. Fabricant F. "Front burner": Meringue-based cakes, fried turkey tails and much more. *The New York Times,* October 9, 2013. Available at http://www.nytimes.com/2013/10/09/dining/meringue-based-cakes-fried-turkey-tails-and-more.html, accessed September 7, 2014.
46. Barclay E. Samoans await the return of the Turkey Tail. *NPR, The Salt,* May 9, 2013. Available at http://www.npr.org/blogs/thesalt/2013/05/14/182568333/samoans-await-the-return-of-the-tasty-turkey-tail, accessed September 7, 2014.
47. United States Department of Agriculture, ChooseMyPlate. *Nutrients and Health Implications.* Available at http://www.choosemyplate.gov/food-groups/protein-foods-why.html, accessed September 7, 2014.
48. Food and Agriculture Organization of the United Nations, Fisheries and Aquaculture Department, Reduction of bycatch and discards. Available at http://www.fao.org/fishery/topic/14832/en, accessed September 7, 2014.
49. Plummer B. Just how badly are we overfishing the oceans? *The Washington Post,* October 29, 2013. Available at http://www.washingtonpost.com/blogs/wonkblog/wp/2013/10/29/just-how-badly-are-we-overfishing-the-ocean/, accessed September 7, 2014.
50. Finley C, Oreskes N. Maximum sustained yield: A policy disguised as science. *ICES J. Marine Sci.* 2013;70(2):245–50.
51. Trebon website. Available at http://www.trebon-info.com/fish-farming/, accessed September 7, 2014.
52. Burden D. *The Organic Aquaculture Quandry.* Agricultural Marketing Resource Center. November, 2009. Available at http://www.agmrc.org/commodities__products/aquaculture/the_organic_aquaculture_quandary.cfm, accessed September 16, 2014.
53. Simon DR. *Meatonomics: How the Rigged Economics of Meat and Dairy Make you Consume Too Much.* Berkeley: Conari, 2013.
54. United States Food and Drug Administration, Animal & Veterinary, Genetic Engineering, updated June 10, 2014. Available at http://www.fda.gov/AnimalVeterinary/DevelopmentApprovalProcess/GeneticEngineering/GeneticallyEngineeredAnimals/ucm113605.htm.
55. Monterey Bay Aquarium, Seafood Watch. Available at http://www.seafoodwatch.org/cr/cr_seafoodwatch/sfw_aboutsfw.aspx, accessed September 7, 2014.
56. Simon DR. *Meatonomics: How the Rigged Economics of Meat and Dairy Make you Consume Too Much.* Berkeley: Conari, 2013, 78.
57. Bureau of Labor Statistics, U.S. Department of Labor. Available at http://www.bls.gov/oes/current/oes513023.htm, accessed December 6, 2013.

58. United States Department of Agriculture, Economic Research Service, Amber Waves April 2006. Available at http://webarchives.cdlib.org/sw1tx36512/http://www.ers.usda.gov/AmberWaves/April06/Findings/Chicken.htm, accessed September 7, 2014.
59. Agriculture Council of America, Agriculture fact sheet. Available at http://www.agday.org/media/factsheet.php, accessed September 7, 2014.
60. United States Department of Justice, Competition and Agriculture: Voices from the Workshops on Agriculture and Antitrust Enforcement in our 21st Century Economy and Thoughts on the Way Forward Justice, May 2012.
61. Imhoff D. *The CAFO Reader: The Tragedy of Industrial Animal Factories.* California: University of California Press, 2010.
62. GRACE Communications Foundation, Sustainable Table: Food processing and slaughterhouses. Available at http://www.gracelinks.org/279/food-processing-slaughterhouses, accessed on September 7, 2014.
63. Hendrickson M, Heffernan W. Concentration of agricultural markets. February 2002. Available at http://www.foodcircles.missouri.edu/CRJanuary02.pdf, accessed September 7, 2014.
64. WH Group, Corporate Profile. Available at http://www.wh-group.com/en/about/profile.php, accessed September 7, 2014.
65. Gara T. Smithfield-Shuanghui Deal: The Goldman Sachs connection. *Wall Street J.* May 29, 2013. Available at http://blogs.wsj.com/corporate-intelligence/2013/05/29/how-america-owned-chinese-pigs-before-the-tables-turned/, accessed September 7, 2014.
66. United States Department of Agriculture, Economic Research Service. U.S. farm structure: Declining—but persistent—small commercial farms. *Amber Waves,* September 2010. Available at http://webarchives.cdlib.org/sw1tx36512/http://www.ers.usda.gov/AmberWaves/September10/Features/USFarm.htm, accessed on September 7, 2014.
67. Pew Commission on Industrial Farm Animal Production, Putting meat on the table: Industrial farm production in America. Available at http://www.ncifap.org/_images/PCIFAPFin.pdf, accessed on September 7, 2014.
68. Environmental Working Group, Farm subsidy database. Available at http://farm.ewg.org/progdetail.php?fips=00000&progcode=corn, accessed September 7, 2014.
69. National Sustainable Agriculture Coalition, 2014 Farm Bill Drill Down: Conservation -Working Lands Programs, February 10, 2014. Available at http://sustainableagriculture.net/blog/2014-farm-bill-working-lands/, accessed September 7, 2014.
70. GRACE Communications Foundation, Sustainable Table: Animal feed, 2013. Available at http://www.gracelinks.org/260/animal-feed, accessed September 7, 2014.
71. PBS Frontline, Modern Meat: Antibiotic Debate Overview. Available at http://www.pbs.org/wgbh/pages/frontline/shows/meat/safe/overview.html, accessed September 7, 2014.
72. GRACE Communications Foundation, Sustainable Table: Antibiotics. Available at http://www.gracelinks.org/257/antibiotics, accessed on January 13, 2013.
73. GRACE Communications Foundation. Sustainable Table: Hormones. Available at http://www.gracelinks.org/258/hormones, accessed September 7, 2014.
74. Center for Food Safety, Food Safety Fact Sheet, Ractopamine, February 2013. Available at http://www.centerforfoodsafety.org/files/ractopamine_factsheet_02211.pdf, accessed September 7, 2014.
75. Opensecrets, Center for Responsive Politics. Available at http://www.opensecrets.org/industries/background.php?cycle=2014&ind=A, updated September 2013, accessed September 7, 2014.
76. United States Environmental Protection Agency, Water, Pollution Prevention, CAFO Regulations. Available at http://water.epa.gov/polwaste/npdes/afo/CAFO-Regulations.cfm, accessed September 7, 2014.
77. Clifton PM. Protein and coronary heart disease: The role of different protein sources. *Curr. Atheroscler. Rep.* 2011;13(6):493–498.

78. Torgan C. Red meat-heart disease link involves Gut microbes. *NIH Research Matters,* April 22, 20–13. Available at http://www.nih.gov/researchmatters/april2013/04222013meat.htm#, accessed September 7, 2014.

79. U.S. Department of Agriculture and U.S. Department of Health and Human Services. *Dietary Guidelines for Americans,* 7th edition, Figure 3-4. Washington, DC: U.S. Government Printing Office, December 2010, 25, 26, 40.

80. United States Department of Agriculture, Agriculture Research Service, National Nutrient Database for Standard Reference, Release 27. Available at http://ndb.nal.usda.gov/ndb/, accessed September 7, 2014.

81. Wang Y, Beydoun MA. Meat consumption is associated with obesity and central obesity among US adults. *Int. J. Obesity.* 2009;33(6):621–628.

82. Romaguera D, Ängquist L, Du H et al. Food composition of the diet in relation to changes in waist circumference adjusted for body mass index. *PLoS One.* 2011;6(8):e23384.

83. Corpet DE. Red meat and colon cancer: Should we become vegetarians, or can we make meat safer? *Meat Sci.* 2011;89(3):310–316.

84. Kirby D. *Animal Factory: The Looming Threat of Industrial Pig, Dairy, and Poultry Farms to Humans and the Environment.* New York, NY: St. Martin's Press, 2010.

85. Centers for Disease Control and Prevention, Estimates of Foodborne Illness in the United States. Available at http://www.cdc.gov/foodborneburden/, updated January 8, 2014, accessed September 7, 2014.

86. Centers for Disease Control and Prevention. General Information. *Escherichia coli.* Available at http://www.cdc.gov/ecoli/general/, accessed November 19, 2014.

87. Rothschild M. *FDA Oks Drug Offering Hope for HUS Treatment.* September 24, 2011. Available at http://www.foodsafetynews.com/2011/09/fda-approves-drug-offering-hope-for-hus-treatment/, accessed September 7, 2014.

88. Centers for Disease Control and Prevention, General Information, Salmonella, last updated March 9, 2015, http://www.cdc.gov/salmonella/general, accessed May 10, 2015.

89. Flynn D. Egg rule violations cost Sparboeit's McDonald's account. *Food Safety News,* November 19, 2011. Available at http://www.foodsafetynews.com/2011/11/egg-rule-violations-cost-sparboe-mcdonalds-account, accessed September 7, 2014.

90. Editorial. The U.S.D.A inspects it's inspectors. *The New York Times,* June 17, 2013. Available at http://www.nytimes.com/2013/06/18/opinion/the-usda-inspects-its-inspectors.html?emc=eta1&_r=1, accessed September 7, 2014.

91. Nixon R. Shipping continued after computer inspection system failed at meat plants. *The New York Times,* August 17, 2013, http://www.nytimes.com/2013/08/18/us/computer-system-failed-at-meatpacking-plants-but-shipping-continued.html?emc=eta1, accessed September 7, 2014.

92. United States Food and Drug Administration, Food Code, 2009 summary of changes. Available at http://www.fda.gov/food/guidanceregulation/retailfoodprotection/foodcode/ucm188119.htm, accessed September 7, 2014.

93. Regulations.gov, FSIS mandatory labeling for mechanically tenderized beef products. Available at http://www.regulations.gov/#!docketDetail;D=FSIS-2008-0017, accessed September 7, 2014.

94. Centers for Disease Control and Prevention, About BSE. Available at http://www.cdc.gov/ncidod/dvrd/bse, updated February 21, 2013, accessed September 7, 2014.

95. Centers for Disease Control and Prevention, Fact Sheet: Variant Creutzfeldt-Jakob Disease, last updated February 22, 2013. Available at http://www.cdc.gov/ncidod/dvrd/vcjd/factsheet_nvcjd.htm.

96. United States Food and Drug Administration, Bovine Spongiform Encephalopathy, last updated July 9, 2014. Available at http://www.fda.gov/animalveterinary/guidancecomplianceenforcement/complianceenforcement/bovinespongiformencephalopathy/default.htm, accessed September 7, 2014.

97. Heuer H, Schmitt H, Smalla K. Antibiotic resistance gene spread due to manure application on agricultural fields. *Curr. Opin. Microbiol.* 2011;14(3):236–243.

98. United States Food and Drug Administration, NARMS Retail Meat Report, 2011. Available at http://www.fda.gov/AnimalVeterinary/SafetyHealth/AntimicrobialResistance/National AntimicrobialResistanceMonitoringSystem/ucm334828.htm, last updated February 5, 2013, accessed September 7, 2014.

99. Pew Charitable Trusts. Record-High Antibiotic Sales. February 6, 2013. Available at http://www.pewhealth.org/other-resource/record-high-antibiotic-sales-for-meat-and-poultry-production-85899449119, accessed December 6, 2013.

100. Raymond R. Antibiotics and animals raise for food: Lies, damn lies and statistics. *Food Safety News,* January 7, 2013. Available at http://www.foodsafetynews.com/2013/01/antibiotics-and-animals-raised-for-food-lies-damn-lies-and-statistics/.

101. United States Food and Drug Administration. Enrofloxacin for Poultry, updated July 14, 2014. Available at http://www.fda.gov/AnimalVeterinary/SafetyHealth/RecallsWithdrawals/ucm042004.htm, accessed September 7, 2014.

102. Love DC, Halden RU, Davis MF, Nachman KE. Feather meal: A previously unrecognized route for reentry into the food supply of multiple pharmaceuticals and personal care products. *Environ. Sci. Technol.* 2012;46(7):3795–3802.

103. United States Food and Drug Administration, FDA secures full industry engagement on antimicrobial resistance strategy, June 30, 2014. Available at http://www.fda.gov/AnimalVeterinary/NewsEvents/CVMUpdates/ucm403285.htm, accessed September 7, 2014.

104. GRACE Communications Foundation. *Sustainable Table: Air Quality.* Available at http://www.gracelinks.org/266/air-quality, accessed September 7, 2014.

105. Sutton MA, Bleeker A, Howard CM, Erisman JW, Abrol YP, Bekunda M, Datta A et al. *Our Nutrient World. The Challenge to Produce More Food and Energy with Less Pollution. Key Messages for Rio + 20.* Available at http://unep.org/gpa/documents/GPNM/GEFProject/KeyMessagesRio20Summit.pdf, accessed May 10, 2015.

106. McMichael AJ, Powles JW, Butler CD et al. Food, livestock production, energy, climate change, and health. *Lancet.* 2007;370(9594):1253–1263.

107. United States Environmental Protection Agency, Greenhouse gas emissions. Available at http://www.epa.gov/climatechange/emissions/index.html, last updated March 18, 2014, accessed September 7, 2014.

108. GRACE Communications Foundation. *Sustainable Table: Energy Use and Climate Change,* 2013. Available at http://www.gracelinks.org/982/energy-use-climate-change, accessed September 7, 2014.

109. United States Department of Agriculture. Quality System Verification Programs, updated January 23. Available at http://www.ams.usda.gov/AMSv1.0/QualitySystemsVerification Programs, 2014, accessed September 7, 2014.

110. Engber D. The Branding—and rebranding, and re-rebranding—of pink slime. *Slate,* October 25, 2012. Available at http://www.slate.com/articles/news_and_politics/food/2012/10/history_of_pink_slime_how_partially_defatted_chopped_beef_got_rebranded.html, accessed September 7, 2014.

111. Moss M. Safety of beef processing method is questioned. *The New York Times,* December 30, 2009.

112. USDA. Agricultural Marketing Service. National Organic Program. Available at http://www.ams.usda.gov/AMSv1.0/nop, accessed September 14, 2014.

113. Rinehart L, Baier A. Pasture for Organic Ruminant Livestock: Understanding and Implementing the National Organic Program (NOP) Pasture Rule, p. 2. USDA. Available at http://www.ams.usda.gov/AMSv1.0/getfile?dDocName=STELPRDC5091036, accessed September 16, 2014.

114. Johnson R. Organic Agriculture in the United States: Program and Policy Issues. *Congressional Research Service,* p. 5, updated November 25, 2008. Available at http://

nationalaglawcenter.org/wp-content/uploads/assets/crs/RL31595.pdf, accessed September 16, 2014.

115. Laskawy T. Battle for the soul of organic dairy farmers goes on behind the scenes. *Grist,* January 30, 2010. Available at http://grist.org/article/battle-for-the-soul-of-organic-dairy-farmers-goes-on-behind-the-scenes/, accessed September 7, 2014.

116. United States Department of Agriculture. Agricultural Marketing Service, National Organic Program, last modified October 17, 2012. Available at http://www.ams.usda.gov/AMSv1.0/ams.fetchTemplateData.do?template=TemplateC&leftNav=NationalOrganicProgram&page=NOPConsumers&description=Consumers.

117. United States Department of Agriculture. Agriculture Marketing Service, FAQs COOL Labeling Provisions Final Rule, September 20, 2013. Available at http://www.ams.usda.gov/AMSv1.0/getfile?dDocName=STELPRDC5105133, accessed September 7, 2014.

118. *American Meat Institute v. United States Department of Agriculture*, July 29, 2014, 13–5281 (D.C. Cir. 2014).

Grains

9

INTRODUCTION

Grain is used in the United States as food for humans and animals, and also as a means of creating biofuel, a gas alternative. The term "grain" describes "a small, dry, one-seeded fruit of a cereal grass."[2] Cereal grasses include wheat and wheat varieties such as spelt, as well as barley, rye, triticale (a wheat/rye hybrid), oats, millet, rice, sorghum, and corn. Other plants that have edible seeds or fruits, such as amaranth, buckwheat, quinoa, and "wild rice," are not technically cereals, as these plants are not members of the grass family, but are used in a similar manner in our diet, and thus commonly considered grains.

In this chapter, we develop the historical context of our grain consumption and production. We use grain to highlight how technological developments, such as packaging, helped encourage interest in U.S. government regulation of food. We review U.S. Dietary Guidance related to grains, using whole grains as a means to highlight the FDA's role in regulating health claims for food. Finally, we touch briefly on agricultural policies, and the use of surplus grain. High-fructose corn syrup serves as an illustration of the FDA's role in food additives and the use of "generally recognized as safe" (GRAS).

GRAINS IN OUR DIET

Agricultural Transition

That our dietary choices include grains at all is the result of what is commonly termed "the agricultural transition"; namely a shift among our ancestors from a predominantly mobile lifestyle to a predominantly localized lifestyle that placed a greater emphasis on growing plants and grains for consumption.

Recent archeological research suggests that such domestication of grains may have occurred independently in at least 13 and up to 24 areas, and with differing paths regarding cultivation, animals, and a mix of reliance wild versus domesticated food supply. Additionally, this process may have taken several thousand years—the introduction of cultivating and harvesting grains to the human diet may have been a lengthy, meandering transition rather than a sudden revolution.[3]

Although grain consumption is often considered to have begun 20,000–10,000 years ago, evidence of sorghum seeds has been found indicating consumption as long as 100,000 years ago.[4] Arguments over the timing of the transition to a grain-centric diet may also include arguments as to if humans have (or have not) evolutionarily adapted to a grain-centric diet.[5]

Grains are a significant source of the macronutrient carbohydrate, and do provide some other nutrients. However, grains are not, nutritionally speaking, essential to the human diet. Some argue that the grains are nutritionally significant in our diet simply because the calories and nutrients provided by grains are inexpensive.[6] Others have commented that the only vitamin notably provided by grain is thiamine.[7] Furthermore, the relative contribution of nutrients from grains to our diet has changed over time due in part to changes in our dietary pattern.

As this book focuses on food policy, rather than specifics regarding the physiological basis of dietary needs, discussing whether grain consumption is nutritionally a good or bad dietary pattern based on evolutionary development is beyond our scope.

However, it is important to realize that as long as sufficient calories are available from other sources, grains are not essential to the human diet. And in particular, since uniform milling and processing became available only within the past 200 years, the consumption of refined, uniformly small particulate grains (i.e., flour) is a very recent dietary pattern for humans. Refined grains are certainly not essential to our diet, even though as of 2005, 85.3% of cereal grains consumed in the United States are highly processed refined grains.[5]

Historical Grain Production in the United States

George Washington apparently personally conducted the first agricultural survey, in part to answer the question: how much grain can the United States export? The reported data showed an emphasis on wheat, flax, and corn production due to demand from European markets.[8] Thus from the very start of the country, the production of grain had an important economic role as a recognized potential national tool of trade.

However, among grains, the significance of any particular grain as a tool of trade has varied over time and of course, by the level of geopolitical entity. By 1930, wheat constituted 27% of major grain production and corn 54% in the United States. Although corn was the predominate crop grown, wheat was more economically significant as a source of income at the county level and to the small farming communities; 70% of wheat was exported out of the county in which it was grown, but only 16–17% of corn.[9]

Conducting an agricultural survey was the job of the Bureau of the Census until 1997, when the USDA's National Agricultural Statistics Service (NASS) took over the work. Every 5 years, NASS conducts an agricultural survey that provides information concerning all areas of farming and ranching operation.[10]

The NASS agricultural survey helped inform a 2013 report from the USDA's Economic Research Service that large farms now dominate crop production in the United States. Farming in the United States is trending toward a "U" shape, with less mid-size farms, but increased numbers of smaller farms and larger farms. In crop production (which would include grains), most farms in the 1980s were less than 600 acres, but now most crop farms are at least 1100 acres, and many are 5–10 times that size.[11]

The consolidation of food production in the United States into larger producers also means a concentration of decision-making. It is important to realize that the choices made by any food producer, small or large, are not inherently or necessarily concerned with nutrition. Farmers have always considered characteristics not necessarily related to nutritional traits, such as speed and consistency of growth and size in their choices of varietal production. Furthermore, seeking profitability is a natural behavior of any farmer, whether small or large.

However, the concentration of farming into a smaller number of larger producers means that the choices around crop characteristics and production become concentrated among a smaller number of producers, and those choices in turn affect a larger and larger number of people. This significant market concentration gives "Big Ag" the ability to spend money on research and lobbying to influence public policy as well as the content of information disseminated to consumers, which can transform nutritional public policy into availability of dietary choices targeted to consumers based on profitability.

United States Agricultural Transformation

Chapter 2 of this book describes the birth of the United States Department of Agriculture. In the same year of 1862, President Lincoln signed three other Acts that significantly transformed the United States through the redistribution of public lands, namely the Morrill Land Grant College Act, the Pacific Railway Act, and the Homestead Act.[12]

Generally, these laws supported the development of the West. The timing of these enactments speaks to the preceding political clashes over this western development; land reform and Homestead Act-like variations had been proposed and argued over since the 1840s; the enactment of these three Acts became politically feasible only after the secession of Southern states into the Confederacy.[13]

The promotion of agriculture by these Acts was also a reversal of direction for a country that had become increasingly urbanized and industrialized. In the 50 years from 1810 to 1860, estimates of the percentage of the labor force working on farms dropped from a range of 74–83% to a range of 53–56%.[14] The rise in urbanization and accompanying poverty fed into the increasing pace of arguments over land reform and western settlement. Our modern nostalgia over America's farming history is attributable to the land-use policies signed in 1862.

The Morrill Land Grant College Act gave public lands to the states and territories for establishing educational institutions specializing in agriculture research and training. The Pacific Railway Act provided federal support for building the first transcontinental railroad. In addition to loans, the railroads received public lands along the track, almost 128 million acres. The railroad could sell these lands to settlers, a guaranteed profit to the railroads, but the lands would not be useful for agriculture and settlement without the railroads to provide access and distribution. Additionally, access to some land was clouded in uncertainty as to where the routes would actually be struck, which benefited land speculators.[15] Nonetheless, this policy of developing a mass transportation and distribution system in turn influenced food production, access, distribution, and in the long run, food policy.

The third act, the Homestead Act, put land ownership and potential self-sufficiency through farming within the reach of average citizens or future citizens, including immigrants, African Americans, and women. In 1887 a variation was extended for Native Americans, but at the cost to the Native Americans of breaking up tribal lands and the loss of collective bargaining power.[16]

Under the initial Homestead Act program, a down payment of $10 (along with a $2 dollar commission to the land agent), meant that 160 acres of land could be claimed as owned outright after 5 years and a final payment of $6 as long as improvements were made, such as building a home and farming the land. For those that had recourse to funds, the land could be owned after 6 months upon paying a $1.25 an acre. Over 270 million acres were claimed.[17]

Because the holding period of 5 years was also the same time requirement to become a citizen, the Act also encouraged immigration.[18] The convergence of promises of land with European overpopulation, unemployment, political unrest, crop failures, and famine influenced patterns of settlement such as the migration of Scandinavians to the Midwest.[19]

The increasing settlement in the west prompted by these Acts and later variations[20] exacerbated conflicts between open-range cattle herders and the farmers who wanted to fence off their lands, and this conflict further accelerated demands for lands that had been set aside for the Native Americans.[21] Nonetheless, this promotion of making land available at a low cost for individuals lacking capital was a shift in policy for the federal government from the previous policy of outright selling land, which favored those who had already established capital.[20]

United States Agricultural Grain Production Expansion

At the same time land was opening, farming technology and expertise were advancing. The combination of more land and better farming resulted in a massive increase in wheat production. The production of wheat expanded rapidly in the United States, from just over 100,000,000 bushels in 1859 to just over 468,000,000 in 1890.[9] Contributing factors to this growth were the opening up of western lands through the Homestead Act, increased immigration, and technological developments in production and harvesting methods through technology—including the steel plow, the mechanical reaper, and the development of horse-drawn cultivators and other machinery.

The development of transportation was critical for the expansion of agricultural production. It allowed for distribution of product back to market, and perhaps even more critically, for the transportation of fuel through the colder areas of the Middle and North West that would have been otherwise inhospitable. For example, prior to 1867, Minnesota had no rail connections outside the state, and relied on using the Mississippi river. River transportation lacked the constant reliability of train distribution. The river could be closed during the winter due to ice and a challenge to navigate in warmer weather if the water level dropped.[22]

For wheat, nearly all the increase in both output and acreage was due to the expansion North and West of the Ohio River. However, even as this expansion of wheat-producing land occurred, crops shifted quickly within this area, with a leading edge of wheat racing with the tide of expansion. In Minnesota, the leading wheat counties of 1870 all fell to the back of the class by 1890, while newly established counties exhibited huge productions. Olmsted County produced over 2 million bushels in 1870 but less than 200,000 bushels only 20 years later. Conversely, Otter Tail County, which had produced less than 10,000 bushels in 1870, produced over 2.5 million bushels in 1890.[22]

Minnesota was not unique, but instead a typical example of historical progression in wheat production, namely, that of wheat as a frontier crop, moving rapidly in production to new lands. In other words, wheat has historically been an agricultural gold rush. Since wheat required significant labor only at plant and harvest, investors were drawn to it along with the "free land."[22] This shift in production to the new lands also changed the farm production profiles of the older states, as farmers could not keep up with the production capabilities of the newer areas, and had to shift production to other items.[9]

Packaging Technology: The Creation of a New Market for Crackers and Bread

The technological advances that pushed grain production to new levels were paralleled by the advancements of food-packaging technology—namely, cardboard boxes and waxed paper sleeves—that created new forms of consumer access to grain products and snacks.

The increased production and distribution of grain in full swing by the end of the nineteenth century, along with mechanization in bakery production, furthered a series of bakery mergers into the National Biscuit Company (Nabisco). These mergers consequently fostered the development of nationally distributed brands of crackers and cookies. The development of consumer packaging to promote national distribution also fostered consumer adoption of these products.

Until the 1890s, crackers were made locally in most towns and distributed in large barrels to grocers from which consumers purchased in bulk. In 1899, Nabisco introduced a paperboard package with wax paper lining that, due to an interfolded construction, kept crackers fresh and intact during transportation and provided compact, attractive, stackable boxes that consumers could buy directly for the household with confidence that the contents were sanitary and fresh. The packages were kept small and inexpensive to maximize volume of sales.[23]

Grain-based snacks, then, as now, were available and marketed as a source of cheap food. The innovation in packaging, which allowed direct marketing to the consumer, instead of through the grocer, also influenced the provision of information on the packaging. Packaging provided a bundled shortcut of information, branding, and advertising, directly to the consumer. Nabisco and Heinz, companies that were seeking national distribution, had the most to lose from competitors hawking poor-quality imitations and accordingly provided industry support for the developing federal regulation of food products, packaging, and labeling that arose from the Pure Food Movement.

The wheat "arms race" and interest in varietals also contributed to the growth of industrial production of bread as well as crackers. Until the 1870s, the wheat grown in the United States was mostly "soft" wheat, which, due to lower protein content, is better for items such as crackers and pastries. Hard spring wheat, with higher protein content, was introduced in Minnesota in the mid-1800s, and a hard winter wheat was introduced in Kansas shortly thereafter. The introduction of these hard wheat varieties helped encourage the increased demand for bread and commercial bread bakeries began to appear in the late 1880s.[24]

But even in 1910, 95% of bread in the United States was still made at home. Commercially baked bread was viewed with suspicion, with lack of confidence that the ingredients were wholesome, made in sanitary conditions, or perhaps not even baked well enough to prevent inappropriate organism growth.[25] Although the Pure Food Movement and the 1906 laws are associated with Upton Sinclair's vivid description of the terrible conditions at abattoirs, the bakeries of the time were not much better. Flour was stretched with other ingredients, and the production facilities were often in cellars, which easily allowed for vermin and the contamination, not just from dirt, but also from sewage lines. Breathing in flour without ventilation lead to lung diseases, and workers were sometimes even sleeping in the bakeries.[26]

By 1907 cities had begun bakery reform, albeit with a focus on the carelessness of dirty immigrant workers, and so the marvels of a sanitary bakery—bread produced (and later wrapped) without the touch of human hands—were technological advances that fostered the adoption of buying bread, as well as the growth of national brands.[26]

The hygienic attractiveness of perfectly formed, perfectly white bread, also evoked society's dawning modernity and interest in advanced technologies, including those of nutrition science, discussed in more detail in Chapter 11.

DIETARY GUIDANCE, WHOLE GRAINS, AND THE FDA

Grains in the U.S. Diet

The composition of our diet, and the relative percentages associated with grain, has changed over time. In the United States, prior to 1800, corn was indeed the king. Due to the difficulties in wheat production in the North East and the South and the expense of transportation, wheat was generally consumed by the wealthier. Yeast-risen wheat bread was also difficult to prepare for home cook. The cast-iron wood-fired cooking stove was an invention of the 1800s, and prior to that, corn bread and other quick breads, such as hearth-based flat breads, were more viable for the home cook. The combination of improved milling technology in the late 1790s, followed by advances in wheat production technology starting in the 1830s, and then distribution improvements in transportation meant the cost of wheat dropped such that it became more economically available to a greater number of households.[24]

Rising prosperity, and reduced sugar prices, meant a significant shift from cornmeal to wheat flour, and increased sugar usage. At the beginning of the twentieth century, grains comprised the bulk of the American diet. In 1909, 300 pounds of grains per capita provided the major source of energy (39%), protein (37%), and carbohydrate (57%) in the American diet. Additionally, Americans consumed a greater amount of a broader variety of grains, such as buckwheat, rye, and barley than at the end of the twentieth century, and this is reflected in the overall relative contribution of nutrients by grain. For example, in 1909, grain comprised a greater source of vitamin E, magnesium, and copper in the American diet than in 1997. And in the early 1900s thiamin was provided in similar amounts by meat, poultry, and fish as a group compared to grains as a group, at 32–31%, respectively. By the early 2000s, however, grain was the main contributor of thiamin to our diet, at 59%. This high level of thiamin, however, is due not to consumption of whole grains, but instead to the fact that we enrich refined grain products with thiamin.[27]

By the close of the century, grains were no longer the major source of protein, having been replaced by meat, poultry, and fish (38%) due to a combination of increased consumer income and greater efficiencies in meat production. The increased use of sugar meant that added sugars now provided just as much (and even slightly more) carbohydrate as grains at 41–38%, respectively. Grains also no longer provided the major source of energy. Fats and sugars accounted for 38% of the total energy available for consumption in 1997.[28] By 2004, energy from grains in our diet had decreased to an average 24%.[27]

Reliance on wheat flour dropped after 1920 as a more diversified diet (eggs, dairy, fruits, and vegetables) became more widely available. The per capita food supply of flour and cereal products reached an all-time low of 133 pounds in 1972, which was less than half of the available 300 pounds per capita supply in 1909.[28] (Although, in 1972, the per capita supply may not have been low solely in response to a low demand; 1972 will be discussed again later in this chapter.)

1972 aside, the per capita supply had been trending downward. The 1980s saw a reversal of that trend. Wheat consumption rose again, for a variety of reasons including consumer perception that wheat products were a healthy alternative to an animal-based diet, and conversely, due to an increase in fast-food consumption. By 1998, the per capita flour and cereal products supply had reached 195 pounds, driven by consumer demand for variety in snack items and store bought and baked breads as well as the reliance of fast-food products on buns and tortillas.[28]

Four-fifth of this increase in per capita supply happened post 1980, the period following the release of the first *Dietary Guidelines for Americans*. The 1980 Dietary Guidelines encouraged consumers to "select foods which are good sources of fiber and starch such as whole-grain breads, cereals, fruit, vegetables, beans, peas, and nuts."[28] But by 2011, the per capita use of flour had dropped again, attributable by the USDA to a general interest in lower carbohydrate diets.[29]

From Grains to Whole Grains: U.S. Dietary Guidance 2000–2005

As noted above, the advances in technology that fostered increased consumer demand for wheat flour and wheat flour products also changed the characteristics of the flour we were consuming.

The refining process of milling removes the bran and the germ from the whole grain, leaving just the starchy endosperm, which makes white flour that lasts longer on the shelf and produces baked goods of a finer texture. However, whole grains provide dietary fiber along with thiamin, riboflavin, niacin, folate, iron, magnesium, and selenium.[30] Also, whole grains, similarly to fruit and vegetables, contain numerous phytochemicals, which may contribute to an overall health-promotion effect.[31] Our policy decisions to enrich white bread may have improved our health over certain specific significant metrics at the start of enrichment but in the long run, may have also helped mask the overall dietary consequences of consuming more refined grains.

By 2000, our dietary guidance began to stress the importance of whole grains. Although prior Guidelines encouraged the consumption of grains as part of a dietary pattern (the 1995 Guidelines recommended to "choose a diet with plenty of grain products, vegetables, and fruits"),[32] the 2000 Dietary Guidelines were the first to include a distinct guideline for grain consumption separately from fruit and vegetables.

The 2000 Dietary Guidelines recommended 6–11 servings of grains based on age, gender, and activity level, in addition to the general guidelines of "choose a variety of grains daily, especially whole grain." Guidelines for grain consumption were offered in three calorie ranges—1600 for children, women, and older adults, 2200 for older children, teen girls, active women, most men, and 2800 for teen boys and active men. Example serving size for grains (identified as bread, cereal, rice, and pasta) were identified as one slice of bread, one ounce of ready-to-eat-cereal, or half cup of cooked cereal, rice, and pasta.[33]

As whole grains in our diet became of greater interest, U.S. government dietary guidance began to include more explanatory information. The 2000 Dietary Guidelines, in addition to presenting numbers of servings for grains, also made the recommendation to include several servings of whole grain foods, noting that eating "whole grains, as part of the healthful eating patterns described by these guidelines, may help protect you against many chronic diseases."[34]

The 2000 Dietary Guidelines were also the first guidelines to recognize that health benefits associated with grains were more correctly attributed to whole grain consumption,[28] although at the time, the focus of the benefit of whole grains was on the contribution of dietary fiber.[34] Helpful examples of how to increase intake of whole grain foods, such as by looking for an ingredient like brown rice, whole wheat, or oatmeal first in the listed contents, were given, but the Dietary Guidelines did not otherwise attempt to define whole grains.

By 2005, the Dietary Guidelines recommendations had become more specific. "Servings" was changed to "ounce-equivalents." Additional diet patterns were added, expanding from 3 patterns ranging from 1600 to 2800 calories, to 12 patterns ranging from 1000 to 3200 calories. The recommendation for grain consumption across this expanded range of diet patterns is 3–10 "ounce-equivalents," which created a slight downward trend for recommended grain consumption at some calorie levels.

The 2005 Dietary Guidelines were also the first to make a specific recommendation to consume whole grains, recommending that half of the grains consumed daily should be whole grains. The consumer-facing brochure "Finding Your Way to a Healthier You: Based on the *Dietary Guidelines for Americans*, 2005" detailed this recommendation as at least 3 oz of whole grains every day.[35]

Whole grains can be eaten as a complete food, such as popcorn, in which the grain kernel is eaten in entirety. Whole grains, however, can also be cracked, split, flaked, or ground for use as an ingredient in bread, cereals, and other items.

In the material meant for professional educators and policymakers (vs. consumers), the 2005 Dietary Guidelines defined "whole grain" as the entire grain seed (or kernel) of the plant, consisting of endosperm, bran, and germ. However, the definition also allowed for the kernel to be broken in processing. Per the 2005 Guidelines, if whole grains are not eaten as a complete food, but are instead processed in some way that breaks the kernel, the resulting food can still be identified as "whole grain" as long as the food retains the same relative proportions of these components as existing in the intact grain.[36]

Defining "whole" in this manner is significant. This definition of "whole" grain allows processing that in actuality breaks the "wholeness" of the grain. Under this definition, a product identified as "whole grain" can contain grain that is milled (which separates the components of the endosperm, bran, and germ) as long as those components are mixed back in the same proportions for use in the product. This allows for the creation of consumer-friendly "whole grain" food products of greater palatability than otherwise would be possible.

Perhaps not surprisingly, the definition adopted by the Dietary Guidelines committee had been earlier proposed by General Mills in an application to the FDA about making a health claim about whole grains on a food package label. Health claims are discussed in more detail later in this chapter.

Eat This versus Eat That? Public Health Dilemma of Whole Foods versus Enrichment

In retrospect, the 2005 U.S. Dietary Guidelines adoption of the recommendation to eat whole grains may seem like a nutritional food policy slam-dunk. *What was there to question?* However, because of our historical food policy reliance on enrichment, advising people to eat whole grains as a substitute for enriched refined grains could have unintended detrimental nutritional consequences. For example, increasing intake of whole grains at the expense of enriched, folate-fortified refined grains could decrease the intake of some micronutrients (e.g., iron, folate, or zinc) to undesirably low levels.

The decision to emphasize a guideline of making half a day's servings of grains whole (which for most people would approximate three servings) was based on an analysis of dietary patterns using 1994–1996 Continuing Survey of Food Intake by Individuals (CSFII) composites. This research indicated that substituting three servings of whole grains for three servings of enriched, folate-fortified refined grains would not adversely affect nutrient intake levels.[37]

Further, keep in mind that the type and amount of micronutrients found in whole grains vary significantly dependent on the specific type of whole grain consumed. For example 100 g of oats has 26.2% iron, but 100 g of brown rice has only 8.2% iron. Bulgur and barley are much higher in fiber than brown rice. Rye is a better source of lignan and sterols than wheat, oats, or barley. Corn contains vitamin A, which is lacking in brown rice, oats, and sorghum.[38]

Instead of enriching refined grains, why not encourage and recommend a full whole grain diet?

Although enrichment of refined grains supports the ease and palatability of a diet completely lacking in whole grains, when refined but enriched grains are eaten in conjunction with whole grains it does smooth out the nutritional impact of consumer dietary preferences of either direction made by the consumer.

2010 Dietary Guidelines

With the change from MyPyramid to MyPlate (see Chapter 3), grains were demoted from the category of primary food consumption to at least share the stage with vegetables. The MyPlate graphic even shows a slight emphasis of vegetables over grains as the primary food group.[39]

The 2010 Dietary Guidelines recommend that an adult consumes 45–65% of daily calories from carbohydrates.[40] Although we naturally think of grains as a source of the macronutrient carbohydrate, fruit, vegetables, and dairy products also provide carbohydrates in conjunction with other important micronutrients. Accordingly, the percentage of calories recommended for carbohydrate consumption is not meant to apply as a one-to-one ratio to grain consumption.

U.S. dietary guidance aims at nutrient density within caloric ranges, and thus the USDA publishes "food patterns" providing suggestions organized by the basic food groups, including grains.[41] The food patterns are provided for 12 daily calorie ranges, ranging from 1000 to 3200 calories, as calorie needs vary by age, gender, and amount of activity. The current MyPlate, based on a food pattern for moderately active adults, recommends consuming 5–8 "ounce-equivalents" of grains depending on age and gender. Recall from above that general dietary guidance for whole grains is to "make half your grains whole." Given the MyPlate recommendation of 5–8 "ounce-equivalents" of grains, 3 "ounce-equivalents" is a good goal for whole grain consumption for most people.[39]

The USDA translates "ounce-equivalent" into consumer-friendly terms by identifying "1 ounce-equivalents" of common foods. One slice of bread, one cup of ready-to-eat cereal, or half cup of cooked rice, pasta, or cereal are all considered "1 ounce-equivalent" of grains. One large bagel is not the same as one slice of bread, however, as the large bagel provides the equivalent of 4 oz of grains instead of the equivalent of 1 oz. Likewise, one large (12 inch) tortilla is the equivalent of 4 oz of grains.[39]

The 2010 Dietary Guidelines advises that the most direct way to make "half your grains whole" is to consume 100% whole grain foods (such as brown rice) since 100% whole grains foods are naturally a one-to-one substitution, and 100% whole grain is the basis of the "ounce-equivalent" calculation. Guidelines uses bread slices (100% whole grain, partly whole-grain, and refined) to illustrate how to meet the whole grains recommendation (see Figure 4-1 therein). The accompanying text describes partial whole grain food products that can help meet whole grain consumption goals as those containing at least 51% of weight as whole grains or those that provide 8 g of whole grains per ounce-equivalent.[42]

Compared to the Dietary Guidelines, MyPlate may offer more screen-time and visual suggestions to whole grains still in an unprocessed state. Even though the Dietary Guidelines are updated and published every 5 years, the flexibility of online content and adaptive nature for immediate promotion of new ideas and trends must be considered in studying underlying policy in governmental dietary guidance.

Whole Grains and Consumption Recommendations on Labels

Product label messaging around whole grains also provides an example of how the food industry can cull information from U.S. dietary guidance to enhance product promotion and marketing.

In 2005, the Whole Grains Council, a trade association and a program of Oldways, a 501c3 nutrition education organization, created several versions of an eye-catching black and gold stamp for food packaging to help guide consumers toward whole-grain food products. One such version indicates the food contains at least 8 g of whole grain per serving, and another version indicates that a product is 100% whole grain.[43]

Can such industry-developed icons provide a useful shortcut for consumers trying to assess if they are meeting the recommended amount of whole grains consumption? *What is the U.S. dietary recommendation for the amount of whole grains consumption?*

In a mid-2013 website posting, the Whole Grains Council stated that current U.S. dietary advice is to consume 48 g of whole grains, but failed to provide a citation or link to such a statement.[44] This idea that 48 g of whole grain consumption is the current U.S. dietary recommendation is also occasionally repeated elsewhere.[45]

However, "48 g" is not a current U.S. dietary recommendation for whole grain consumption; "48 g" is not stated anywhere in the 2010 Dietary Guidelines. It is a historical artifact, stemming from a 2004 report released by the Dietary Guidelines Advisory Committee. At the time, there was a need to quantify the then current dietary guidance lingo of "servings" to translate the amount of foods eaten in the dietary interview component of the National Health and Nutrition Examination Survey NHANES into the metrics used by the 2005 Pyramid Serving database. (This database is now referred to as the USDA Food Pattern Equivalents Database.)

The 2005 Pyramid Serving database used for translation of dietary interviews defined one serving of grains as "the grams of grain product containing 16 g of flour," reflecting the idea that a grain product would include other ingredients than flour.

Applying this schematic of 16 g of flour directly to the recommendation of three or more servings of whole grains results in a total of 48 g of flour. Obviously, this definition and calculation is specifically useful only for flour-based food products.[46] Additionally, the Food Guide Pyramid was replaced with MyPyramid very shortly thereafter, along with a change over from "servings" to cups and ounces, and the database used for translating dietary interview information was updated from servings to cups and ounces in 2006.

The 2004 report from the Dietary Guidelines advisory committee was the first proposal of official U.S, dietary goals for whole grain consumption. It was released a few months after General Mills had filed a health claim with the FDA to allow referencing a health benefit to whole grain consumption on food package labels.

Buoyed by the combination of the industry push for a whole grain health claim, and the subsequent first-ever U.S. specific dietary guidance recommendation for whole grains, the Whole Grain Council ran very quickly with the idea of a daily goal of three servings of whole grain = 48 g. Thus, this relatively official looking symbol identifies half servings (8 g) and full servings (16 g) of whole grains.[46]

The key point here is to recognize that packaging nutrition claims may be based on outdated standards and furthermore, that even science-based standards may have components that have been pushed by industry. The recommendation of "48 g" for whole grain consumption is an artifact from overeager industry adoption of preliminary information in 2004, and was still being repeated in 2013 by registered dieticians. Having made the stamp, the interested parties would naturally desire that the recommendation become truly official and continuing pushing for adoption. *What do you think of the Whole Grains stamp? Is it a helpful shortcut for consumers even if "48 g" isn't current official U.S. dietary guidance?*

Increasing Whole Grain Consumption as a Public Health Goal

Healthy People 2010 is the first version of Healthy People to include a specific target for whole grain consumption[28] and a whole grain consumption objective continues in Healthy People 2020, albeit a modest objective of increasing the mean daily intake of 0.3 ounce-equivalent per 1000 calories (as measured in the 2001–2004 NHANES), to a mean daily intake of 0.6 ounce-equivalent.[47]

Whole grain consumption, however, has already been influenced through industry adoption and interpretation of recommendations made in prior editions of the U.S. Dietary Guidelines and educational tools. As food producers are generally eager to establish a market share in perceived consumer desires, the publication of any specific objectives or recommendations made in Healthy People or the Dietary Guidelines may lead to product reformulation, enthusiastic marketing, and the adoption of quickly identifiable, consumer-friendly "front of the package" icons, as discussed above with the Whole Grains stamp.

This interest in producing and promoting food products that help consumers meet the objectives of dietary guidance is not limited to industry. The health and nutrition professionals who developed the 2000 Dietary Guidelines commented on the essential need for partnership from the food industry, noting that consumers must have access to affordable and acceptable grain foods in convenience and away-from-home settings to increase consumption levels.[28]

To achieve public health goals of increasing whole grain consumption across the population, whole grain foods need to be available. Although consumers can readily buy some minimally processed whole grain foods, such as brown rice, steel cut oats, or popcorn, other whole grain foods may be too costly for some consumers, less easily procured, and more challenging to prepare at home. Specialty grains like quinoa may be expensive. And though most home cooks are familiar with wheat flour, whole kernels of wheat (wheat berries) may not be easily found outside of specialty stores and are less commonly used in standard cooking in the United States.

The current dietary guidance message regarding whole grains is to make half your grains whole but without any strong specific emphasis on variety, or type of whole grains. In response to an increased demand for whole grains, more "whole white" wheat varieties are being planted. Although some are trademarked products, in general "whole-white wheat flour" is simply flour produced from hard white spring wheat, rather than the hard red form. The pigmentation in the red wheat carries an astringent flavor in addition to the color, thus the white variety produces whole-wheat flour that is lighter in color and milder in taste.[48] Public health dietary guidance, as it creates an interest in nutritional improvements, will in turn help drive product development, but we have limited information as to if such products will in fact support the intended health improvement goals. Even if whole-wheat flour is an improvement over refined flour, it may not be as healthful as a still relatively unprocessed whole form of grain.

Given that enrichment of food products as a public health policy in the United States was already entrenched as a means of approximating the nutrition available from whole grains, it is not surprising that the objective of increasing whole grain consumption in the United States would manifest via contributions from processed food products, with increased availability of whole grains in ready-to-eat food products, meals away from home, and convenient cooking options.

When steps in preparation have been done ahead of time for the consumer, the consumer is reliant on available information about a food product when considering purchase. The available information includes cost, knowledge gleaned from research, and marketing and advertising information from the manufacturer, which includes the packaging of the food product. Additionally, a consumer may also consider the factual information present on the ingredient list and nutrition label.

Whole Grains: Food Package Labeling

Although food product labeling is not an issue specific to grains, much of the extensive array of packaged products available for us to purchase in supermarkets are grain-based, such as bread, tortillas, cakes, muffins, pastries, crackers, cookies, granola, and other similar snack bars and chips, frozen dough, popcorn and rice cakes, breakfast cereals (hot and cold), rice-based beverages, pasta, and grain-based entrees or side dishes that are fast to prepare such as polenta, rice pilafs, macaroni dishes, or those already made and ready-to-eat after being heated up, such as lasagnas and pizzas. Food labeling has played a critical role in the promotion of grains as a food product, and in whole grains as a dietary concept.

Food labeling in general is introduced in Chapter 4, which provides more details on the proposed update to the Nutrition Facts box. Under the charge of the Nutrition Education Labeling Act of 1990 (NLEA), the FDA was given the authority to require nutrition labeling on food packages. The Nutrition Facts box presents a ratio of nutrient per serving of food to compare with an ideal consumption total, described as Percent of Daily Value.

However, the Nutrition Facts box only provides details for specific, mandated nutrients. Dietary fiber was considered by the IOM in 2002 to play such a significant role in health promotion that it is treated specifically as a nutrient, and a percentage per serving applicable to a day's targeted consumption is described—the Daily Value. The mandated presence (or lack thereof) of a specific nutrient on the Nutrition Facts label influences the implementation of dietary guidance, even when such implementation is done through the efforts of the food industry to maintain a consumption market. The Nutrition Facts label is silent when it comes to whole grains. Because whole grains are not considered nutrients, a Daily Value has not been established for whole grains.

An additional complicating factor for using the Nutrition Facts box in monitoring whole grain consumption is that Nutrition Facts information is provided in grams whereas current 2010 U.S. dietary guidance regarding grain consumption is in ounces (or "ounce-equivalents"). The recommendation for moderately active adults, depending on age and gender, is to consume 5–8 "ounce-equivalents" of grains and to "make at least half of your grains whole grains."[49] Consumers can assess adequacy of whole grain intake through using the "half your grains whole" approach in combination with one-to-one substitutions such as brown rice for white rice.

However, health-seeking consumers may still aim to select foods and food products based on assessing the whole grain content of a food, and using dietary fiber as a proxy for whole grain is of limited value. To assess the whole grain content of a food product, the consumer must read the fine print of the list of ingredients and be able to discern the relevant descriptive language.

The lack of detailed information in the nutrient box as to the amount of whole grains per serving allows manufacturers to more easily leverage consumer interest into sales through the use of front-of-package marketing. Front-of-package marketing may attract a consumer but not help much with assessing actual nutritional or whole grain content. As the case in point, in 2008, one of the authors walked into a Starbucks and noticed a sign on the pastry case: "Looking for whole grains? Try this muffin with whole grain topping!" Sprinkling of topping aside, the muffin was otherwise very much a refined grain product.[50]

This "front of the package information" can be a shortcut for a purchasing decision. Seeing the term "multigrain," for example, may lead someone to assume that an item contains whole grains—but this isn't necessarily the case. And the trending interest in gluten-free foods provides the opportunity for foods that inherently do not contain gluten to assert their virtues as gluten-free, a "nudging" reminder to consumers (see Box 9.1 for more information on gluten labeling).

Seeing the "front of the package" information that a children's breakfast cereal is a good source of fiber and contains whole grains may be enough to sway a purchase, even

BOX 9.1 GLUTEN

Whether a grain contains gluten, a type of protein, is another increasingly important distinction used in describing grains in our diet. Wheat and related varieties, barley, rye, and triticale contain gluten. Oats do not inherently contain gluten, but are often contaminated in processing on shared equipment. Amaranth, buckwheat, corn, millet, quinoa, rice, sorghum, teff, and wild rice are gluten free.

An estimated three million Americans suffer from celiac disease, a chronic inflammatory autoimmune disorder in which the consumption of gluten triggers the production of antibodies that attack and damage the intestinal lining. In addition to the physical discomfort of gastrointestinal distress, the damage reduces the ability to absorb nutrients, which in turns leads to health complications. For those with celiac disease, avoiding gluten is a must. Food products must be accurately labeled.

In 2004, Congress passed the Food Allergen Labeling and Consumer Protection Act (FALCPA), discussed in more detail in Chapter 13, which in part required a definition of the term "gluten-free" for food label use. In 2007, the FDA published a proposed rule. In August 2013, the Food and Drug Administration (FDA) finalized the definition of the term "gluten-free" on food labels, which means that all manufacturers that want to use the phrase "gluten-free" on their products must now adhere to strict guidelines as of August 2014.

As stated by the FDA, the term gluten-free now refers to foods that are (i) inherently gluten-free and (ii) foods that do not contain any ingredient that is a gluten-containing grain or derived from a gluten-containing grain that has not been processed to remove the gluten. The use of an ingredient that normally contains gluten but that has been processed to remove gluten is acceptable as long as there is no more than 20 parts per million (ppm) of remaining gluten.

Additionally, any unavoidable presence of gluten in the food must be less than 20 ppm. If a food label bears the claim "gluten-free," "free of gluten," "without gluten," or "no gluten," but fails to meet the FDA requirements, the food would be considered misbranded and therefore subject to regulatory enforcement action by the FDA.

> *U.S. Food and Drug Administration, Food Facts, Gluten and Food Labeling: FDA's Regulation of "Gluten-Free" Claims. Updated June 5, 2014. Available at http://www.fda.gov/Food/ ResourcesForYou/Consumers/ucm367654.htm, accessed June 17, 2014; U.S. Food and Drug Administration, For Consumers, What is Gluten-Free? Available at http://www.fda.gov/forconsumers/ consumerupdates/ucm36069.htm, accessed September 20, 2014.*

though the ingredient label on the back reveals that the first ingredient is sugar, followed by whole grain corn flour, then wheat flour, along with oat fiber, and soluble corn fiber.

Although the ingredients must be listed in descending order of proportion, there is no simple way to discern how much whole grain is included, and it is clear that the fiber content has been added independently of the whole grain ingredients.

On the other hand, if the same sugar-sweetened cereal would have been purchased previously anyway, perhaps any improvement in the nutritional profile, even if small, is a positive trend toward improving the public's diet overall.

FDA and Nutrient Claims

FDA's specific regulations as to labeling and health claims are quite detailed and extensive, and beyond the scope of this book. Guidance for the food industry is published on the FDA website.[51] Generally, allowable claims for food labeling consist of health, structure/function claims, and nutrient claims. Health claims, which are subject to FDA review and authorization, will be discussed in more depth later in this chapter.

Structure/function claims describe the role of a nutrient or dietary ingredient that is believed to impact our normal physiological processes. The statement "calcium builds strong bones" is an example of a structure/function claim.[52] Unlike health claims, structure/function claims for conventional foods are not subject to FDA review, and the manufacturers do not need to provide notice to the FDA of the intended use.[53]

Looking at the differences in how dietary fiber and whole grains can be described on package labels is illustrative of nutrient claims. The NLEA allows labels to characterize the level of nutrients in a food such as high or low, or to compare to other foods such as more or reduced, for example, "low in sugar" or "reduced fat." However, such descriptive characterizations are limited to those that are allowed under the FDA's authorizing regulations. Most of the FDA's authorizing regulations apply only to those substances that have an established Daily Value.

To keep the public from being misled by claims on food packaging, the FDA regulates the use of certain statements, such as the phrase "healthy" or "good source." The FDA considers the use of "healthy" as an implied nutrient content claim in part because a consumer may associate the word "healthy" with current dietary advice, and because it characterizes a food item as having "healthy" levels of total fat, saturated fat, cholesterol, and sodium.[52]

For a food to be labeled a "good source" of a particular nutrient, the amount customarily consumed must contain 10–19% of the FDA's established reference daily intake (RDI) or daily reference value (DRV).[54] The term RDI is used for vitamins and minerals considered to be essential and the term DRV is used for food components, such as fat and fiber (see Chapter 4 for more details on RDIs and DRVs).[55] These established dietary baselines for nutrients and food components are necessary to make a descriptive claim about the level of that item within a food.

Whole grains, although a significant item of interest in current dietary guidance, do not have an established RDI or DRV. Because we do not apply an RDI or DRV to "whole grains," it is impossible to have a reference amount that is "10–19%" to meet the needs of the regulation cited above. Accordingly, the phrase "good source" cannot be used in conjunction with whole grains on a food product label. The label may only state that it is "made with whole grains."

Conversely, there is an established DRV of 25 g for dietary fiber,[56] and thus, a food can be labeled as a "good source" of dietary fiber. A breakfast cereal can state "A good source of fiber" and "Made with whole grains," but is not a "good source of whole grains."

In 2006, the FDA issued draft guidance for the food industry on whole grains and labeling, which included a comment that food manufacturers can make factual statements about whole grains on the label of their products, such as "10 g of whole grains," "½ ounce of whole grains," and "100% whole grain oatmeal."[57]

As of 2014, this draft guidance has not yet been finalized. It seems that in 2013, the FDA communicated that it wanted to finalize the guidance and was still open to comments. The Center for Science in the Public Interest (CSPI) is suggesting disclosure of the percentage of whole grains within a serving. General Mills suggested that the FDA proceed with using the definition of whole grains that was included in the 2006 draft guidance, to not limit whole grain statements to foods that provide a good source of fiber, and to use a "grams per serving" based recommendation. Other parties are arguing that whole grain statements should only be on foods that do contain a good source of fiber, or that the addition of extra bran and/or germ should count for purposes of identifying a product as whole grain, or, most recently, in December of 2013, a proposal that whole grains shouldn't be limited to cereal grains but instead expanded to include whole seeds.[58] *Do you think the definition of whole grains should be expanded to include whole seeds?*

FDA and Health Claims

The developing interest in the 1980s in the connection between nutrition and health—and the potential to tap into an associated consumer market—created a surge in health claims on packaging. In particular, health claims became most extensively promoted in concentrated market areas, such as cereals.[59] Just like packaging technology, health claims greatly contributed to the marketing, sales, and consumption of grain-based food products.

Per the FDA, a "health claim characterizes the relationship between a food substance (food, food component, or dietary supplement ingredient) and reduced risk disease or health-related condition."[52] A health claim connects a food (or component nutrient) with the possibility of disease risk reduction, and must be limited to claims about disease risk reduction. A health claim cannot invoke the possibilities of diagnosis, cure, mitigation, or treatment of disease.

Manufacturers can also make claims that do not address risk disease reduction, known as structure/function claims. "Calcium builds strong bones" is an example of a structure/function claim. Structure/function claims were authorized in 1994 under the Dietary Supplement Health and Education Act, and unlike health claims, structure/function claims for foods are not subject to FDA review. Manufacturers do not need to provide notice to the FDA of the intended use of a structure/function claim.[60] *Do you think the structure/function claims are useful to consumers? Is the difference between health claims and structure/function claims appreciable?*

The Food and Drug Administration is interested in health claims as it is in charge of regulating the truthfulness of food labeling. Due to the FDA's long-standing role in combating fraud in "patent medicines," the general policy prior to the mid-1980s had been to classify any food product making health claims as an unapproved drug, which would then be subject to the systemic and lengthy drug review process. However, the Federal Trade Commission (FTC) is in charge of advertising that is not part of the food package label. Even though the FDA did not have authority over advertising campaigns, such campaigns could trigger the FDA into starting the "unapproved drug" review process and this strongly dampened the use of health claims in advertising. As a policy this helped prevent the public from being misled or subjected to harm, but also kept useful information from being disseminated.[59]

The strengthening evidence on diet and connections to health, particularly chronic disease and cancer, however, was too tantalizing for the food industry to ignore as a means to boost sales. And groups interested in improving public health, such as the National Cancer Institute, wanted to give consumers this information. In the mid-1980s, the Kellogg Company, partnering with the National Cancer Institute, began a labeling and marketing campaign on the box of a high-fiber breakfast cereal product advising that dietary fiber could help prevent some types of cancer. Furthermore, the FTC supported these messages. Accordingly, the FDA backed off from taking any action against the Kellogg campaign, and other companies quickly followed suit with similar claims. The use of health claims on food products escalated. In the first half of 1989, 40% of new food products were labeled with general and specific health claims.[59]

The legislative arm of the government was also interested in promoting such claims and hence, passed the 1990 Nutrition Labeling and Education Act (NLEA). The NLEA charged the FDA, among other things, with issuing regulations regarding the use of health claims on food labels.[61]

Although the NLEA authorized the use of health claims in general, the FDA had to issue a regulation allowing a health claim before manufacturers could place the health claim on the label. Furthermore, the FDA would only issue such a regulation if it had sufficient scientific evidence and agreement among scientific experts as to effects and dosage.

As a baseline to keep the value of health claims from being trivialized, the FDA published a position in 1993 noting that a food making a health claim must contain a minimum of specified nutritive elements (such as calcium, vitamin C, etc.).[62] These general baseline requirements also include caps on less-nutritive components. For example, the regulations now state that food products can use the term "healthy," but only if "disqualifying nutrients" (currently total fat, saturated fat, cholesterol, and sodium) do not exceed certain levels. Additionally the food must contain at least 10% of the RDI or DRV for vitamin A, vitamin C, iron, calcium, protein, or fiber for a serving customarily consumed prior to the addition of any nutrients.[63]

1997 Food and Drug Administration Modernization Act and First Amendment Challenges to Health Claims

The 1997 Food and Drug Administration Modernization Act (FDAMA) streamlined the health claim approval process to allow a food manufacturer to propose a health claim to the FDA based on an authoritative statement from approved federal scientific entities, such as the National Institutes of Health (NIH), the National Academy of Science (NAS), and the Centers for Disease Control and Prevention (CDC). The streamlined proposal review process was faster, but it did not make approval by the

FDA any more likely to happen. The FDA kept rejecting proposed health claims, which left industry chaffing about the continued restrictions.[64]

First Amendment challenges were raised. (Commercial speech has some protection under the First Amendment. See the discussion in Chapter 12 as to if calorie labeling constitutes being compelled to express an opinion.) In a 1999 case, *Pearson v. Shalala,* the U.S. Court of Appeals ruled that if health claims could be presented in a manner that was not deceptive (although potentially misleading), the FDA could not ban such claims outright, but must instead consider if suitability could be achieved with appropriate disclaimers.[65]

Accordingly, the FDA added a new category: "qualified" health claims. "Qualified" claims are supported by scientific evidence but do not meet the significant scientific agreement standard, and require an additional disclaimer regarding the scientific evidence. Summarizing, there are three paths to a health claim: the NLEA authorized health claim (significant scientific agreement standard), the FDAMA authoritative statement path, and the qualified health claim path.

To date, one qualified health claim has been made in regard to whole grains. In 2012, ConAgra petitioned for a qualified health claim concerning whole grains and type 2 diabetes. In the fall of 2013, the FDA approved a qualified health claim for whole grain and type 2 diabetes in these forms: "Whole grains may reduce the risk of type 2 diabetes, although the FDA has concluded that there is very limited scientific evidence for this claim," or "Whole grains may reduce the risk of type 2 diabetes. FDA has concluded that there is very limited scientific evidence for this claim."[66]

Could it be argued that the above approved qualified health claims for whole grains are opinions of the FDA? Are there statements that could be considered more neutral? The next section describes two FDAMA whole grain health claims that have been successfully made.

Whole Grains and Health Claims

Two FDAMA authoritative statement whole grain health claims have been successfully made, one claim in 1999 by General Mills for whole grain foods and one claim in 2003 by Kraft for whole grain foods with moderate fat content.[67]

Significantly, the 1999 General Mills claim provided the FDA with a definition of whole grain foods. This same definition was then used in the 2005 Dietary Guidelines. *Does it matter that our dietary guidance is based on a definition provided by industry?*

Both industry health claims incorporate a modicum of perceived healthful characteristics, such as limits on fat and cholesterol, which are required for eligibility of the food product to state the claim on the label; however, unlike the 2012 "qualified" health claim from ConAgra, both of the earlier claims also establish a threshold for the amount of whole grains required in the product.

General Mills used the FDMA's fast-track process by citing the 1989 National Academy of Sciences' statement: "diets high in plant foods—fruits, vegetables, legumes, and whole-grain cereals—are associated with a lower occurrence of coronary heart disease and cancers of the lung, colon, esophagus, and stomach."

In the notification, General Mills also handily provided the FDA a suggested definition of "whole grain foods" as foods containing 51% or more whole grain ingredients by weight per reference amount customarily consumed, and further proposed that compliance to this standard could be assessed through comparison to the amount of dietary fiber in whole wheat, wheat being the predominant grain in the U.S. diet.

The FDA did not reject or modify the claim. The lack of response meant that as of July 8, 1999, foods that fit the criteria put forth in the General Mills notification could boast the following health benefit claim: "Diets rich in whole grain foods and other plant foods and low in total fat, saturated fat, and cholesterol, may help reduce the risk of heart disease and certain cancers."[68]

The 2003 notification from Kraft is much the same as the one from General Mills, although it removed the reference for cancer, added a restriction on trans-fat, and modified the total fat criterion by adjusting up to a "moderate" level, although still limited to no more than 6.5 g of total fat.[69] The acceptance of the Kraft terms by the FDA means that future health claims can be proposed for foods in this area without reference to a diet low in total fat (see Table 9.1).

TABLE 9.1   FDAMA health claims (health claims authorized based on an authoritative statement by federal scientific bodies), whole grains

APPROVED CLAIMS	FOOD REQUIREMENTS	CLAIM REQUIREMENT	MODEL CLAIM STATEMENTS
Whole Grain Foods and Risk of Heart Disease and Certain Cancers (Docket No.1999P-2209)	Contains 51% or more whole grain ingredients by weight per RACC, and Dietary fiber content at least: • 3.0 g per RACC of 55 g • 2.8 g per RACC of 50 g • 2.5 g per RACC of 45 g • 1.7 g per RACC of 35 g	Required wording of the claim: "Diets rich in whole grain foods and other plant foods and low in total fat, saturated fat, and cholesterol may reduce the risk of heart disease and some cancers."	NA
Whole Grain Foods with Moderate Fat Content and Risk of Heart Disease (Docket No. 03Q-0547)	Contains 51% or more whole grain ingredients by weight per RACC, and Dietary fiber content at least: • 3.0 g per RACC of 55 g • 2.8 g per RACC of 50 g • 2.5 g per RACC of 45 g • 1.7 g per RACC of 35 g	Required wording of the claim: "Diets rich in whole grain foods and other plant foods and low in total fat, saturated fat, and cholesterol may help reduce the risk of heart disease."	NA

Source: Adapted from U.S. Food and Drug Administration, Guidance for Industry: A Food Labeling Guide (11 Appendix C: Health Claims) January 2013. Available at http://www.fda.gov/Food/GuidanceRegulation/GuidanceDocumentsRegulatoryInformation/LabelingNutrition/ucm064919.htm, last updated November 22, 2013, accessed November 19, 2014.

Accepting the General Mills notification also meant a *de facto* adoption of the General Mills definition of whole grain, although only promulgated as draft guidance by the FDA. This definition has some weaknesses. For example, the criteria of weight do not indicate if the reference is dry or wet weight. The weight of a whole grain may be influenced by a high water rate, or a mix of ingredients may prevent the weight criteria from being met. Raisin bread that does not contain any refined flour at all, consisting completely of whole grain flour, might still not meet this 51% weight guideline.[70]

Keep in mind that this defining criteria is only for making health claims on the package label. The ingredients on the nutrition label would still reveal the usage of whole grains and the lack of refined flour, just not in an easily marketable format.

The definition of whole grain presented by General Mills also conflates the idea of whole grain and dietary fiber, by suggesting the use of the level of dietary fiber in whole wheat as a reference point that could be used to assess compliance with the requirement that the food contains 51% or more of whole grains. This makes sense historically, as dietary guidance initially focused on fiber as the compelling healthful benefit of whole grains.

However, as research and dietary guidance evolved (see the 2000 Dietary Guidelines), the contributions of other components of whole grains became more apparent. Accordingly, other whole grain producers may not have been pleased with getting hamstrung by the fiber requirement. Brown rice and quinoa have less dietary fiber than whole wheat.

AGRICULTURAL POLICIES: GRAIN PRODUCTION AND SURPLUS

Agricultural policies and the current grain commodity support system are topics of critical contention. Detailed critiques of economic pros and cons of such policies, as well as topics such as crop insurance supporting diverse and drought-tolerant planting, are crucial, but beyond the scope of this chapter. The policies and politics around grain subsidies are contentious, and this chapter is limited to some brief highlights on this issue. Commodity and grain subsidies are keeping some family farms in the business, but the transformation of our rural and farming landscape from small to large—and larger—farms is well documented elsewhere. In general, our agricultural policies have kept a larger focus on commodities rather than the structure and shape of rural farming.

This section touches on how the interests of agribusiness and grain producers can drive farm policies, illustrated by the historical use of surplus grain in international food aid as well as a tool of political negotiation, the abandonment of a national grain reserve, and the absorption of surplus crops through our food supply and the development of biofuels. Developing regional and producer industry concentrations created political support that kept farm subsidies extended past the original needs of the Great Depression. Commodity support continued in World War II, even though the economic engine of the war itself increased prices and eliminated surpluses. Furthermore, the structure of the price supports (more for a greater volume of production) favored larger farmers. In other words, agribusiness became a key policy player earlier than we might think.[71]

Agricultural Policy and Developing Agribusiness

The legislation of the 1930s, outlined in Chapter 2, developed supply management policy using price support and production controls. Price support means that the government is acting to maintain a market price, for example, by buying grain and taking it off the market by placing it into reserve. Production controls are a means to affect supply by setting limitations on production. In the United States, historically production controls meant that acreage was restricted but volume (which established the levels of price support) was not. Over time, U.S. agricultural policies have fluctuated from price support to fixed income support, with production controls eventually fading.[71]

Grain Surplus: Export Subsidies to Aid

Farmers naturally seek efficiencies in all areas, and together with the technological advantages, productivity boomed post war. Between 1945 and 1970, wheat productivity per acre almost doubled and corn productivity more than doubled. Additional surplus control mechanisms were introduced to deal with this bounty, such as the 1954 Agricultural Act and the Agricultural Trade and Development Assistance Act of the same year (also known later as "Food for Peace"), which added export subsidies.[72]

The 1954 legislation set up a means by which the United States could provide economic assistance to other countries through supplying grain, mostly wheat. This allowed the United States to dispose of agricultural surplus while still maintaining the same domestic policies of supply management. The idea caught on, and other countries joined in on exporting. Thus this legislation, and the eventual immense amount of global aid, ushered in a reshaped world economy around grain, as well as changes in our global diet. An estimated 80% of wheat exported from the Unites States prior to 1996 was done so under some form of government assistance. Additionally, food aid paved the way for the development of commercial markets, which in particular helped the grain traders and foster the growth and concentration of agribusiness.[71]

Another form of international food aid could be policy that encourages self-sufficiency, perhaps through improvements in traditional cropping. However, some, such as the U.S. Grains Council, a private nonprofit company that develops export markets, argue that self-sufficiency is "expensive and elusive,"[73] suggesting instead the idea of using the market for trade efficiencies, that is, to develop a trading market for food items of which countries can produce on an off season compared to the intended market, or are niche producers or just really efficient at producing. Quinoa provides a good example for discussing some of the complexities of food policy promoting trade efficiencies.

Bolivia and Peru are "efficient" producers of quinoa. Quinoa, a local traditional dietary staple, grows in the cool, arid climate of the Andean highlands, but is not grown much elsewhere as it is difficult to grow, even in the Andean climate. The market for quinoa in the United States dramatically exploded over a 5-year period from 2007, when the United States imported 7.3 million pounds, to 2012, when imports reached 57.6 million pounds. Interestingly, the price of quinoa also tripled during the same period.[74]

Nutritionally, quinoa is a powerhouse of a grain (as well as gluten-free). In addition to touting the nutritional benefits, some retail channels promote fair trade arrangements, thus providing additional value-adds for shoppers seeking to "vote with their fork" and to feel good on multiple levels about their dietary choices.

In 2013, however, those shoppers struggled with the news that, due to their demand and willingness to pay high prices, quinoa has become too expensive for poor Bolivians. This is not the outcome a fair trade shopper is envisioning. However, this disparity in quinoa access manifested in urban markets, such as La Paz and Lima, where quinoa may not have been as broadly consumed. It was also reported at the time that local quinoa producers and highlanders were still able to include the traditional grain in their diet, and, due to boosted income levels, were able to increase other high-quality nutritional items, such as produce, in their diet. However, the increased interest and expanded market in quinoa is leading to the development of varieties that might more easily be grown outside the Andean highlands. If quinoa production expands outside the Andean highlands, the world may be better off nutritionally, but the sustainability of the rise in income of local Andean producers may be a challenge.[75]

Grain Surplus: Reserves and Political Negotiation

The early 1970s saw another shift in agricultural policies. In 1972, after a disastrous crop failure, the Soviet Union decided to maintain livestock production by buying grain on the world market and the United States decided to sell an enormous amount of grain to the Soviets—both from surplus and from commercial markets. Although the United States government may have intended the sale to be a multi-year process, the involvement of the commercial grain traders could have also unduly influenced the process; in just 1 month the Soviets bought a quarter of the 1972 U.S. wheat crop, along with large amounts of corn and soybeans.[76]

Wheat prices doubled. Because the Soviets had already secured their grain early at a low price, the grain traders still profited and the American consumer took the hit, although it appeared first as rapidly escalating farm income. The 1973 emergency policy response to address the unexpected severe consumer inflation was to empty all government stockpiles to market. Government wheat stocks sank from 714 million bushels in 1972 to 19 million as of 1974, while domestic prices nonetheless still rose 15% in 1973 and 1974.[72] Price supports were adjusted to reflect market conditions and the agricultural reduction requirement was eliminated,[71] thus encouraging and supporting continued overproduction.

The combination of high grain prices in the 1970s along with available government subsidies caused another "gold rush" in wheat as there had been 100 years prior, with outside investors plowing marginal land and reaping a quick profit, then selling off land once price dropped; fragile land was damaged that could have been publically conserved instead.[77]

The U.S. grain economy became complicated with dependencies on foreign policy, export issues, and a reluctance to institute trade barriers. Weather conditions and policy shifts by other governments or large corporations created instability. The extent of our surplus and ability to export seemed like it should be useful in political negotiation by the federal government, but often is not as useful as planned. For example, in 1975, President Ford tried to embargo sales to the Soviets during his tenure to help curb domestic inflation and with an eye toward some political concessions as to Soviet oil exports, but reelection political pressure forced dropping the embargo after 2 months, leading to a trade agreement with the Soviets that failed to garner the desired concessions.[78]

In 1977, in response to the decade's early turmoil, the Farmer-Owned Grain Reserve program was instituted.[79] This program created an incentive for farmers to store grain but did not leave the government holding the stock. Under the reserve program, the government would subsidize loans and storage payments, and the farmer agreed to limitations as to marketing the grain.[80]

As the next decade began, leveraging grain for political concerns was tried again. President Carter instituted a grain embargo withholding any excess sales of grain to the Soviet Union as a means of expressing disapproval over the Soviet occupation of Afghanistan. President Carter's embargo initially had success. However, for any such embargo to succeed, consider that other countries and, in particular, transnational corporations must agree to play along. In 1980, export firms such as Cargill did informally agree for a period of time to play along. As Paarlberg commented at the time of the 1980 embargo, "multinational firms have a well-deserved reputation for discretion and ingenuity when it comes to servicing the needs of their largest and wealthiest customers." In other words, secret sales were basically expected, and, as Paarlberg pointed out, the USDA is not necessarily a natural enforcer, considering the agency's mission of promoting agriculture.[78]

Discussion about the power of corporate lobbyists in shaping our food policies is touched on in other chapters of this book, and other resources address this in more detail. Corporate influence is hardly news anymore. Nonetheless, it is helpful to be aware that corporate influence on food policy is not limited just to lobbying. Cargill is the largest grain trader in the world, and as such, has been able to accordingly influence agricultural grain policy, even if simply called on as an expert, as in 1971 when a Cargill vice president took leave to help Nixon shape trade policy in Asia. Later, Cargill and other such entities participated in committees shaping the World Trade Organization and Freedom to Farm policies, in 1995 and 1996, respectively.[81]

Although it is beyond the scope of this text to discuss the World Trade Organization, it is worth mentioning that this organization is part of a mid-1990s emphasis on free trade. Domestically, the interest in free trade resulted in the elimination of the Farmer Reserve program in the 1996 Farm Bill. The official name of the 1996 Farm Bill was "Federal Agricultural Improvement and Reform Act," but, in line with its moniker, "Freedom to Farm Act," the 1996 Farm Bill moved away from supply management as a policy by generally reducing government intervention. Production controls and the grain reserve program were both eliminated.[82]

The United States no longer has a grain reserve. Agribusinesses, such as Cargill do, however. *Does it matter that a corporation holds grain but our nation does not?* In 1976, the U.S. General Accounting Office concluded that a grain reserve would act as a buffer against unpredictable shocks, such as the adverse weather conditions in 1972 and 1974, and would help avoid decisions regarding domestic price increases and foreign allocations.[72] *In the case of adverse weather now, who will benefit from our policies— farmers (small or large), consumers, or Cargill?*

Grain Surplus: Food and Fuel

By the end of the 1970s, United States corn producers also found another market capable of absorbing excess corn supply—beverages, and processed, packaged food—through the development of high-fructose corn syrup.

Glucose, fructose, and sucrose, among others, are specific types of sugars. Sucrose, which we are familiar with in the form of table sugar, consists of a 50/50 ratio of glucose to fructose. Corn syrup is made from corn starch, and consists almost entirely of glucose. High-fructose corn syrup (HFCS) is exactly "as described on the tin"; it is corn syrup that is higher in fructose than regular corn syrup. Introduced into the U.S. marketplace in the late 1960s and early 1970s, HFCS is made by converting a portion of the glucose found in corn syrup to fructose through enzymatic isomerization, most commonly in a ratio that turns the syrup into 55% fructose and 45% glucose, or "HFCS-55." Formulations also exist with different ratios, such as "HFCS-43" (43% fructose) and "HFCS-90" (90% fructose). HFCS turned out to be highly useful for the processed food and beverage industry; the liquid nature makes easier and cheaper to transport and store, as it can be pumped directly from delivery vehicles to storage, and is stable in an acidic environment which is helpful for industrial-level baking.[83]

Although our policies subsidizing corn are a contributing factor to the inexpensiveness of HFCS, the fact that sugar is also a highly protected industry is sometimes overlooked. Given a combination of subsidies for corn with tariffs on imported sugar, cane sugar can be cheaper abroad while corn syrup remains cheaper in the United States. Domestically in the United States, the relative price of sugar compared to corn price tends to be at multiples higher than elsewhere in the world.[84] The opportunity to reap greater profit by using HFCS domestically was naturally attractive to the food industry.

High fructose corn syrup, food additives, and GRAS

The Food Additives Amendment of 1958 tasked the FDA with reviewing the safety of food additives. This Amendment provides a legal definition of food additives for the purpose of imposing a premarket approval requirement by the FDA. Food additives are defined as "all substances, the intended use of which results or may reasonably be expected to result, directly or indirectly, either in their becoming a component of food, or otherwise affecting the characteristics of food." The legal definition of food additives also excludes items whose use is GRAS; government approval is not needed for such items.[85] Additionally, some items are explicitly exempted in Section 201(s) of the Act, including pesticide chemicals, pesticide chemical residue, and color additives, because those items are subject to different legal regulations for premarket approval.

If a substance was used in food prior to January 1, 1958, a food can be considered GRAS "through experience based on common use in food," defined as a history of consumption by a substantial number of users. The other option is to have qualified scientific experts evaluate the safety of the item.[86]

HFCS was identified GRAS by the FDA in 1983, and this status was affirmed in 1996, in part because the 45% glucose/55% fructose ratio is quite close to the 50/50 ratio of table sugar. In other words, although HFCS has much more fructose than does regular corn syrup, this formulation of HFCS does not have a hugely greater amount of fructose than does table sugar. HFCS-43 and HCFS-55 were listed as safe in the FDA's final ruling, but HFCS-90 was not included, primarily because the ratio of glucose and fructose is not approximately equal. The ruling goes on to state "additional data on the effects of fructose consumption that is not balanced with glucose consumption would be needed to ensure (safety)." For HFCS-90 to be found safe, it would need to proceed through the petition process in Section 170.35 of the regulation.[87]

It important to understand the parameters under which this regulation considers additives to be safe, namely "there is a reasonable certainty in the minds of competent scientists that the substance is not harmful under the intended conditions of use."[85]

In other words, recognizing a food additive as GRAS is specifically tied to the conditions of intended use. But what happens if we change our "intended use" of the additive? For example, we may be consuming more HFCS than originally thought, and in particular, more via a beverage form. In 1970, less than 1% of all caloric sweeteners in the United States food supply consisted of HFCS, but by 2000, this had increased to 42%, with soft drinks a high majority.[88]

Furthermore, we may be consuming a product that is in actuality different from the one that the FDA determined was GRAS. It is possible that the 55% ratio has only been assumed to be an accurate representation of the fructose content in the syrup being commonly used in our food products and beverages. A 2010 study testing HFCS in beverages found a mean ratio of 59% fructose, with several major products apparently using a ratio of 65%.[89] This would appear to be an example of a potential consumption imbalance of fructose compared to glucose, the matter over which the FDA has stated that data would be needed to evaluate the safety of HFCS-90.

From a health perspective, the additional fructose might make a difference. Glucose and fructose are handled differently within our body. Glucose provides "satiety" signals to the brain that fructose does not, and fructose, unlike glucose, is rapidly taken up the liver, bypassing a key step in the glucose breakdown process. Research has shown that consumption of high amounts of fructose has been associated with insulin resistance.[89] Conversely, a 2014 article summarizes the issue as "excessive fructose intake can have deleterious metabolic effects," but when fructose is ingested together with glucose, the extent of such effects is much less clear.[90]

Corn producers have proposed in recent years to change the name to "corn sugar." As of 2012, the FDA rejected this proposal, in part because "corn sugar" has historically meant dextrose, in part because there is more fructose in HFCS than there is table sugar, and in part because "corn sugar" itself is safe to consume by people with fructose intolerance, which would not be the case if it were HFCS.[91]

Grain surplus: Biofuel

Biofuel also absorbs grain crops in the United States. In 2007, Congress passed a law that required oil companies to blend ethanol into gasoline. By 2010, about 85% of gas in the United States contained 10% ethanol, and the percentage of ethanol-blended gas on the market has continued to increase.[92]

Some argue that the choice to produce biofuel as an alternate source of energy is problematic, particularly with land usage conversions, as we are choosing fuel over food for those that are hungry. One argument is that if acreage that was used to produce food for humans were swapped into biofuel production, the loss in human food production would impact global food stocks and accordingly, hunger. Biofuel in the United States, however, is produced almost exclusively from corn feedstock. Nonetheless, the increased demand for corn feedstock can still impact food for humans as the increased price for corn feedstock ripples through our supply chain. The U.S. Congressional Budget Office (CBO) reported that about 20% of the increase in domestic corn prices between 2007 and 2008 was due to domestic ethanol demand and accounted for about 10% of the rise in food prices over the same period.[93]

Historically, we used most of our corn to feed our livestock. In 2010, for the first time, fuel topped the list of how we use corn. Fuel stayed on top until 2013, when it dropped slightly below livestock feed, and declining slightly more in 2014, with 43% of corn going to fuel and just slightly more at 45% to livestock feed.[94]

As demand increases, so does the price of the supply. As the price of corn (or any grain) increases, farmers (small and large) will seek to maximize their chance of profit by growing more of the desirable crop. Individual decisions to pursue maximum profit can impact issues of collective concern, such as sustainability, when land less suited to agriculture or the particular crop is converted for that production, such as hilly, grassy areas of Iowa being turned to corn. Of even more concern is land that was set aside for conservation—wetland, prairies—turned into corn production.[95] Other countries use grass, sugarcane, and byproducts to produce biofuel. Although land conversions are still an issue, biofuel sourced from these alternatives may be a more efficient means of energy conversion and less of a food crop replacement threat.

Global Hunger/Food Security/Global Production

Our diet has globalized toward grain. Can we continue to grow enough grain? Arguments have been made that there is slowing productivity growth in agriculture, in particular for cereal yields.[96] In other words, the amount of grain harvested per unit of land is leveling off. What will happen as our world population continues to grow?

Bread and grains as a dietary staple have historically been very closely tied to political stability. An 1801 bakery strike in New York City caused enough unrest that wealthy citizens formed a larger baking company to keep the peace.[97] Although food riots have not contributed to political unrest in the United States post the Great Depression, food riots have been prominent on the global stage, in 2008–2009 when wheat and rice prices rose over 200% from 2004–2005 prices, and again in early 2011 when food index prices generally rose again to new heights, known as the Arab Spring.[98]

FAO's Cereal Price Index (CPI) shows a record high at 265 in April 2011, with an overall annual value in 2013 of 240.9.[99] International food prices generally peaked in 2011, and although prices have declined slightly, they remain well above historical averages. Grain (cereal) prices increased again mid-2012, dropping in the fall of 2013.[96]

The Future of Grains in Our Diet

The current global human population world is dependent on the continued success of the agricultural transition and the calories provided by grain. The UN projects a world population of more than nine billion people in 2050, some two billion more than today. Additionally, these extra two billion people will not be evenly distributed; it is believed that population growth is most likely to occur in areas already limited in resources. The favorable news, however, is that this population growth is a deceleration from the period of 1970–2010. Nonetheless, to meet the increase in energy needs, we may need to double our global crop production by 2050.[100] Some countries are addressing the problem of input restraints (crop yield, land, water, weather) through purchasing or long-term leasing land from other countries, then sending the harvest home.

Wheat and rice together directly provide about 40% of our current global energy intake. Grain production has increased over time due to the development of high-yielding varieties and the spread of agricultural technologies, including greater use of irrigation and fertilizer. Historically, grain production has increased in tandem with people production, and crop yield has been argued to be a more sustainable means of achieving food self-sufficiency than clearing more land. However, there is significant concern that we have reached "peak grain," that is, the yield rates in agriculture have already reached their maximum and now are slowing down.[101]

Weather has always affected agriculture output. However, climate change may be escalating the impact of weather on our crops, through increased drought and changing weather patterns. For example, among wheat, corn, and rice, wheat is most susceptible to heat. And corn is an open pollinator, meaning that it needs rainfall in a certain window or the crop will be lost. The summer of 2012 brought a one-two-punch of the hottest July in U.S. history along with the worst drought in 50 years. Corn and soybean yields plummeted, among others, and the National Climatic Data Center calculates that the 2012 drought cost us $30 billion.[102]

The location of current wheat production in the United States is shifting, just as it did in the 20 years between 1860 and 1880 when wheat raced through the newly open counties in Minnesota; 85% of the U.S. durum wheat crop, a variety prized for pasta rather than bread, has historically been grown throughout North Dakota.[103] However, as rainfall patterns change, the growing zone has been shifting farther west as the eastern part of the state is now too wet.[104]

If our wheat crops are not as productive under changing climate conditions, we will need to develop new varieties of wheat or eat something else. Perhaps our dietary pattern will have to rely less on grains, or incorporate grains that thrive better in changing conditions. *How can/will our dietary guidance be retooled? Could there be a greater emphasis on whole grains of all kinds instead of on wheat flour products?*

CONCLUSION

Nutritionally speaking, grains are not essential to our diet. However, grains are an essential means of feeding the global population, both directly and indirectly through the use as animal feed. Wheat, in particular, became a staple of the American diet in the mid-1800s. Land expansion and development of rail transportation encouraged the production of wheat and other grains. The high "packagability" of grains and grain-based products contributed to the rise of national brands and increasing federal regulation over packaging. With such prominence in our food supply in the form of packaged goods on grocery store shelves, grain products provide good illustrations of federal food and labeling regulations, particularly the use of health claims on packaging and discussions over GRAS.

As we increasingly adopted refined grains in our diet as a cheap way to feed ourselves, we shored up the nutritional deficiencies caused by refinement through enriching the resulting food products. This in turn encouraged continued reliance on grain as a staple in our national diet for public health such that when official U.S. Dietary Guidance began to promote whole grain consumption, it could recommend only consuming half the daily grain as whole.

The United States has become an expert at grain production. What to do with the surplus? Grain is used as international food aid, as a tool of political negotiation, as an industrial ingredient throughout our food supply, and as biofuels. However, given the stress of climate change, reduced water resources, and other inputs, we may not continue to easily have a surplus. Looking forward, we may need to change where we grow grain, the grains themselves, our diet, and our dietary guidance.

REFERENCES

1. Bates KL. America the Beautiful. Available at http://www.gilderlehrman.org/history-by-era/art-music-and-film/resources/"america-beautiful"-1893, accessed November 20, 2014.
2. The American Heritage Dictionary of the English Language. Available at http://www.ahdictionary.com/word/search.html?q=grain, accessed April 12, 2014.
3. Fuller DQ. An Emerging Paradigm Shift in the Origins of Agriculture. *General Anthropol.* 2010;17:2. Available at http://www.academia.edu/369943/An_Emerging_Paradigm_Shift_in_the_Origins_of_Agriculture, accessed April 12, 2014.
4. Mercader J. Mozambican grass seed consumption during the middle stone age. *Science.* 2009;326(5960):1680–1683.
5. Cordain L, Eaton B, Sebastian A, Mann N, Lindeberg S, Watkins BA, O'Keefe JH, Brand-Miller J. Origins and evolution of the Western diet: Health implications for the 21st century. *Am. J. Clin. Nutr.* 2005;81(2):341–354. Available at http://ajcn.nutrition.org/content/81/2/341.full, accessed April 12, 2014.
6. Drewnowski A. Concept of a nutritious food: Toward a nutrient density score, *Am. J. Clin. Nutr.* 2005;82(4):721–732.

7. Hansen RG, Wyse BW, Sorenson AW. Nutritional Quality Index of Food. Westport, CT: AVI Publishing Co, 1979.

8. Brandt R. New timeline traces history of US agriculture stats. *Western Farm Press*, May 29, 2012. Available at http://westernfarmpress.com/management/new-timeline-traces-history-us-agriculture-stats, accessed April 12, 2014.

9. Bohannan CD. *Fifteenth Census of the United States: Census of Distribution*. Agricultural Commodity Series, Distribution of Grain, Issue 5, Bureau of the Census, United States.

10. United States Department of Agriculture, Census of Agriculture, About The Census. Available at http://www.agcensus.usda.gov/About_the_Census/index.php, last modified April 10, 2012.

11. MacDonald J, Korb P, Hoppe R. Farm size and the organization of U.S. Crop Farming, Economic Research Report No. (ERR-152), August 2013. Available at http://www.ers.usda.gov/publications/err-economic-research-report/err152/report-summary.aspx#.U03QeeZdVvA, accessed April 12, 2014.

12. Rasmussen WD. Lincoln's agricultural legacy, United States Department of Agriculture, National Agricultural Library. Available at http://www.nal.usda.gov/lincolns-agricultural-legacy, accessed April 12, 2014.

13. Arrington BR. Free Homes for Free Men: A political history of the Homestead Act, 1774–18632012. Available at http://digitalcommons.unl.edu/cgi/viewcontent.cgi?article=1049&context=historydiss, accessed April 12, 2014.

14. Weiss TJ. U.S. Labor force estimates and economic growth, 1800–1860. In: Gallman RE, Wallis JJ (eds.), *American Economic Growth and Standards of Living before the Civil War*, Chicago: University of Chicago Press, 1992, 19–78. Available at http://www.nber.org/chapters/c8007.pdf, accessed April 12, 2014.

15. U.S. History Scene, The Homestead Act of 1862: Dreams and Realities. Available at http://www.ushistoryscene.com/uncategorized/1862homesteadact/, accessed April 12, 2014.

16. U.S. History Scene, The Homestead Act of 1862: Dreams and Realities. Available at http://www.ushistoryscene.com/uncategorized/1862homesteadact/#footnote_11_2085, accessed April 12, 2014.

17. National Archives, The Homestead Act Anniversary, May 20, 1862. Available at http://www.archives.gov/legislative/features/homestead-act/, accessed April 12, 2014.

18. Bell B. Historian, Homestead National Monument of America. *America's Invitation to the World: Was the Homestead Act the First Accommodating Immigration Legislation in the United States?* Available at http://www.nps.gov/home/historyculture/upload/Immigration-White-Paper.pdf, accessed April 12, 2014.

19. Harvard University Library Open Collections Program. *Aspiration, Acculturation, and Impact Immigration to the United Sates*, 1789–1930, Scandinavian Immigration. Available at http://ocp.hul.harvard.edu/immigration/scandinavian.html, accessed April 12, 2014.

20. Frail TA, Gambino M. Document Deep Dive: How the Homestead Act Transformed America, *Smithsonian Magazine*, May 2012. Available at http://www.smithsonianmag.com/history-archaeology/How-the-Homestead-Act-Transformed-America.html, accessed April 12, 2014.

21. Arrington BR. Free Homes for Free Men: A political history of the Homestead Act, 1774–18632012. Available at http://digitalcommons.unl.edu/cgi/viewcontent.cgi?articl=1049&context=historydiss, accessed April 12, 2014.

22. Jarchow ME. King Wheat, *Minnesota History*. 1948;29,1–28. Available at http://collections.mnhs.org/MNHistoryMagazine/articles/29/v29i01p001-028.pdf, accessed April 12, 2014.

23. Twede D. The UNEEDA biscuit package: A historical landmark, Michigan State University. *8th Conference on Historical Research in Marketing and Marketing Thought*, Kingston, Ontario, May 22–25, 1997. East Lansing: MSU Department of Marketing, 1997: 75–82.

24. United States Department of Agriculture, Economic Research Service, Wheat's role in the U.S. Diet. Available at http://www.ers.usda.gov/topics/crops/wheat/wheats-role-in-the-us-diet.aspx#.UVcK9FsjqFd, last updated June 19, 2013.

25. Waldo FJ. Does baking sterilize bread? Being an inquiry, on bacteriological and other grounds, as to how far baking affects the vitality of organisms in dough. *Lancet.* 1894;3612(2):906–908.

26. Bobrow-Strain A. *White Bread: A Social History of the Store-Bought Loaf.* Boston: Beacon Press, 2012, 38–39, 41.

27. United States Department of Agriculture, Center for Nutrition Policy and Promotion. *Nutrient Content of the U.S. Food Supply, 1909–2004, A Summary Report,* February 2007. Available at http://www.cnpp.usda.gov/publications/foodsupply/foodsupply1909-2004report.pdf, accessed April 12, 2014.

28. Kantor LS, Variyam JN, Allshouse JE, Putnam JJ, Lin B-W. Choose a variety of grains daily, especially whole grains: A challenge for consumers. *J. Nutr.* 2001;131(2):473S–486S.

29. United States Department of Agriculture, Economic Research Service. Wheat's role in the U.S. Diet, last updated June 19, 2013. Available at http://www.ers.usda.gov/topics/crops/wheat/wheats-role-in-the-us-diet.aspx#.UVcK9FsjqFd, accessed September 6, 2014.

30. United States Department of Agriculture, Choose My Plate.gov. Why is it important to eat grains, especially whole grains? Available at http://www.choosemyplate.gov/food-groups/grains-why.html, accessed April 12, 2014.

31. American Cancer Society, Phytochemicals, last revised January 17, 2013. Available at http://www.cancer.org/treatment/treatmentsandsideeffects/complementaryandalternativemedicine/herbsvitaminsandminerals/phytochemicals.

32. United States Department of Agriculture, Center for Nutrition Policy and Promotion, 1995 Guidelines, last updated August 15, 2013. Available at http://www.cnpp.usda.gov/dgas-1995guidelines.htm.

33. United States Department of Agriculture, Center for Nutrition Policy and Promotion, 2000 Guidelines, last updated August 15, 2013. Available at http://www.cnpp.usda.gov/dgas2000guidelines.htm.

34. United States Department of Agriculture, Center for Nutrition Policy and Promotion, Nutrition and Your Health: Dietary Guidelines for Americans, 2000. Available at http://www.cnpp.usda.gov/dgas2005guidelines.htm, accessed April 23, 2014.

35. Dietary Guidelines for Americans. *Finding Your Way to a Healthier You: Based on the Dietary Guidelines for Americans, 2005.* Available at http://www.health.gov/dietaryguidelines/dga2005/document/html/brochure.htm, accessed April 12, 2014.

36. Dietary Guidelines for Americans, 2005. Available at http://www.health.gov/dietaryguidelines/dga2005/document/default.htm, accessed April 12, 2014.

37. United States Department of Agriculture, Agricultural Research Service. *Report of the Dietary Guidelines Advisory Committee on the Dietary Guidelines for Americans, 2000.* Available at http://www.cnpp.usda.gov/Publications/DietaryGuidelines/2000/2000DGCommitteeReport.pdf, accessed April 12, 2014.

38. De Moura FF, Lewis KD, Falk MC. Applying the FDA definition of whole grains to the evidence for cardiovascular disease health claims. *J. Nutr.* 2009;139(11):2220S–22263.

39. United States Department of Agriculture, MyPlate. Available at http://www.choosemyplate.gov/index.html, accessed April 23, 2014.

40. United States Department of Agriculture, Center for Nutrition Policy and Promotion, Dietary Guidelines for Americans, 2010. Available at http://www.cnpp.usda.gov/dgas2010guidelines.htm, accessed September 6, 2014.

41. United States Department of Agriculture, Center for Nutrition Policy and Promotion, USDA Food Patterns. Available at http://www.cnpp.usda.gov/USDAFoodPatterns.htm, accessed September 6, 2014.

42. U.S. Department of Agriculture and U.S. Department of Health and Human Services. *Dietary Guidelines for Americans, 2010,* 7th edition, Washington, DC: U.S. Government Printing Office, December 2010. Available at http://www.health.gov/dietaryguidelines/dga2010/DietaryGuidelines2010.pdf, accessed April 23, 2014.

43. Whole Grains Council, an Oldways Program. Available at http://wholegrainscouncil.org/whole-grain-stamp, accessed April 23, 2014.

44. Whole Grains Council, an Oldways Program. Available at http://wholegrainscouncil.org/newsroom/blog/2013/07/danish-whole-grain-consumption-up-72, accessed September 6, 2014.

45. Nestle, Whole Grain. Available at https://www.nestle-cereals.com/whole-grain, accessed September 6, 2014.

46. Marquart L, Jacobs DR, McIntosh GH, Poutanen K, Reicks M (eds.). *Whole Grains and Health.* Iowa: Blackwell, 2007.

47. Healthy People 2020, Food and Nutrient Consumption. Available at http://healthypeople.gov/2020/topicsobjectives2020/objectiveslist.aspx?topicId=29, accessed April 23, 2014.

48. Whole Grains Council, an Oldways Program, Whole Grains 101. Available at http://wholegrainscouncil.org/whole-grains-101/whole-white-wheat-faq, accessed April 23, 2014.

49. United States Department of Agriculture, My Plate.gov. Available at http://www.choosemyplate.gov/food-groups/grains-counts.html, accessed April 23, 2014.

50. Whole Grains Council, an Oldways Program, Newsroom Blog. Available at http://wholegrainscouncil.org/newsroom/blog/2008/10/just-ask-for-whole-grains-at-starbucks, accessed April 23, 2014.

51. U.S. Food and Drug Administration, Guidance for Industry: A Food Labeling Guide, last updated November 22, 2013. Available at http://www.fda.gov/Food/GuidanceRegulation/GuidanceDocumentsRegulatoryInformation/LabelingNutrition/ucm064908.htm#nutrcontent.

52. U.S. Food and Drug Administration, Label Claims for Conventional Foods and Dietary Supplements, last updated February 26, 2014. Available at http://www.fda.gov/Food/IngredientsPackagingLabeling/LabelingNutrition/ucm111447.htm.

53. U.S. Food and Drug Administration, Structure/Function Claims, last updated February 26,2014. Available at http://www.fda.gov/Food/IngredientsPackagingLabeling/LabelingNutrition/ucm2006881.htm.

54. 21CFR101.54, 2014.

55. Nielsen SS (ed.). *Food Analysis.* Berlin: Springer, 2010, 39.

56. 21CFR101.9, 2014.

57. U.S. Food and Drug Administration, Draft Guidance: Whole Grain Label Statements, updated March 18, 2014. Available at http://www.fda.gov/food/guidanceregulation/guidancedocumentsregulatoryinformation/labelingnutrition/ucm059088.htm.

58. Regulations.gov, Guidance for Industry and FDA Staff: Whole Grains Label Statements, Docket Folder Summary. Available at http://www.regulations.gov/#!docketDetail;D=FDA-2006-D-0298, accessed May 4, 2014.

59. John E, Pappalardo JK. Public policy issues in health claims. *J. Public Policy Mark.* 1991;10(1):33–53.

60. U.S. Food and Drug Administration, Structure/Function Claims, last updated December 2013. Available at http://www.fda.gov/Food/IngredientsPackagingLabeling/LabelingNutrition/ucm2006881.htm.

61. U.S. Food and Drug Administration, Guidance for Industry: Evidence-Based Review System for the Scientific Evaluation of Health Claims—Final, last updated February 25, 2014. Available at http://www.fda.gov/food/guidanceregulation/guidancedocumentsregulatoryinformation/labelingnutrition/ucm073332.htm.

62. Department of Health and Human Services, Food and Drug Administration, Final rule. Food Labeling: General Requirements for Health Claims for Food. *Fed. Regist.* 1993;58:3.

63. 21CFR101.14, 2014.

64. Fortin ND. *Food Regulation: Law, Science, Policy, and Practice.* New Jersey: Wiley, 2009, 125.

65. *Pearson v. Shalala*, 164 F.3d 650 1999.
66. Response from FDA regarding the ConAgra qualified health claim petition, September 2013. Available at http://www.fda.gov/downloads/Food/IngredientsPackagingLabeling/ LabelingNutrition/UCM368051.pdf, accessed May 4, 2014.
67. U.S. Food and Drug Administration, Qualified Health Claims, last updated June 7, 2013. Available at http://www.fda.gov/Food/IngredientsPackagingLabeling/LabelingNutrition/ ucm2006874.htm, accessed September 6, 2014.
68. U.S. Food and Drug Administration, Health Claim Notification for Whole Grain Foods, last updated April 18, 2013. Available at http://www.fda.gov/Food/Ingredients PackagingLabeling/LabelingNutrition/ucm073639.htm.
69. U.S. Food and Drug Administration, Health Claim Notification, last updated April 18, 2013. Available at http://www.fda.gov/Food/IngredientsPackagingLabeling/LabelingNutrition/ ucm073634.htm.
70. Whole Grains Council, Government Guidance, Limitations of FDA Whole Grain Health Claims. Available at http://wholegrainscouncil.org/whole-grain-stamp/government-guidance, accessed May 10, 2014.
71. Winders B. *The Politics of Food Supply, U.S. Agricultural Policy in the World Economy.* New Haven: Yale, 2012,83
72. United States General Accounting Office, *Grain Reserves: A Potential U.S. Food Policy Tool.* March 1976.
73. U.S. Grains Council, Key Issues, Relationships that Advance Food Security and Economic Growth. Available at http://www.grains.org/index.php/key-issues/food-security/issues/ 3619-relationships-that-advance-food-security-and-economic-growt, accessed May 12, 2014.
74. Depillis L. Quinoa should be taking over the world. This is why it isn't. *The Washington Post,* July 11, 2013. Available at http://www.washingtonpost.com/blogs/wonkblog/ wp/2013/07/11/quinoa-should-be-taking-over-the-world-this-is-why-it-isnt/, accessed May 12, 2014.
75. Aubrey A. Your Love of Quinoa is Good News for Andean Farmers. *NPR: The Salt.* July 17, 2013. Available at http://www.npr.org/blogs/thesalt/2013/07/16/202737139/is-our-love- of-quinoa-hurting-or-helping-farmers-who-grow-it, accessed May 12, 2014.
76. Penn MJ. America Gets the Shaft, *The Harvard Crimson,* November 16, 1973. Available at http://www.thecrimson.com/article/1973/11/16/america-gets-the-shaft-pthe-1972/, accessed May 12, 2014.
77. Malone LA. Reflections on the Jeffersonian ideal of an agrarian democracy and the emergence of an agricultural and environmental ethic in the 1990 Farm Bill. *Faculty Publication,* 1993. Available at http://scholarship.law.wm.edu/facpubs/597, accessed May 12, 2014.
78. Paarlberg RL. Lessons of the grain embargo. *Foreign Affairs,* Fall 1980. Available at http:// www.foreignaffairs.com/articles/34274/robert-l-paarlberg/lessons-of-the-grain-embargo, accessed May 10, 2014.
79. Meyers WH, Ryan ME. The farmer-owned reserve: Is the experiment working? *Am. J. Agric. Econ.* 1981;63:2.
80. Chambers RG, Foster WE. Participation in the farmer-owned reserve program: A discrete choice model, *Am. J. Agri. Econ.* 1983;65:1.
81. Hauter W. *Foodopoly: The Battle over the Farming in America.* New York: The New Press, 2012.
82. Food and Agriculture Organization of the United Nations. *The Review of the 1996 Farm Legislation in the United States.* Available at http://www.fao.org/docrep/w8488e/ w8488e04.htm, accessed May 12, 2014.
83. White JS. Straight talk about high-fructose corn syrup: What it is and what it ain't. *Am. J. Clin. Nutr.* 2008;88(6):1716S–1721S.

84. Goone A.U.S. Sugar Protectionism, *Dartmouth Business Journal,* May 11, 2011. Available at http://dartmouthbusinessjournal.com/2011/05/u-s-sugar-protectionism/, accessed May 13, 2014.

85. 21CFR3, 2014.

86. U.S. Food and Drug Administration. Generally Recognized as Safe (GRAS), last updated February 12, 2014. Available at http://www.fda.gov/food/ingredientspackaginglabeling/gras/default.htm.

87. Department of Health and Human Services, 21 CFR Parts 182 and 184. Direct Food Substances Affirmed as Generally Recognized as Safe; High Fructose Corn Syrup, Food and Drug Administration, Final Rule. Available at http://www.gpo.gov/fdsys/pkg/FR-1996-08-23/html/96-21482.htm, accessed May 13, 2014.

88. Bray GA, Nielsen SJ, Popkin BM. Consumption of high-fructose corn syrup in beverages may play a role in the epidemic of obesity. *Am. J. Clin. Nutr.* 2004;79:(4):537–543.

89. Ventura EE, Davis JN, Goran MI. Sugar content of popular sweetened beverages based on objective analysis: Focus on fructose content. *Obesity.* 2011;19(4):868–874.

90. van Buul VJ, Tappy L, Brouns FJ. Misconceptions about fructose-containing sugars and their role in the obesity epidemic. *Nutr. Res. Rev.* 2014;27(1):119–130.

91. U.S. Food and Drug Administration, Response to Petition from Corn Refiners Association to Authorize "Corn Sugar" as an Alternate Common or Usual Name for High Fructose Corn Syrup. Available at http://www.fda.gov/aboutFDA/CentersOffices/OfficeofFoods/CFSAN/CFSANFOIAElectronicReadingRoom/ucm305226.htm, accessed September 6, 2014.

92. U.S. Energy Information Administration, Ethanol blending provides another proxy for gasoline demand, October 7, 2013. Available at http://www.eia.gov/todayinenergy/detail.cfm?id=13271, accessed May 14, 2014.

93. Congressional Budget Office, Ethanol, Food Prices, and Greenhouse-Gas Emissions. Available at http://www.cbo.gov/publication/24883, April 8, 2009, accessed May 14, 2014.

94. United States Department of Agriculture, Economic Research Service, last updated May 14, 2014. Available at http://www.ers.usda.gov/media/866543/cornusetable.html.

95. Plumer B. Corn and soy wiping out America's grasslands at fastest pace since the 1930s. *The Washington Post,* February 20, 2013. Available at http://www.washingtonpost.com/blogs/wonkblog/wp/2013/02/20/biofuel-craze-wiping-out-americas-grasslands-at-fastest-rate-since-the-dust-bowl/, accessed May 14, 2014.

96. Food and Agriculture Organization of the United Nations. *The State of Food and Agriculture,* 2012. Available at http://www.fao.org/docrep/017/i3028e/i3028e.pdf, accessed May 14, 2014.

97. Bobrow-Strain A. *White Bread: A Social History of the Store-Bought Loaf.* Massachusetts: Beacon, 2012, 35.

98. Winders B. *The Politics of Food Supply, U.S. Agricultural Policy in the World Economy.* New Haven: Yale, 2012, Preface.

99. Food and Agriculture Organization of the United Nations, World Food Situation, FAO Food Price Index, last release update May 8, 2014. Available at http://www.fao.org/worldfoodsituation/foodpricesindex/en/.

100. Alexandratos N, Bruinsma J. World agriculture towards 2030/2050: The 2012 revision. ESA Working paper No. 12-03. Rome, FAO. Available at http://www.fao.org/docrep/016/ap106e/ap106e.pdf, accessed April 12, 2014.

101. Ray DK, Mueller ND, West PC, Foley JA. Yield trends are insufficient to double global crop production by 2050. *Plos One.* June 2013. Available at http://www.plosone.org/article/info%3Adoi%2F10.1371%2Fjournal.pone.0066428#pone.0066428-FAO3, accessed April 12, 2014.

102. National Oceanic and Atmospheric Administration, National Climatic Data Center. Available at https://www.ncdc.noaa.gov/billions/events, accessed May 14, 2014.

103. Boehm D, Sayler T. Wheat industry yields rich history. *Prairie Grains Magazine,* 46 2002. Available at http://www.smallgrains.org/springwh/June02/rich/rich.htm, accessed May 14, 2014.
104. Ashbrook T. The great greening of the global North, *WBUR,* November 12, 2013. Available at http://onpoint.wbur.org/2013/11/19/climate-change-corn-belt-north-dakota, accessed May 14, 2014.
105. U.S. Food and Drug Administration. Food Facts. Gluten and Food Labeling: FDA's Regulation of "Gluten-Free" Claims. Updated June 5, 2014. Available at http://www.fda.gov/Food/ResourcesForYou/Consumers/ucm367654.htm, accessed June 17, 2014.
106. U.S. Food and Drug Administration, For Consumers. What is Gluten-Free? Available at http://www.fda.gov/forconsumers/consumerupdates/ucm36069.htm, accessed September 20, 2014.
107. U.S. Food and Drug Administration, Guidance for Industry: A Food Labeling Guide (11 Appendix C: Health Claims) January 2013. Available at http://www.fda.gov/Food/GuidanceRegulation/GuidanceDocumentsRegulatoryInformation/LabelingNutrition/ucm064919.htm, last updated November 22, 2013, accessed November 19, 2014.

Fruits and Vegetables

10

Eat real food. Not too much. Mostly plants.
Michael Pollan[1]

INTRODUCTION

This chapter examines the meaning of the words "fruits" and "vegetables" and the recommendations for their consumption. We look at how "food group" advice treats fruits, vegetables, and legumes, and review the health benefits of fruit and vegetable consumption. We examine dietary recommendations for fruits and vegetables during the past century, including the genesis and validity of the campaign to have Americans fill half their plates with fruits and vegetables. We look at the promotion of fruits and vegetables by industry and the genesis of the catch phrases "Fruits & Veggies—More Matters®," "5 A Day,"[2] and "Make half your plate fruits & vegetables." We consider how government promotes consumption of fruits and vegetables through dietary recommendations, agricultural subsidies, and nutrition assistance programs, and question why government does not do more to help Americans achieve Healthy People 2020 fruit and vegetable consumption objectives.

Is It a Fruit or Vegetable?

In the introduction chapter, we discuss the importance of defining "food." In this chapter, we consider the relevance of defining "fruits" and "vegetables." The Supreme Court weighs in on defining these terms, as does the U.S. Department of Agriculture (USDA), and the crafters of the Farm Bill. Botanically, tomatoes are classified as fruits. However, according to culinary and common usage and, surprisingly, as a matter of law, tomatoes are deemed to be vegetables.

Not all U.S. states are in agreement as to how to classify tomatoes. Currently, three states recognize the tomato as an official state symbol: the Creole tomato is the state vegetable of Louisiana, the generic tomato is the state fruit of Tennessee, and the S. Arkansas pink tomato is both the state fruit and vegetable of Arkansas.[3] In 2005 a bill designating the tomato as the official New Jersey state vegetable was introduced, but for various reasons unrelated to nomenclature, the bill was never approved.

So, how does one respond to the question, "Is a tomato a vegetable or a fruit when even the States cannot agree on how to classify them?" Predictably, we look to the USDA for the answer. Unexpectedly, however, it is the U.S. Supreme Court that has the final say.

The U.S. Department of Agriculture

The USDA defines horticulture as the branch of agriculture concerned with *intensively cultivated plants* that are *used by people for food* [emphases added], medicinal purposes, and aesthetic gratification. Thus, horticultural crops are differentiated from other crops by the level of management employed in their production and by their subsequent use.[4] This chapter is limited to plants that are grown for human consumption—fruits, vegetables, and tree nuts. Beyond the scope of this book is a discussion of crops used as herbs and spices, for medicinal purposes, and for aesthetic purposes (floriculture).

The USDA describes vegetables as "herbaceous plants of which some portion is eaten raw or cooked during the main part of a meal," whereas fruits are "plants from which a more or less succulent fruit or closely related botanical structure is commonly eaten as a dessert or snack." Despite the fact that the edible portion of the tomato, squash, and cucumber is defined botanically as a fruit, they are considered by the USDA and, in fact, most of us as vegetables because they are customarily eaten during the main part of the meal.[4] But, the story does not end here.

The U.S. Supreme Court

The idea that plants could be delineated as fruit or vegetable by common usage rather than by botany was made legal doctrine at the end of the nineteenth century by the unanimous U.S. Supreme Court decision in *Nix vs. Hedden*, 149 U.S. 304 (1893).[4]

The Tariff Act of March 3, 1883, required a 10% tax to be paid on vegetables entering the United States, but allowed fruit to enter tax-free. The litigants (Nix, et al.) were merchants who imported tomatoes and other crops from the West Indies. They sued a New York port tariff collector (Hedden) to recoup back duty paid on tomatoes, claiming that tomatoes are botanically fruits and therefore should not be taxed. The Supreme Court disagreed, and unanimously decided in favor of the tariff collector. The justices found that, under the customs regulations, the tomato should be classified as a vegetable, based on the ways in which it is used, and the general perception of the tomato as a vegetable. Thus, the tomato was declared a vegetable as a matter of law. In handing down its ruling, the Court said:

> Botanically speaking, tomatoes are the fruit of a vine, just as are cucumbers, squashes, beans, and peas. But in the common language of the people, whether sellers or consumers of provisions, all these are vegetables which are grown in kitchen gardens, and which, whether eaten cooked or raw, are, like potatoes, carrots, parsnips, turnips, beets, cauliflower, cabbage, celery, and lettuce, usually served at dinner in, with, or after the soup, fish, or meats which constitute the principal part of the repast, and not, like fruits generally, as dessert.[5]

Taking our cue from the Supreme Court, tomatoes are vegetables in this book. But, how do we classify legumes?

Legumes

Edible legumes include beans, peas, lentils, peanuts, and soybeans. Because of their rich protein content, they are a mainstay of many vegetarian diets. While lacto-ovo vegetarians may rely on dairy products and eggs for the preponderance of their protein, the richest source of protein for vegans is legumes.

A legume is a fruit or a seed of plants in the family *Leguminosae*. Legumes that are cultivated for their seeds are also known as pulses, examples of which are dry peas, lentils, and peanuts (which, despite their name, are not nuts) and dry beans like pinto beans, kidney beans, and navy beans. Legumes used to be considered an "incomplete protein." See Box 10.1 for a discussion on "incomplete proteins."

Based on their protein content, legumes have traditionally been part of the protein-rich meats, fish and poultry food group. See Chapter 3, "Dietary Guidance," for more information on the history of "food group" guidance in the United States. For purposes of our discussion in this chapter, we will start with the USDA's "Basic Four" guidance initiated in 1956. The "Basic Four" plan recommended a minimum number of servings of foods from four food groups as the basis for a nutritionally adequate diet. The "Basic Four" consists of milk, meat, cereal and breads, and vegetables and fruits (legumes were included in the meat category).

The "Basic Four" was indeed basic, perhaps too basic, as it lacked guidance on fats, sugar, sodium, and maximum number of servings from each food group, instead making only minimum recommendations. It recommended a minimum of two servings from the milk group, a minimum of two servings from the meat group, a minimum

---

BOX 10.1   PROTEIN-COMBINING

In 1979, Frances Moore Lappé introduced the "protein-combining" theory in her seminal book, *Diet for a Small Planet* (see Chapter 5). According to Lappé, two *plant* sources of protein with different amino acid profiles could complement each other, thus producing a complete amino acid pool that is necessary for protein synthesis. The route to this "protein complementarity" was eating the two "incomplete protein" sources at the same meal. For example, only when legumes were combined with a complementary protein source, such as rice (*arroz con gondules*) would the meal provide all the essential amino acids in the right proportions to support protein synthesis. "Protein combining" has lost favor, even with Lappé, who retracted the theory in subsequent editions of her book. But, for years after the theory was proposed, "protein combining" was included in introductory nutrition textbooks and was taught in introductory nutrition college courses throughout the country. With dismay, the second author of this book remembers teaching her introductory nutrition students how to "combine proteins."

BOX 10.2   ENERGY DENSITY OF FOOD

The *mass* of an object is the amount of matter it contains while its *volume* is the amount of three-dimensional space it takes up. "One-half of the plate should be fruits and vegetables" means that fruits and vegetables should make up one-half of the space on the plate (*not* one half of the weight of the food on the plate). *Density* equals mass divided by volume. When the mass stays the same but the volume increases, the density decreases.

Energy density is the number of calories (energy) in a given amount (volume) of food. Fruits and vegetables in their natural state tend to be low in energy density due to their high water content (mass) which takes up a lot of space (volume). For weight control, good choices would be healthy foods like fruits and vegetable, particularly when fresh. They deliver a small amount of energy relative to their high volume. By choosing foods that are low in calories, but high in volume, in theory, one can eat more and feel fuller on fewer calories. Fruits and vegetables in their natural state tend to be low in energy density, which is why they are known as "low-calorie foods."

In terms of mass, volume and density explain why one cup of grapes has a lower energy density than one cup of raisins.

of four servings from the vegetables and fruit group, and a minimum of four servings from the cereals and bread group. Lacking guidance as to what constituted a serving and lacking any upward bounds that would provide proportionality resulted in guidance that heavily tipped the scales toward animal products. In terms of energy density (mass and volume), the recommended intake of animal products far outweighed foods from the plant kingdom. See Box 10.2 regarding the energy density of food.

The "Basic Four" grouping lasted until the mid-1980s, with legumes classified as a component of the meat, fish, and poultry group. In 1984, with the introduction of the six food-group Food Guide Pyramid, the Fruits and Vegetables group was split into two separate entities. Legumes maintained their status within meats, fish, and poultry, but also became a component of the vegetable group. With the 2011 launch of the five food-group MyPlate (replacing MyPyramid), fruits were reunited with vegetables into a single fruits and vegetables group.

Currently, legumes are classified as components of meats, fish, and poultry *and* fruits and vegetables. Legumes occupy a unique niche because they are the only foods that are considered components of more than one food group. In addition to being the only plant products in the protein-rich meat, fish, and poultry group, legumes are also included in fruits and vegetables. Legumes are grouped with fruits and vegetables based on their botany and nutrient content.

In a single meal, however, nutritionists count legumes as representing meat, fish, and poultry *or* fruits and vegetables, but not both. A typical serving of cooked lentils (about one cup plus 2½ tablespoons) from the fruits and vegetables group contains roughly the same amount of protein as a typical 3-oz serving of boneless chicken breast from the meats, fish, and poultry group.

In our meat-centric society, legumes are often referred to as "meat alternates." In that same vein, vegans might refer to meat as "legume alternates."

Benefits of Eating Fruits and Vegetables

Fruits contain fiber and the sugars glucose, fructose, and sucrose. Most fruits are relatively low in calories due to high water and fiber content and low content of fat. They are rich sources of vitamin C, folate, and potassium. Vegetables provide fiber and have varying amounts of starch and only small amounts of sugar. They are important sources of at least 20 micronutrients, including potassium, folate, vitamins A and E, and, in the special case of legumes, protein (a macronutrient).

The benefits of fruits and vegetables go beyond their nutritional contribution to the diet. Adults whose fruit and vegetable consumption meets the recommended daily intake will likely be consuming amounts that are associated with a decreased risk of such chronic conditions as cardiovascular diseases,[6–8] type 2 diabetes,[9] and certain types of cancer.[10] In addition, fruit and vegetable eaters *may* be better able to achieve and sustain a healthy body weight than those who consume less produce.[11,12] Consuming less than five servings a day of fruits and vegetables is associated with progressively shorter survival and higher mortality rates.[13]

FRUIT AND VEGETABLE DIETARY RECOMMENDATIONS

Since at least the 1940s, the official U.S. government-sanctioned recommendation regarding the intake of fruits and vegetables has always been to eat more of them. Until the 1970s, the over-riding message in food guidance for Americans was to eat enough of a balanced diet to eliminate hunger and prevent malnutrition. Many recommendations called for eating *at least* a certain amount of food from each of the food groups. However, with the advent of the first edition of the Dietary Goals (1977), the message shifted from "eat more" to "eat less" *except fruits and vegetables*.[14] The goal of the changed thrust is to eat a balanced diet and avoid overweight (or if overweight, reduce it) to decrease the risk of developing diet-related chronic diseases, often associated with excess body fat.

Fruits and vegetables are perennial players in federal dietary guidance. See Table 10.1 for a snapshot of U.S. dietary recommendations for fruit and vegetable consumption since 1977.

The 1990 and 1995 editions of the Guidelines contain identical recommended daily intakes for fruits and vegetables: two to four servings of various fruits and three to five servings of various vegetables. The minimum recommended intake from 1990 through 2000 was five servings per day, which led to the 5 A Day campaign discussed further.

The 2000 Guidelines recommend a broader range of fruit and vegetables intake, five to nine servings each day, with recommendations covering a range of daily energy intakes of 1200–3200 calories. The recommendation for someone consuming 2200

TABLE 10.1   Official U.S. dietary recommendations for fruits and vegetables, 1977–2014

1977 *Dietary Goals for the United States*
U.S. Senate Select Committee on Nutrition and Human Needs
Increase the consumption of complex carbohydrates and "naturally occurring" sugars
  from about 28% of energy intake to about 48% of energy intake
Increase consumption of fruits and vegetables

1988 The Surgeon General's Report on Nutrition and Health
HHS Public Health Service
Increase consumption of vegetables (including dried beans and peas) and fruits

1980–2014 *Dietary Guidelines for Americans*
USDA and HHS

- 1980: Select foods that are good sources of fiber and starch, such as *fruits and vegetables, beans, peas, and nuts.* Consuming fruits and vegetables is suggested as a strategy for attaining three of the seven guidelines presented in the first edition of the *Dietary Guidelines for Americans.* (Henceforth, nuts are included as a component of the Meat, Fish & Poultry food group.)
- 1985: Choose foods that are good sources of fiber and starch, such as *fruits, vegetables, and dry beans and peas.*
- 1990: Choose a diet with plenty of vegetables and fruits. Eat more vegetables, including dry beans and peas, fruits.
- 1995: *Choose a diet with plenty of vegetables and fruits.* Most calories should come from grain products, vegetables, and fruits. Plant foods provide fiber, and a variety of vitamins and minerals essential to health.
- 1995: *Choose a diet with plenty of vegetables and fruits.* Most calories should come from grain products, vegetables, and fruits. Plant foods provide fiber, and a variety of vitamins and minerals essential to health.
- 2000: Choose a variety of fruits and vegetables daily.
- 2005: Increase daily intakes of fruits and vegetables.
- 2010: Increase vegetable and fruit intake. Eat a variety of vegetables, especially dark-green and red and orange vegetables and beans and peas.
- 2015:_____?

calories was seven servings per day, namely three servings of fruit and four servings of vegetables. Paradoxically, however, the "advice for today" section still advised to "Enjoy five a day—eat at least 2 servings of fruit and at least 3 servings of vegetables each day."

In the 2005 edition of the Guidelines, the recommended range of fruit and vegetable intake was increased to 5–13 servings (2½ to 6½ cups) each day, for daily energy intakes of 1200–3200 calories. For someone consuming 2200 calories, the recommendation amounts to 10 one-half cup servings per day, namely 4 servings (2 cups) of fruits and 6 servings (3 cups) of vegetables.[15]

The 2010 Guidelines represent a shift from recommending specific quantities of fruits and vegetables (as presented in previous editions of the Guidelines) to directional intake of fruits and vegetables. Key Recommendations (described as the most important in terms of implications for improving public health) include "increase vegetable and fruit intake," and "eat a variety of vegetables," especially dark-green and red and orange vegetables and beans and peas. However, specific quantities are still recommended in Appendix 7, "USDA Food Patterns." At the 2200 daily calories level, the recommended daily intake remains the same as for 2005: two cups of fruit and three cups of vegetables.[16]

Information in the 2015 *Dietary Guidelines for Americans* is not available at the time of this writing. When the Guidelines are released, compare the 2015 recommendations for fruits and vegetables to those of previous years.

Fruits and Vegetables Messaging

In 1988, the California Department of Health Services embarked upon 5 A Day for Better Health—a campaign designed to increase fruit and vegetable consumption. This social marketing program was distinctive in approach and penetration. The campaign presented a simple, positive, behavior-specific message—eat five servings of fruits and vegetables every day. That message was promoted widely through use of mass media, partnerships with produce and supermarket industries, and an extensive use of point-of-purchase messages.[17]

The 5 A Day message worked its way into federal dietary guidance. In 1990, the 2000 edition of Healthy People (HP 2000) was released. One population health objective was to "increase complex carbohydrate and fiber-containing foods in the diets of adults to five or more daily servings for vegetables (including legumes) and fruits."[18] "Eat 5–9 servings of fruits and vegetables a day" was a catchphrase based on recommendations in the 2000 edition of the Dietary Guidelines and USDA's Food Guide Pyramid.

5 A Day for Better Health (1991–2007)

In 1991, a partnership was formed between the National Cancer Institute (NCI) and the Produce for Better Health Foundation (PBH) to promote U.S. consumption of fruits and vegetables. PBH leverages resources of private industry and the public sector to influence policy makers, motivate key consumer influencers, and promote fruits and vegetables directly to consumers.[19] PBH describes its mission as "increase[ing] fruit and vegetable consumption for better health,"[20] The role of NCI as a convening member of this partnership speaks to the idea that increased fruit and vegetable consumption might help reduce the incidence of certain types of cancer.

Leveraging the HP 2000 objective of "five or more daily servings for vegetables and fruits" and the groundwork laid by the 1988 California campaign, the PBH named its program "5 A Day for Better Health." By 1994, over 700 industry organizations and 48 states, territories, and the District of Columbia were licensed to participate.[17] The PBH and NCI also launched a targeted campaign to address health disparities. In 2003, the PBH and NCI committed to reaching African-American men aged 35–50 with the message to eat nine servings of fruits and vegetables a day through a multi-year communications effort.[21]

The agency authority changed in 2005 when the CDC replaced NCI as the lead federal agency and national health authority for the 5 A Day program. This change in agency authority was consistent with the notion that fruit and vegetable consumption leads not just to less cancer, but to better health overall. The change in agency also meant that data could be more easily and consistently collected to track progress on the fruit and vegetable consumption objective.

Although PBH is a 501(c)(3) nonprofit organization, it is *largely funded by industry.* Recall from Chapter 7 that Dairy Management, Inc.™ is another food industry group that is associated with the federal government (specifically, the USDA). Just as the mission of Dairy Management Inc.™ is to promote consumption of dairy products, the mission of PBH is to increase consumption of fruits and vegetables. In other words, both dairy and produce consumption is promoted by industry and taxpayer dollars.

Fruits & Veggies—More Matters and Half Your Plate (2007–)

The program messaging from PBH is updated periodically to reflect new dietary recommendations. Accordingly, in 2007, the Fruits & Veggies—More Matters™ program replaced 5 A Day. The new slogan reflected the increased amount of fruits and vegetables recommended in the 2005 Dietary Guidelines, from 5 *servings* in 2000 to 2 to 6½ *cups* in 2005. Depending on individual calorie needs, these recommendations based on cups translated to 4 to 13½ "servings" per day of fruits and vegetables. Adopting a more open-ended message would address the ambiguity as to if "5" meant servings or cups, as 5 "servings" a day would actually now be considered insufficient intake for most people. In 2012, the program messaging was updated again to incorporate the advice to make one-half one's plate fruits and vegetables.[22]

As discussed in Chapter 3, "Dietary Guidance," the New American Plate developed by the American Institute for Cancer Research portrays a plate, one-half of which should be filled with fruits and vegetables, including dry peas and beans, one-quarter with starchy foods (potatoes, pasta, and bread), and the remainder with animal products (dairy, fish, chicken, and beef). The "half-plate" theme for fruits and vegetables was echoed at an Advisory Committee meeting for the 2010 Dietary Guidelines when an officer of the United Fresh Produce Association asked the committee to consider operationalizing the recommended intake of fruits and vegetables by using the easy-to-understand concept "half the plate be fruits and vegetables."[23] PBH adopted the half-your-plate concept messaging; however, the 2010 *Dietary Guidelines for Americans* do not specifically make the "half plate" recommendation. Nevertheless, as indicated in Table 10.2, PBH has been successful in keeping its fruit and vegetable theme consistent with the current Dietary Guidelines.

From 2011 through (at least) 2015, the USDA encouraged Americans to "Make Half Your Plate Fruits and Vegetables," as a supporting message of MyPlate, USDA's nutrition guide. The MyPlate icon (see Chapter 3, "Dietary Guidance") illustrates the amount of space of each food group for healthy eating. According to the USDA, excluding dairy, fruits, and vegetables should be the source of one-half of the *volume* of food consumed (the amount of space the food takes up).

From Servings to Cups

The units used to describe desired amounts of food in official dietary guidance have changed over time from servings to cups.

TABLE 10.2   Produce for better health foundation messages are consistent with each new edition of the dietary guidelines, 1990–2010

DIETARY GUIDELINES FOR AMERICANS		PRODUCE FOR BETTER HEALTH	
PUBLISHED	DGA RECOMMENDATION FOR F&V INTAKE PER DAY	YEAR	PBH SLOGAN
1990	At least 5 servings	1991	Five a Day
1995	At least 5 servings	1991	Five a Day
2000	At least 5 servings (5–9) (DGA starts using the "five a day" slogan)	1991	Five a Day
2005	For 2000 calorie intake, 4½ cups (9 servings). (For 1000–3200 calorie intake, 2–6½ cups.) (5–13 daily servings) Recommends that the number of daily servings of fruits and vegetables should reflect one's sex, age, and physical activity level.	2007	Nine A Day Fruits & Veggies— More Matters™
2010	For 2000 calorie intake, 2 cups fruit and 2½ cups vegetables. Operationalized through USDA's MyPlate nutrition guide by the recommendation to make half your plate fruits and vegetables for meals and snacks. Slogan: Half Your Plate	2011	Fill half your plate with fruits and veggies
2015	TBA	2015 or later	TBA

1990–2000: Servings

The 1990 edition of the Guidelines recommended two to four servings of fruits and three to five servings of vegetables. The minimum recommended intake was five servings per day. Thus, the 5 A Day mantra owes its genesis to the 1990 Dietary Guidelines. The 1995 recommendation was identical.

The 2000 Guidelines recommend a broader range of fruit and vegetables intake, five to nine servings each day, depending on calorie intake as discussed earlier in this chapter. Taking its cue from the Five A Day campaign, in the "advice for today" section of the 2000 Guidelines, people are advised to "Enjoy five a day—eat at least 2 servings of fruit and at least 3 servings of vegetables each day." Thus, government used the slogan developed by the produce industry.

2005–2010: Cups and ounces

The 2005 edition of the Guidelines featured several improvements over previous editions, one of which was describing recommendations using cups and ounces to help consumers better understand the quantity of food recommended. The recommended intake increased as well to 2½ to 6½ cups (5–13 servings) because the daily energy

TABLE 10.3   "Cup equivalents" for selected fruits and vegetables

FRUIT OR VEGETABLE		AMOUNT EQUIVALENT TO ONE CUP
Apple	½	Large, 3¼ inch diameter
Grapes	32	Whole, seedless
Orange juice, 100%	½	Cup
Raisins	½	Cup
Strawberries	8	Large
Watermelon	1	Wedge, small, 1 inch thick
Broccoli	3	Spears, 5-inches long, raw or cooked
Celery	2	Large stalks, 11–12 inches long
Corn	1	Large ear, 8–9 inches long
Lettuce	2	Cups shredded or chopped, raw
Spinach	2	Cups, raw
Beans and Peas (such as black, garbanzo, kidney, pinto, black eyed peas, split peas)	1	Cup whole or mashed, cooked

Source: Adapted from USDA. ChooseMyPlate.gov. What Counts as a Cup of Fruit? Available at http://www.choosemyplate.gov/food-groups/fruits-counts.html, accessed November 1, 2014; USDA. What Counts as a Cup of Vegetables? Available at http://www.choosemyplate.gov/food-groups/vegetables-counts.html.

intakes were expanded to 1200–3200 calories. For a reference 2200-calorie intake, that amounts to two cups of fruit (4 servings) and three cups of vegetables (6 servings) per day.[24] At this juncture, PBH replaced Five A Day with its new slogan, Fruits & Veggies—More Matters.

Changing the recommended portions from "servings" to "cups" may have been a step in the right direction, but did it go far enough?

The 2010 edition of the Dietary Guidelines dropped the use of servings and retains the concept of cups, ounces, and "cup equivalents." The recommendations for a reference 2200-calorie intake remained at the same level as 2005: two cups of fruit and three cups of vegetables per day.[25] "Cup equivalents" were developed to provide the weight (roughly 3 oz) of many recommended foods. Some examples of "cup equivalents" are listed in Table 10.3. Note that none of the examples in Table 10.3 measures one cup.

HEALTHY PEOPLE 2020: OBJECTIVES FOR FRUIT AND VEGETABLE CONSUMPTION

Healthy People 2020 objectives for fruit and vegetable consumption are within the Nutrition and Weight Status (NWS) topic. NWS-2.2 is to increase the proportion of school districts to require schools to make fruits or vegetables available whenever other food is offered or sold from 6.6% of districts (as of 2006) to 18.6%. NWS-3 is to increase the number of States that have state-level policies to incentivize food retailers to provide foods "encouraged by the Dietary Guidelines." Although the objective to increase the number of states with such policies from 8 to 18 does not actually describe fruits and vegetables as "encouraged foods," the data source to measure improvement is CDC's State Indicator Report on Fruits and Vegetables. NWS-14's goal is to increase the mean daily fruit consumption in people aged 2 years and older from 0.5 cup equivalent of fruit per 1000 calories (as per NHANES 2001–2004 data) to 0.9 cup equivalent per 1000 calories. NSW-15.1's goal is the same structure as NSW-14, but applied to increasing overall vegetable intake from 0.8 cup equivalent to 1.1 cup equivalent, with an emphasis on increasing dark green vegetables, orange vegetables, and legumes from 0.1 cup equivalent to 0.3 cup equivalent per 1000 calories.[26]

According to the 2010 Dietary Guidelines, the recommended daily intake of fruits and vegetables for someone consuming 2000 calories a day is two cups of fruits and 2½ cups of vegetables. On a simple calculation of "per 1000 calories" basis, this would translate into 1 cup of fruit and 1¼ cup of vegetables, meaning that even our Healthy People objectives may be in a slight shortfall in comparison to Guideline recommendations. (Note, however, that the dietary guidelines for fruit and vegetable intake do not increase or decrease in this direct manner; this simple calculation is done here only as an illustrative exercise.)

Methods to Achieve Our Fruit and Vegetable Consumption Population Health Objectives

In addition to dietary guidance and nutrition messaging aimed at changing individual behavior discussed above, policies and initiatives also exist that promote fruit and vegetable consumption at a structural level by increasing access.

For example, incentive programs have increased the number of low-income shoppers purchasing fresh produce at farmers' markets. In the New York City Health Bucks program, for each $5 of SNAP benefits that a customer spends at a farmers' market, they will receive one Health Buck coupon worth $2 for fresh fruits and vegetables redeemable at any farmers' market in New York City. Philadelphia has adopted a similar approach with its Philly Food Bucks program.[27] Detroit and other parts of Michigan have a Double-Up Food Bucks program.[28] Boston's Bounty Bucks[29] offers similar kinds of benefits.

SNAP incentive programs at farmers' markets increase fruit and vegetable consumption and have a positive impact on local economies.[30] Community supported agriculture (CSA) is another means of marketing directly from the farmer to the consumer. Consumers purchase a "membership" or a "subscription" in the CSA and in return receive a box or bag of seasonal produce that is delivered to a convenient drop-off location each week throughout the farming season. The CSA farmers benefit from this arrangement because they limit their marketing activities to early in the year before the planting and harvesting season, and they receive payment early in the season, which helps their cash flow. Benefits to consumers include access to just-harvested produce and the potential exposure to unfamiliar vegetables and new ways of cooking.[31] However, as farmers often ask their members to pay for the entire season in advance, low-income

---

BOX 10.3   CDC'S STRATEGIES FOR INCREASING FRUIT AND VEGETABLE CONSUMPTION

1. Improve access to retail stores that sell high-quality fruits and vegetables or increase the availability of high-quality fruits and vegetables at retail stores in underserved communities.
2. Establish policies to incorporate fruit and vegetable activities into schools as a way to increase consumption.
3. Promote food policy councils as a way to improve the food environment at state and local levels.
4. Start or expand farm-to-institution programs in schools, hospitals, workplaces, and other institutions.
5. Start or expand farmers' markets in all settings.
6. Start or expand CSA programs in all settings.
7. Ensure access to fruits and vegetables in workplace cafeterias and other food service venues.
8. Ensure access to fruits and vegetables at workplace meetings and events.
9. Support and promote community and home gardens.
10. Include fruits and vegetables in emergency food programs.

Centers for Disease Control and Prevention, State Indicator Report on Fruits and Vegetables, 2013, Atlanta, GA: Centers for Disease Control and Prevention, U.S. Department of Health and Human Services, 2013. Available at http://www.cdc.gov/nutrition/downloads/State-Indicator-Report-Fruits-Vegetables-2013.pdf, accessed August 29, 2014.

people may be excluded from participating in CSAs. One solution is to subsidize participation by grants and contributions from other CSA participants.

Are there other models for low-income participation? Is subsidization a sustainable model?

The CDC has identified 10 approaches designed to increase the consumption of fruits and vegetables through improving access. The strategies are listed in Box 10.3.

Federal Polices Supporting Increased Access to Fruits and Vegetables

Departments of Agriculture, HHS, and Defense all have programs that focus specifically on increasing consumption of fruits and vegetables to promote the health of Americans.

The USDA supports fruit and vegetable consumption through nutrition assistance programs. The USDA provides funds for the purchase and distribution of fruits and vegetables to schools, food banks, and other programs, and use of fruits and vegetables by the Child Nutrition Programs (NSLP, SBP, CACFP, SFSP). Participants can purchase

TABLE 10.4   USDA food and nutrition services programs with components designed to increase fruit and vegetable consumption

PROGRAM	ABBREVIATION	URL
Food Assistance		
Supplemental nutrition assistance program	SNAP	www.fns.usda.gov/snap
Women, Infants, Children, and Seniors		
Special Supplemental Nutrition Program for Women, Infants and Children (WIC)	WIC	www.fns.usda.gov/wic/ women-infants-and-children-wic
WIC Farmers' Market Nutrition Program	FMNP	www.fns.usda.gov/fmnp
Senior Farmers' Market Nutrition Program	SFMNP	www.fns.usda/gov/sfmnp
Child Nutrition		
National School Lunch Program (includes afternoon snacks)	NSL	www.fns.usda.gov/nslp/ national-school-lunch-program-nslp
School Breakfast Program (SBP):	SBP	www.fns.usda.gov/sbp
Summer Food Service Program	SFSP	www.fns.usda.gov/ summer- food-service-program-sfsp
Child and Adult Care Food Program	CACFP	www.fns.usda.gov/cnd/care/
Fresh Fruit and Vegetable Program	FFVP	www.fns.usda.gov/cnd/ffvp
Department of Defense Fruit and Vegetable Program	DoDFresh	http://www.fns.usda.gov/fdd/ dod-fresh-fruit-and-vegetable- program

fruit and vegetables using benefits from SNAP, WIC, and the farmers' market nutrition programs (FMNP and SFMNP). The USDA also provides nutritional education by producing and disseminating messages and materials that encourage consumption of fruits and vegetables.[32] Table 10.4 lists USDA's programs that feature components designed to increase fruit and vegetable consumption.

Supplemental Nutrition Assistance Program

From 2011 through 2013, USDA made $4 million available to FNS for the expansion of its program to improve access to fresh produce and healthy foods by Supplemental Nutrition Assistance Program (SNAP) recipients at America's farmers' markets and direct marketing farmers (roadside stands). The program expands the availability of wireless point-of-sale equipment for farmers' markets and roadside stands not currently accepting SNAP benefits. Funds may be used to purchase or lease equipment or pay for wireless access.[33]

SNAP-Ed is SNAP's nutrition education component. Its goal is to "improve the likelihood that persons eligible for SNAP will make healthy food choices within a limited budget and choose physically active lifestyles consistent with the current *Dietary Guidelines for Americans* and USDA food guidance." Thus, SNAP-Ed's activities include messages to increase fruit and vegetable consumption, which is consistent with the *Dietary Guidelines for Americans* and its associated food guidance system, *MyPlate*. The program has not been vigorously evaluated.[34]

Farmers' Market Nutrition Programs

As described in Chapter 6, some WIC food packages were adjusted to provide more whole fruits and vegetables, but less fruit juice. The WIC Farmers' Market Nutrition Program (FMNP) provides additional benefits aimed at increasing fruit and vegetable consumption. Eligible WIC participants are issued FMNP checks or coupons *in addition to* their regular WIC benefits. These vouchers are used to buy eligible foods (locally grown unprepared fresh fruits, vegetables, and herbs) at farmers' markets and/ or roadside stands that have been approved by the State agency to accept FMNP coupons. The Federal FMNP benefit level of $10 to $30 per year may be supplemented with State, local, or private funds. Fruit and Vegetable Checks issued by WIC may be used at approved grocery stores, but only some States also allow them to be used at farmers' markets.

The Senior Farmers' Market Nutrition Program awards grants to States, U.S. Territories, and federally recognized Indian tribal governments to provide low-income seniors with coupons that can be exchanged for eligible foods (fruits, vegetables, honey, and fresh-cut herbs) at farmers' markets, roadside stands, and CSA programs.

Child Nutrition Programs

The Healthy, Hunger-Free Kids Act of 2010, under the Child Nutrition Reauthorization Bill, authorizes funding for Federal school meal and child nutrition programs and improves access to "healthy" food for low-income children. The bill reauthorizes child nutrition programs for 5 years and includes $4.5 billion in new funding for these programs over 10 years. One of the many objectives of the program is to increase federal reimbursement for school lunches, an undertaking that will offer more access to fruit and vegetables.

Fresh Fruit and Vegetable Program

The nationwide Fresh Fruit and Vegetable Program (FFVP) provides a variety of free fresh fruits and vegetables throughout the school day, exclusive of breakfast and lunch, to children attending eligible elementary schools with the highest free and reduced-price NSLP participation. The goal of the FFVP is to improve the overall diet of children and create healthier eating habits to impact their present and future health. The FFVP was designed to expand the variety of fruits and vegetables children experience as a means to increase their fruit and vegetable consumption and to help schools create school environments that provide healthier food choices. Fruit juice is not permitted in the program. Participating schools receive $50.00 to $75.00 per student for the school year.[35] The FFVP was permanently authorized in the 2008 Farm Bill and provided with $40 million funding in FY 2009, adjusted annually by changes in the Consumer Price Index. For SY 2013–2014, $165.5 million was available to the program.[36]

Students who were part of the FFVP consumed on average one-third of a cup (0.32 cups) more fruits and vegetables on FFVP days than students in comparable schools not participating in the program. FFVP appears to have been especially effective in improving fruit consumption. Comparing students in FFVP schools with their counterparts in nonparticipating schools, there was neither any evidence of a statistically significant difference in total energy intake, nor consistent evidence of differences in intake of foods besides fruits and vegetables. Combined, these findings provide weak evidence that FFVP fruit and vegetable consumption was *in addition to*, rather than in place of, other foods.[37]

Department of Defense Fresh Fruit and Vegetable Program

The Department of Defense Fresh Fruit and Vegetable Program (DoD Fresh) allows schools to use their USDA Foods entitlement dollars to buy fresh produce. Entitlement dollars used for the DoD Fresh program will result in less funds available for the school district to purchase other USDA Foods, but this is compensated for by the advantage of greater buying power utilizing the military's contracted volume purchasing price. Other advantages of the program include an emphasis on high quality and the potential to purchase produce that is sourced from within the state of service or adjacent states, including use of the Farm to School vendors. About 15–20% of the produce DoD provides to schools is currently designated as local. Schools in 46 states, the District of Columbia, Puerto Rico, the Virgin Islands, and Guam participate, with more than $100 million worth of anticipated produce purchases during SY 2012–2013.[38]

Community Development

The Departments of Agriculture, Treasury, and HHS partnered in 2009 to create the Healthy Food Financing Initiative (HFFI). HFFI's aim is to improve healthy food access, and, through improved food retail channels, support the economic backbone of communities with local jobs. Assistance is provided to community development financial institutions, other nonprofits, and businesses with plausible strategies for addressing the healthy food needs of communities. The initiative makes available a mix of federal tax credits, below-market rate loans, loan guarantees, and grants to attract private sector capital. Modeled on the successful Pennsylvania Fresh Food Financing Initiative, HFFI is an example of how food system activism can start local and have national results.

The 2014 Farm Bill formally established HFFI at the USDA and authorized up to $125 million in funding.[39]

National Council of Fruit and Vegetable Nutrition Coordinators

The National Council of Fruit & Vegetable Nutrition Coordinators serves as an organized voice in decisions that affect national fruit and vegetable health promotion campaigns. To the extent possible, each state, territory, and the District of Columbia have one person designated as the Fruit and Vegetable Nutrition Coordinator. Coordinators manage a wide range of activities, including working on programs to improve access to fruits and vegetables for low-income populations. The Council is a voluntary membership-based initiative. Coordinator positions may be underwritten through universities, a variety of federal and state funding sources, and state agencies, such as health departments. All coordinators are members of the Association of State and Territorial Public Health Nutrition Directors.[40]

How Much Fruit and Vegetables Are We Eating?

Given all the resources government and industry have allotted for increasing consumption of fruits and vegetables, have Americans actually increased their intake of foods in the Fruits and Vegetables Food Group? To answer this question, we look to reports of fruit and vegetable consumption from three major national surveys—the National Health and Nutrition Examination Survey (NHANES),[41] the Behavioral Risk Factor Surveillance System (BRFSS),[42] and the Youth Risk Behavior Surveillance System (YRBSS).[43]

Despite recommendations from the public and private sectors, Americans are currently *not* eating more fruits and vegetables now than in the recent past. According to individual 24-hour recalls, fruit and vegetable consumption did not increase between 1988–1994 (NHANES III) and 1992–2002 (NHANES).[44] Indeed, CDC data suggest that from 2000–2009, average fruit consumption actually declined and vegetable consumption remained the same, with both levels falling considerably below the desired targets.[45]

The 2007 BRFSS and YRBSS indicate that only a third of adults met the recommendation for fruit consumption and slightly more than one-quarter (27%) met the recommended servings of vegetables. High school students reported an even worse diet—just under a third (32%) were eating at least two servings of fruit daily and only 13% were eating at least three servings of vegetables each day.[46]

The 2009 BRFSS results indicate that about one-third (32.5%) of adults consumed fruit two or more times per day and one-quarter (26.3%) consumed vegetables three or more times per day, far short of national recommendations. Note, however, that the number of times per day that fruit is consumed is used as a proxy for the quantity consumed. A person who reports having eaten fruit two or more times per day is presumed to have eaten two fruit servings.

In terms of change, from 2000 to 2009, there was no significant change in the proportion of adults who met the vegetable target, while there was a small but statistically significant decline in fruit consumption from 34.4% to 32.5% during the same period.[45] Note that the BRFSS revised its methodology in 2011. Starting that year, estimates of fruit and vegetable intake cannot be compared to previous years' estimates. Thus, 2011 became the new baseline for assessing changes in fruit and vegetable consumption.

The 2010 Dietary Guidelines suggest a range of intake from 2½ to 6½ cups of fruits and vegetables daily, depending on energy requirements.[47] Based on 2008 data, the PBH reported in 2010 that Americans were eating, on average, less than one-half of the recommended quantity of fruit per day and slightly more than one-half of the recommended quantity of vegetables.[48] According to CDC data, in 2013, adults in the United States had a median intake of fruit and vegetables only about 1.1 and 1.6 times per day, respectively.[49]

Are We Spending Enough? Are We Producing Enough?

A 2010 PBH report compared the 2005 Dietary Guidance priorities on food group recommendations to USDA spending on food groups. Fruits and vegetables together constitute the highest priority at 41.4% of recommended food servings, whereas meat accounts for 8.3% of total servings—the lowest share of servings. However, funding was allocated conversely: 54.7% of USDA commodity spending was allocated to meat and only 9.8% to fruits and vegetables, and commodity administrative costs were similar at 59.9% for meat and 6.3% for fruit and vegetables. The allotment of research spending still heavily favored meat at 53.6%, but 24.1% went to fruits and vegetables. Nutrition assistance programs, conversely, did spend more money on fruits and vegetables (35.9%) than on any other group. However, even with the boost from the nutrition assistance program spending, total USDA overall spending on fruit and vegetables was just under 20%. Although $3.6 billion sounds like a great deal of money, this 3.6 billion spent on fruit and vegetables would have to be double to align USDA food group spending with Dietary Guidelines food group priorities.[48]

In 2005, the USDA's Economic Research Service reported that the 2003 food supply provided 1.4 daily servings of fruit per capita and 3.7 daily servings of vegetables,[50] which was a slight decrease since 2000 in the amount of vegetables and the same amount of fruit.[51] The situation has not improved. The United States does not produce enough fruit and vegetables for each of us to be able to eat the USDA's 2010 Dietary Guidelines recommended servings of fruit and vegetables without relying on imported produce. We have only six cups of fruit available to meet the weekly recommendation of 14 cups of fruit and just over 11 cups of vegetables available to meet the weekly recommendation of just over 24 cups of vegetables.[52]

Sales of fruits, vegetables, and tree nuts account for nearly one-third of U.S. crop cash receipts and one-fifth of U.S. agricultural exports. However, despite their relatively large share of crop receipts, as of 2007, these "specialty crops" occupy only about 3% of U.S. harvested cropland.[53] See Appendix to this chapter for a list of specialty crops.

If we need more fruits and vegetables on our plates, *why are fruits and vegetables planted on only 3% of U.S. farmland?* The only significant federal funding of fresh produce is that of apples.[54] In short, the incentives of our agricultural policies are at odds with our dietary guidance.

Since 1996, farm bills traditionally have treated fruits and vegetables as specialty crops, which lack the subsidy incentives for production that are available to commodity crops. Although some fruits and vegetables are truly consumed only as specialty items, one might ask what is so special about apples, onions, and carrots?

Conversely, farmers that grow commodity crops are largely restricted from planting fruits and vegetables on acreage reaping government subsidies. The 2014 Farm Bill has kept this planting restriction. Growers of fresh fruits and vegetables (especially, vegetables) support this restriction, arguing that because specialty crop growers do not get subsidies, it would be unfair if other farmers who do get subsidies started to compete with them by growing specialty crops. On the other hand, to expand the production of healthy food, the Union of Concerned Scientists (UCS), a 501(c)(3) charitable organization that encourages innovative and environmentally sustainable ways to produce high-quality, safe, and affordable food, recommends the federal government to eliminate restrictions that provide disincentives for farmers to plant fruits and vegetables.[55] In either case, Americans would eat better if more of our agricultural land supported the production of fruits and vegetables rather than the commodity crops.

Given the health benefits of fruit and vegetable consumption, and the continuing mandate to eat more, should there be restrictions on planting fruit and vegetables? What would happen if farmers who get subsidies for a commodity crop also grew fruit or vegetables on the subsidized acreage?

In lieu of subsidies, the USDA offers support to fruit, vegetable, and tree nut farmers by directly purchasing crops and then transferring the crops for use in domestic nutrition and food assistance programs. Improving the diet of those vulnerable to hunger and malnutrition does help achieve population health goals.

Also, specialty crop farmers, like commodity farmers, are eligible for federal crop insurance. Crop insurance, authorized in 1938 as part of the first farm bill to mitigate the effects of the Great Depression and the Dust Bowl, has become the most important crop subsidy program in the United States. Government costs have escalated. During FY 2000–2007, crop insurance costs ranged from $2.1 billion to 3.9 billion, but escalated to $7 billion in 2009. Rising crop prices meant premiums were higher, which in turn caused increased premium subsidies to farmers and greater expense reimbursements to private insurers. Poor weather resulting in crop loss and high prices cost the federal government $11.3 billion in FY2011 and $14.1 billion in FY2012. With better weather and smaller crop losses, costs in FY2013 dropped back to $6 billion. Many specialty crop producers depend on crop insurance as the only *safety net* for their operation, and about 73% of specialty crop acreage is insured.[56]

2014 Farm Bill Specialty Crop Highlights

The 2014 Farm Bill does increase our investment into fruits and vegetables from $676 million mandatory funding to $739 million, with an additional $257 million as discretionary. Authorized USDA funding for the FFVP program is increased from $101 million to $150 million. Funding is also increased to $75 million on the Specialty Crop Block Grants, focusing on local markets and safe food production. Funding is maintained for the Seniors Farmer Market Nutrition program through FY2018. The majority of support for the specialty crop industry comes through purchases made under USDA's nutrition assistance programs, with $406 million for "Section 32 purchases" (surplus purchased for use in domestic nutrition assistance programs). The 2014 Farm Bill also introduced the Food Insecurity Nutrition Incentive program to establish projects that incentivize SNAP participants to buy fruits and vegetables.[57]

Food Safety

Although produce was previously considered an unlikely culprit, produce was either the first or second leading source of foodborne illness in the United States between 2006 and 2008. Nonprocessed foods viewed as healthy—spinach, lettuce, cantaloupes—have become significant carriers of pathogens. Commodity produce as a group (fruits, nuts, and five vegetables) is the leading cause of foodborne illness in the United States, and of all items, more illnesses were attributed to leafy greens than any other item.[58]

The major source of microbial contamination of fresh produce is associated with human or animal feces. The use of irrigation water that contains animal or human feces and deposition of feces by grazing livestock and wild animals can contaminate the crops with toxin-producing *Escherichia coli* 0157:H7 (*E. coli* 0157:H7), an organism that lives in the intestines of humans and animals. Although most strains of *E. coli* are harmless, *E. coli* 0157:H7 produces toxins that can cause severe acute hemorrhagic diarrhea, and in a small number of cases, hemolytic uremic syndrome and possibly kidney failure.[59]

The Food Safety Modernization Act of 2010 (FSMA), P. L. 111–353, aims to shift U.S. food safety practice from responding to outbreaks to preventing contamination.[60] The FSMA is the first major change in U.S. food safety laws since 1938, when the Federal Food, Drug, and Cosmetic (FDC) Act was passed by Congress. The law's standards for fresh produce call for identifying routes of microbial contamination of produce and set requirements to prevent or reduce the introduction of pathogens.

These are some of the provisions required by P.L. 111–353:

- Farmers are required to keep records documenting that the agricultural water they use has been tested and contains not more than the allowable limits of *E. coli*. If it does, the farmer must immediately discontinue using that water.
- There are specific requirements for sprouts, which present a unique risk because the warm, moist, and nutrient-rich conditions required to produce sprouts are the same conditions that are also ideal for the growth of pathogens.
- Personnel who handle covered produce and their supervisors are required to receive training that includes principles of food hygiene and food safety, and health and personal hygiene.

In 2013, the FDA published in the Federal Register proposed standards for implementing the produce section of FSMA.[61] Final standards are expected by 2015.

Finally, a chapter about fruits and vegetables could only be complete if it considered vegetarianism.

Vegetarianism

It is a vegetarian diet composed solely of vegetables? Of course it is not; a vegetarian food plan is based on foods from plants, which, in addition to vegetables, includes fruits, grains, legumes, peanuts, and plant oils. *Shouldn't we all have diets based primarily on foods from the plant kingdom?* According to official food guidance from the federal government, notably the USDA, DHHS, and from not-for-profit organizations devoted to health, the answer is *Yes*.

If planned well, vegetarian diets are nutritionally adequate. Additionally, they may help in the prevention and treatment of essential hypertension, hypercholesterolemia, heart disease, and type 2 diabetes. Vegetarians have lower overall cancer rates and lower body mass indexes (BMIs). Long-time (more than 5 years) adherence to all kinds of vegetarian diets is associated with lower BMI, although rigorous prospective longitudinal studies are still needed to determine cause and effect.[62-64]

There are several kinds of vegetarian diets, characterized as to the extent to which animal products are avoided.

- Vegan. Diet is based on grains, fruits, vegetables, legumes, seeds, and nuts. Excluded from the vegan diet are all animal products and sometimes honey. People choose to be vegan for health, religious, economic, environmental, or ethical reasons.
- Lactovegetarian. Diet includes all vegan foods, plus dairy products. Excluded from the lacto-vegetarian diet are eggs, meat, fish, and poultry.
- Lacto-ovo-vegetarian. Diet includes all vegan foods, plus dairy products and eggs. Excluded from the lacto-ovo vegetarian diet are meat, fish, and poultry.
- Flexitarian (semi-vegetarian). Diet includes mostly plants, but also incorporates dairy, eggs, meat, fish, and poultry occasionally, when it is more convenient, or on special occasions. This eating style is for those who are moving their meat-centric diets in a more vegetarian direction, as well as vegetarians who are adding some animal products back into their meals.[65]

U.S. Dietary Guidelines and Vegetarianism

It is not a new idea that we should place more emphasis on food from plants rather than animals. As discussed in Chapter 3, "Dietary Guidance," the USDA has been recommending diets that rely more on plants than animals on and off for the past century. However, it was not until 1977 that such a recommendation became official. In that year, the Senate Select Committee on Nutrition and Human Needs *intimated* that Americans should *approach vegetarianism*. The 1977 Dietary Goals recommended decreasing the percent of calories from fat (from about 40–30% of energy intake) and increasing the percent of energy from complex carbohydrates and "naturally occurring sugars" (from about 28% to about 48%) while maintaining the percent of energy from protein (12%). The only way to achieve that goal would require changing the source of protein from animal products to plants. Yet, the Dietary Goals never uses the word "vegetarian" or any of its derivatives (such as "vegetarianism"), nor does the phrase "plant-based diet" appear anywhere in the report.

In 1980, a U.S. Senate committee on appropriations directed the establishment of a committee to review scientific evidence and recommend revisions to the Dietary Guidelines. The Dietary Guidelines Advisory Committees (DGACs) gave short shrift to vegetarian diets until 2010. In 2010 the value of vegetarianism was discussed in the DGAC Report to the USDA and DHHS,[66] and the DGAC report recommends a shift in food intake patterns to a more "plant-based diet."[66] In *Diet for a Small Planet* (1971), Frances Moore Lappé called this "eating lower on the food chain."[67] Ultimately, plant-based diets were sanctioned in the 2010 Dietary Guidelines.

The 2010 Dietary Guidelines include lengthy descriptions of how and why to incorporate plant foods in one's diet, with recommendations to consume 2½ cups of fruits and 6½ daily. The 2010 Guidelines also provide lacto-ovo vegetarian and vegan eating plans and an appendix that contains the quantity of food from each food group necessary for lacto-ovo vegetarians and vegans at the 1000–3200 calorie levels.

In January 2011, the same month the 2010 Guidelines were published, PBH operationalized the DGAC's recommendation for 4½ cups of fruits and vegetables by suggesting that (excluding dairy), one-half of every meal and snack we eat should come from the Fruit and Vegetable Group. In June 2011, the USDA released the MyPlate icon that depicted just what the PBH recommended 6 months earlier—that fruits and vegetables comprise one-half of the plate, as indicated in MyPlate (see Figure 10.1).

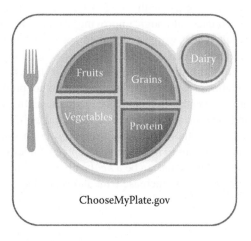

FIGURE 10.1 USDA's MyPlate icon depicts fruits and vegetables as one half of an ideal plate of food. (From U.S. Department of Agriculture.)

Note that the adjectival phrase "plant-based diet" appears in the DGAC report, but is conspicuous by its absence in the 2010 Dietary Guidelines. It would appear that in the first decade of the twenty-first century, the USDA and DHHS were not yet ready to endorse vegetarianism. Doing so would be tantamount to recommending the avoidance of food industry products that the USDA supports. *Given the trajectory of the Dietary Guidelines, do you think the phrase "plant-based diet" or the adjective "flexitarian" will be included in the 2015 and subsequent editions of the Dietary Guidelines? Do you think they should be included? Why?*

CONCLUSION

Most of our calories (energy) should be supplied by food from the plant kingdom, whether fruit, vegetable, or legume. Numerous federal programs encourage fruit and vegetable consumption through increasing access to these items and through nutrition education. Yet our agricultural policies still seem at odds with our dietary guidance, and even our dietary guidance seems tentative in the promotion of a more plant-based diet.

In the mid-1950s, the USDA opened the floodgates on influence and education to the food industry when its diet recommendations were streamlined to the "basic four" food groups of meat, milk, cereal and bread, and fruits and vegetables. The authors of this book do not know the extent to which food industry could have influenced the development of the "basic four" (*that* would make an instructive expose), but, as described in Chapter 7, the dairy industry did seize the opportunity to create nutrition education materials focusing on their products, turning school teachers into unwitting food industry spokespeople. It took the fruit and vegetable "industry" almost 40 years to catch up to dairy in this footrace with the "5 A Day" campaign.

Our agricultural policies, although improving, remain at odds with our dietary guidance, and the result is that we grow less and spend less when it comes to fruits and vegetables. The Healthy People 2020 objectives concerning fruit and vegetable consumption are striving to bring mean consumption of fruit and vegetables close to the 2010 Dietary Guidelines recommendations, yet we do not produce enough fruit and vegetable for each of us to be able to meet these recommendations.

Looking forward, how do we bring our agricultural polices in closer alignment with our dietary guidance without succumbing to relying on starchy vegetables with longer shelf-life such as potatoes to meet the need for vegetables in our diet? What will it take to ensure that each American can be a *healthy person* with the resources (knowledge, skills, money, and access) to increase fruit and vegetable consumption?

APPENDIX

Plants USDA Classifies as "Specialty Crops"

Fruits

Apple
Apricot
Avocado
Banana
Blackberry
Blueberry
Breadfruit
Cacao
Citrus
Cherimoya
Cherry
Coconut

Coffee
Cranberry
Currant
Date
Feijou
Gooseberry
Grape (including raisin)
Guava
Kiwi
Macadamia
Mango
Nectarine

Olive
Papaya
Passion fruit
Peach
Pear
Persimmon
Pineapple
Pomegranate
Quince
Raspberry
Strawberry
Suriname cherry

Tree Nuts

Almond
Beechnut
Brazil nut
Bush nut
Butternut
Cashew
Chestnut

Coconut
Filbert (hazel nut)
Ginko nut
Hickory nut
Lichee nut
Macadamia nut
Nangai nut

Pecan
Pine nut
Pistachio
Shea nut
Walnut

Vegetables

Artichoke
Asparagus
Bean (snap or green, Lima, dry, edible)
Beet, table
Broccoli (including broccoli raab)
Brussels sprouts
Cabbage (including Chinese)
Carrot
Cauliflower
Celeriac

Celery
Chive
Collards (including kale)
Cucumber
Eggplant
Endive
Garlic
Horseradish
Kohlrabi
Leek
Lettuce
Melon (all types)

Mushroom (cultivated)
Mustard and other greens
Okra
Onion
Opuntia
Parsley
Parsnip
Pea (garden, English or edible pod)

(Continued)

Vegetables

Pepper	Salsify	Taro
Potato	Spinach	Tomato (including tomatillo)
Pumpkin	Squash (summer and winter)	Turnip
Radish (all types)	Sweet corn	Watermelon
Rhubarb	Sweet potato	
Rutabaga	Swiss chard	

Culinary Herbs and Spices

Ajwain	Cinnamon	Mace
Allspice	Clary	Mahlab
Angelica	Cloves	Malabathrum
Anise	Comfrey	Marjoram
Annatto	Common rue	Mint (all types)
Artemisia (all types)	Coriander	Nutmeg
Asafetida	Cress	Oregano
Basil (all types)	Cumin	Orris root
Bay (cultivated)	Curry	Paprika
Bladder wrack	Dill	Parsley
Bolivian coriander	Fennel	Pepper
Borage	Fenugreek	Rocket (arugula)
Calendula	Filé (gumbo, cultivated)	Rosemary
Chamomile	Fingerroot	Rue
Candle nut	French sorrel	Saffron
Caper	Galangal	Sage (all types)
Caraway	Ginger	Savory (all types)
Cardamom	Hops	Tarragon
Cassia	Horehound	Thyme
Catnip	Hyssop	Turmeric
Chervil	Lavender	Vanilla
Chicory	Lemon balm	Wasabi
Cicely	Lemon thyme	Water cress
Cilantro	Lovage	

Medicinal Herbs

Artemissia	Ephedra	Goldenseal
Arum	Fenugreek	Gypsywort
Astragalus	Feverfew	Horehound
Boldo	Foxglove	Horsetail
Cananga	Ginkobiloba	Lavender
Comfrey	Ginseng	Liquorice
Coneflower	Goat's rue	Marshmallow

(Continued)

Medicinal Herbs

Mullein	Senna	Urtica
Passion flower	Skullcap	Witch hazel
Patchouli	Sonchus	Wood betony
Pennyroyal	Sorrel	Wormwood
Pokeweed	Stevia	Yarrow
St. John's wort	Tansy	Yerba Buena

Source: Adapted from National Institute of Food and Agriculture. Available at http://www.csrees.usda.gov/funding/pdfs/definition_of_specialty_crops.pdf/, accessed September 6, 2014.

REFERENCES

1. Pollan M. *In Defense of Food: An Eater's Manifesto.* New York: Penguin Press, 2008.
2. Centers for Disease Control and Prevention. *5 A Day Works!* Atlanta: U.S. Department of Health and Human Services, 2005. Available at http://www.cdc.gov/nccdphp/dnpa/nutrition/health_professionals/programs/5aday_works.pdf, accessed October 31, 2014.
3. State Symbols USA. Available at http://www.statesymbolsusa.org/Lists/AmericanFood Symbols.html, accessed August 28, 2014.
4. USDA Definition of Specialty Crops. Available at http://www.ams.usda.gov/AMSv1.0/getfile?dDocName=STELPRDC5082113, accessed August 28, 2014.
5. United States Supreme Court. *NIX v. HEDDEN*, 149 U.S. 304 (1893).
6. Hu FB. Plant-based foods and prevention of cardiovascular disease: An overview. *Am. J. Clin. Nutr.* 2003;78(3Suppl):S544–551.
7. Habauzit V, Morand C. Evidence for a protective effect of polyphenols-containing foods on cardiovascular health: An update for clinicians. *Ther. Adv. Chronic. Dis.* 2012;3(2):87–106.
8. Fung T, Chiuve S, McCullough M et al. Adherence to a DASH-style diet and risk of coronary heart disease and stroke in women. *Arch. Intern. Med.* 2008;168(7):713–720.
9. Montonen J, Knekt P, Jarvinen R, Reunanen A. Dietary antioxidant intake and risk of type 2 diabetes. *Diabetes Care.* 2004;27(2):362–366.
10. World Cancer Research Fund, American Institute for Cancer Research. *Food, Nutrition, Physical Activity, and the Prevention of Cancer: A Global Perspective.* Washington, DC: American Institute for Cancer Research, 2007.
11. Tohill BC, Seymour J, Serdula M, Kettel-Khan L, Rolls BJ. What epidemiologic studies tell us about the relationship between fruit and vegetable consumption and body weight. *Nutr. Rev.* 2004;62(10):365–374.
12. Rolls BJ, Ello-Martin JA, Tohill BC. What can intervention studies tell us about the relationship between fruit and vegetable consumption and weight management? *Nutr. Rev.* 2004;62(1):1–17.
13. Bellavia A, Larsson SC, Bottai M, Wolk A, Orsini N. Fruit and vegetable consumption and all-cause mortality: A dose-response analysis. *Am. J. Clin. Nutr.* 2013;98:454–9.
14. Nestle M. *Food Politics: How the Food Industry Influences Nutrition and Health, First Edition Revised.* Berkeley: The University of California Press, 2013.
15. Health.gov, Dietary Guidelines for Americans, 2005. Available at http://www.health.gov/dietaryguidelines/dga2005/document/html/AppendixA.htm#appA2, accessed August 28, 2014.

16. Health.gov, Dietary Guidelines for Americans, 2010. Available at http://www.health.gov/dietaryguidelines/dga2010/DietaryGuidelines2010.pdf, accessed August 28, 2014.

17. Foerster SB, Kizer KW, Disogra LK, Bal DG, Krieg BF, Bunch KL. California's "5-day—for better health!" campaign: An innovation population scale dietary change. *Am. J. Prev. Med.* 1995;11(2):124–31.

18. U.S. Department of Health and Human Services, Public Health Service. *Healthy People 2000: National Health Promotion and Disease Prevention Objectives,* September 6, 1990.

19. Produce for Better Health Foundation, About PBH. Available at http://pbhfoundation.org/about, accessed September 6, 2014.

20. Produce for Better Health Foundation. 5 A Day Research. Available at http://pbhfoundation.org/about/res/5aday_res, accessed September 6, 2014.

21. U.S. Department of Health and Human Services. NIH National Cancer Institute, NCI Office of Media Relations, last updated April 4, 2003. Available at http://www.cancer.gov/newscenter/newsfromnci/2003/9aday, accessed September 6, 2014.

22. Fruits and Veggies More Matters. Available at http://www.fruitsandveggiesmorematters.org/myplate-and-what-is-a-serving-of-fruits-and-vegetables, accessed August 28, 2014.

23. USDA. Dietary Guidelines Advisory Committee Meeting, January 29–30, 2009. Statement by DiSigora LK. Available at http://www.unitedfresh.org/assets/files/GR/FINAL_Lorelei_testimony_DG_2010.pdf, accessed November 19, 2014.

24. Health.gov, Dietary Guidelines for Americans, 2005. Available at http://www.health.gov/dietaryguidelines/dga2005/document/html/AppendixA.htm#appA2, accessed August 28, 2014.

25. Health.gov, Dietary Guidelines for Americans, 2010. Available at http://health.gov/dietaryguidelines/dga2010/dietaryguidelines2010.pdf, accessed November 4, 2014.

26. Healthy People 2020. Available at http://www.healthypeople.gov/2020/topicsobjectives2020/objectiveslist.aspx?topicId=29, accessed August 28, 2014.

27. The Food Trust. What We Do: Philly Food Bucks. Available at http://thefoodtrust.org/what-we-do/farmers-markets/philly-food-bucks, accessed August 28, 2014.

28. Fair Food Network. Double Up Food Bucks. Available at http://www.doubleupfoodbucks.org, accessed August 28, 2014.

29. The Food Project. Boston Bounty Buck. Available at http://thefoodproject.org/bounty-bucks, accessed August 28, 2014.

30. Fair Food Network. Available at http://www.fairfoodnetwork.org/resources/double-up-food-bucks-5-year-report, accessed August 28, 2014.

31. Mulvaney D (ed.). *Green Food: An A to Z Guide.* Thousand Oaks, CA: Sage Publications, 2011.

32. Altman JM. Increasing fruit and vegetable consumption through the USDA Nutrition Assistance Programs: A progress report. United States Department of Agriculture Food and Nutrition Service, March 2008.

33. USDA. Food and Nutrition Service. Feasibility of Implementing Electronic Benefit Transfer Systems in Farmers' Markets. Report to Congress. June 2010. Available at http://www.fns.usda.gov/multimedia/Kohl—Feasibility.pdf, accessed August 29, 2014.

34. Conway C, Kennel J, Zubieta AC. Assessing optimal nutrition education dosage and long-term behavior change retention in SNAP-Ed participants. N.d. Ohio SNAP-Ed. Available at http://www.csrees.usda.gov/nea/food/fsne/pdfs/ohio_improv_SnapEd_exec_sum.pdf, accessed August 29, 2014.

35. USDA. Fresh Vegetable Program: A Handbook for Schools. December 2010. Available at http://www.fns.usda.gov/cnd/ffvp/handbook.pdf, accessed August 29, 2014.

36. USDA. Fresh Fruit and Vegetable Program (FFVP): Allocation of Funds for School Year (SY) 2013/2014. Food and Nutrition Service. Document # SP 33-0213, April 2, 2013. Available at http://www.fns.usda.gov/ffvp-allocation-funds-school-year-2013-14, accessed August 29, 2014.

37. Bartlett S, Olsho L, Klerman J, Patlan KL et al. Evaluation of the Fresh Fruit and Vegetable Program (FFVP): Final Evaluation Report. USDA. Food and Nutrition Service. March 23, 2013. Available at http://www.fns.usda.gov/sites/default/files/FFVP.pdf, accessed August 29, 2014.

38. USDA. DoD Fresh Fruit and Vegetable Program, last updated April 25, 2014. Available at http://www.fns.usda.gov/fdd/dod-fresh-fruit-and-vegetable-program.

39. Healthy Food Access Portal. Available at www.healthyfoodaccess.org/sites/default/files/updated-hffi-fact-sheet.pdf, accessed August 29, 2014.

40. National Council of Fruit & Vegetable Nutrition Coordinators. Available at http://www.astphnd.org/newsletter.php?issue_id=15, accessed August 29, 2014.

41. CDC. About the National Health and Examination Survey, last updated February 3, 2014. Available at http://www.cdc.gov/nchs/nhanes/about_nhanes.htm.

42. CDC. Behavioral Risk Factor Surveillance System, last updated June 18, 2014. Available at http://www.cdc.gov/brfss.

43. CDC. Youth Risk Behavior Surveillance System, last updated July 9, 2014. Available at http://www.cdc.gov/healthyyouth/yrbs.

44. Casagrande SS, Wang Y, Anderson C, Gary TL. Have Americans increased their fruit and vegetable intake? The trends between 1988 and 2002. *Am. J. Prev. Med.* 2007;32(4):257–263.

45. CDC. State-specific trends in fruit and vegetable consumption among adults—United States, 2000–2009. *Morb. Mortal. Wkly. Rep.* 2010;59:1125–1130.

46. CDC. *State Indicator Report on Fruits and Vegetables 2009.* Available at http://www.cdc.gov/nutrition/downloads/StateIndicatorReport2009.pdf, accessed June 9, 2014.

47. U.S. Department of Agriculture and U.S. Department of Health and Human Services. *Dietary Guidelines for Americans*, 7th edition, Washington, DC: U.S. Government Printing Office, December 2010. Available at http://www.health.gov/dietaryguidelines/dga2010/DietaryGuidelines2010.pdf, accessed August 28, 2014.

48. Produce for Better Health Foundation. Available at http://www.pbhfoundation.org/pdfs/about/res/pbh_res/2010gapanalysis.pdf, accessed August 29, 2014.

49. CDC. *State Indicator Report on Fruits and Vegetables 2013.* Available at http://www.cdc.gov/nutrition/downloads/State-Indicator-Report-Fruits-Vegetables-2013.pdf, accessed August 29, 2014.

50. USDA. Economic Research Service, Understanding Economic and Behavioral Influences on Fruit and Vegetable Choices. Available at http://www.ers.usda.gov/amber-waves/2005-april/understanding-economic-and-behavioral-influences-on-fruit-and-vegetable-choices.aspx#.VADo9CiRndk, accessed August 29, 2014.

51. Drewnoswki A, Darmon N. Food choices and diet costs: An economic analysis. *J. Nutr.* 2005;135(4):900–904.

52. Palma MA, Jetter KM. Will the 2010 Dietary Guidelines for Americans be any more effective for consumers? *Choices: The Magazine of Food, Farm and Resource Issues* 2012;1:27.

53. USDA, NASS. *2007 Census of Agriculture,* Specialty Crops, vol. 2, November 2009.

54. Russo M, Smith D. Apples to Twinkies 2013: Comparing Taxpayer Subsidies for Fresh Produce and Junk Food. U.S. PIRG. http://www.uspirg.org/sites/pirg/files/reports/Apples_to_Twinkies_2013_USPIRG.pdf, accessed August 29, 2014.

55. O'Hara JK. Ensuring the harvest: Crop insurance and credit for a healthy farm and food future. April 2012. Union of Concerned Scientists. Available at http://www.ucsusa.org/assets/documents/food_and_agriculture/Ensuring-the-Harvest_summary.pdf, accessed September 6, 2014.

56. Dennis A. Shields. *Federal Crop Insurance*, December 12, 2013, Congressional Research Service. Available at http://fas.org/sgp/crs/misc/R40532.pdf, accessed August 29, 2014.

57. Johnson R. *Specialty Crop Provisions in the 2014 Farm Bill,* July 10, 2014, Congressional Research Service. Available at http://nationalaglawcenter.org/wp-content/uploads//assets/crs/R43632.pdf, accessed July 10, 2014.

58. Painter JA, Hoekstra RM, Ayers T, Tauxe RV, Braden CR, Angulo FJ, Griffin PM. Attribution of foodborne illnesses, hospitalizations, and deaths to food commodities by using outbreak data, United Sates, 1998–2008. *Emerg. Infect. Dis.* 2013;19(3):407–15.

59. Hoar B, Atwill ER, Carlton L et al. Buffers between grazing sheep and leafy crops augment food safety. Calif. Agric. 2013;67(2):104–109.

60. P.L. 111-353. FDA Food Safety Modernization Act. Available at http://www.gpo.gov/fdsys/pkg/PLAW-111publ353/pdf/PLAW-111publ353.pdf, accessed August 29, 2014.

61. Regulations.gov. *Standards for the Growing, Harvesting, Packing and Holding of Produce for Human Consumption.* Available at http://www.regulations.gov/#!documentDetail;D=FDA-2011-N-0921-0001, accessed August 29, 2014.

62. Brathwaite N, Fraser HS, Modeste N, Broome H, King R. Obesity, diabetes, hypertension, and vegetarian status among Seventh Day Adventists in Barbados: preliminary results. *Ethnicity. Dis.* 2003;13(winter):34–39.

63. Newby PK. Plant foods and plant-based diets: Protective against childhood obesity. *Am. J. Clin. Nutr.* 2009;89(5):1572S–875S.

64. Craig WJ. Health effects of vegan diets. *Am. J. Clin. Nutr.* 2009;89(5):16275S–1633S.

65. Blatner DJ. *The Flexitarian Diet.* New York: McGraw-Hill, 2009.

66. Dietary Guidelines Advisory Committee. *Report of the Dietary Guidelines Advisory Committee on the Dietary Guidelines for Americans, 2010, to the Secretary of Health and Human Services.* Washington, DC: U.S. Department of Agriculture, Agricultural Research Service, 2010.

67. Lappé FM. *Diet for a Small Planet.* New York: Ballantine Books, 1971.

68. USDA. ChooseMyPlate.gov. What Counts as a Cup of Fruit? Available at http://www.choosemyplate.gov/food-groups/fruits-counts.html, accessed November 1, 2014.

69. USDA. What Counts as a Cup of Vegetables? Available at http://www.choosemyplate.gov/food-groups/vegetables-counts.html.

70. Centers for Disease Control and Prevention. State Indicator Report on Fruits and Vegetables, 2013. Atlanta, GA: Centers for Disease Control and Prevention, U.S. Department of Health and Human Services, 2013. Available at http://www.cdc.gov/nutrition/downloads/State-Indicator-Report-Fruits-Vegetables-2013.pdf, accessed August 29, 2014.

Enrichment and Fortification* 11

Everything is toxic ... it is the dose which makes an item non-toxic.
Paracelsus (Swiss medical scientist Theophrastus
Bombastus von Hohenheim) (1493–1541)

INTRODUCTION

This chapter provides an overview of the history and current status of policies, guidelines, and regulations related to nutrient fortification in the United States, which emphasizes the addition of nutrients during the processing of food.

The FDA's authority to set standards of identity for food products under the 1938 Federal Food, Drug, and Cosmetic Act allows the FDA to regulate food fortification by enforcing label claims. Thus, the FDA can set a standard of identity for "enriched bread," specifying what nutrients must be added in what quantities to bread labeled as "enriched." With the exception of the period of 1943–1946, when fortification was mandatory for all flour sold for interstate commerce, the FDA's policy is that fortification is not mandatory for any food product. A baker chooses whether or not to produce enriched bread.

Fortification through the 1940s addressed obvious population health issues. The additions of iodine, vitamin D, thiamin, riboflavin, niacin, and iron to the U.S. food supply sought to correct, prevent, and/or correct outright nutrient deficiencies in the overall population. One important distinction, however, is that iodine and vitamin D were public health efforts initiated primarily through health organizations, whereas the enrichment of flour and bread was precipitated by national defense interests. Starting in the 1970s, we see several trends—the burgeoning influence of the food industry on fortification, fortification efforts trending toward health promotion instead of deficiency correction, and the justification of widespread fortification based on the needs of subpopulations.

---

* Angelica Santana, MPH, RDN, contributed primary research and prepared draft versions of portions of this chapter.

As a public health initiative in the United States, food fortification is highly dependent on food industry cooperation, and this in turn risks coopting by industry. The food industry was highly involved with the proposed increase in iron fortification; and calcium fortification for the most part is an industry-based initiative. The goal of folic acid fortification is to fortify the diets of women immediately before and during the first months of pregnancy to reduce the incidence of neural tube defects (NTDs), and thus is a departure from previous uses to correct widespread nutrient deficiencies among the population as a whole.

After a background discussion on fortification policy, this chapter provides a chronological discussion of fortification trends: the 1920s iodization of salt, a public health initiative; the 1930s fortification of milk with vitamin D, a joint public health and industry promotion; the enrichment of flour and bread in the 1940s to support national defense; the iron fortification controversy of the 1970s, a political solution to the problem of hunger; the addition of calcium to numerous products in the 1980s, a case of fortification driven by industry over optimal health rather than deficiency with limited regulation; and the addition of folic acid to the enrichment mix in the 1990s, arguably a public health intervention being used to solve a medical problem.

BACKGROUND ON FORTIFICATION POLICY

Food fortification is one of the most effective public health initiatives to prevent nutritional deficiencies. Widespread micronutrient deficiency exists worldwide. While deficiencies in micronutrients such as riboflavin, folic acid, vitamin C (ascorbic acid), and calcium are common, the conditions creating the greatest global public health burden are iron deficiency anemia, iodine deficiency disorders (IDD), and vitamin A deficiency, which result in blindness, goiter, impaired cognitive development.[1] The Food and Agriculture Organization (FAO) and the World Health Organization (WHO) have established general principles for the addition of vitamins and minerals to foods. However, each country determines its own policy or regulations, and fortification approaches can vary widely throughout the world. Financing of fortification also varies, from partnerships of government and private philanthropy to consumer-supported, as in the United States.[2] Fortification does not need to be centralized. Powders can be provided for home use and smaller regional programs that target areas of risk can be effective. However, the centralization of a food industry may encourage both the initiation and ongoing evaluation of voluntary or marketplace-based fortification programs.

Global fortification programs and technical fortification issues are beyond the scope of this book, but briefly, technical issues include identifying bioavailable forms of a nutrient, factors affecting nutrient bioavailability, determining a maximum safe intake level for the nutrients, ensuring the nutrient remains stable during storage and processing, considering—how the added nutrients might change the food, and selecting a carrier food suitable for cultural factors, food availability and access, and food system logistics. For example, if a population does not widely use salt, another more commonly used condiment would be a better carrier for iodine.[3]

Additional fortification policy concerns also include assessing the possible introduction of unanticipated nutrients or levels of nutrients to foods through exposure during processing, cooking, or through unexpected levels of consumption. For example, the iodine content in milk can be greatly influenced by the use of iodine sanitizers in the dairies; a change in milling processing methods drastically reduced the level of niacin in our grain supply; cooking food in iron pans can contribute dietary iron; and a reliance on breakfast cereals as a dietary mainstay can incur an excess of vitamin consumption.[4,5]

Definitions

To start, we explain some terms used in discussions of fortification: micronutrients (minerals and vitamins), fortification, enrichment, generally recognized as safe (GRAS), and "standard of identity." Understanding the difference between whole and refined grains is also helpful.

Minerals are elements that originate in the earth and cannot be made by living organisms. Plants obtain minerals from the soil; most of the minerals in our diet come directly from plants or indirectly from animal sources. The mineral content of edible plants may vary, because the mineral content of the soil varies geographically. Minerals are also present in drinking water, which also varies with geographic locale. *Vitamins* are organic (carbon-containing) compounds that are produced in plants. We obtain vitamins directly from plants or indirectly from animal sources. In general, vitamins and minerals are nutrient essentials that cannot be produced endogenously (synthesized in the body); their only source is exogenous (our diet). The one exception is vitamin D, which is synthesized in the skin when exposed to ultraviolet light. The discoverer of the dietary factors that cause the classic deficiency diseases designated these dietary factors *vital amines* or *vitamines* (*vita* for life, and *amine* for an essential nitrogenous substance). *Enrichment* in the United States refers to a food product that has nutrients added to replace those lost in processing, such as those added to refined flour. *Fortified* refers to the addition of vitamins not originally found in the food, such as vitamin D added to milk or soy beverages.[6] In this chapter, the broad term "fortification" is used to refer to both enrichment and fortification. *GRAS* is an acronym for "generally recognized as safe." The 1958 Food Additives Amendment to the Federal Food, Drug, and Cosmetic Act of 1938 regulates intentional additives to foods. A food additive is subject to approval by the FDA before the food is marketed, excepting items that are considered "generally recognized as safe." Although it broadly covers any substance that becomes a component of food or affects the characteristics of food, it excludes substances that have been scientifically shown to be safe under the conditions of their intended use, or, for items used in food prior to January 1, 1958, through experience based on common use in food.[7] *Standard of identity* refers to federally set requirements that describe what ingredients a food product of a specific identity (such as "enriched bread") must contain to be labeled as such and marketed in interstate commerce. Standards of identity are issued and regulated by the FDA under the authority of the 1938 Federal Food, Drug, and Cosmetic Act. Think back to the discussion in the introduction chapter of the relevance of how we define "food."

Botany of Whole Grains; Impact of Refining on Nutrients

One needs to be acquainted with some elementary botany to appreciate why refined grains and cereals are enriched with four B-vitamins and iron. In the beginning, all grains are whole. The entire seed of a plant is its whole grain. This seed (or kernel) is composed of three key parts: the bran, the germ, and the endosperm.

- The bran is the tough outer covering of the kernel. The bran contains antioxidants, B vitamins, and fiber.
- The endosperm is the germ's food supply, a kick-starter for the seed so it can send roots down for water and nutrients, and send sprouts up for sunlight's photosynthesizing power. The endosperm is the largest portion of the kernel and contains carbohydrates, proteins, and some vitamins and minerals.
- The germ is the embryo, which could germinate into a new plant. It contains many B vitamins, some protein, minerals, and wheat germ oil.

During the refining process, the bran and germ are separated from the endosperm. The endosperm alone is ground into flour; the product is now known as white flour. White flour is less coarse than whole wheat (because it has lost its fiber-rich bran) and it has a longer shelf-life (because it has lost the oils in the germ that over time can become rancid). About 25% of the grain's protein and at least 11 other key nutrients are lost in refinement. In contrast, the entire wheat kernel is used in producing whole meal or whole grain flour, a grinding process that has no appreciable nutrient loss. It is important to note that not all nutrients lost in processing are returned in enrichment, and some nutrients are in fact added in excess. For details see Table 11.1.

TABLE 11.1   Nutrient percentage of refined and enriched flour compared to whole wheat

NUTRIENT	WHOLE	REFINED	ENRICHED
Folate	100	59	416
Riboflavin (vit B_2)	100	19	230
Thiamin (vit B_1)	100	27	176
Iron	100	30	120
Niacin (vit B_3)	100	20	93
Selenium	100	48	
Copper	100	38	
Potassium	100	26	
Zinc	100	24	
Fiber	100	22	
Magnesium	100	16	
Vitamin B_6	100	13	
Vitamin E	100	7	

Source: Adapted from Whole grains: The inside story, *Nutrition Action Health Letter,* May 2006. Available at https://www.cspinet.org/nah/05_06/grains_can.pdf, accessed September 26, 2014. Based on data from the Oldways, Preservation Trust and the Whole Grains Council.

Vitamin(s): A Brief History

Our desire and ability to fortify foods with nutrients is dependent on our understanding of the characteristics of foods. Our understanding of nutrients began developing in the 1800s, and by the end of that century it was accepted that there were four essential components to a nutritious diet—proteins, carbohydrates, fats, and minerals. Lacking an understanding of the dietary connection to recurring diseases such as pellagra observed among people who lived in close quarters (*families who shared similar diets*) fostered the idea that such diseases were infectious or familial in nature.

Yet more than one researcher expressed the idea of "unsuspected dietary factors" and "unknown substances indispensable to human nutrition." Milk was a popular subject for animal diet experiments, and by 1912, researchers found that "astonishingly small amounts" of milk in a diet for rats that consisted otherwise of protein, starch, cane sugar, lard, and minerals would support normal growth. Given this diet without milk meant the rats did not thrive.[9]

Researchers had also compiled enough information to sustain a new deficiency model of disease. In 1912, Casimir Funk proposed the term "vitamine" to describe the substance in foods (of the nature of organic bases), that, when insufficient in the human diet, would lead to a "deficiency disease."[10] Scientists continued isolating, describing, and then synthesizing vitamins (the "e" was dropped by Jack Cecil Drummond's suggestion as not all vitamins contain an anime group), with the majority of efforts completed by the end of the 1930s.

This new body of scientific knowledge, along with the ability to inexpensively synthesize vitamins, led to the efforts to improve population health through selective additions of nutrients in food to address specific diet related deficiencies. These efforts were not without controversy, with ongoing discussions about sufficient scientific justification.

Recommended Dietary Allowances

In 1940, the Committee on Food and Nutrition of the National Research Council (now the Food and Nutrition Board of the Institute of Medicine, FNB), was established to advise on nutrition for the U.S. population. This committee developed the first Recommended Dietary Allowances (RDAs). Committee member Harriet Stiebeling was an early pioneer in dietary micronutrient requirement research, having used the term "dietary allowances" in 1933 for quantitative estimates of individual dietary intake for several vitamins and minerals. Stiebeling proposed dietary allowances that were 50% more than the average requirement to account for variability in the population, and thus cover most normal individuals. The RDAs were presented and accepted at the 1941 National Nutrition Conference for Defense.[11]

As with other results of scientific research, new and adaptable technologies arose. Some food producers eagerly adopted the idea of adding nutrients as a means of enticing customers. After the release of the RDAs, Russell Wilder, the FNB Chairman, noted that the RDAs were quickly misinterpreted and leveraged by the new vitamin industry to promote the idea that any diet providing less than 100% daily RDAs was ipso facto nutritionally inadequate.[12]

Fortification and the FDA

The FDA plays an integral role in U.S. fortification policy, but has struggled in governing the tension between consumer, public health, and industry interests. The regulatory power of the FDA has developed over time. Although the 1906 Food and Drug Act prohibited interstate commerce of misbranded food items, it did not give the FDA authority to determine what actually constituted misbranding. The 1938 Food, Drug, and Cosmetic Act corrected this by giving the FDA the ability to authorize standards of identity, quality, and fill-of-container for foods, thus allowing the FDA to in turn assess and enforce misbranding.[13] The FDA's authority to set standards of identity is the authority that allows the FDA to regulate food fortification through enforcing label claims.

The FDA can set a standard of identity for "enriched bread," specifying what nutrients must be added in what quantities to bread labeled as "enriched." If a manufacturer chose to add ingredients not including in the standard of identity, but yet still labeled the bread "enriched" (such as an early example of marketing bread as enriched by adding other ingredients, such as peanut flour)[14] the FDA could enforce misbranding.[15] See Box 11.1 regarding the FDA's policy on fortification.

In Chapter 4, we comment on how the FDA's developing authority over nutrition reflected the move of the agency to the (now) Department of Health and Human Services in 1953. Broadening regulatory authority over misbranding to cover nutrition was not a political certainty, but the FDA was concerned about the expanding intrusion of vitamins into the food supply, especially through marketplace offerings such as fortified sugar that were at odds with a scientifically advisable diet for the population. In 1961, the FDA brought a fortified sugar maker to court due to the manufacturer's claim that the fortified sugar was more nutritious than other sugar. The FDA argued that such claims were misleading and, furthermore, sugar was not an appropriate vehicle for

---

BOX 11.1   FDA's POLICY ON FORTIFICATION

With the exception of the period of 1943–1946, when fortification was mandatory for all flour sold for interstate commerce, the FDA's policy, standing since the 1940s, is that fortification is not mandatory for any food product. Any standard of identity set for an enriched food product has a corresponding standard of identity for the same food product without the added nutrients. Prior to 1990, individual states could regulate fortification for foods sold in state, and many did so with fervor. When the FDA released the post-war standard of identity for enriched bread in 1952, 26 states had already made flour and bread enrichment mandatory, and full federal preemption over standards of identity in general did not occur until the 1990 National Labeling Education Act.[16]

fortification.[17] The court held that FDA had no legal authority to prohibit food fortification unless it can be shown to be unsafe, stating that "no federal agency has the power to determine what foods should be included in the American diet; this is the function of the marketplace...v."[16]

Leery of industry-driven nutrient fortification, the FDA tried twice in the 1960s to tamp down fortification by proposing limitations on which foods could be fortified and which nutrients could be added to foods, but the proposals were pushed back each time by Congress.

In the mid-1970s, the FDA tried again to propose comprehensive regulations; in response Congress specifically limited the FDA's authority through the Vitamins and Minerals Amendments ("Proxmire Amendments"), which prevented FDA from establishing standards limiting potency of vitamins and minerals in food supplements or regulating them as drugs based solely on potency.[13]

The FDA's proposed regulations were eventually published in 1980 as a policy statement by the FDA, codified and revised from time to time in 21 CFR 104.20, "Nutritional Quality Guidelines for Foods." The objective of the policy is "to establish a uniform set of principles that will serve as a model for the rational addition of nutrients to foods," and to discourage indiscriminate addition of nutrients. The FDA also states that it does not consider it appropriate to fortify fresh produce and meat, sugars, or snack foods, including candies and carbonated beverages.[18] Although the policy statement reflects the standards the FDA urges manufacturers to follow if they elect to add nutrients, it is technically not enforceable.[16]

The FDA's policy speaks to when it is appropriate to fortify foods, for example, to address known nutrient-deficiency diseases, to restore nutrients lost in storage, handling, and processing, to address nutritional inferiority of products that are replacing traditional foods. The policy also includes recommendations as to nutrients that are appropriate to add, using the same standards as used for nutrition labeling. The policy provides guidance that the nutrients should be stable, physiologically available, and not at a level that would encourage excessive intake.

Finally, the policy statement further reinforces the scope of authority over labeling with a reminder that claims and statements on the label regarding nutrients cannot be false or misleading, and products must also meet the definition and standard of identify for the same item unfortified. The USDA's Food Safety and Inspection Service (FSIS) has followed the FDA in this policy for the products regulated by the FSIS, with some limited additions of nutrients that contribute to meat processing, such as vitamin C which accelerates curing. The FSIS also allows label statements about nutrients approved by FDA, such as calcium-fortified egg noodles.[16]

The FDA's position is that if a manufacturer fortifies food in a manner that is inconsistent with this stated policy, such fortification may be misleading but also potentially harmful. However, beyond regulating label claims, or the ability to remove dangerous products from the market, the FDA does not have much authority over industry-sponsored fortification, particularly when beyond the scope of standard of identity regulations. (See Chapters 4 and 9 for more on label claims.)

For fortification, a provision for "reasonable overages" in the standards of identity for enriched food provides flexibility for the manufacturer and accounts for loss of nutrients over the shelf-life of a product.[4] It is important to note that this also means that consuming a food shortly after production might expose the consumer to a higher level of nutrients than what would be expected based on the label.

Fortification and Population Health

Understanding the role of fortification in population health outcomes may be more complex than imagined. Fortification may essentially encourage an inferior diet by the improvement of only a small portion of the nutrient profile (such as is the case with enriched white bread). In addition to the technical issues of nutrient bioavailability and amount, fluctuations in overall dietary patterns and dietary guidance may have further unintended consequences. For example, consuming dairy products and eggs can inhibit iron absorption. A Danish study conducted from 1984 to 1994 found that after a period of mandatory iron fortification ended, iron stores were unchanged in premenopausal women, whereas iron stores and the prevalence of overload in postmenopausal women had increased. The increase in iron stores after the period of fortification had ended was unexpected. The reason was thought to be changes in dietary habits.[19]

Fortification policy also must address the fact that regulated fortification may encroach on individual liberties and autonomy over one's physical self. Consumption of fortified foods may often incur without an individual's specific knowledge, especially with servings of enriched grain products. At the best, excess unintended consumption of nutrients for nontargeted recipients may simply lack benefit, and at the worst, may be harmful.

Other policy approaches to solving micronutrient deficiency exist and should be evaluated (see Table 11.2).

1920s IODIZED SALT: A PUBLIC HEALTH INITIATIVE

Background

Iodine is an essential mineral component of thyroid hormones, which regulate human growth and development. Thyroid hormone is also important for myelination of the central nervous system. Iodine deficiency disorders include mental retardation, hypothyroidism, enlargement of the thyroid gland (goiter), and varying degrees of other growth and developmental abnormalities. Simple goiter, characterized by an enlarged thyroid gland, was recognized in classical times and depicted in artifacts from the early Christian era. For centuries much of Europe experienced both endemic goiter and cretinism. In the late nineteenth century, research conducted in Europe suggested an association between iodine deficiency and goiter.[20] It was found that without iodine, the body could not properly synthesize thyroid hormones, which often resulted in an unsightly neck goiter or in more serious cases, the mental retardation of cretinism. Iodine deficiency is now accepted as the most common cause of preventable brain damage in the world.[21]

TABLE 11.2 Micronutrient deficiency approaches

METHOD	REQUIREMENTS	COMMENTS
Dietary education	Educational messaging and programming that is made accessible to all population members. Consistent bioavailability of nutrient within suggested dietary items.	Difficult to change dietary habits. Dietary items must be accessible (physically and financially) for education to be effective. Rest of diet may suffer if overconsumption of suggested items; industry may piggy-back onto educational messages. Education-centered initiatives are least intrusive on personal autonomy as to dietary choices.
Supplementation	Target-specific demographic. Consistent formulations of supplements. Adherence by individual.	Can target specific demographic, but requires reaching targeted demographic. Can be conducted outside of physician channel. One approach consists of small powder packets to be sprinkled on individual serving of food. Independent of diet, as powder can be sprinkled on the most basic food item. Risk of toxicity if combined with fortification.
Dietary-based food supplementation	Target-specific demographic at known risk of deficiency (WIC).	Program evaluation may require biometric testing.
Regulated fortification of foods	Standards for dosage and serving size. Industrial methods. Cost for the most part is spread across marketplace and consumers.	Reaches people through all socioeconomic and demographic groups given consistent dietary trends. Risk of excessive intake by nondeficient population members. Risk of undertake by targeted population. Financial constraints may impact how fortified food is consumed; including overconsumption. Changes in dietary trends may reduce effectiveness of fortification program (sea salt vs. iodized salt, olive oil vs. vitamin A margarine, breakfast cereal vs. nongrain based breakfast). If fortified products dominate the marketplace, fortification becomes the most intrusive as to personal choice.

(Continued)

TABLE 11.2 (Continued)   Micronutrient Deficiency Approaches

METHOD	REQUIREMENTS	COMMENTS
Voluntary fortification of food products by industry	Government population level dietary recommendations that lead to a market. Cost is borne for the most part directly by consumer making specific choice.	If nutrient is not regulated, the dosage may vary depending on manufacturer and food group. Industry overfortification may lead to toxicity. Marketplace may retain more options for consumer to choose between fortified and nonfortified product. Will create a market.
Food-processing methods and food modifications (soil enhancements, breeding, GMO)	Centralized production may be most efficient but not necessary.	Impact on nutrient profile may be unintentional.
Secular increase of income	Requires funding either through government/ taxpayer programs.	Increase in income leading to improved variety in dietary choices, leading to improved micronutrient intake. May enhance personal autonomy (SNAP).

Dietary Source of Iodine

Iodine is found in the ocean and soil close to the ocean. Inland and mountainous areas do not contain iodine compounds in the soil. Soil can also become depleted of iodine due to flooding, heavy rainfall, or glaciation. The older an exposed soil surface, the more likely the iodine has been leached away by erosion. Mountainous regions, such as the Himalayas, the Andes, and the Alps, and flooded river valleys, such as the Ganges, are among the most severely iodine-deficient areas in the world. Before the 1920s, iodine deficiency was common in the United States in what was then referred to as the *goiter belt*—the Great Lakes, Appalachia, and northwestern U.S. regions. These areas are characterized by being mountainous, inland, or greatly shaped by glaciation, such as the Great Lakes.

Natural sources of dietary iodine include animals and plants from the ocean. Marine animals that concentrate iodine from seawater are good sources of iodine, along with certain seaweeds. The iodine content of most plant foods depends on the iodine content of the soil used for cultivation. The variance in growing conditions makes it difficult to quantify if daily requirements have been met through plant food consumption. In the UK and northern Europe, iodine levels in dairy products tend to be lower in summer when cattle are allowed to graze in pastures with low soil iodine content.[22] However, the use of iodine supplements in feed and iodophor sanitizers can also contribute to the iodine content of milk.[23]

Recommended Intake of Iodine

The recommend intake in the United States is 150 µg daily for anyone over the age of 14 who is not pregnant. Iodized salt in the United States contains about 400 µg iodine per teaspoon of salt.

As indicated in Table 11.3, iodine requirements are greater during pregnancy and lactation than during any other stage of the lifecycle. This increased requirement is the result of increased thyroid hormone production, increased renal iodine excretion, and fetal iodine requirements. The American Thyroid Association (ATA) recommends that, in addition to dietary sources, women take a 150 µg iodine supplement daily during pregnancy and lactation. The ATA also recommends that all prenatal vitamin/mineral preparations contain 150 µg of iodine.[24] Although most people in the United States consume sufficient iodine in their diets from iodized salt, dietary sources, and food additives, an additional 150 µg/day is unlikely to result in excessive iodine intake.

Iodine Deficiency

Thyroid enlargement (goiter) is an early sign of iodine deficiency. The thyroid enlarges in response to persistent stimulation by thyroid stimulating hormone. When deficiency is mild, this adaptive response may produce enough hormone for the body, but in more severe cases of deficiency, hypothyroidism results. Adequate iodine intake will generally reduce the size of a goiter, but the reversibility of the effects of hypothyroidism depends on an individual's stage of development. The damages of iodine deficiency are most pronounced in the developing brain. Most dietary iodine absorbed in the body eventually appears in the urine; thus, urinary iodine excretion is used to assess recent dietary iodine intake (see Table 11.4).

Fortification

In 1921, two Ohio doctors at the American Medical Association meeting presented their clinical trial findings demonstrating the effectiveness of sodium iodide treatments for the prevention of goiter in schoolgirls, suggesting treatment by physicians for people and

TABLE 11.3 Iodine: Recommended intake

The RDAs for iodine:
- 90 µg for children 1–8 years
- 120 µg for children 9–13 years
- 150 µg for adolescents (14–18 years) and for nonpregnant adults
- 220 µg per day for pregnant women
- 290 µg per day during lactation

The UL is 1100 µg per day for adults.

Source: Adapted from National Insitute of Health, Iodine Fact Sheet for Health Professionals. Available at http://ods.od.nih.gov/factsheets/Iodine-HealthProfessional/, accessed May 10, 2015.

TABLE 11.4   Median population urinary iodine values and iodine nutrition status

MEDIAN URINARY IODINE CONCENTRATION (ng/mL)	ESTIMATED CORRESPONDING IODINE INTAKE (µg/DAY)	IODINE NUTRITIONAL STATUS
<20	<30 (insufficient)	Severe deficiency
20–49	30–74 (insufficient)	Moderate deficiency
50–99	75–149 (insufficient)	Mild deficiency
100–199	150–299 (adequate)	Adequate
200–299	300–449 (above requirements)	More than adequate (likely to provide adequate intake for pregnant/lactating women but may pose a slight risk of more than adequate intake in the overall population)
>299	>449 (excessive)	Possible excess. Risk for adverse health consequences (for example, iodine-induced hyperthyroidism)

Source: Adapted from American Thyroid Association. How do you diagnose iodine deficiency? June 4, 2012. Available at http://www.thyroid.org/iodine-deficiency, accessed September 25, 2014; Anderrson M. et al. (eds.). *Iodine Deficiency in Europe: A Continuing Public Health Problem*, WHO, UNICEF, 2007. Available at http://whqlibdoc.who.int/publications/2007/9789241593960_eng.pdf, accessed September 25, 2014.

iodized salt for animals. Treatment ideas rapidly evolved, with another physician suggesting the use of iodized salt in humans a few months later, and by 1922, a University of Michigan pediatrician had persuaded the Michigan State Medical Society to set up an Iodized Salt Committee.[2]

Michigan became the first state to institute a public campaign to provide dietary iodine via salt. An extensive educational campaign that involved schoolteachers, industry, and the medical and public health communities helped increase consumer awareness and demand for iodized salt, which in turn encouraged voluntary industry fortification. Consumption of salt was not stratified by demographics and salt could be supplemented at a cost low enough that the manufacturer could pass along the cost to the consumer without significant market disruption. The relevance of Michigan should not be overlooked; Michigan was then the largest dietary salt producer in the country. By marketing iodized salt nationally, the industry made the shift to iodization financially feasible, which was a small cost to maintain national market share.[21] Over 5 years, from 1922 to 1927, the goiter rate in Michigan fell almost 75% from 38.6%–9.0%.[25]

Although iodized salt provided a great public health benefit, there was also a corollary cost; a too sudden influx of iodine in cases of chronic iodine deficiency can cause fatal iodine-induced hyperthyroidism. Excess deaths due to iodization would have occurred in older age groups, as the period of deficiency prior to influx would have been longer. Between 1921 and 1926 the death rate due to goiter nearly doubled from 2.1 to 4 per 100,000, and remained elevated for another 10 years, with an estimated 10,000 excess deaths over 1925–1942.[21]

By 1924, most table salt in the United States was iodized, and continued to be so. The rapid consumer adoption and acceptance of iodized salt was greatly fostered by educational campaigns that incited demand while largely preserving consumer choice, as noniodized salt was still available in most states. Iodization of salt remains voluntary in the United States,[2] and is regulated by the FDA only under misbranding. Iodized salt must be labeled "This salt supplies iodide, a necessary nutrient" and un-iodized salt labeled "This salt does not supply iodide, a necessary nutrient."[26]

Iodine deficiency in the United States has been nonexistent or rare for the past 80 years and the "goiter belt" virtually eliminated. However, trends in consumption can become an issue (sea salt, for example, contains some iodine, but not as much as iodized salt) and the food-processing industry can be complicating factor, as can be seen in the dairy industry. Iodine is a natural constituent of milk, and ranges dependent on the cow's diet. Before 1970 in the United States, values of iodine in milk were generally less than 100 μg/kg and no higher than 165 μg/kg. The concentration of iodine in milk increased after 1970 such that there was concern of excess iodine exposure. The increase was attributed to supplemental iodine in cow feed and the use of iodophor sanitizers in dairies.

The Food and Nutrition Board (FNB) recommended in the 9th edition (1980) of the *Recommended Dietary Allowances* that iodine used in the food system as disinfectants, alginates, coloring dyes, and dough conditioners be replaced or reduced. In a twist on public health education, educational campaigns were conducted among California farmers about reducing iodine supplements and sanitizers and in 1980 the California dairy industry voluntarily eliminated excess iodine feed supplements.[23]

Conversely, the dairy industry in Australia switched sanitizing chemicals in the 1990s and subsequently iodine deficiency began to rise. It turned out that the main source of iodine in Australia had been from milk containing iodine residue from iodophor sanitizers. The ubiquity of iodized salt in the United States made the iodine in milk excess, whereas in Australia, iodized salt had not been adopted, making the iodine in milk the major source of iodine. Consequently, Australia adopted mandatory fortification of bread with iodized salt.[27]

The potential impact of industry food processing changes, the call for decreasing salt consumption to decrease cardiovascular disease risk, and the consumption trend starting in the mid-2000s toward using sea salt, which is not iodized, should all serve as a reminder of the need for continued population monitoring of iodine status.

Populations at Risk of Iodine Deficiency

People eating foods grown in iodine-deficient soils are at risk of iodine deficiency unless they have access to iodized salt or foods grown elsewhere. People with marginal iodine status who eat a diet overemphasizing goitrogenic foods, such as soy and cassava, cabbage, broccoli, cauliflower, and other cruciferous vegetables are at risk, as are pregnant women and people who do not use iodized salt.[28] Iodine deficiency during pregnancy has been associated with increased incidence of miscarriage, stillbirth, birth defects, and if severe, hypothyroidism and neurocognitive deficits.[27]

Iodine deficiency was perceived to be limited to endemic areas delimited by geography until a 1994 ground-breaking report quantified the global population at risk of iodine deficiency as about 30% with 110 countries being iodine deficient. The World Health Organization (WHO) recommended global iodization of all salt used for human consumption, which led to wider implementation of systemic iodine fortification programs.[27] By 2011, the number of iodine-deficient countries had dropped from 110 to 32, but just under 30% of school-aged children worldwide lack sufficient iodine.[29] Not all countries have a means to report iodine status, and this too remains a problem, along with the converse problem of excessive iodine consumption in a number of countries.[27]

Iodine Status in the U.S. Population

In the United States, women aged 20–39 years show iodine status bordering on insufficiency.[30] Overall, however, the median iodine status in the U.S. population is considered sufficient, even though it has decreased by approximately 50% since National Health and Nutrition Examination Survey (NHANES) I (1971–1974).[31] Health recommendations calling for reduced salt and egg intake may have contributed to the decrease as well as structural changes in our food supply. For example, in addition to the removal of iodine in the dairy industry, bread manufacturers replaced iodate dough conditioners and the use of an iodine-based dye used in fruit-flavored breakfast cereals declined.[28]

Estimating dietary intake of iodine is complicated, as food sources have substantial variation, and the USDA's database of food composition does not currently include iodine.[31] According to a 2012 report, the USDA and FDA are collaborating on tools to provide iodine content of foods and dietary supplements. Having these tools would improve nutritional monitoring in the U.S.[20]

How to reconcile recommendations to decease sodium consumption as a means of preventing and treating cardiovascular disease with the recommendation to consume iodized salt?

1930s VITAMIN D FORTIFIED MILK: A JOINT PUBLIC HEALTH AND INDUSTRY PROMOTION

Background

Vitamin D is a secosteroid hormone that is produced by the skin when irradiated by ultraviolet B rays. Vitamin D is naturally present in very few foods, and technically, vitamin D is not a "true" vitamin, as dietary intake is not necessary when individuals have adequate sun exposure.

The major biologic function of vitamin D is to maintain normal blood levels of calcium and phosphorus. Vitamin D aids in the absorption of calcium, helping to form and maintain strong bones. It is used, alone or in combination with calcium, to increase bone mineral density and decrease fractures. Generally, vitamin D deficiency causes bone disorders, due to impaired mineralization, such as osteoporosis, osteomalacia, myopathy, and rickets.[32] Rickets can result in growth retardation, muscle weakness, skeletal deformities, hypocalcemia, tetany, and seizures. Rickets was the most common nutritional disease of children in the mid-to-late nineteenth and early twentieth century, estimated in 1921 to impact 75% of all New York City infants.[33]

The first scientific description of rickets was provided in the seventeenth century by Whistler (1645) and Glisson (1650); cod-liver oil was noted in folklore and in German medical literature as a remedy in the early 1800s, and by 1861, the relationship of lack of sun exposure and faulty diet (improved by cod-liver oil) to rickets was noted by Trousseau of France. By the end of the 1800s, it was known that sun exposure was preventative of rickets in children that had poorer diets. Dietary studies on rickets began with Mellanby's 1919 experiments with cod-liver oil and puppies.[34] McCollum extended this research and found that certain fats contained a specific substance that regulated the metabolism of bones. As this was the fourth discovery in the sequence of newly discovered vitamins, the substance was named vitamin D.[34]

Sunlight Exposure and Dietary Sources

Sunlight exposure is the natural physiological source of vitamin D. Sun-exposed skin forms previtamin D_3. In turn, this form of vitamin D is changed in the liver to D_2 (25-hydroxyvitamin D, [25OH-D]), the main circulating form of the vitamin. The kidneys convert D_2 into the nutrient's physiologically active form, D_3(1,25-dihydroxyvitamin D).

Vitamin D is found naturally in only a few foods. Fish-liver oils and fatty fishes (salmon, tuna, mackerel, etc.) are among the best sources, and small amounts of vitamin D are found in egg yolks, cheese, beef liver, and some mushrooms.[35] In the United States, most dietary intake of vitamin D comes from fortified milk and breakfast cereals. Milk products, such as cheese and ice cream, are not normally fortified, however, which may be an area of confusion for consumers who may believe all dairy products are good sources of vitamin D.[36]

Recommended Intake of Vitamin D

No dietary or supplemental intake of vitamin D is needed in healthy people who have sufficient sunlight exposure. However, season, time of day, length of day, cloud cover, smog, pollution, aging, sunscreen use, and skin pigmentation can all influence vitamin D production. Although geographic latitude would seem a simple way to describe risk, latitude alone is not a consistent predictor of vitamin D status. Seasonal sun exposure opportunities allow for production and storage of vitamin D, even in the far north latitudes.[35] Nonetheless, people may not have enough sun exposure for numerous reasons, including form of dress, occupation, or being homebound. The recommended dietary intake for vitamin D is meant to achieve nutritional sufficiency to maintain bone health and normal calcium metabolism when sunlight exposure is minimal.[35]

Vitamins D, A, and E are measured in International Units (IU). IUs refer to a biological activity specific to the vitamin, so the conversion rate is specific to each vitamin. For vitamin D, 1 IU = 0.025 µg, or 40 IU to 1 µg.[37] The most recent (2010) RDAs for vitamin D are 600 IU (15 µg) for anyone aged 1–70 years, and 800 IU (20 µg) for

TABLE 11.5   RDA for vitamin D, male and female

AGE	RDA
0–12 months[a]	400 IU (10 µg)
1–70 years (includes pregnancy and lactation through age 50)	600 IU (15 µg)
>70 years	800 IU (20 µg)

Source: Available at http://www.iom.edu/Reports/2010/Dietary-Reference-Intakes-for-calcium-and-vitamin-D.aspx.

[a] Adequate Intake (AI). AI is used when there is insufficient evidence to develop an RDA; and is set at a level that is assumed to be nutritionally adequate.

anyone over 70 years. For children and adolescents, the RDA prior to 2010 was 200 IU; the increase to 600 IU is significant.[35]

The Tolerable Upper Intake Level (UI) is 4000 IU (100 µg) per day in North America for individuals 9 years of age and older, and ranges from 1000 to 3000 IU for infants and children less than 9 years of age; as dietary intake increases above this amount, so does the risk for adverse consequences.[38] Excessive sun exposure does not create toxicity; the formation of vitamin D_3 is limited by certain mechanisms and not all D_3 is converted to active forms.[35] See Table 11.5 for intake recommendations for vitamin D.

Vitamin Deficiency

Vitamin D status is determined by measuring the amount of D_2 (25OH-D) in blood serum.[39] Because D_2 has a half-life of 15 days, it is a good reflection of dietary intake, exposure to sunlight, and the mobilization vitamin D out of tissue stores. However, this measurement of vitamin D status does not indicate the amount of vitamin D that remains stored in body tissues.[35] Recommendations regarding vitamin D intake are partly based on studies examining the dose–response relationship between vitamin D intake and the rise in serum 25OH-D. See Table 11.6 for standards to measure deficiency.

Populations at Risk of Deficiency

Older adults, people with limited sun exposure, and people with limited ability to absorb nutrients are at risk of vitamin D deficiency. Our skin is less efficient at producing vitamin D as we age, and our kidneys decrease in ability to activate vitamin D. People who are homebound, night shift workers, or who wear protective clothing all have limited sun exposure. People who have had postbariatric surgery or Crohn's or celiac disease have limited nutrient absorption, and obese people, because greater amounts of body fat take more vitamin D into storage, preventing active circulation, are at risk.[35]

Two populations are of particular interest: people with dark skin and infants. People with dark skin have less ability to produce vitamin D from the sun. However, it

TABLE 11.6   IOM 25OH-D standards for measuring vitamin D sufficiency

SERUM 25OH-D nmol/L	INTERPRETATION	EXPLANATION
<30	At risk for deficiency	Leads to rickets in infants and children and osteomalacia in adults
30–49	At risk for inadequacy	Considered inadequate for bone and overall health in healthy individuals
50–125	Desirable level	Generally considered adequate for bone and overall health in healthy individuals
>125	Approaching excess	Emerging evidence links potential adverse effects to such high levels, particularly >150 nmol/L

Source: Adapted from Institute of Medicine, *Dietary Reference Intakes for Calcium and Vitamin D,* Washington, DC: National Academies Press, 2011. Available at http://www.ncbi.nlm.nih.gov/books/NBK56070/pdf/TOC.pdf, accessed October 1, 2012.

is unclear if the lower serum level of vitamin D for such individuals is significant for health outcomes, as African-Americans have reduced rates of bone fracture and osteoporosis compared with Caucasians.[35]

Infants provide an example of conflicting health policy outcomes. Although health policy promotes breastfeeding, infants that are solely breastfed may be at increased risk of vitamin D deficiency. After birth, vitamin D status for an exclusively breastfeeding infant may be dependent on both sunlight exposure and vitamin D intake from the mother. Even in healthy mothers, breast milk contains relatively small amounts of vitamin D. Thus, vitamin D deficiency in infants may occur who are exclusively breastfed if sunlight exposure is also limited.

In the United States, infant formula is a special class of food. Infant formula is the only food that by law must contain vitamin D. The 1980 Infant Formula Act and subsequent legislation mandated that all infant formulas must be fortified with vitamin D between 40 and 100 IUs per 100 kilocalories. Manufacturers are required to assure these levels in their products (see Box 11.2).

Fortification

The encouragement of sensible sun exposure and the fortification of milk with vitamin D resulted in almost complete eradication of rickets in the United States,[40] and rickets largely disappeared in the UK due to mass programs distributing cod-liver oil to children.[34] However, cases of rickets in dark-skinned infants who are solely breastfed by mothers themselves deficient in vitamin D are still periodically seen.[34,35] Deficient calcium intake postweaning is also a problem.

Cod-liver oil had been long recommended for rickets. Milk was recognized to be an important dietary component for children. In 1925, it was discovered that irradiated milk was better than regular milk in stimulating bone calcification in children (the irradiation converts inactive ergosterol into physiologically active vitamin D_2) and the

---

BOX 11.2   THE INFANT FORMULA ACT OF 1980

In 1978, a major manufacturer of infant formula reformulated two of its soy products. Federal regulation for infant formula at the time specified minimums for some nutrients, but not for salt. The reformulated products contained an amount of chloride that was inadequate for infant growth and development. By mid-1979, a substantial number of infants who had been exclusively fed the chloride-deficient formula had been diagnosed with hypochloremic metabolic alkalosis, a syndrome associated with chloride deficiency.

Congress determined that greater regulatory control over the formulation and production of infant formula was needed, and enacted the Infant Formula Act of 1980 (Pub.L. 96–359). The Infant Formula Act amended the Federal Food, Drug, and Cosmetic Act, and in short, an infant formula is deemed adulterated unless it provides certain required nutrients.

In passing the Infant Formula Act of 1980 and the 1986 amendments, Congress created infant formulas as a special class of foods. In the legislative history that accompanied the 1986 amendments, Senator Metzenbaum explains the increased regulatory scrutiny when he stated "there is simply no margin for error in the production of baby formula. An infant relies on the formula to sustain life and provide the proper nourishment at a time of rapid physical and mental development."

Available at http://www.fda.gov/ohrms/dockets/
ac/02/briefing/3852b1_01.htm;
see Spark, A., Nutrition in Public Health: Principles,
Policies, and Practice, *Boca Raton, FL: CRC Press, 2007.*

---

invention of a milk irradiator for industrial use followed in 1929. Thus cod oil, milk, irradiated milk, and combinations thereof were all available in the mid to late 1920s. Vitamin D_2 was isolated in 1932, and the medical profession began encouraging the production of vitamin D milk. Milk was widely available, inexpensive, palatable to young children,[41] and rich in calcium and other nutrients.

The dairy industry was also eager for the opportunity to promote the healthfulness of their product. The promotion of vitamin D milk was a voluntary cooperative effort between the medical profession and the dairy industry—several large dairies sought the American Medical Association seal of approval, which was offered to manufacturers who passed advertising and content tests and conformed with the Food and Drug Act. The AMA issued reports announcing the sale of vitamin D fortified milk by dairies.[2] Government agencies making large-scale purchases also helped foster the supply of fortified milk in general, as dairies found it technically difficult to produce both fortified and unfortified milk.[2] Today, although voluntary, most milk in the United States is fortified with vitamin D.

Vitamin D is GRAS; however, it is fat-soluble and potentially toxic if consumed in excess. FDA regulations limit fortification amounts and foods to which vitamin D can be added,[42] with maximum limits generally supplying between 10% and 25% of DV.[36] Regulations authorizing vitamin D fortification of soy beverages and products were published in 2009, which may have contributed to the inclusion of soy beverages as an alternative to milk in the revised WIC packages. If a food product manufacturer wishes to add vitamin D to any other food or in a manner not already prescribed, the manufacture must petition the FDA to amend the regulations.[36]

Excessive intake of vitamin D can have adverse health outcomes, including raising blood levels of calcium, which in turn leads to vascular and tissue calcification and damage to the heart, blood vessels, and kidneys. Toxicity is most likely to occur from high intake of dietary supplements rather than from foods, particularly given the limitations on vitamin D food fortification as of 2014. However, it is important to note that the fortification process must be tightly controlled, as incorrectly implemented food fortification can be a serious food safety issue. As discussed in Chapter 13, food safety related diseases can spread more rapidly and widely among a population given the consolidation of our food system and increasing national distribution. Likewise, food that has been unintentionally excessively fortified could result in an unexpectedly wide exposure to toxicity, even with the historically smaller distribution and delivery patterns of milk. In the late 1980s and early 1990s a Boston home-delivery dairy distributed over-fortified milk to 11,000 households. Due to problems with the dispensing equipment, the milk contained highly variable amounts of vitamin D, from 70–600 times the then state regulated limit of 500 IU per quart.[43]

Vitamin D Status in the U.S. Population

During the first decade of the twenty-first century, the vitamin D status of the U.S. population was under intense investigation to determine if a downward trend was apparent. Indeed, mean 25OH-D concentrations decreased by approximately 10% between NHANES III (1988–94) and the periods 2001–2002 and 2003–2006. Decreases were seen in all groups stratified by gender or race/ethnicity.[38]

NHANES data reveals that concentrations of 25OH-D generally decreased with increasing age, but no consistent pattern was observed with regard to gender. Less than 2% of children (1–11 years) were at risk for vitamin D deficiency. More females (10%) than males (6%) were at risk for deficiency. The likelihood of being vitamin D-deficient was significantly influenced by race/ethnicity; non-Hispanic blacks had the lowest 25OH-D concentrations and non-Hispanic whites had the highest. Specifically, more non-Hispanic blacks (31%) were at risk for deficiency than Mexican Americans (11%) or non-Hispanic whites (4%). The high rates of low serum 25OH-D in non-Hispanic blacks is perplexing because clinical data indicate there fewer fractures and greater bone density in this group. Further research is needed to explain why non-Hispanic blacks have better bone health despite serum vitamin D levels that would suggest otherwise.[8]

Nevertheless, serum 25OH-D data suggest that requirements are being met for most people in the United States.[44] NHANES data from 2003 to 2006 suggest that two-thirds of the population had sufficient Vitamin D, about one-quarter were at risk of adequacy, 8% were at risk for deficiency, and less than 1% above safe levels. See Table 11.7 for population level Vitamin D status.

TABLE 11.7 Prevalence estimates of deficiency and excess in the population >1 year of age as determined by NHANES analyses of 25OH-D during the 2001–2006 survey cycle

PREVALENCE ESTIMATES (2001–2006)	25OH-D[a]	INTERPRETATION	EXPLANATION
	VITAMIN D STATUS BASED ON SERUM CONCENTRATIONS OF 25-HYDROXYVITAMIN D (25OH-D)		
8%	<30	At risk for deficiency	Leads to rickets in infants and children and osteomalacia in adults
24%	30–49	At risk for inadequacy	Considered inadequate for bone and overall health in healthy individuals
67%	50–125	Desirable level	Generally considered adequate for bone and overall health in healthy individuals
<1%	>125	Approaching excess	Emerging evidence links potential adverse effects to such high levels, particularly >150 nmol/L

Source: Adapted from Looker, A.C. et al., Vitamin D status: United States 2001–2006, NCHS data brief, no 59. Hyattsville, MD: National Center for Health Statistics, 2011. Available at http://www.cdc.gov/nchs/data/databriefs/ db59.pdf, accessed September 25, 2014.
[a] Serum concentrations of 25OH-D in nanomoles per liter (nmol/L).

Optimal Vitamin D Intake

Continued research on vitamin D has led to a more complex view of potential interaction in the body beyond skeletal health. Research has shown that vitamin D deficiency is associated with increased risk of cardiovascular diseases, cancer, asthma, and other chronic conditions.[45] Some argue that basing vitamin D recommendations solely on bone health is too simplistic. In 2011, the US Endocrine Society suggested that the current RDAs may not be sufficient for achieving the full potential of nonskeletal vitamin D health benefits, but like the IOM, they found that evidence was not yet decisive.[46] Our understanding of the relationships between vitamin D status and nonskeletal health outcomes—positive and negative—need further development.

What constitutes the optimal intake of vitamin D remains a matter of some disagreement. Defining necessary dietary intake is a challenge as the relative contributions from sun, diet, mobilization of tissues store are unknown and likely also variable.[39] Because sun exposure affects vitamin D status, serum 25OH-D levels are generally higher than would be predicted on the basis of dietary and supplement intakes alone of vitamin D.[35]

One researcher has argued that the nutritional value for vitamin D should be defined as the equivalent that an adult would acquire through exposing full skin surface to summer sunshine, which would range up to 250 µg per day,[47] which is 10,000 IU—a much greater amount than the range recommended by the IOM since 2010.

Part of the challenge in evaluating conflicting recommendations is appreciating the distinction between population level recommendations versus clinical guidelines used to evaluate specific patients at risk of deficiency. Conceptual language plays a role, too;

using "deficient" as a label of nutritional status instead of assessing vitamin D status as a measure of risk of health outcome may inappropriately escalate concern.[48]

The RDAs for vitamin D and calcium were updated in 2010, the last prior update being 1997. The expanding research on vitamin D after the late 1990s intrigued everyone—industry, scientists, and the public. In 2010, the media described vitamin D as what "promises to be the most talked-about and written-about supplement of the decade."[49] By 2014, the idea that vitamin D could be widely insufficient in the U.S. population was well established.[39] Research is still needed to bring consensus as to definitions of vitamin D status, threshold levels of deficiency, and nonskeletal health outcomes.

1940s B VITAMIN ENRICHMENT OF BREAD TO SUPPORT NATIONAL DEFENSE

Background

For the first few decades of the twentieth century, the American diet relied heavily on refined flours, a recent innovation, as described in Chapter 9. During the refining process, grains were stripped of many nutrients, including the B-vitamins thiamin, riboflavin, and niacin, and the mineral iron. As a result of an overall diet trending toward the affordable and desirable refined grains and corn as staples, these nutrients were not readily found elsewhere in the diet. Nutrient-deficiency diseases such as pellagra became common. As pellagra in particular brought a high toll to the public's health, this section will briefly highlight thiamin and riboflavin, but focus on niacin. Iron is addressed in a later section of this chapter.

Thiamin, riboflavin, and niacin are all water-soluble B vitamins. Excess is passed out in the urine, which means that a continuous dietary supply is necessary.

Thiamin, also known as B$_1$, supports the conversion of carbohydrates into energy and is involved in muscle contraction and nerve conduction. Insufficient thiamin can lead to muscle weakness, fatigue, psychosis, and nerve damage. Alcoholics are particularly at risk, as their ability to absorb thiamin from foods decreases. Beriberi is a thiamin-deficiency disease.[50] *Riboflavin*, also known as B$_2$, works in conjunction with the other B vitamins, and supports physical growth, red blood cell production, and also helps release energy from carbohydrates. Insufficient riboflavin causes symptoms of anemia, oral sores, skin disorders, sore throat, and mucus membrane swelling.[51] *Niacin*, also known as B$_3$, supports the digestive system, skin and nerves, and contributes in the conversion of food to energy. Pellagra is a niacin-deficiency disease.[52]

Dietary Source of B1, B2, B3 (Nonenriched Foods)

Thiamin is found in beef liver, pork, egg, legumes and peas, nuts and seeds, and whole grains.[53] Riboflavin is found in milk, dairy foods, eggs, green leafy vegetables, lean meats, legumes, and nuts.[51] Niacin is found in poultry, fish, dairy products, eggs, lean meats, legumes, and nuts.[52]

Recommended Intake of B1, B2, B3

Thiamin, riboflavin, and niacin were among the first group of vitamins to be given RDAs, as described later in this chapter. The recommendations for niacin are based on preventing deficiency. Niacin can be synthesized by the body from dietary niacin, but can also be synthesized from tryptophan (in a more inefficient process) so the term "niacin equivalent" is used to cover both sources. Niacin in cereal grains is only about 30% available (alkali treatment increases availability); niacin in the coenzyme nicotinamide adenine dinucleotide/nicotinamide adenine dinucleotide phosphate (NAD/NADP) in meats appears to have greater bioavailability. Niacin in the free form is found in some foods, including beans and liver, and is highly available. Niacin added during enrichment or fortification is in the free form.[54]

Pellagra can be prevented by about 11 mg of niacin equivalent, but as pellagra is considered to be the manifestation of a severe deficiency, the recommendations from the FNB are instead based on the excretion of niacin breakdown products (metabolites) as an assessment of niacin nutritional status; 12–16 mg/day of niacin equivalents has been found to normalize the urinary excretion of niacin metabolites in healthy young adults.[55]

See Table 11.8 for current recommended dietary intake values for thiamin, riboflavin, and niacin.

Deficiency: Niacin and Pellagra

Pellagra, a consequence of niacin deficiency, is characterized by the manifestation of the "Four Ds"—diarrhea, dermatitis, dementia, and death, with the symptoms appearing in that order. Joseph Goldberger, a U.S. Public Health Service physician, conducted groundbreaking epidemiological studies from 1914 until his death in 1929 that demonstrated that pellagra was not infectious but instead associated with poverty and the consequential poor diet. Despite compelling evidence, his hypothesis remained controversial and unconfirmed until 1937.

Although pellagra was likely present in the South post–Civil War, it surged into an observable epidemic at the dawn of the twentieth century. From 1906 to 1940, there were an estimated three million cases of pellagra, with about 100,000 deaths.[56]

A brief contextual discussion illustrates the blossoming of a nutrition deficiency epidemic. Early on, pellagra was considered to be infectious or a result of food spoilage. The association of poverty and a corn-based diet was recognized, but was confounded by the fact that the corn available to the poor was often spoiled corn rejected by others. For example, a 1906 pellagra outbreak in a mental hospital for African-Americans in Alabama was considered due to spoiled corn.[34] Pellagra also had a known and relatively consistent geographic distribution in the South.

Then, in 1909, pellagra made a surprising leap past the recognized Southern geographic boundaries of the disease, and manifested at a mental hospital in Peoria,

TABLE 11.8 Recommended dietary allowances (RDA) and tolerable upper limits (UL) in milligrams for thiamin, riboflavin, and niacin

AGE	RDA	UL
Thiamin		
0–6 months	0.2 mg[a]	ND
7–12 months	0.3 mg[a]	ND
1–3 years	0.5 mg	ND
4–8 years	0.6 mg	ND
9–13 years	0.9 mg	ND
14–18 years	1 mg females, 1.2 mg males	ND
19–71+ years	1.1 mg females, 1.2 mg males	ND
Pregnancy at any age	1.4 mg	ND
Lactation at any age	1.4 mg	ND
Riboflavin		
0–6 months	0.3 mg[a]	ND
7–12 months	0.4 mg[a]	ND
1–3 years	0.5	ND
4–8 years	0.6	ND
9–13 years	0.9	ND
14–18 years	1 mg females, 1.3 mg males	ND
19–71+ years	1.1 mg females, 1.3 mg males	ND
Pregnancy at any age	1.4 mg	ND
Lactation at any age	1.6 mg	ND
Niacin		
0–6 months	2 mg[a]	ND
7–12 months	4 mg[a]	ND
1–3 years	6 mg	10 mg[b]
4–8 years	8 mg	15 mg[b]
9–13 years	12 mg	20 mg[b]
14–71+ years	14 mg females, 16 mg males	30–35 mg[b]
Pregnancy at any age	18 mg	30–35 mg[b]
Lactation at any age	17 mg	30–35 mg[b]

Source: Adapted from the IOM Dietary Reference Intakes, Vitamin Table. Available at http://www.iom .edu/Home/Global/News%20Announcements/~/media/Files/Activity%20Files/Nutrition/DRIs/ DRI_Vitamins.ashx, accessed September 14, 2014.

Note: ND, not determined.

[a] Adequate intake.
[b] The UL for niacin is based on flush reaction.

Illinois. Just a year later, more than 13 states had reported more than 3000 cases in total. Within 2 years, pellagra had been reported in all but nine states nationwide. About 25,000 cases, with a mortality rate of 40%, occurred between 1907 and 1912.[34,57]

Pellagra's surge to epidemic status in the South was multifactored. The South was struggling to regain prosperity through cotton production in the face of dropping cotton prices, resulting in a labor system of sharecroppers given a small portion of land in exchange for the production of cotton. The efforts of the sharecroppers to survive crushing poverty lead to plots being mostly fully farmed in cotton, with very little crops grown for their own consumption. Molasses, corn meal, and pork fat became a consistent diet for sharecroppers and low-paid urban factory workers. Access to fresh meat, a good source of niacin, was further limited due to a regional lack of slaughterhouses and cold storage.[58] Thus many people had a diet overly reliant on corn as source of niacin.

Corn must be prepared appropriately to maximize niacin bioavailability. Reliance on corn as a dietary staple became even more problematic after the invention of the Beall degerminator, which removed the germ—and consequently, the niacin—from the corn kernel. The degermed corn meal lasted longer in storage, which was attractive to consumers, and the millers could process the germ separately to produce corn oil and expand profits. Widespread adoption of degermed corn meal may have been the last straw for nutritionally compromised diets and the trigger for the epidemic surge of pellagra beyond the South.[58]

Population at Risk

As like other B vitamins, people who have malabsorptive disorders issues (such as celiac or Crohn's disease) or who drink too much alcohol can be at risk of deficiency. The homeless population is at risk of malnutrition and niacin deficiency, as is anyone with a diet heavily emphasizing corn and lacking in animal protein. Anyone with a physical condition or who takes medication that interferes with the tryptophan-niacin synthesis can have increased dietary niacin needs and susceptible to niacin deficiency.[55]

Fortification

Although the United States did not enter World War II until after December 7, 1941, anticipatory national defense planning was already underway. In 1940, the United States had enacted a draft bill, and the Committee on Food and Nutrition of the National Research Council (now the Food and Nutrition Board of the IOM, referred to as FNB) was established to advice on nutrition for the population. The public health achievements that resulted were based on a concern of national defense rather than improving the lives of Americans, nonetheless the physical examination conducted as part of the draft did provide supporting evidence for the need to address population health.

Based on the combined judgment of nutrition authorities and the best available evidence, the FNB developed a table of RDAs detailing specific intakes for calories, protein, iron, calcium, and vitamins A, C, D, thiamin (B1), riboflavin (B2), and nicotinic acid (B3).[11] Nicotinic acid was renamed to niacin due to the baking industry's desire to avoid association with the tobacco chemical nicotine.[58]

Armed with the RDAs, the FNB developed specifications for enrichment of foods and presented the results to the FDA. The 1938 Federal Food, Drugs, and Cosmetic Act gave the FDA the authority to regulate standards of identity for foods, and this meant the FDA could regulate food fortification by specifying what nutrients (and in what quantities) could be added to which foods by describing the "standard of identity" for foods.[15]

The enrichment proposal was to add thiamin, riboflavin, iron, and the "pellagra-preventing factor" niacin to flour in amounts consistent with or higher than those would be found in stone-ground wheat. The public was also assured that the cost of flour and bread was not anticipated to rise. Enrichment would occur either in a change in milling that would retain the vitamins but remove most of the bran, or by addition of the vitamins to the flour post milling.[14] The addition of vitamins, supported by industry, was the answer. See Box 11.3 regarding this nutrient mix.

The FNB issued a press release in January, 1941 urging millers and bakers to begin producing enriched products immediately as the FDA standards were expected soon and would be close enough to the FNB proposal that waiting was not needed. The anticipation of war made this an emergency. Voluntary enrichment commenced, an early example of corporate self-regulation as discussed in Chapter 12, "Nanny State."

The FNB also discussed enriching sugar, but decided against it as a diversion of vitamins to sugar could mean a lack of vitamin supply for enriched bread.[14] Although today we see vitamins everywhere, the technology to make vitamins was then new. And, as discussed in Chapter 9, the almost universal adoption of store-bought bread was fairly new as well. Further public acceptance of tinkering with bread was not necessarily a given, and the idea of adding things to foods could be a suspect to the generation that had fought for pure and wholesome food regulations to keep inappropriate additions *out* of food.

In May, 1941, President Roosevelt called a National Nutrition Conference for Defence, at which the RDAs received national approval. The United States entered the war in December, 1941, and in early 1942 the Army and the Navy announced they would only procure enriched flour and bread, which created additional demand, acceptance, and interest by the public in enriched bread, which in turn, fostered industry's compliance. By 1942s end, three-quarters of the white bread sold in the United States was voluntarily fortified with thiamin, riboflavin, niacin, and iron.[2]

BOX 11.3   WHY WAS RIBOFLAVIN INCLUDED IN THE ENRICHMENT MIX?

Colleagues on the nutrition faculty in the City University of New York shed some light on this mystery. Niacin, folate, and iron metabolism are dependent on riboflavin coenzymes. Milk is an excellent source of riboflavin, but riboflavin is easily destroyed by exposure to light. Milk stored in glass bottles, which was the most common method of packaging milk prior to the introduction of milk carton packages, could lose up to 50% of riboflavin after light exposure. The loss of riboflavin from milk due to the available packaging at the time combined with the loss from refined grain products would thus put the population at risk for decreased bioavailability of all four nutrients—riboflavin, niacin, folate, and iron.

Although now 2 years past the call for voluntary enrichment, the FDA still had not released official standards. The industry complained that some bakers were taking advantage of the wave patriotic consumption by marketing additions of any type (soy flour, peanut flour) as "enriched" on the label. Concerned that voluntary enrichment would falter, the War Food Administration issued "Food Distribution Order No. 1" effective January 18, 1943. In addition to requiring enrichment of all commercially baked white bread, Order No. 1 regulated the baking, packaging, and distribution of bread; this consisted of elements related to rationing supplies of all kinds. Excess bread that failed to sell at the store could not be returned to the bakery. Wrapping should be economized, and thus bread could not be sold presliced.[59] Order No. 1 was repealed in 1946, ending mandatory fortification.

Health Outcome

Niacin status of the U.S. population is not currently available,[38] which leads to the conclusion that a risk of niacin deficiency is no longer considered a public health issue in the United States. Fortification is often credited with ending the pellagra epidemic. However, as Marion Nestle points out, pellagra death rates were declining before the 1943 mandated niacin fortification.[60] Voluntary fortification had started as early as 1941, and the war also brought dietary improvements through increased economic power. The democratizing effect of food rationing was also significant. Since food was rationed for everyone, the poor had increased access to previously unavailable foods.[58] *Is it possible that the pellagra epidemic could have been reversed in the absence of fortification by a multi-factored approach of public awareness campaigns, agricultural and diet diversification, financial support for people to buy the appropriate food items, and a reversal in grain milling techniques?*

1970s IRON: A POLITICAL SOLUTION FOR HUNGER

A proposal to significantly increase the amount of iron included in the standard of identity for enrichment caused significant controversy in the 1970s. The use of fortification as a political solution collided with a scientific concern over validity of measurement. The resulting conflict introduced the promotion of ideal nutrient status health outcomes versus widespread deficiency disease prevention.[15]

Background

Iron is an essential mineral used by the body to make proteins, including the oxygen-carrying hemoglobin and myoglobin. How much dietary iron is absorbed depends on variables: the amount of stored iron in the body, the rate of red blood cell production, the type and amount of iron consumed, as well as interactions with other substances. Vitamin C can enhance iron absorption, while calcium may inhibit iron absorption.[61]

Iron that is absorbed but not immediately used is stored by ferritin (a protein). Unabsorbed iron is picked up by cells lining the intestinal tract (enterocytes), which will eventually slough off and leave the body. However, once iron is absorbed and stored, blood loss is the only physiological means of excreting excess. Most people are protected from overload due to mechanisms regulating absorption, but too much iron can be toxic.[61] Men (and postmenopausal women) are at a higher risk of accumulating excess iron.

NHANES uses a marker of iron stores status called "body iron." Body iron is in a positive balance (≥ 0 mg/kg) when there is residual storage iron or in a negative balance (<0 mg/kg) when there is functional iron deficiency. The latter represents a deficit in iron required to maintain a normal hemoglobin concentration. Body iron can be estimated from a ratio of serum ferritin (the major iron-storage protein compound) and that of the soluble transferrin receptor (which helps deliver iron into cells). The body iron model produced lower estimates of iron-deficiency prevalence and a better prediction for anemia among nonpregnant women than using the ferritin method, and also appeared to be less affected by inflammation.[62]

Dietary Source

Dietary sources for iron include dried beans, dried fruits, egg yolks, liver, lean red meat, oysters, dark red poultry meat, salmon, tuna, and whole grains. Iron from meat, fish, seafood, and poultry is easier for the body to absorb than iron from vegetables. Mixing sources of iron at a meal will increase absorption of iron from vegetable sources. Water drawn from wells may also be a significant source of iron.

Recommended Iron Intake

Table 11.9 lists the iron recommendation for nonvegetarians. The RDAs for vegetarians are 1.8 times higher than for people who eat meat, as the iron from plant-based food is not as bioavailable. For infants from birth to 6 months, the FNB established an AI for iron that is equivalent to the mean intake of iron in healthy, breastfed infants.[63]

Iron Deficiency

Iron deficiency ranges from mild (observed by decreased serum ferritin) to functional tissue deficiency (observed by increase of serum transferrin receptor concentration, even though hemoglobin levels may still be normal) and iron deficiency anemia (IDA), in which iron stores are exhausted and levels of hemoglobin are low.[61] Public health is concerned with the impact of iron deficiency prior to the clinical manifestation of IDA.[64]

Iron deficiency can be due to lack of food, poor-quality food, impaired iron transport or absorption, as well as a consequence of chronic blood loss due to conditions such

TABLE 11.9   Recommended dietary allowances (RDAs) for iron

AGE	MALE	FEMALE	PREGNANCY	LACTATION
Birth to 6 months	0.27 mg[a]	0.27 mg[a]		
7–12 months	11 mg	11 mg		
1–3 years	7 mg	7 mg		
4–8 years	10 mg	10 mg		
9–13 years	8 mg	8 mg		
14–18 years	11 mg	15 mg	27 mg	10 mg
19–50 years	8 mg	18 mg	27 mg	9 mg
51+ years	8 mg	8 mg	–	–

Source: Adapted from National Institutes of Health, Iron Fact Sheet for Health Professionals, reviewed April 8, 2014. Available at http://ods.od.nih.gov/factsheets/Iron-HealthProfessional/, accessed September 25, 2014.
[a] Adequate intake (AI).

as gastrointestinal parasites. Iron deficiency during infancy and toddler years may have lasting effects on psychomotor development, cognitive performance, and social adjustment.[61] Iron deficiency is considered the most prevalent global nutrition problem.[45]

Populations at Risk from Iron Deficiency

Pregnant women need greater amount of iron due to increased red blood cell production and to supply the fetus and placenta. Infants and young children have increased needs for iron due to rapid growth. Preterm infants, low-birth-weight infants, or those whose mothers have iron deficiency are at risk of iron deficiency. All infants risk iron deficiency at 6–9 months unless diet includes bioavailable iron from food or fortified formula, as maternal reserves of iron transmitted in breastmilk become depleted.

Iron is added to infant formula. Most European infant formulas contain 4–8 mg/L of iron, whereas U.S. formula typically contained 10–12 mg/L. Some researchers comment that anemia rates are not significantly different between the United State and Western Europe, and instead of interpreting this as confirmation that more iron is better, we should be asking why more is better if a lesser amount seems to be sufficient.[65]

Frequent blood donors have an increased risk of iron deficiency. People with cancer are at risk due to anemia of chronic disease, chemotherapy-induced anemia, chronic blood loss, and deficiencies of other nutrients. People who have gastrointestinal disorders or have had gastrointestinal surgery that reduces iron absorption or incurs blood loss in the gastrointestinal tract are at risk. People with heart failure may experience poor nutrition, iron malabsorption, defective mobilization of iron stores, and cardiac cachexia.[61]

Fortification

In 1909, the amount of iron in our food supply was at 14.3 mg per capita per day, but by the early 1930s had trended downward to a low of approximately 12.5 mg. War-time bread and flour enrichment added 4 mg of iron per capita per day to our food supply. Post war, the supply of iron in our diet stabilized at roughly 14.5 mg per capita per day until the mid-1960s when it began again to rise. By 2000 our iron supply was at about 23 mg per capita per day.[66] *With such an increase, one might wonder why there is still iron deficiency at all in the U.S. population. Are there issues with the supply not being evenly distributed?*

The post-war escalation of iron in our food supply is in part attributable to industry fortification of breakfast cereals in the 1970s. In 1969, just 14% of ready-to-eat breakfast cereals were fortified but by 1979, 92% of such cereals were fortified.[15] As our grain consumption increased in the 1980s, our iron intake rose again.

Iron fortification was also federally sponsored; a response to a society shocked by the malnutrition, poverty, and hunger in American in the mid to late 1960s. The Special Supplemental Nutrition program for Women, Infants, and Children (WIC) arose from this era. WIC directly addresses specific nutrient deficiency risks in a vulnerable population by providing vouchers for the purchase of specific foods, some of which are required to be fortified with certain nutrients. Iron is one of the WIC nutrients of interest.[15]

Nutrition, poverty, and hunger were significant political and social issues at the end of the 1960s. The April 1967 Senate Poverty Subcommittee hearings included television coverage of Robert Kennedy touring sharecroppers' shacks, and the public and media attention incited further studies, new committees, hearings, and reports. Politically, conservatives and liberals clashed over the extent of the problem and solutions; the USDA was caught having to defend both farm subsidies and food programs and was reluctant to escalate food aid.[67] Popular media continued reporting on poverty and hunger, reaching an apex with the shocking 1968 CBS series *Hunger in America*, further engaging our social conscience with a need to act.

Congress had charged the CDC with investigating malnutrition, and in January 1969, the CDC released the preliminary results of the Ten-State Nutrition Survey, which included both dietary intake recall and blood testing. The survey results showed that generally the amount of available food was a greater problem than food quality.[68] The only nutrient varying by race or income was vitamin A. However, dietary intake of iron was generally low across all demographic categories, and for children under the age of 36 months, iron was the only nutrient for which mean intakes were generally below RDA.[68]

Additionally, high prevalence of low hemoglobin and hematocrit values was found throughout all segments of the population, and unexpectedly, this included many adolescent and adult males. Given a cutoff for males of 14 g/100 mL, 22% of white men and 53% of black men aged 45–49 had low hemoglobin.[68] Iron deficiency anemia, if defined by low levels of hemoglobin, seemed widespread.[68]

Some physicians argued instead that the unexpectedly high result among males reflected a problem with the survey instead of true iron deficiency.[69] One concern was that the cutoff indicating deficiency was too high. However, the heightened political and media attention on malnutrition and hunger meant that some subtleties of the data were lost in the call to act. An identifiable concrete nutritional issue could not be explained away as ignorance and indifference as to diet. The CDC concluded that solving this nutritional deficiency would likely require making foods available in the market place with a higher iron-to-calorie ratio than available at the time,[68] a different policy solution than encouraging or promoting the consumption of foods that are naturally iron-rich.

The pressure to act was reinforced by a White House conference on Food, Nutrition, and Health in early December 1969; by the end of the month, the FNB recommended in a short (just over two pages) report that the best way to increase dietary intake of iron was to triple the amount regulated by the FDA standards for the bread and flour enrichment mix.[70] A Committee on Iron Nutritional Deficiency was appointed to consider practical measures and invited industry representative to a workshop in January of 1970. Industry quickly sent a proposal to the FDA to quadruple the amount of iron in flour and bread on April 1, 1970—long before the January meeting notes were published that July.[71]

The FDA published a somewhat less aggressive proposal at the end of 1971 that still received hundreds of comments and caused considerable activity in the scientific community.[72] Objections were raised on the consequence of iron overload (hemochromatosis) particularly on those individuals suffering from hereditary hemochromatosis.[4] Additionally, the launch of NHANES I (1971–1974) provided additional information on population dietary intake and nutritional status, showing that low dietary iron intake (40%–56% below the RDA) in females aged 10–54 did not necessarily translate into serum iron deficiency.[73]

After 3 years of study, and a change in commissioners, the FDA announced in October of 1973 that it would approve industry requests to increase iron enrichment from a range of 13–16.5 to 40 mg/lb with the increase to begin April 15, 1974. The declining use of iron cookware was cited as one of the reasons that enrichment was necessary to increase iron in our diet.[72] Objections continued such that the FDA agreed to push back the effective date for the increase and held a hearing on the matter in April 1974 instead.[74] From that point, the iron increase proposal seemed to have entered a holding pattern[73] until November, 1977, when yet a different Commissioner proposed to withdraw the idea, concluding that increases of iron are not proved to be needed, safe, or effective. One reason for the reversal was apparently that a Swedish study found that 5% of the population had consistently elevated serum iron levels, and 2% preclinical hemochromatosis, a higher incidence than previously thought.[4]

However, by May, 1980, increasing iron had a foothold on the FDA agenda again,[75] and in 1981, the FDA increased the iron standard for enriched flour to a set level of 20 mg per pound.[15] Critics commented that no new data had been reviewed and the move seemed to be in response to industry (in particular, the Pennwalt Corporation, a major supplier of vitamin-iron enrichment mixes) but this time, the increase did not draw the same fervor as the earlier proposal.

Current Iron Status in the U.S. Population

NHANES has reported on iron status since inception in 1971. Generally, women have been at risk for iron deficiency, and men at risk from iron excess. The prevalence of anemia did substantially decrease in the 1970s and 1980s.[15]

Over 1999–2002, women of childbearing age had a 13% prevalence of low serum ferritin, while men had a 29% prevalence of high serum ferritin.[38] Starting in 2003, NHANES limited iron status reporting (at least that of a certain measurement) to children aged 1–5 years and women aged 12–49 (childbearing age). Over 2003–2006, less than 10% of children (8% of boys and 5% of girls) and 10% of women had iron deficiency as measured by negative body iron balance.[38]

Using the body iron store measure, racial/ethnic groups show variation in iron status, with higher rates of iron deficiency in Mexican-American children (11%), Mexican-American women (13%), and in non-Hispanic black women (16%) when compared to other race/ethnic groups in these age brackets. Among pregnant women, depleted iron stores deficiency is more common in Mexican-American (23.6%) and non-Hispanic black women (29.6%) than in non-Hispanic white women (13.9%).[76]

Iron Toxicity

The risk of iron toxicity is greater among whites of Northern European descent, due to an increased prevalence in this group of hereditary hemochromatosis caused by gene mutation. The rate of homozygosity (having two copies of a mutated gene) is 4.4 per 1000 among white persons in the United States with much lower rates found in other racial/ethnic groups.[77]

In 2002, Backstrand commented that from 1979 to 1992, age-adjusted rates of hemochromatosis-related deaths increased by 60%, and, even though methodological issues may have contributed to the increase, the trend was consistent with the rise of iron in our food supply. As hemochromatosis is a more prevalent problem than NTDs, for which we have adopted folic acid fortification, Backstrand called out for increased regulation and scientific scrutiny, commenting that industry-initiated iron fortification may have contributed to a much more prevalent negative health outcome than NTDs.[15]

In 2003, the National Heart Lung and Blood Institute and the National Human Genome Research Institute sponsored a 5-year, multisite study, HEIRS (Hemochromatosis and the Iron Overload Screening), to study the risk of iron overload and chronic disease associated with the known susceptible genotypes to make informed decisions regarding primary care clinical and genetic screening for hemochromatosis.[78]

1980s CALCIUM: INDUSTRY-DRIVEN FORTIFICATION FOR OPTIMAL HEALTH

Background

Calcium, a mineral, plays an integral role in nerve conduction, muscle contraction, blood clotting, and, as the major mineral of the skeleton, structural support. Calcium is critical for normal skeletal development of skeleton and teeth and optimizing bone mass accretion prior to adulthood.[79] The body stores calcium in the skeleton and very tightly regulates serum calcium, such that fluctuations in dietary intake will not cause serum calcium to fluctuate.[80]

Calcium is considered to help prevent osteoporosis, a disease of low bone mass, and low calcium intake has been associated with increased risk for bone fractures as well as osteoporosis.[79] Osteoporosis is reported to affect 10 million Americans and cause 2 million fractures per year.[81] However, as noted in Chapter 7, despite the strong case for milk as the best source of calcium, osteoporotic fractures occur at higher rates in milk-consuming Western countries than countries in Asia and sub-Saharan Africa with less dairy consumption.[82] The connection is unclear, but one reason for this difference may be that high intakes of protein lead to decreased calcium absorption.[83]

Dietary Sources of Calcium (Nonfortified)

Dairy products are naturally high in calcium: 8 oz of milk contains an average of 288 mg of calcium. Canned fish such as sardines or salmon, due to the presence of consumable soft small bones, also contain calcium. Some vegetables, such as turnip greens and kale, also have calcium, as does spinach, although the bioavailability of calcium in spinach is poor. Grains provide a small amount of calcium, but because people eat grains often, grains can be a significant contributor of calcium in our diet. A cup of kale contains an average of 97 mg of calcium.[80]

The amount of calcium we absorb varies. As we increase calcium intake our absorption becomes less efficient, such that consuming smaller periodic doses may be more effective than a single larger dose. We are also less efficient at absorbing calcium as we age. Conversely, some phytic and oxalic acid containing foods can inhibit absorption of calcium, but these interactions should balance out in a varied diet. Vitamin D improves calcium absorption.[80]

Recommended Intake

Chapter 7 on milk and Chapter 3 on dietary guidance provide context about dairy recommendations over time. The RDAs for calcium are calculated as required for bone health and to maintain adequate rates of calcium retention in healthy people. See Table 11.10 for the most recent (2010) RDAs for calcium.

Calcium Deficiency

As noted above, circulating levels of serum calcium are tightly regulated within the body. A shortfall in consumption will not be noticed immediately. Medical problems or treatments, however, can cause significant hypocalcemia, causing numbness and tingling in the fingers, muscle cramps, convulsions, and abnormal heart rhythms. Hypocalcemia, if untreated, is fatal.

Inadequate calcium over the long term can lead to poor health outcomes of public health concerns, namely osteopenia and osteoporosis, without ever-reaching hypocalcemic status. The best health outcome is felt to be most likely when bone mass is

TABLE 11.10   Calcium: Recommended dietary allowances (RDA) and tolerable upper limits (UL) in milligrams

AGE	RDA	UL
0–6 months	200[a]	1000
7–12 months	260[a]	1500
1–3 years	700	2500
4–8 years	1000	2500
9–13 years	1300	3000
14–18 years[b]	1300	3000
19–50 years[b]	1000	2500
51–70 years	1000 (male) 1200 (female)	2000
71+ years	1200	2000

Source: Adapted from Committee to Review Dietary Reference Intakes for Vitamin D and Calcium, Food and Nutrition Board, Institute of Medicine, *Dietary Reference Intakes for Calcium and Vitamin D*. Washington, DC: National Academy Press, 2010.

[a] Adequate intake.
[b] Includes pregnant and lactating.

optimally developed during young adulthood, maintained through the middle years, and loss is minimized during the elderly years.

Populations at Risk of Calcium Deficiency

Postmenopausal women, women with amenorrhea, or the female athlete triad (amenorrhea, eating disorders, and osteoporosis), individuals with lactose intolerance or cow's milk allergy, and vegans are at higher risk of calcium deficiency.

Fortification

In 1984, an expert panel of the National Institutes of Health held a consensus conference to determine optimal calcium intake for disease prevention as well as growth, and released a statement advising that calcium was a major player in preventing and managing osteoporosis (the other being estrogen).[84] The food industry, perceiving an increased market for nutrition, and having no inhibiting FDA regulation, began introducing a variety of calcium-fortified products.[2]

The National Dairy Board (NDB) was authorized under the Dairy Act of 1983 to promote milk and other dairy products. Accordingly, the dairy industry ran with marketing dairy as a source of calcium. Between 1984 and 1985, the NDB spent about $20 million on the "Dairy Foods, Calcium the Way Nature Intended" campaign.[85]

Ten years later, in 1994, the NIH held a consensus conference regarding optimal calcium intake, and recommended calcium intakes higher than the then current RDAs for many age and gender groups. The NIH report also concluded that although the preferred source of calcium is through calcium-rich foods (specifically dairy products), private industry should help promote optimal calcium intake by developing a wide variety of calcium-fortified foods.[86] Calcium fortified orange juice, for example, provides a convenient source of calcium in a similar fluid form for people who do not drink milk.

Upon updating the DRIs in 1997, the IOM, recognizing the increase in calcium-fortified food products, recommended surveillance of such products and monitoring the effect on calcium intake, but the food industry already had at least a 10-year's head start.[87]

Although some fortified products may certainly help people achieve adequate calcium intake, one must pause for reflection at seeing Hershey's chocolate syrup with calcium on store shelves.[88] As Hershey's chocolate syrup is often used on ice cream, or in milk, one might wonder what critical role this fortified syrup plays in addressing nutritional deficiencies. Other nutrients besides calcium have been added to the syrup, such as zinc, biotin, vitamin e, and vitamin b5, apparently replacing the need to use crushed multivitamins as an ice cream topping to meet one's daily nutritional needs.

Recall that the limitation the FDA places on fortifying candy and other such substances is only a policy. The best the FDA could do with calcium-fortified chocolate syrup was to enforce labeling regulations prohibiting the use of plus, +, or fortified. Thus we have chocolate syrup "with calcium."[89]

Current Calcium Status in the U.S. Population

The NHANES survey provides an estimate of calcium intake in the United States through both food and dietary supplements. At a population level, information from the 2003–2006 surveys indicated that certain groups (boys and girls aged 9–13, girls aged 14–18, women aged 51–70, and both men and women 71+) had greater than 50% prevalence of inadequate calcium intake. Generally, females are less likely than males to get sufficient calcium from food. Many people take calcium supplements (including almost 70% of older women) and use dietary supplements containing calcium. However, even when considering total calcium intakes including supplements, some age groups are still at high risk of calcium inadequacy, namely females aged 4 years and older, males aged 9–18 years and older than 51 years.[80]

Calcium and Osteoporosis

Although the necessity of calcium to maximize bone strength is well accepted, there remains some doubt as to if the level of calcium intake recommended for adults is effective in reducing risk of osteoporosis. Some countries have much lower average daily calcium intake while also having a lower incidence of bone fractures. Harvard researchers found that large prospective long-term dietary studies did not reveal an association between calcium intake and fractures. Likewise, combined randomized trials comparing calcium supplements with a placebo did not show any protective effect of calcium against fractures. The interaction between calcium and vitamin D remains an important factor.[90]

1990s FOLIC ACID: A PUBLIC HEALTH INTERVENTION TO HELP SOLVE A MEDICAL PROBLEM?

Background

Folate, a water-soluble B vitamin, is essential for many biological functions. In general, it helps produce and maintain new cells. Folate is required for the metabolism of homocysteine and helps maintain optimal levels of this amino acid in the body.[91] This is especially vital during periods of rapid cell division and growth, such as in infancy and pregnancy.[92] Folate is important for cell division and growth since it is a cofactor in nucleic acid synthesis.[93]

Lucy Wills, a leading English hematologist, discovered folate. Wills conducted work in India in the late 1920s and early 1930s on macrocytic anemia during pregnancy. Wills decided to investigate possible nutritional treatments by first studying the effects of dietary manipulation on a macrocytic anemia in albino rats. Pregnant rats fed the same diet as Bombay Muslim women became anemic, dying before giving birth, but the addition of yeast to the diet prevented anemia. The unknown nutritional compound in the yeast was initially referred to as *Wills factor*. Folate was identified as the corrective substance in yeast extract in the late 1930s, was extracted from spinach leaves in 1941, and first synthesized in 1945.[94]

Dietary Source of Folate

Folate is found naturally in dark, leafy green vegetables such as spinach and broccoli, whole grains, legumes, egg yolks, and citrus fruits. The typical Western diet contains about 0.2 mg natural folate per day. Natural food folate is approximately 50% less bioavailable than its synthetic form, folic acid. There is some loss from harvesting, processing, and cooking of foods and some folates remaining bound and unavailable within the plant material.[95] Also folic acid is absorbed more rapidly and effectively across the intestine than folate.[96]

Recommended Intake

Table 11.11 lists the current RDAs for folate as micrograms of dietary folate equivalents (DFEs). DFEs reflect the higher bioavailability of folic acid (85%) over food folate (50%).

- 1 µg DFE = 1 µg food folate
- 1 µg DFE = 0.6 µg folic acid from fortified foods or dietary supplements consumed with foods

TABLE 11.11   Recommended dietary allowances (RDAs) for folate

AGE	MALE	FEMALE	PREGNANT	LACTATING
Birth to 6 months[a]	65 μg DFE[a]	65 μg DFE[a]	–	–
7–12 months[a]	80 μg DFE[a]	80 μg DFE[a]	–	–
1–3 years	150 μg DFE	150 μg DFE	–	–
4–8 years	200 μg DFE	200 μg DFE	–	–
9–13 years	300 μg DFE	300 μg DFE	–	–
14–18 years	400 μg DFE	400 μg DFE	600 μg DFE	500 μg DFE
19+ years	400 μg DFE	400 μg DFE	600 μg DFE	500 μg DFE

Source: Adapted from National Institutes of Health, Folate, dietary supplement fact sheet, last reviewed December 14, 2012. Available at http://ods.od.nih.gov/factsheets/Folate-HealthProfessional/, accessed September 26, 2014.

[a] Adequate intake (AI).

- 1 μg DFE = 0.5 μg folic acid from dietary supplements taken on an empty stomach

Folate Deficiency

A serum folate concentration is commonly used to assess folate status, with a value above 3 ng/mL considered adequate, but serum folate is sensitive to recent dietary intake. Erythrocyte folate concentration might be a better indicator of long-term folate status, and a concentration above 140 ng/mL is considered to be adequate. The primary clinical sign of folate deficiency is megaloblastic anemia.[96]

Populations at Risk of Deficiency

Like other B vitamins, people who have malabsorptive disorders issues (such as celiac or Crohn's disease) or who consume too much alcohol can be at risk of deficiency.[96] Women who lack adequate folate intake have an increased risk of giving birth to an infant with NTDs,[96] defects of the brain and spinal cord that occur when the neural tube fails to close early in embryonic development. The two most common NTDs are anencephaly and spina bifida. In anencephaly, much of the brain does not develop. Babies with this condition are either stillborn or die shortly after birth. In spina bifida, the fetal spinal column does not close completely during the first month of pregnancy, typically causing some paralysis of the legs.[92]

According to the Centers for Disease Control and Prevention (CDC), 3000 pregnancies are affected by spina bifida or anencephaly annually.[33] The neural tube closes early in embryonic development (17–30 days of gestation). As 50% of pregnancies in the United States are unplanned and as NTDs can occur even when no family history of the disorder, NTDs can occur before a woman has access to prenatal care or takes supplements. The medical costs for treating U.S. children with NTDs are estimated to exceed $200 million per year.[97]

Fortification

The relationship between folate deficiency and NTDs was hypothesized as early as 1965.[98] Since this time, various studies have strengthened this hypothesis; however, the exact biochemical pathway of folate that works to prevent NTDs remains unclear.[95] In the 1970s, it became possible to add folic acid to vitamins, and there was some discussion then about adding folic acid to flour, but the vitamin approach was retained. Interestingly, lowering the RDA for folate intake from 40 to 200 μ was being considered in the late 1980s. In 1990, the National Academy of Sciences recommended additional studies on folate intake prior to pregnancy, but suggested that women of childbearing years follow dietary guidelines and increase consumption of fruits, vegetables, and whole grains. Nutrition from food sources was being suggested as a recommendation over supplements, and the vitamin industry was not pleased.[99]

The 1990 Nutrition Labeling Education Act required the FDA to examine the validity of 10 specific nutrient–disease associations as the basis for allowing health claims on foods and dietary supplements without waiting for industry to make a petition. Folate and NTDs was one of the mandated nutrient–disease associations, which reportedly surprised FDA scientists at the time. Furthermore, undertaking this review was challenging, because under the NLEA, the FDA can only use publically available information and cannot rely on unpublished studies. At the time, the available public evidence was inconclusive, leaving the FDA struggling to fulfill this mandate, when a British randomized controlled trial study reported that women with a previous history of an NTD pregnancy reduced their recurrence risk by 70% by taking 4 mg (or 4000 μg) of folic acid daily, a much larger dose than the RDA.[99]

Based on the British study and other evidence, the CDC issued a formal recommendation in September, 1991 that women who had already had an NTD-impacted pregnancy should take 4 mg of folic acid supplement daily. But at the time, a 4-mg folic acid tablet was not available, and a 1 mg folic acid tablet was available only by prescription. Over-the-counter (OTC) vitamins contained smaller amounts of folic acid, but an attempt to get the recommended 4 mg through OTC vitamins would likely expose women to toxic or teratogenic levels of vitamins A and D. The choices were to reformulate prescription prenatal vitamin and/or OTC supplements, make folic acid a separate standalone vitamin, or fortify the food supply for everyone.[99]

The FDA rejected a health claim for folic acid and NTD prevention in November, 1991, in part because the results of the British study were still being evaluated, and in part because the level of supplement in the study was considered a pharmacologic dose. Media and public health authorities lambasted the FDA for not being willing to generalize from the British study, which caused the FDA to push unpublished studies to make their results public more quickly, which would then allow the FDA to use the data under the NLEA mandate.[99]

One such study was a Hungarian randomized control trial that found a 100% reduction in risk of a first occurrence of an NTD-affected pregnancy among women who took a prenatal vitamin containing 800 μg of folic acid daily. The second study, a case control, suggested that 400 μg of folic acid daily from multivitamin/mineral supplements was associated with a reduced risk of NTD but also suggested that a diet adequate in folate (meaning more than 250 μg daily) was also protective. With this new evidence, the recommendation changed from women who already had a NTD pregnancy to all women of childbearing age.[99]

The FDA was now under a time crunch to complete final regulations by the 1993 NLEA deadline. Now able to cite published data, the FDA found scientific agreement on the relationship between folic acid and NTDs, but still could not reconcile concerns over the safety of high folic intake across the population. One major concern was the risk of adverse effects from undetected vitamin B12 deficiency, particularly among the elderly, and it was not necessarily clear that overall, social costs of reduced NTDS would outweigh the social cost of increased neurologic disabilities.[97]

Public support for a fortification effort began to grow while the FDA's folic acid subcommittee struggled with the nutrient-focus of health claims versus promotion of a healthy diet. In October of 1993, the FDA finally proposed that health claims for folic acid and NTDs could be permitted if the food contained 0.04 mg (400 μg) or more of folic acid per serving and if the claim was limited to NTD risk reduction rather than NTD prevention. The FDA also proposed limitations on foods to which folic acid could be added, as well as limitations on levels, to resolve safety issues for the rest of the population. Children, for example, may be high consumers of enriched products and breakfast cereals, a population that does not need the high levels of folate.

The FDA also made an effort to recognize a possible wider public health benefit from folic acid fortification, namely a possible decline in rates of cardiovascular disease.[100] Folic acid is known to help break down homocysteine, which at high levels is a risk factor for cardiovascular disease (CVD).[92] The FDA hoped that the possible wider population benefit of reduced risk in CVD would make the fortification program more acceptable to those that felt the policy exposed too many people to risks while only benefiting a small segment of the population.[101] Unfortunately, homocysteine is now considered a marker of CVD disease instead of a cause.[102] Other possible health associations have been studied and there is evidence that adequate folate intake might reduce the risk of some forms of cancer. Research has also indicated that folate may be a helpful addition to treatment for depression.[96]

The FDA's caution in proposing a health claim was due to the recognition that once a health claim was authorized, food manufacturers might indiscriminately add folic acid to products to leverage the marketability of a health claim, which could expose many people to a dangerous excess of folate. Yet the FDA also did not want to discourage women having access to multiple servings of fortified foods.[99]

Less in this case is more, and addressing these concerns required a multiple pronged approach:

- Folate health claims were authorized under certain conditions.
- Fortification policy of adding folate to the standard of identity regulations for enrichment.
- An amendment to food additive regulations that would prohibit folic acid from being added to other foods than breakfast cereals and those products falling under the enrichment standard of identity.[103]

In 1996 the FDA added folate to the standards of identity for enriched breads, flours, corn meals, pastas, rice, and other grain products. Effective January 1, 1998, these products labeled as enriched must contain 140 mg of folic acid per 100 g of enriched cereal-grain product.[104]

Folate Status in the U.S. Population

The exact blood folate concentration required to reduce risk for folate-sensitive NTDs is not known. However, several studies have found the greatest reduction in NTDs with blood folate concentrations much higher than deficiency levels, implying that elimination of folate deficiency may not be enough to prevent folate-status related NTDs.[98]

Results from the NHANES for 1999–2000 showed an increase in folate status among women of childbearing age.[33] Average serum folate concentrations increased significantly after the implementation of folic acid fortification. Medium blood folate levels among women of childbearing age increased from 4.8 to 13.0 ng/mL between 1994 and 2000.[93] Before fortification began, folate deficiency, as determined by blood folate levels, was approximately 12% for women of childbearing age and decreased to less than 1% after 1998.[105]At current fortification levels, U.S. adults who do not consume supplements or who consume an average of ≤400 μg folic acid per day from supplements are unlikely to exceed the upper limit in intake for folic acid.[106]

Health Outcomes Post Folic Acid Fortification

According to the CDC, post folic acid fortification, the reported prevalence of spina bifida declined 31%, and the prevalence of anencephaly declined 16%. An estimated 1000 more babies are born without NTDs since fortification.[107] During October 1998 to December 1999, the birth prevalence of spina bifida in the United States declined 23% compared with 1995–1996. NTD declines were not significant from 1999 to 2005 but rates for 2005 were 17.96 per 100,000 live births, the lowest ever reported according to the CDC.[33]

However, racial–ethnic disparities in the rates of NTDs remain.[108] The prevalence of NTD-affected pregnancies had been higher among Hispanic women than women in other racial/ethnic populations before fortification and remained higher post fortification.[33] Folic acid intake from fortified foods and supplements has been found to be lower among Hispanic women.[109] Additionally, Hispanic women are less likely to have heard about folic acid, to know it can prevent birth defects, or take vitamins containing folic acid before pregnancy.[108]

A policy statement from the American College of Genetics (ACG) notes that the amount of decrease in NTD risk has been disappointing. Some groups have accordingly called for an increase in the amount of folic acid fortification.[110] The ACG, however, does not recommend increasing folic acid fortification, pointing out that it is possible that the remaining occurrences of NTDs are folate-resistance, and that there is no evidence yet that increasing folic acid will prevent additional cases. Furthermore, the actual content of folic acid in fortified food has been reported to be 50% higher than as stated on the label.[110] Many products have been found to contain higher levels of folic acid than required by the FDA regulation, which may result in a higher folic acid intake from fortified foods than expected.[109]

Folic acid fortification is a shift in public health policies that stressed the necessity of widespread nutrient deficiency as a prerequisite for fortification. NTD prevention is aimed at the population of women who actually become pregnant. However, reaching this population through primary and prenatal care is difficult due to the numerous reasons, including the early manifestation of NTDs in gestation, the fact that many pregnancies are unplanned, and the lack of consistent, universal access to primary healthcare. Women may not have access to prenatal care and counseling until it is too late to address NTDs. Furthermore, it was felt that many women of childbearing age would not adhere to these recommendations.[98]

Consider the statement above concerning the lack of adherence in light of the discussion in Chapter 12 on individual liberty. It is important to note that many women may feel comfortable assessing their own individual risk of pregnancy. Women that felt their risk of pregnancy was quite small might reasonably choose not to adhere to a supplement regime, or may reasonably choose to simply eat a more nutritious diet. Folate fortification covers the entire population of women who risk becoming pregnant to reach the much smaller actual target population, women who do become pregnant. Additionally, even women that do become pregnant may spend quite a lengthy period of time of their childbearing years not being pregnant. In other words, the number of person-years spent being exposed to folic acid may be far greater than the period of time in which the folic acid is truly preventative. Folate fortification also exposes the rest of the American population to higher levels of this vitamin.

Some argue that the original clinical trials that produced the convincing evidence on folic acid and risk reduction for NTDs used supplementation, not fortification, and that supplementation is really the preferred policy option, instead of fortification, particularly in the context that we are using fortification, which is a public health intervention, to help solve a medical problem.[110] Others argue that any food fortification policy that includes folic acid should also include vitamin B1 to address the masking deficiency issue noted above.[111]

CONCLUSION

Fortification is part of our food environment, and has been for so over the lifetime of most Americans. The goiter belt has vanished. The iodization of salt is a text book example of a successful public health campaign, in part because the campaigns incited demand but preserved consumer choice at the time. Looking forward we need to be conscious of trends in diet and food processing that might impact the level of iodine in our diet.

For rickets, we can look back and be thankful that as children we were given fortified milk and formula instead of tablespoons of cod-liver oil, but looking forward we need to recognize the possibility of reemergence as children may incur less sun exposure today due to the use of sunscreen and greater time spent indoors.

Pellagra has become relatively unheard of as a population level health concern in countries where the population has widespread access to a nutritionally appropriate diet. Measuring the niacin status of the population is apparently no longer a public health concern in the United States, as niacin is not one of the 58 biochemical indicators reported by the CDC from NHANES data in *CDC's Second Nutrition Report*.[38] Looking forward we must be aware of subpopulations at risk and dietary trends that may impact the level of consumption of enriched grain products.

The 1970s era increase of iron fortification in the enrichment mix presents an interesting example of an unexpected initial pushback from the medical community followed by relative indifference, perhaps due to the development of a more circumscribed primary care role in handling iron toxicity.

Calcium fortification has made alternative sources of this nutrient available to the population, such as orange juice, and, as such fortification has not saturated the marketplace completely, consumers are able to still choose nonfortified products. Looking forward, however, this may be an area in which increased regulation and monitoring by the FDA may be useful.

Folic acid fortification is a notable shift in public health policies that stressed the necessity of widespread nutrient deficiency as a prerequisite for fortification. The FDA successfully regulated folic acid fortification to avoid rampant introduction by industry into foods besides breakfast cereals and those using the enrichment mix. Looking forward, this approach should inform any new fortification initiatives, while monitoring of folic acid fortification and issues with vitamin B1 deficiency should continue.

With the new proposed Nutrition Facts label (see Chapter 4) our expressed nutrients of public health interest may change. By the time you read this chapter you may be able to discern if there is any consequential interest in new trends of nutrient fortification.

Our use of historical context is aimed at highlighting the potential mutability of food policies that may seem to be a "given," such as fortification. In the case of folic acid, fortifying the food supply was a specific choice made over the alternative of reformulating dietary supplements. In the case of grain enrichment, a conscious choice was made to add nutrients postprocessing instead of the alternative of changing the milling process to retain more nutrients.

When enriched flour and bread were made mandatory during World War II, the Office of Health and Welfare and Defense Activities produced a "twenty questions" consumer leaflet.[112] Today, fortified products are generally accepted by consumers, but without question. *Should we?*

REFERENCES

1. Chakravarty I, Sinha RK. Prevalence of micronutrient deficiency based on results obtained from the national pilot program on control of micronutrient malnutrition. *Nutr. Rev.* 2002;60(May Pt 2):S53–S58.
2. Bishai D, Nalubola R. The history of food fortification in the United States: Its relevance for current fortification efforts in developing countries. *Econ. Dev. Cult.* 2002;51:37–53.

3. Allen L, de Benoist B, Dary O, Hurrell R (eds.). *Guidelines on Food Fortification with Micronutrients*. Geneva: World Health Organization/Food and Agriculture Organization, 2006.

4. Quick JA, Murphy EW. *The Fortification of Foods: A Review*. Agriculture Handbook No. 598. Washington, DC: U.S. Department of Agriculture, Food Safety and Inspection Service, 1982.

5. Cook JD, Reusser ME. Iron fortification: An update. *Am J Clin Nutr* 1983;38:648–659. http://pdf.usaid.gov/pdf_docs/Pnaaq794.pdf.

6. Eat right. What's the difference between the terms "enriched" and "fortified" on food labels? Academy of Nutrition and Dietetics. Available at http://www.eatright.org/Public/content.aspx?id=6442453536, accessed September 26, 2014.

7. United States Food and Drug Administration. FDA's approach to the GRAS provision: A history of processes. Last updated April 23, 2013. Available at http://www.fda.gov/Food/IngredientsPackagingLabeling/GRAS/ucm094040.htm, accessed September 26, 2014.

8. Institute of Medicine. *Dietary Reference Intakes for Calcium and Vitamin D*. Washington, DC: National Academies Press, 2011. Available at http://www.ncbi.nlm.nih.gov/books/NBK56070/pdf/TOC.pdf, accessed October 1, 2012.

9. Semba R. The discovery of the vitamins. *Int. J. Vit. Nutr. Res.* 2012;82(5):310–315.

10. Funk C. The etiology of the deficiency diseases. *J. State. Med.* 1912;20:341–368.

11. National Research Council. *Recommended Dietary Allowances*. Washington, DC: The National Academies Press, 1941. Available at http://www.nap.edu/catalog.php?record_id=13286, accessed September 26, 2014.

12. Wilder RM. Misinterpretation and misuse of the recommended dietary allowances. *Science*. 1945;(101):2621.

13. U.S. Food and Drug Administration. Significant dates in U.S. food and drug law history. Last updated March 25, 2014. Available at http://www.fda.gov/aboutfda/whatwedo/history/milestones/ucm128305.htm, accessed September 26, 2014.

14. Wilder R, Williams R. Enrichment of flour and bread. A history of the movement. *Bulletin of the National Research Council* 1944, Number 110. Appendix C, Resolution and Actions of the Food and Nutrition Board of National Research Council, National Academy of Sciences, Washington, DC.

15. Backstrand J. The history and future of food fortification in the United States: A public health perspective. *Nutr. Rev.* 2002;60(1):15–26.

16. Committee on Use of Dietary Reference Intakes in Nutrition Labeling. Overview of Food Fortification in the United States and Canada. *Dietary Reference Intakes: Guiding Principles for Nutrition Labeling and Fortification,* Chapter 3. Washington, DC: The National Academies Press, 2003.

17. *United States v. 119 Cases ...* "New Dextra Brand Fortified Sugar," 231 F. Supp. 551 (D. Fla. 1963), aff'd per curiam, 334 F 2d 238 (5th Cir. 1964).

18. 21CFR104.20, revised as of April 1, 2014.

19. Milman N, Byg KE, Ovesen L, Kirchhoff M, Jürgensen KSL. Iron status in Danish women, 1984–1994: A cohort comparison of changes in iron stores and the prevalence of iron deficiency and iron overload. *Eur. J. Haematol.* 2003;71(1):51–61.

20. Swanson CA, Zimmerman MB, Skeaff S, Pearce EN, Dwyer JT, Trumbo PR et al. Summary of an NIH workshop to identify research needs to improve the monitoring of iodine status in the United States and to inform the DRI. *J.Nutr.* 2012;142(6):1175S–1185S.

21. Feyrer J, Politi D, Weil DN. The cognitive effects of micronutrient deficiency: Evidence from salt iodization in the United States, NBER Working Paper Series, Working Paper 19233, National Bureau of Economic Research, July 2013. Available at http://www.econ.brown.edu/faculty/David_Weil/Politi%20Feyrer%20Weil%20w19233.pdf, accessed September 25, 2014.

22. Higdon J. *An Evidence-Based Approach to Vitamins and Minerals: Health Benefits and Intake Recommendations.* New York: Thieme Medical Publishers, 2003.
23. Bruhn JC, Franke AA, Bushnell RB, Weisheit H, Hutton GH, Gurtle GC. Sources and Content of Iodine in California Milk and Dairy Products. *J. Food. Protection.* 1983; 46(1):41–66. Available at http://drinc.ucdavis.edu/research/vol46.pdf, accessed September 25, 2014.
24. Public Health Committee of the American Thyroid Association, Becker DV, Braverman LE, Delange F, Dunn JT, Franklyn JA, Hollowell JG et al. Iodine supplementation for pregnancy and lactation-United States and Canada: Recommendations of the American Thyroid Association. *Thyroid.* 2006;16(10):949–51.
25. Langer PL. History of goitre. *Endemic Goitre.* Geneva, Switzerland: World Health Organization, 1960:9–25. WHO Monograph Series No. 44.
26. 21 CFR (G) 100.155. Salt and Iodized Salt, April 1, 2012.
27. Li M, Eastman CJ. The changing epidemiology of iodine deficiency. *Nat. Rev. Endocrinol.* Advance online publication 3 April 2012. Available at http://www.iccidd.org/cm_data/2012_Li_Teh_changing_epidemiology_of_iodine_deficiency_NatRevEnd.pdf, accessed September 25, 2014.
28. National Institutes of Health, Iodine Fact Sheet for Health Professionals, reviewed June 14, 2011. Available at http://ods.od.nih.gov/factsheets/Iodine-HealthProfessional/, accessed September 25, 2014.
29. Zimmermann MB, Andersson M. Update on iodine status worldwide. *Curr. Opin. Endocrinol. Diabetes. Obes.* 2012;19(5):382–387.
30. Caldwell KL, Pan Y, Mortensen ME, Makhmudov A, Merrill L, Moye J. Iodine status in pregnant women in the National Children's Study and in U.S. women (15–44 years), National Health and Nutrition Examination Survey 2005–2010. *Thyroid.* 2013;23(8):927–937.
31. Perrine CG, Pan Y, Mortensen ME, Makhmudov A, Merrill L, Moye J. Intakes of dairy products and dietary supplements are positively associated with iodine status among U.S. children, *J. Nutr.* 2013;143:1155–1160.
32. Dobnig H. A review of the health consequences of the vitamin D deficiency pandemic. *J. Neurol. Sci.* 2011;311(1-2):15–8.
33. Centers for Disease Control and Prevention. Achievements in Public Health, 1990–1999: Safer and Healthier Foods. *MMWR.* 1999;48(40);905–913. Available at http://www.cdc.gov/mmwr/preview/mmwrhtml/mm4840a1.htm, accessed September 25, 2014.
34. Rajakumar K. Vitamin D, cod-liver oil, sunlight, and rickets: A historical perspective. *Pediatrics.* 2003;112(2);e132–e135. Available at http://pediatrics.aappublications.org/content/112/2/e132.full, accessed September 25, 2014.
35. National Institutes of Health, Vitamin D Fact Sheet for Health Professionals, reviewed June 24, 2011. Available at http://ods.od.nih.gov/factsheets/VitaminD-HealthProfessional/, accessed September 25, 2014.
36. Calvo MS, Whiting SJ, Barton CN. Vitamin D fortification in the United States and Canada: Current status and data needs. *Am. J. Clin. Nutr.* 2004;80(suppl):1710S–06S.
37. National Institutes of Health, Office of Dietary Supplements. Dietary Supplement Ingredient Database. http://dietarysupplementdatabase.usda.nih.gov/ingredient_calculator/equation.php.
38. U.S. Centers for Disease Control and Prevention. Second National Report on Biochemical Indicators of Diet and Nutrition in the U.S. Population 2012. Atlanta, GA: National Center for Environmental Health, April 2012. Available at http://www.cdc.gov/nutritionreport/report.html, accessed September 25, 2014.
39. Cashman KD, Kiely M. Recommended dietary intakes for vitamin D: Where do they come from, what do they achieve and how can we meet them? *J. Hum. Nutr. Diet.* 2014;27(5):434–42.

40. Holick MF. Resurrection of vitamin D deficiency and rickets. *J. Clin. Invest.* 2006;116:2062–72.
41. Relation of vitamin D milk to rickets. *JAMA.* 1952;148(14):1227.
42. 21 CFR 184.1(b)(2).
43. Blank S, Scanlon KS, Sinks TH, Lett S, Falk H. An outbreak of hypervitaminosis D associated with the overfortification of milk from a home delivery dairy. *Am. J. Public Health.* 1995;85(5):656–659.
44. Looker AC, Johnson CL, Lacher DA, Pfeiffer CM, Schleicher RL, Sempos CT. Vitamin D status: United States 2001–2006. NCHS data brief, no 59. Hyattsville, MD: National Center for Health Statistics, 2011. Available at http://www.cdc.gov/nchs/data/databriefs/db59.pdf, accessed September 25, 2014.
45. Tulchinsky TH. Micronutrient deficiency conditions: Global health issues. *Public Health Rev.* 2010;32:243–255.
46. Holick MF, Binkley NC, Bischoff-Ferrari HA, Gordon CM, Hanley DA, Heaney RP, Murad MH, Weaver CM. Evaluation, treatment, and prevention of vitamin D deficiency: An Endocrine Society clinical practice guideline. *J. Clin. Endocrinol. Metabol.* 2011;96(7):1911–1930.
47. Vieth R. Critique of the considerations for establishing the tolerable upper intake level for vitamin D: critical need for revision upwards. *J. Nutr.* 2006;4(136):1117–1122.
48. McKenna MJ, Murray BF. Vitamin D dose response is underestimated by endocrine society's clinical practice guideline. *Endocr. Connect.* 2013;2(2):87–95.
49. Brody J. What do you lack? Probably vitamin D. *New York Times*, July 26, 2010. Available at http://www.nytimes.com/2010/07/27/health/27brod.html, accessed September 25, 2014.
50. Medline Plus, Beriberi, updated August 10, 2012. Available at http://www.nlm.nih.gov/medlineplus/ency/article/000339.htm, accessed September 25, 2014.
51. Medline Plus, Riboflavin, updated February 18, 2013. Available at http://www.nlm.nih.gov/medlineplus/ency/article/002411.htm, accessed September 25, 2014.
52. Medline Plus, Niacin, updated February 18, 2013. Available at http://www.nlm.nih.gov/medlineplus/ency/article/002409.htm, accessed September 25, 2014.
53. Medline Plus, Thiamin, updated February 18, 2013. Available at http://www.nlm.nih.gov/medlineplus/ency/article/002401.htm, accessed September 25, 2014.
54. Institute of Medicine (US) Standing Committee on the Scientific Evaluation of Dietary Reference Intakes and its Panel on Folate, Other B Vitamins, and Choline. Dietary Reference Intakes for Thiamin, Riboflavin, Niacin, Vitamin B6, Folate, Vitamin B12, Pantothenic Acid, Biotin, and Choline. Washington (DC): National Academies Press (US), 1998. 6, Niacin. Available at http://www.ncbi.nlm.nih.gov/books/NBK114304/, accessed November 2, 2014.
55. Oregon State University. Linus Pauling Institute. Micronutrient Information Center. Available at http://lpi.oregonstate.edu/infocenter/vitamins/niacin/#rda, accessed November 1, 2014.
56. Bollet AJ. Politics and pellagra: The epidemic of pellagra in the US in the early twentieth century. *Yale J. Biol. Med.* 1992;65:211–221.
57. Lanska DJ. Stages in the recognition of epidemic pellagra in the United States: 1865–1960. *Neurol.* 1996;47:829–34.
58. Reilly B. *Disaster and Human History: Case Studies in Nature, Society and Catastrophe.* Jefferson, NC: McFarland & Co, 2009.
59. Ed. Bread: The Staff of Life falls victim to the war. *St. Petersburg Times*, p. 6, January 15, 1943.
60. Nestle M. Folate fortification and neural tube defects: Policy implications. *J. Nutr. Educ.* 1994;26(6):287–293.
61. National Institutes of Health, Iron Fact Sheet for Health Professionals, reviewed April 8, 2014. Available at http://ods.od.nih.gov/factsheets/Iron-HealthProfessional/, accessed September 25, 2014.

62. Cogswell ME, Looker AC, Pfeiffer CM, Cook JD, Lacher DA, Beard IL, Lynch SR, Grummer-Strawn LM. Assessment of iron deficiency in US preschool children and non-pregnant females of childbearing age: National Health and Nutrition Examination Survey 2003–2006. *Am. J. Clin. Nutr.* 2009;89:1–9.
63. Institute of Medicine. Food and Nutrition Board. Dietary Reference Intakes for Vitamin A, Vitamin K, Arsenic, Boron, Chromium, Copper, Iodine, Iron, Manganese, Molybdenum, Nickel, Silicon, Vanadium, and Zinc: A Report of the Panel on Micronutrients. Washington, DC: National Academy Press, 2001.
64. Expert Scientific Working Group, Summary of a report on assessment of the iron nutritional status of the United States population. *Am. J. Clin. Nutr.* 1985;42:1318–1330.
65. Quinn EA. Too much of a good thing: evolutionary perspectives on infant formula fortification in the United States and its effect on infant health. *Am. J. Human. Biol.* 2014;(26)1:10–17.
66. Gerrior S, Bente L, Hiza H. *Nutrient Content of the U.S. Food Supply, 1909–2000.* (Home Economics Research Report No. 56). U.S. Department of Agriculture, Center for Nutrition Policy and Promotion, 2004. Available at http://www.cnpp.usda.gov/publications/foodsupply/foodsupply1909-2000.pdf.
67. Raphael C. *Investigated Reporting, Muckrakers, Regulators, and the Struggle over Television Documentary.* Illinois: University of Chicago Press, 2005:47.
68. Centers for Disease Control. Ten-State Nutrition Survey, 1968–1970: Highlights. *DHEW (HSM) Publication 72–487*, 4th edition, Washington, DC: US Dept of Health, Education, and Welfare, 1972.
69. Crosby WH. Serum ferritin and iron enrichment. *JAMA.* 1974;290:1435–1436.
70. Recommendation for Increased Iron Levels in the American Diet, November 1969, Food and Nutrition Board, National Academy of Sciences National Research Council, Washington DC.
71. Summary of Proceedings. Workshop on Measures to Increase Iron in Foods and Diets, January 22–23, 1970. Committee on Iron Nutritional Deficiencies, Food and Nutrition Board, National Academy of Sciences National Research Council, Washington D.C.
72. Harold M. Schmeck, Jr., FDA. Weighs role of iron in bread. *The New York Times*, February 14, 1972.
73. Crosby WH. Improving iron nutrition. *West. J. Med.* 1975;122(6):499–450.
74. FDA delays plan to increase iron in bread. *Abilene Reporter News*, Tuesday February 12, 1974.
75. American diets deficient in iron. *The New York Times*, May 30, 1980.
76. National Institutes of Health. Iron fact sheet for health professionals, reviewed April 8, 2014. Available at http://ods.od.nih.gov/factsheets/Iron-HealthProfessional/, accessed September 25, 2014.
77. U.S. Preventive Services Task Force, Screening for Hemochromatosis, Recommendation Statement, August 2006. Available at http://www.uspreventiveservicestaskforce.org/uspstf06/hemochromatosis/hemochrs.htm#summary, accessed September 25, 2014.
78. McClaren CE, Barton JC, Adams PC, Harris EL, Acton RT, Press N et al. Hemochromatosis and the Iron Overload Screening (HEIRS) study design for an evaluation of 100,000 primary care-based adults. *Am. J. Med. Sci.* 2003;325:53–62.
79. Bailey RL, Dodd KW, Goldman JA, Gahche JJ, Dwyer JT, Moshfegh AJ et al. Estimation of total usual calcium and Vitamin D intakes in the United States. *J. Nutr.* 2010;(140)4;817–822.
80. National Institutes of Health, Calcium Dietary Supplement Fact Sheet, reviewed November 21, 2013. Available at http://ods.od.nih.gov/factsheets/Calcium-HealthProfessional/, accessed September 25, 2014.
81. Benjamin RM. Bone health: Preventing osteoporosis. *Pub. Health Rep.* 2010;125(3):368–370.

82. Prentice A. Diet, nutrition and the prevention of osteoporosis. *Pub. Health Nutr.* 2004; 7:227–243.

83. Committee to Review Dietary Reference Intakes for Vitamin D and Calcium, Food and Nutrition Board, Institute of Medicine. *Dietary Reference Intakes for Calcium and Vitamin D.* Washington, DC: National Academy Press, 2010.

84. NIH Consensus Development Program. Osteoporosis. *NIH Consens. Dev. Conf. Consens. Statement* 1984;5(3):1–6. Available at http://consensus.nih.gov/1984/1984Osteoporosis043 html.htm, accessed September 25, 2014.

85. USDA. *Marketing US Agriculture: The 1988 Agriculture Yearbook.* Washington, DC: US Government Printing Office, 1988. Available at http://naldc.nal.usda.gov/naldc/download. xhtml?id=IND50000192&content=PDF, accessed September 25, 2014.

86. NIH Consensus Development Program. Optimal calcium intake. *NIH Consens. Dev. Conf. Consens. Statement* 1994;12(4):1–31. Available at http://consensus.nih.gov/1994/1994opti malcalcium097html.htm, accessed September 25, 2014.

87. Institute of Medicine, Standing Committee on the Scientific Evaluation of Dietary Reference Intakes, Food and Nutrition Board, Institute of Medicine. *Dietary Reference Intakes for Calcium, Phosphorus, Magnesium, Vitamin D, and Fluoride.* Washington, DC: National Academy Press, 1997.

88. Hershey. The Hershey Company. Hershey's Syrup with Calcium. Available at http:// www.thehersheycompany.com/brands/syrup/hersheys-syrup-with-calcium.aspx, accessed September 25, 2014.

89. United States Food and Drug Administration, Warning Letter. The Hershey Company February 14, 2012. Available at http://www.fda.gov/ICECI/EnforcementActions/Warning Letters/ucm314829.htm, accessed September 25, 2014.

90. Harvard School of Public Health. The Nutrition Source. Calcium and Milk: What's best for your bones and health? Available at http://www.hsph.harvard.edu/nutritionsource/cal-cium-full-story/, accessed September 25, 2014.

91. Smith AD, Kim Y-I, Refsum H. Is folic acid good for everyone? *Am. J. Clin. Nutr.* 2008;87(3):517–533.

92. Mosley BS, Cleves MA, Siega-Riz AM, Shaw GM, Canfield MA, Waller DK et al. Neural tube defects and maternal folate intake among pregnancies conceived after folic acid for-tification in the United States. *Am. J. Epidemiol.* 2009;169(1):9–17.

93. Hoffbrand AV, Weir DG. Historical review. The history of folic acid. *Br. J. Haematol.* 2001;113:579–589.

94. Shelke N, Keith L. Folic acid supplementation for women of childbearing age versus sup-plementation for the general population: A review of the known advantages and risks. *Int. J. Family Med.* 2011;2011:173705.

95. National Institutes of Health. Office of Dietary Supplements. Folate dietary supplement fact sheet, last reviewed December 14, 2012. Available at http://ods.od.nih.gov/factsheets/ Folate-HealthProfessional/, accessed November 1, 2014.

96. Backstrand JR. The history and future of food fortification in the United States: A public health perspective. *Nutr. Rev.* 2002;60(1):15–26.

97. Crider KS, Bailey LB, Berry RJ. Folic acid food fortification—Its history, effect, concerns, and future directions. *Nutrients.* 2011;3:370–384.

98. Junod SW. Folic acid fortification: Fact and folly. Available at http://www.fda.gov/ AboutFDA/WhatWeDo/History/ProductRegulation/SelectionsFromFDLIUpdateserieson FDAHistory/ucm091883.htm, accessed September 26, 2014.

99. Verhoef P. New insights on the lowest dose for mandatory folic acid fortification? *Am. J. Clin. Nutr.* 2011;93(1):1–2.

100. Lawrence MA, Chai W, Kara R, Rosenberg IH, Scott J, Tedstone A. Examination of selected national policies towards mandatory folic acid fortification. *Nutr. Rev.* 2009;67:S73–S78.

101. University of Oxford. Study closes debate on folic acid and heart disease, February 22, 2012. Available at http://www.ox.ac.uk/media/news_stories/2012/120222.html, accessed September 26, 2014.

102. Food standards: Amendment of standards of identity for enriched grain products to require addition of folic acid, Final Rule, March 5, 1996. Available at http://www.gpo.gov/fdsys/pkg/FR-1996-03-05/html/96-5014.htm, accessed September 26, 2014.

103. Bentley TG, Weinstein MC, Willett WC, Kuntz KM. A cost-effectiveness analysis of folic acid fortification policy in the United States. *Public Health. Nutr.* 2009;12(04):455–467.

104. Centers for Disease Control and Prevention. *Second National Report on Biochemical Indicators of Diet and Nutrition in the U.S. Population.* Atlanta, GA: National Center for Environmental Health, April 2012.

105. Yang Q, Cogswell ME, Hamner HC, Carriquiry A, Bailey LB, Pfeiffer CM et al. Folic acid source, usual intake, and folate and vitamin B-12 status in US adults: National Health and Nutrition Examination Survey (NHANES) 2003–2006. *Am. J. Clin. Nutr.* 2010;91:64–72.

106. Centers for Disease Control and Prevention. Folic acid. Data and Statistics, reviewed July 7, 2010. Available at http://www.cdc.gov/ncbddd/folicacid/data.html, accessed September 26, 2014.

107. Dowd JB, Aiello AE. Did national folic acid fortification reduce socioeconomic and racial disparities in folate status in the US? *Int. J. Family Med.* 2008;37(5):1059–1066.

108. Hamner HC, Mulinare J, Cogswell ME, Flores AL, Boyle CA, Prue CE et al. Predicted contribution of folic acid fortification of corn masa flour to the usual folic acid intake for the US population: National Health and Nutrition Examination Survey 2001–2004. *Am. J. Clin. Nutr.* 2009;89(1):305–315.

109. Toriello HV. Policy statement on folic acid and neural tube defects. American College of Medical Genetics. 2005. *Genet. Med.* 2011;13(6):593–596.

110. Lawrence M. Fortification. Folic acid and spina bifida. Is it safe? Is it wise? Is it right? [Commentary]. *World Nutr.* 2013;4(3):95–111.

111. Selhub J, Paul L. Folic acid fortification: Why not vitamin B12 also? http://www.ncbi.nlm.nih.gov/pubmed/21674649 *Biofactors.* 2011;37(4):269–271.

112. Legislative Record of the Ninety-Third Legislature of the State of Maine, 1947- House. *Daily Kennebec Journal, Augusta Maine*, April 22, 1947:1168.

113. National Insitute of Health, Iodine Fact Sheet for Health Professionals. Available at http://ods.od.nih.gov/factsheets/Iodine-HealthProfessional/, accessed May 10, 2015.

114. American Thyroid Association. How do you diagnose iodine deficiency? June 4, 2012. Available at http://www.thyroid.org/iodine-deficiency, accessed September 25, 2014.

115. Anderrson M, de Benoist B, Darnton-Hill I, Delange F (eds.). *Iodine Deficiency in Europe: A Continuing Public Health Problem.* WHO, UNICEF, 2007. Available at http://whqlibdoc.who.int/publications/2007/9789241593960_eng.pdf, accessed September 25, 2014.

116. Spark A. *Nutrition in Public Health: Principles, Policies, and Practice.* Boca Raton, FL: CRC Press, 2007.

117. The IOM Dietary Reference Intakes, Vitamin Table. Available at http://www.iom.edu/Home/Global/News%20Announcements/~/media/Files/Activity%20Files/Nutrition/DRIs/DRI_Vitamins.ashx, accessed September 14, 2014.

118. National Institutes of Health. Folate, dietary supplement fact sheet, last reviewed December 14, 2012. Available at http://ods.od.nih.gov/factsheets/Folate-HealthProfessional/, accessed September 26, 2014.

"The Nanny State"

12

Taxes, Zoning, Ordinances, and Corporate Self-Regulation*

> *In 1790, the nation which had fought a revolution against taxation without representation discovered that some of its citizens weren't much happier about taxation with representation.*
> Lyndon B. Johnson[1]

INTRODUCTION

Local, regional, and federal governmental regulatory powers can influence "who eats what, when, how, and with what impact." Food policy can be shaped through federal, regional, and local regulations of taxation, zoning, and ordinances. Conversely, food policy can also be shaped through self-regulation in the food system by industry in the absence of such public rules.

Governmental regulation of food production has become generally accepted as appropriate in order to protect the common goods of trade and food safety. However, regulatory efforts at modifying our consumption of food can be quite controversial. The attempt by the New York City Department of Public Health to limit the size of sodas sold in certain venues in the city was such a controversy that brought out cries of "Nanny

---

* Jeanine Kopaska-Broek, MPH, contributed primary research and prepared draft versions of portions of this chapter.

State," an overbearing government imposing undue restrictions on the personal choices of citizens. Interestingly, the introduction of smaller-sized single-serving cans by soda manufacturers—an example of potential industry self-regulation that could result by default in limiting the choices available to us—did not stir up the same controversy.

We attempt to provide context as to the tension between American socio-political ideas of individual liberty and regulation of food choices with a brief discussion of John Stuart Mills' "harm principle." By using historical examples, this chapter aims to show that governmental regulation of food is not inherently in opposition to American values and is indeed an appropriate area of public health intervention. We explore how food policy conveyed through taxes, zoning, and ordinances could support public health objectives, such as addressing obesity, and we conclude with a discussion of the potential role of industry self-regulation.

THE HARM PRINCIPAL, INDIVIDUAL LIBERTY, AND PUBLIC HEALTH

When studying, creating, or indeed living under, food policy regulations, it might be helpful to be aware of some of the underlying political philosophical ideas.

John Stuart Mill's highly influential treatise *On Liberty* introduced a perspective that became a significant element of political thinking in the United States; namely that the only justification for infringing an adult individual's liberty is to prevent harm to others.

Nicknamed the harm principle, it is a useful baseline for discussing how society can meaningfully, and appropriately, balance social goods and individual liberty. Mill's thoughts are briefly highlighted due to prevalence in political thought and rhetoric, but of course there are other approaches which are not able to be presented due to the limited scope of this chapter.

The harm principle has been described as the foundational principle for public health, as it is a clear statement of when it is permissible for governmental action that restricts individual liberty.[2] It has been used specifically as a framework for addressing childhood obesity.[3]

Generally, the harm principal is relatively straightforward:

"That the only purpose for which power can be rightfully exercised over any member of a civilized community, against his will, is to prevent harm to others."[4]

This curtailing of free will to prevent harm to other people affects our liberties that we might otherwise enjoy. Free speech, for example, is one of those liberties, but the principle of harm restricts our freedom to shout "Fire" without justification in a crowded movie theatre.[5]

Mill continues by discussing the parameters of what action may be appropriate:

"His own good, either physical or moral, is not sufficient warrant. He cannot rightfully be compelled to do or forbear because it will be better for him to do so, because it will make him happier, because, in the opinions of others, to do so would be wise, or even right. These are good reasons for remonstrating with him, or reasoning with him, or persuading him, or entreating him, but not for compelling him, or visiting him with any evil in case he do otherwise."[4]

In other words, when it comes to food—as long as no harm is caused to others, an adult person should be perfectly free to eat as desired, even to the point of self-harm, without fear of being compelled otherwise or punished by someone else's opinion on the matter.

Compulsion and punishment are not appropriate tools to use. However, *remonstrating, reasoning, persuading, and entreating* are acceptable according to Mill. Even in a highly liberty-sensitive political sphere, "nudging"[5] to modify individual behavior may be considered an appropriate public health intervention, although it is nonetheless often still accompanied by a justification on how such behavior is externalized into a cost that harms everyone else in society. For example, in the case of public health and obesity prevention, health care costs are often cited as the harm that affects everyone.

The concept of harm is both vague and complex. When considering policy that must encroach on individual liberties based on a justification that such liberties present harm to others, it may be helpful to be critical about the details. *What exactly is the harm? Who suffers the harm? Is the harm direct or indirect? If the harm is one of monetary cost, is it of direct immediate consequence to individuals, or is an abstract future social cost?*

Mill's Principles in Application: Posting Calorie Information; Soda Size; Trans-Fat Ban

The public health interest in regulating food consumption choices in modern society is predominantly to prevent a future danger, as a poor health outcome due to diet tends to develop over time. In this scenario, Mill would argue that warning is better than forcibly preventing exposure. Posting calorie information would be a perfectly acceptable regulatory action.

But what about limiting options to purchase beverages by measures of quantity? Does using government regulation to limit the amount we could purchase constitute "forcibly preventing exposure?" Or is it an example of *reasoning and/or persuading*? Could not reasoning and persuading include modifying the content of cues provided in our food environment as part of shifting a social norm, such that a 16-ounce serving of soda becomes a "normal" single serving?

A law limiting soda size servings to 16 ounces could be considered an acceptable incursion of government regulation, as long as individuals can (1) purchase as many as they like are (2) not punished for doing so, or (3) prevented or punished from purchasing any other available size, or (4) punished for the consequences to their own self of drinking as much soda as they like.

What about modifying the food environment? How does limiting choices infringe on personal liberty? How is that different from "market-place" conditions, which inherently limit choices? Is limiting the possibility of choice for something which individual consumers do not actually normally participate in choosing still an infringement on personal liberty?

This last question is related to the topic of trans-fat oil being banned in New York City restaurants. When ordering French fries from a restaurant, does it matter to our individual agency of choice that the potatoes are fried in a different type of oil than in the past?

By the writing of this book, the NYC ban of trans-fat is no longer the controversial topic it was when proposed in 2006 and implemented in 2008. The banning of trans-fat in NYC restaurants modified the food environment by changing the ingredient make-up of processed foods and methods of restaurant cooking, thereby theoretically improving the health profile of an individual's diet.

Eating one batch of French fries cooked in trans-fat oil does not put us in immediate harm. Because the harm is not immediate, at first blush, it may seem the classic Millsian approach of warning rather than forcibly preventing exposure would be the preferable option. However, the burden on the individual to evaluate the risk of eating French fries at any given restaurant and over the course of a life-time would be substantial, and realistically not something most individuals would desire to do.

Improved health seems like something an individual would naturally desire and seek to choose, if available. To that extent, it seems fair to suggest that the modification to remove the "choice" to consume trans-fat was a reasonable anticipation of what the individual would most likely prefer. However, removing some of the burden of making health-related decisions about a specific food and a specific risk (cardiovascular) also may impact criteria that an individual might use about overall dietary choices. New Yorkers no longer have to moderate their intake of French fries in order to reduce trans-fat consumption. But their overall health, not just cardiovascular, might still be improved through moderation of French fry consumption.

Consumers are now left blissfully eating French fries without having to reject French fries in favor of salad a certain number of times a week specifically to avoid trans-fat. But in reality, not to have to make the decision of salad over fries to avoid trans-fat is indeed probably what we would all prefer. We like to make food decisions for health reasons, but also taste, mood, enjoyment, etc.

Individual Liberty: Is Regulation of Food "Un-American?"

American culture emphasizes individual liberty. Having liberty implies agency, or control over our actions and choices. This makes food policy that is simply the provision of information, albeit perhaps information as a hopeful nudge toward certain behavior, generally easier for Americans to accept. The example noted above of posting calories on a menu is simply information transmission.

Encroachments on individual liberties that are not supported by a clear-cut common social good, even if to address laudable public health goals such as obesity and diabetes reduction, could still be perceived as contrary to the tenets of American political philosophy.

However, even during the initial foment of revolution, the emerging Americans recognized that individual food choices could impact the community as a whole. While still under British rule, curtailing personal consumption of sugar and tea was an identified social good strategy toward relieving the community of British taxes. It was recognized that personal choices, even something so personal as diet, had associated social and public consequences.[6]

Post revolution, the American government was looking to relieve some of the debt incurred by the Revolutionary War, and in 1791, Secretary Hamilton proposed an excise tax on whiskey. This suggestion was an expansion of federal authority and controversial for that reason.[7] As an excise tax, it would be paid by the distiller, however, it was recognized that the tax would be passed on to consumers. In an early example of a public health interest supporting federal regulation in the form of a "sin tax," this taxation of an item of personal consumption was supported by groups that felt whisky consumption carried excessive public cost, because the effects of overconsumption were borne not just by the individual but also by the society as a whole.[6]

The Excise Tax was highly controversial; and the same types of arguments we hear now of government overreaching (such as what next, ketchup?) were put forth. The western portion of the country rebelled, but not necessarily over a perceived regulation on consumption choices. Given the lack of infrastructure, the famers in the west (the west being at this time Pennsylvania) relied on converting their food crop harvests to whiskey to send back east because grain would spoil in transit. Furthermore, because the tax was applied equally (per unit), it was regressive as to the quality of the whiskey, and made it even harder for the lower quality whisky from the west to compete with the quality eastern product. The western farmers were being incorrectly targeted by the tax.[8]

President Washington suppressed the rebellion in a manner apparently well received. The events of the taxation, rebellion, and suppression seemed to settle the question as to if the "right of the people" included the right to defy unpopular laws, even though validly made by a legitimately elected and representative government. In any case, since the Founding Fathers themselves supported and passed a law taxing private beverage consumption, it becomes difficult to argue that food policy nudging a private decision of consumption toward behavior to the common public good or social goal is "un-American."[6]

Individual Risk and Social Norms

Individual behavior modification becomes a population level, or public health strategy when the effort is made to change the behavior of many people, thus possibly shifting a cultural norm. This manner of behavior change effort recognizes that even though most individuals might be at a small risk of injury or disease, the fact that there are many such individuals at risk means that reducing the risk for the larger number at lower risk may actually decrease the number of cases of disease (or accident, or injury) more so than by addressing the smaller number of people at higher risk.[9] This is relevant to an overall goal of decreasing social cost.

A widespread shift in behavior reduces the risk across the population. As a new behavior becomes a norm, less education and persuasion is needed to make people change their behavior. The younger author of this book remembers having to constantly remind her parents to use seatbelts when driving, whereas putting on the seatbelt is an automatic behavior within her own generation.

A social norm can be changed through education and other incentives that encourage people to make the desired choice. A social norm may also exist in the context of our physical environment such that it obviates the individual from even having to make the decision.

For example, in our society, in which a great deal of our diet is procured on the shelves of supermarkets, a social norm can be transmitted through the choices available in our food environment. Epidemiologist Geoffrey Rose describes how our food system as a supply industry contributes to a social norm: *"Once a social norm of behavior has become accepted and ... the supply industries have adapted themselves to the new pattern, then the maintenance of that situation no longer requires effort from individuals. The health education phase aimed at changing individuals is, we hope, a temporary necessity, pending changes in the norms of what is socially acceptable."*[9]

Our physical surroundings can provide cues as to a "norm." We see it every day in things such as fashion. It should not be surprising that our physical surroundings can provide cues as to a dietary norm. A drink packaged in a size customarily purchased as a single serving is an example of a social norm. For years, the single serving soda available to purchase in the United States has been relatively standardized to 20 ounce bottles and 12-ounce cans.[10]

Soda happens to also be the subject of a relatively recent local governmental regulatory effort at shifting a dietary social norm via physical cues in our physical environment. In 2012, the Health Board of New York City's Department of Public Health and Mental Hygiene approved Mayor Michael Bloomberg's proposal to restrict the size of sugary drinks sold in certain venues to 16 ounces through an amendment to the New York City Health Code.[11] Highly controversial and erratic in application due to a split in regulatory oversight between New York City and New York State, the proposal was struck down by state appeals court in July of 2013.

The failed New York City soda size ban would have prohibited delicatessens from selling larger sizes even though convenience stores could still sell larger sized options because the New York City Board of Health oversees city restaurants, delis and concession stands while New York State regulates grocery stores.[12] (This issue of split regulatory oversight creating erratic rules also appears at the federal level with food safety responsibilities shared between the United States Department of Agriculture (USDA) and the FDA.)

Regulatory Methods: Reducing Risk through Individual Behavior Modification

Taxation is one way in which local, regional, or federal government can attempt to guide individual choices, and thus behavior (as well as creating some profit to the government from such choice) through "sticks" or "carrots." Taxation can be a food policy "stick" that penalizes the purchase of certain products, such as soda, through applying a higher than usual sales tax.

Food policy regulations could also act as "carrots" rewarding the desired choice. Incentives to consume healthier food choices could include subsidizing the costs of healthy foods or, as in the case of federal food assistance, providing "bonus" money for purchasing healthy foods.

Food policy regulations can also be neutral, that is, neither stick nor carrot, but instead providing information or increasing access for the consumer to make their own choice. Requiring that calories be labeled on a menu is an example of informative food policy.

In 2008, New York City passed a law requiring chain restaurants to post calories directly on the food menu (or board) from which the diner orders. The idea was that posting calorie information could influence consumer behavior. And in fact, one of the authors observed a distinct behavior change occurring on the first day calories were posted at the local Starbuck's—the pastry case, normally ravaged by mid-morning, was left untouched and the lowest calorie item available at the time, multigrain bagels, had flown off the shelf. Over time, she observed the shock wear off and purchases of the more caloric items crept back up. However, the contents of the pastry case also begin to change as Starbucks began introducing slightly less caloric items, and in fact, studies conducted on the effects of calorie posting show mixed results.[13]

Community Level Regulations

Even though above examples are regulations aimed directly at individuals, food policy regulations can also be enacted at the community level through zoning and ordinances. For example, these could encourage grocery stores or farmers markets through tax credits, zoning changes in or small business tax relief. At a federal level, The New Markets Tax Credit program, established by Congress in 2000, has resulted in expanded access to healthy food in underserved communities by encouraging private sector investment. Investors can get federal tax credits with equity investment in eligible locations through Community Development Entities.

Regulatory-based initiatives can drastically improve communities' access to food. Zoning can be used to prohibit undesirable food outlets (such as fast food restaurants), or encourage desirable food outlets like supermarkets by reducing neighborhood entry barriers such as square footage restrictions and parking requirements. In recent years, some municipalities including Los Angeles have passed ordinances to ban new fast food restaurants in designated areas, or to limit such restaurants ability to market unhealthy food to children through toy incentives. Tax incentives and grants are now being used to draw supermarkets into underserved neighborhoods in many cities. The second half of the chapter will provide more detailed examples of local regulations.

Food policy regulations of the sort we have described above can change the choices available to individuals. Although we perceive of food as primarily a matter of individual choice, our choices are nonetheless already inherently constrained by the amount of resources we have to spend on food and also what is available to us for consumption or purchase. This includes the underlying composition of packaged and processed food. For example, it can be a challenge to choose to drink a cane sugar-sweetened soda in the United States. Cane sugar-sweetened soda is not widely available in all venues selling sodas. This is an example of an individual food choice that is constrained by availability, even if not by personal resources due to a premium cost. Our individual liberty to choose is naturally constrained by the choices available to us—our available choices compromise our "food environment."

Although the effect of education on individual consumer behavior may still be unclear, informative food policies, such as calorie labeling, can also influence changes in the available food environment. Food retailers, concerned about losing revenue, adapted their offerings for the perceived burgeoning market now informed by calorie labeling. Treats of smaller caloric amount begin appearing in 2010–2011 at Starbucks and other retailers.[14]

The perception by the food industry that a market for healthier food will be created even through a policy that is only an informative correction of asymmetrical market information can lead to changes in available food choices. The "supply industry" can contribute to a new social norm.

A local shift in one city or region can also foster a broader social norm shift across a larger area. When New York City passes a regulation that impacts the business practices of a national-wide or global retail chain, it can influence the adoption of the requirements into corporate practices elsewhere due to necessary business efficiencies even if other locations lack similar laws.

And, when similar, but not exactly the same, laws are passed in multiple other locations, such as calorie menu labeling, a reversal can occur such that the food companies which originally bitterly contested the local menu calorie labeling laws begin supporting a national version in order to avoid the headache of nationally diverging market inefficiencies.[13,15]

Public Health and Food Environment Modification

Modifying the environment in which we live (the built environment) has been a successful approach for nonfood related public health concerns. For example, public health fostered social good goals through early twentieth century building code requirements that improved living conditions in urban tenements.[16] Public health goals continue to be addressed through similar built-environment regulations such as requiring improved window guards in urban housing to address a high mortality rate for children due to falling.

Although we do not normally think of such safety or housing regulations as compromising individual liberties, such regulations can remove some choices from the market. For example, it would be more difficult today in the United States to choose to save rent by seeking a "cold-water" flat instead of one that provided hot water. These two housing related examples are areas in which we have accepted the role of public health in fostering social goods by limiting individual choices.

Similarly, public health goals can also be met through modifying the food environment in a manner that removes the necessity of choice from the individual. The prevalence of partially hydrogenated vegetable oil in the American diet provides an example. Partially hydrogenated vegetable oil was the first human-made fat introduced to our food supply, with Crisco appearing on grocers' shelves in 1911.

Margarine became popular during World War II due to butter rationing, and in the following decades, the use of partially hydrogenated vegetable oil increased further as it was considered a healthier replacement for saturated fats. However, in the 1990s, nutritional research began revealing health concerns about the relationship between partially hydrogenated oil (also known as trans-fat) and heart disease, and by 2003, the United States government was concerned enough to require that the amount of trans-fat contained in products should be included as a specified line item on nutrition labels.[17] Although this food policy regulation is technically only informative, it also resulted in food environment modification as some manufacturers decided to reformulate their products.

Armed with this label information, individuals could choose their diet more knowledgably. However, the federal labeling law requiring trans-fat information did not apply to foods served in restaurants, which left consumers in the dark when dining out. In 2006, with the rationale that restaurants were a major source of artificial trans-fat, New York City Board of Health passed a requirement barring restaurants of all types from using artificial trans-fats by July 2008.[18]

Going trans-fat free was not controversy-free. Even though this ban concerned ingredients that are in reality usually unimportant to the individual consumer, criticism included the "slippery slope" concern that such a ban would open the door to bans on actual food or ingredients that consumers did hold dear,[19] or even to government regulated portion size. In 2006, the concern over governmentally regulated portion size might have sounded alarmist. As of 2013, however, when New York City's Mayor Bloomberg proposed a size limitation on sugary drinks, government regulation over portion size became a possibility.

Has the trans-fat ban achieved the intended public health goal? A New York City Department of Health study, although limited in scope, indicates dietary improvement can occur through reducing the consumption of trans-fats.[20] The individual consumer's diet is improved without the need for the individual to specifically examine cooking practices at preferred restaurants. The healthier choice is presented as the default—as the social norm—and obviates the need for the customer to choose.

Food Modification and Public Health Goals in United States History

Historically, public health's food-related goals have been most commonly associated with food modification (fortification and enrichment) to address nutritional deficiencies that would otherwise result in disease. As fortification and enrichment is the sole topic of Chapter 11 of this book, we will briefly describe another historical example of food modification related to public health, namely, pasteurization. Although pasteurized milk was a rarity in 1900, it had become the norm by 1936, comprising about 98% of the milk in major cities.[21]

The Food and Drug Administration (FDA) points out: "Pasteurization of milk was adopted in the early twentieth century in the United States as a public health measure to reduce illness from an item commonly considered to be a major food staple, particularly for children."[22] Proponents of raw milk might instead point out that pasteurization was instead promoted primarily because it was critically important to the developing industrialization of milk, not public health, and furthermore that pasteurization does not necessarily prevent foodborne illnesses.[23] In 1985, an outbreak of antimicrobial-resistant salmonellosis affecting 170,000 people was caused by pasteurized milk.[24] Today, the States retain the ability to decide if and how raw milk can be sold in that state, but as of 1987, the FDA has prohibited the sale of raw milk across state lines.[22]

A pasteurized versus raw milk argument can be safely made at a historical distance now that bovine tuberculosis, another disease prevalent in nonpasteurized milk at the turn of the century has been eradicated and forgotten. Bovine tuberculosis can be transmitted through milk to humans. One estimate of the number of deaths circa 1900 likely due to the bovine form was 15,000—mostly children.[25]

In any event, the impact of pasteurization extending the safety and shelf life of milk may have in turn increased the usage of milk in the American diet. As milk became increasingly more accessible and dependable, its role as a food staple also likely solidified. Even though public health may have been the main goal, regulating that milk be pasteurized most likely also increased the amount of milk consumed and sold, thereby supporting industry.

The examples of enrichment and fortification elsewhere in this book and pasteurization in this chapter illustrate that government policy modifying food choices for the consumer in order to achieve public health goals is not new in the United States.

Historically, however, this food modification was mostly predicated on achieving public health goals that were of visible and immediate impact, such as reducing nutrient-deficiency related disease or food-poisoning illnesses. Given that an individual is usually seeking to avoid illness, restrictions that promote safety through limiting the availability of certain foods in a "raw" state (especially foods children are likely to consume, such as milk) are now commonly considered an appropriate balancing of individual choice. See Box 12.1 regarding pasteurization.

BOX 12.1   PASTEURIZATION: NOT JUST FOR MILK

Pasteurization is the process of heating a food in order to kill bacteria that may otherwise cause the food to spoil. Although we commonly associate milk with the pasteurization process, pasteurization is also used for fruit juices and since 2007 in the United States, almonds. Raw almonds that are not sold directly from the grower to the consumer must be pasteurized, either through a steam process or through the use of a chemical, propylene oxide (PPO). The Environmental Protection Agency reviewed PPO in 2006 for reregistration and stated that PPO does not pose a health risk. However, this creates a somewhat deceptive labeling issue, as a consumer may not be aware that "raw" can include PPO treatment or steam.

California Almond Board, Pasteurization, http://www.almondboard
.com/HANDLERS/FOODQUALITYSAFETY/PASTEURIZATION/
Pages/Default.aspx, accessed September 2, 2014.

Food Production versus Food Consumption

Although public health objectives have been initiated previously in the United States through food modification, such objectives have historically been to address immediate dangers of acute illness or death. The move to reduce trans-fat in our food supply is instead an example of modifying food to improve chronic, or long-term health outcomes. Traditionally, such consequences have been seen as ones of personal responsibility associated with poor dietary choices. However, because the push to reduce trans-fat in our food supply is an intervention based at the food production level, just like the efforts at reducing nutritional deficiency related disease and the pasteurization of milk, the blow to our agency of individual choice as consumers is softened.

Although we have come to accept the regulation of food production, regulation of food consumption remains controversial.[6] Box 12.2 "Consumption Controversy"

BOX 12.2 CONSUMPTION CONTROVERSY

The federal government has a public health interest in feeding children to support the nutritional health of the population. As part of ensuring a healthy diet, National School Lunch Program meals must fit nutritional guidelines, and historically, provide a *minimum* of calories. The 2010 Healthy, Hunger-Free Kids Act update to the meal standards was the first major change in 15 years. For the first time in program history, a maximum calorie limit was provided. As press releases make it clear, the changes were intended to help address childhood obesity. The new meals emphasizing fruits, vegetables, and whole grain, became surprisingly controversial when the menus were first served in fall 2012. Students who lacked appreciation of the new emphasis on vegetable and whole grains did not have as many of their accustomed alternatives to fill up on, and struggled with the calorie limitations. For high school students, lunch is limited at 850 calories; which detractors describe as extreme calorie rationing and the intrusion of the nanny-state. In response, the USDA relaxed the limitation on meat and grain portions, but kept the calorie cap.

*Institute of Medicine of the National Academies, Action Taken, http://
www.iom.edu/Reports/2009/School-Meals-Building-Blocks-for-
Healthy-Children/Action-Taken.aspx, accessed September 2, 2014;
The White House, President Barack Obama, Office of the Press
Secretary, December 13, 2010, http://www.whitehouse.gov/
the-press-office/2010/12/13/president-obama-signs-healthy-
hunger-free-kids-act-2010-law, accessed September 2, 2014;
Congressman Steve King, http://steveking.house.gov/
media-center/press-releases/king-supports-the-nutrition-
nanny-challenge-to-usda, accessed September 2, 2014;
United States Department of Agriculture, School Meals Nutrition
Standards, Policy Memos, February 25, 2013, http://www.fns.usda.gov/
sites/default/files/SP26-2013os.pdf, accessed September 2, 2014.*

highlights how government intervention into dietary behavior through restricting food consumption is partially justified through the public health goal of addressing obesity.[5]

TAXATION, ZONING, AND ORDINANCES

This section highlights how regulations like taxation, zoning, and ordinances are being explored as practical solutions for one of today's most pressing contemporary health concerns in the United States—that of obesity and associated chronic diseases. Nationwide, no state achieved the Healthy People 2010 goal to reduce the prevalence of obese adults to 15%.[26] As of 2013, all states are now above 20% and the majority of adults in the United States are now overweight or obese, with a combined prevalence of 68%.[27] The section concludes with a discussion on corporate self-regulation as an alternative approach.

The relationship between socioeconomic status and obesity is complex. Income, sex, and education can all be contributing factors, such that among women, income and education level are inversely related to obesity yet among men, obesity rates are similar across income levels with no discernable significant difference due to education.[28] However, generally, minority populations experience obesity at a disproportionate rate.[29]

Poor quality food environments in areas of low socioeconomic status may contribute to an increased risk of obesity of resident populations.[30] One contributing factor could be that unhealthy diets are simply cheaper, when comparing the cost of energy dense foods to the cost of whole grains, lean proteins, vegetables, and fruits.[31] Geographic access of quality food could be another contributing factor.[32] A 2011 review of 28 studies on weight status, consumption of healthy and unhealthy foods, and environmental factors reported a consistent association between weight status and the food environment. The study also showed an association between an environment (in particular, living in an area of low socioeconomic status) and unhealthy dietary behaviors.[30]

Municipalities across the United State are exploring various regulatory approaches to tinker with local environments, including the modification of zoning codes, the enactment of local ordinances regulating fast food outlets, and the use of taxation as a "stick" (or "carrot") to provide incentives. Such approaches may be aimed directly at the food environment (the topic of this section) but it is worthwhile to point out that these approaches are also being considered for shaping other aspects of neighborhood environments, such as using zoning to limit check-cashing store-fronts or pay-day lenders.

Zoning and Ordinances Impacting Food Stores

An ecological framework addressing distal causes of obesity in the social and physical environments is one approach to obesity prevention.[33] Access to food stores—type, number, quality, access via public-transit and for persons of limited mobility as well as affordable—is an aspect of the local environment. As obesity research continues, researchers may find that healthy food access is not a significant contributing factor to obesity prevention. Nonetheless, the use of local zoning and ordinance to shape access to food of any type is still an idea of significance for discussions concerning social justices or other food and health related issues.

Food Deserts and Grocery Gaps

The term "food desert" is a means of descriptively characterizing areas that lack sufficient access to healthy food items such as produce and whole grains; sufficient referring also to relative affordability across the socioeconomic demographics of the area. Researchers have found that, in general, areas with low-income and minority populations have less access to healthy food, as indicated through lesser numbers of (and greater distance to) supermarket retailers as well as through lesser availability and quality of healthy foods.[34]

In urban areas that lack access to healthy foods, a lack of supermarkets may be coupled with a saturation of convenience and fast food establishments. An analysis of New York City restaurant and food outlet data found an average density of 31 unhealthy food outlets per square kilometer, while the average density of healthy food outlets was only 4 per square kilometer.[35]

Food deserts, however, are not limited to urban environments. One analysis reported that in 418 counties in the United States, nearly all of which are rural, county residents were located 10 or more miles from any supermarket or supercenter.[36]

The term "grocery gap" refers more specifically to a comparative lack of supermarkets in some low-income urban neighborhoods when measured against areas of greater affluence in the suburbs. As the population migrated in the 1960s, 70s, and 80s from urban centers to suburban communities, grocery stores and supermarkets followed this market. The physical facilities of the stores and eventually the business model consequently adapted to the suburban environment, with larger buildings and parking lots, and more efficient homogeneous inventories supplied by national distributors, which consequently made it less likely the chains would return to opening small stores in lower-income neighborhoods.[37] Similarly, with the ability of such stores to offer lower prices, more convenient hours and greater parking access, business was drawn away from the urban core forcing smaller independent grocery stores in urban centers out of business.[38]

Health Impacts of Grocery Gaps and Food Deserts

Several studies have concluded that neighborhood density of, or living in proximity to, small grocery stores or convenience stores is associated with obesity. One study, the Atherosclerosis Risk in Communities study, also found greater prevalence of hypertension in census tracks with convenience stores.[39–41] Another study found that close proximity to supermarkets was associated with lower risk of obesity[39] and with greater consumption of vegetables and fruits in populations including African Americans, Supplemental Nutrition Assistance Program (SNAP) participants, seniors, and adolescents.[42–45] However, this result has not held consistent in similar studies. One study of a multiethic population in Detroit reported an increase of 69 vegetable and fruit servings in neighborhoods with a large grocery store,[46] while other studies found no relationship between supermarket access and fruit and vegetable intake.[47–49]

Attracting Supermarkets to Underserved Communities

Barriers to bringing supermarkets back to urban neighborhoods exist. Operating a supermarket in an urban neighborhood can be more expensive than in a suburban area. Minimal available land development opportunities, along with zoning regulations on square footage, parking spaces, and other issues may discourage new supermarkets from moving into densely populated urban neighborhoods. Further, real estate costs can be high and financing can be difficult to obtain from lenders who are reluctant to risk financing a new business.[50]

However, sophisticated data analysis techniques, such as spatial policy simulations estimating the impacts of introducing supermarkets on the body mass index of the surrounding communities[51] have fostered growing literature on food access and health. Armed with more and better information, community and political support has increased for supermarket initiatives across the country.

This supermarket movement was "kicked off" in 2001, when The Food Trust, a nonprofit food advocacy organization, approached the Philadelphia city council with a series of policy recommendations to encourage supermarket development or improvement of existing food stores in underserved neighborhoods. Garnering community and political partners, including the state of Pennsylvania, the Fresh Food Financing Initiative was launched in 2004 to help finance supermarket projects in underserved neighborhoods through a combination of grants, loans, and tax incentives. This was the first statewide supermarket development program,[52] and as of 2011, the program has provided $72.9 in loans and 11.3 million in grants for 88 supermarket projects across the state of Pennsylvania.[53] No longitudinal evaluation has been conducted yet, although reportedly early data suggests the program has increased access to fresh food in the targeted communities.[54]

Success in Philadelphia fostered a similar approach in New York City with the 2008 Food Retail Expansion to Support Health (FRESH) program. The FRESH program provides zoning and tax incentives for the development of food stores in designated neighborhoods determined to be the most in need of improved food access,[55] based on a Department of City Planning report that found 3 million New Yorkers reside in neighborhoods that scored high on a supermarket need index, characterized by high population density, low access to a household car, low household income, high rates of obesity and diabetes, low consumption of fresh fruits and vegetables, low share of fresh food retail, and capacity for new stores.[56] To support this and other initiatives across the state, then New York State Governor David Paterson launched the New York Healthy Food and Communities Fund, administered in part by The Food Trust, which provides loans and grants for food stores in low or moderate income areas that serve a majority low-income client base.[57] New York City's FRESH website lists 16 sites that have benefited from the FRESH program as of December 2013.[58]

Similar programs have been launched in a number of cities and states across the country, and federally, the Obama Administration proposed the Healthy Food Financing Initiative (HFFI), modeled after Philadelphia's Fresh Food Financing Initiative. The 2014 Farm Bill formally established HFFI and authorized up to $125 million in funding.[59]

The initiative has received mixed reviews, with some skepticism that the evidence supporting the connections between supermarket access, individual shopping behavior and dietary intake of health foods is lacking. Some argue that educational interventions are necessary to increase consumers desire to buy vegetables and fruits.[60] Others suggest that supermarket incentives and increasing access to healthy foods are not enough, and efforts should extend to increasing barriers to nonhealthy foods.

The school food environment suggests similar questions. *For example, will increasing the nutritional content of meals offered in school cafeterias necessarily result in dietary improvement among the students if highly attractive a la carte food options remain available?* Think back to the discussion about individual liberty and choices. In this case, providing the "nudge" of an alternative may not be enough. *Does the fact that the student population is not of adult age influence the appropriateness of removing the choice of a poorer quality food?*

Other Efforts to Increase Access to Healthy Foods

Improving access to healthier food can be done on many levels. Supermarkets are not the only answer. Other efforts focus on corner stores and mobile vending. New York City's Shop Healthy program encourages corner stores to stock healthier items, and to apply for permits to sell produce on the street outside of their stores.[61] Similar programs aimed at improving "corner stores" are being established in communities nationwide.[62]

Another method transforms mobile food vending from tantalizing hot dogs and ice cream to tempting colorful fruits and vegetables. The New York City Green Cart Initiative offers licenses at reduced rates to mobile food vendors to operate carts that sell only fruits and vegetables in designated low-income and produce deficient communities.[63] Other cities, including Chicago, Kansas City, and San Francisco are also developing similar programs.[64]

Zoning and Ordinances Impacting Fast Food Outlets

Since at least 2004, the away from home food market has accounted for approximately half of all food expenditures.[65,66] Foods served at fast food restaurants have increasing portion sizes[67] and energy density.[68] Research has found that fast food consumption is strongly correlated with weight gain, obesity, and insulin resistance.[69]

Several studies have found that low income communities have a higher concentration of fast food establishments than middle or high income communities, and that areas with higher concentrations of ethnic minority groups have more fast food establishments.[70,71] However, the findings from research regarding access to fast food restaurants and obesity is mixed, with some studies finding significant associations between obesity and fast food access, while others find no association.[70,72] Some studies have reported a relationship between fast food availability and lower consumption of vegetables and fruits.[70,72] Some municipalities in the US are experimenting with local ordinances aimed at limiting the concentration or marketing abilities of fast food restaurants in their jurisdictions.

South Los Angeles' fast food restaurant moratorium

In June 2008, the city council in Los Angeles, California unanimously approved an ordinance enacting a one-year moratorium on any new fast food establishment in South and Southeast Los Angeles. The ordinance allowed existing establishments to remain, but disallowed any new stand-alone restaurants from opening. It defined fast food restaurants as "any establishment which dispenses food for consumption on or off the premises, and which has the following characteristics: a limited menu, items prepared in advance or prepared or heated quickly, no table orders and food served in disposable wrapping or containers."[73]

The action was spurred by higher obesity rates and poorer health outcomes in these areas in comparison to wealthier areas of the city. City council members reported that there were approximately 1000 fast food establishments in the 30 square mile South Los Angeles area, which has a population of about 500,000 residents. While other municipalities have enacted similar ordinances for esthetic purposes, Los Angeles was the first to do so for public health.[74]

The two-year moratorium expired in September of 2010, but in December of 2010, the council voted on new city planning regulations intended be adopted as part of the community plan. These regulations would restrict new stand-alone fast-food eateries from opening within a half-mile radius of another, but would still permit fast-food outlets as part of strip malls. Overall, the goal was to encourage mixed use and more variety of services.[75] In mid 2013, the city planning commission suggested exempting a portion of the community from the fast-food moratorium, but after a public meeting, decided to retain the restrictions throughout the area.[76]

Impact of ban on fast food restaurants

The South Los Angeles ban on new fast food restaurants has been controversial. One criticism is over the underlying assumption that the higher concentration of fast food restaurants than other areas of Los Angeles is the signifying problem, rather than other issues. Through analyzing food establishment and consumption data, the RAND Corporation concluded that the more significant issue at hand is actually that South Los Angeles has fewer supermarkets in comparison to wealthier West Los Angeles. The study also found that South Los Angeles residents eat out less often but consume more calories in snacks than do residents of higher income neighborhoods.[77]

Think back to the discussion on harm and individual liberty. *What kind of evidence do you think is needed to justify an ordinance that acts as a restriction?* One concern is that the fast-food ban ordinance lacks clear evidence to support the impact of reducing fast food density on obesity or other health outcomes. A 2011 study using data from the National Longitudinal Study of Adolescent Health found no association between fast food availability and weekly consumption of fast food in young adults, suggesting that limiting access to fast food restaurants may not impact frequency of dining at these establishments.[78] However, another study of young adults found an association between an increase in fast food availability near a resident's home and an increase in fast food consumption among low-income men.[49] *Is waiting for evidence to become available to support such policies problematic? Would working to add positive influencers such as supermarkets or farmers markets be a better tool than banning negative influencers?*

Recall the discussion in the introduction about definitions. Working through definitions can be critically important to policy implementation. Some have pointed out that the ordinance, with its imprecise definition of fast food, may actually limit opportunity for local small food establishments with diverse cultural cuisines to develop in the neighborhood, that it will reduce jobs for local residents, and that it unfairly targets a largely minority community.[79]

Other fast food ordinances

Other types of fast food zoning ordinances are also being utilized; often in connection with schools. In Detroit, an ordinance sets a minimum distance of 500 ft between any fast food establishment and elementary, junior or high schools. In San Francisco, a similar ordinance prohibits any mobile food vendors from selling food within 1500 ft of public middle, junior high, and high schools during daytime hours.[80] Public health law experts also suggest additional possibilities for using zoning strategies to either encourage fast food establishments to improve their menus, or to displace these outlets in favor of restaurants with healthier offerings. Some suggested interventions include using land use restrictions that establish minimum distances of fast food restaurants from playgrounds and other facilities for children, limiting how close together fast food restaurants can be situated, banning drive-through service and charging a fee to fast food restaurants, which can be used to support other anti-obesity efforts.[81]

Fast food toy bans

Many countries have enacted legislation restricting marketing of foods to children, including Canada, Norway, and Sweden, where all television advertising to children under 12 is banned.[82] In the United States, attempts to regulate marketing of unhealthy foods to children have met resistance, and attention has turned toward strategies to regulate the marketplace in which fast foods are sold, rather than the marketing of the foods. The fast food industry practice of including a free toy in children's meals is a practice under scrutiny. Research has shown that, when paired with a collectible toy, preschool children are more likely to find a meal appealing.[83] Advocates hope that by limiting the ease at which children can receive a toy, it will lessen the effect of the toy's value as a marketing tool to entice overconsumption of unhealthy meals.

San Francisco implemented an ordinance in December of 2011 preventing fast food restaurants from including free toys in children's meals unless the meals met predetermined standards as to calories (no more than 600), sodium (less than 640 mg), calories from fat (no more than 35%) and trans-fat (0.5 g or less). In order to include a free toy, meals must meet these standards and also include one half cup of fruit and three-quarters cup of vegetables.[84,85]

McDonald's and Burger King immediately responded by announcing that customers would be able to purchase the toys for 10 cents with a child's meal, which some say will diminish the impact of the new law.[86] Regardless, advocates and organizations such as Public Health Law and Policy are advising municipalities interested in enacting similar legislation to include specific language banning both giving toys away for free and for a nominal fee.[87]

In the prior year of 2010, Santa Clara County (a metropolitan area containing the city of San Jose and larger in population than San Francisco), passed a similar toy ban, albeit with slightly different nutritional requirements.[85] An early analysis of the effects of the Santa Clara ordinance reported some positive impacts to the food environment in fast food restaurants, including the removal of posters marketing the children's toy in 2 of 4 of the restaurants studied, and offering the children's toys separately at an additional cost in two restaurants, but concluded that the ban positively influenced the marketing of healthy options that already existed but did not prompt any change in menu offerings to increase the number of healthy options.[88] In addition to these recent local ordinances that attempt to regulate the fast food marketplace, public health law experts have suggestion additional strategies such as requiring fast food restaurants to sell healthy alternatives to unhealthy meals.[89]

Interestingly, action has also occurred on the corporate side. In July 2013, Taco Bell decided to eliminate toys and meals targeted at children altogether. Although Taco Bell is the first national chain to discontinue toys, the western regional chain Jack in the Box discontinued toys in 2011.[90] *Does it matter if Taco Bell is eliminating children's meals and toys out of sheer business interest, rather than as a nutritional or health policy goal?* How does this connect with the idea of corporate self-regulation (discussed at the end of this chapter)?

Taxation

Sales tax on food

Some states require that an individual pay a sales tax on groceries at the time of purchase. Sales tax on basic necessities such as food or clothing is considered a regressive tax, meaning that the cost of the tax is a greater burden to lower income people. When it comes to groceries, everyone must eat, and will at some point buy groceries. A sales tax on groceries will be applied equally to everyone purchasing the item, whether rich or poor. However, since the poor have less overall income to spend, the cost of the tax disproportionally cuts into the amount of money available to spend by the individual on other necessities. Sales tax on food used to be a more popular funding mechanism among the states, but as of 2011, only 14 states tax groceries.[91] A few states, such as Idaho and Kansas, provide a grocery tax refund,[92] but as of August 2014, a resolution has passed in the Kansas House to abolish the sales tax.[93]

Although grocery taxes are historically an outgrowth of a fiscal policy, grocery taxes become food policy through impact. Grocery taxes could be created specifically as food policy. Such a tax could be designed with the goal of guiding individual behavior toward a choice deemed relevant to a greater social good. For example, a "social good" food policy could be one of taxing foods considered likely to contribute to social expenses (such as government subsidized long term health care for chronic disease such as type-2 diabetes) at a higher rate than foods deemed nutritious and diverting the revenue thus derived toward health intervention and care programs. Taxing food for this purpose would be similar to the adoption of a higher than average sales taxes on cigarettes by certain localities on the idea that a financial stick will cause people to decrease smoking.[94] Just like groceries, the sales tax on cigarettes is also considered a regressive tax, in this case because cigarette smoking is correlated with lower-income. A 2012 study has pointed out that because the cigarette sales tax is quite regressive, more of the revenue should go toward health intervention programs aimed at reducing smoking.[95]

Taxation can be used to guide or influence public behavior while simultaneously raising federal, state, or local revenue.[96] For example, cigarettes and alcohol are usually taxed at higher rates than other food products or beverages. Taxes meant to dissuade the public from unhealthy behaviors are sometimes called "sin taxes" because of general public perceptions about the morality or desirability of such behaviors. Nonetheless, these "sin taxes" are in fact excise taxes, levied on some but not all commodities.[97] Such taxes, sometimes termed "fat taxes" when pertaining to unhealthy foods and beverages, are of interest to both legislators and health researchers due to the ability to conduct parallel efforts to close budget gaps while promoting obesity prevention.

Evidence for taxation strategies to promote public health

The interest in taxing unhealthy beverages or food items is grounded in the success of tobacco taxation programs. Increasing the excise tax rate on cigarettes (and thereby driving up the cost of smoking) effectively reduces smoking rates, with estimates showing a 10% increase in price per pack corresponding with a 3–7% decline in cigarette smoking.[98] Sugar-sweetened beverages are of particular interest to policy makers because of their link to obesity and other health concerns,[99] and based on the success in tobacco, the idea of increasing taxes to help in encouraging decreased consumption seems like it could be effective. Furthermore, the idea of taxing sugared beverages is not a novel idea, as most states in the United States already have small taxes on sugared beverages and snack items.[100] However, these small taxes mean that even when states do tax soda, the rate is not necessarily higher than for other taxed consumable items. A small tax in line with tax rates on other items is not really a "sin" or "fat" tax; as it does not provide a clear "stick" to guide behavior. As of 2011, 40 states tax soda at an average rate of 5.2%, but only in five states is the tax 7.0% or higher.[101]

Sugar-sweetened beverage tax

Attempts have been made across the United States to implement a specific sugar sweetened beverage tax. A federal tax has also been previously considered. In 2008, the Congressional Budget Office undertook a study on imposing a federal tax on sugar-sweetened beverages of 3 cents per 12 ounces. The study found that this tax could generate $50 billion in revenue from 2009–2018, but this tax was not implemented.[102]

In 2008, then New York State Governor David Paterson proposed an 18% tax on sugar-sweetened beverages with a dual purpose of addressing obesity and closing the state's budget gap. The tax proposal was not well received by the public, and was not passed.[103] In 2008, Maine was successful in passing a tax on the syrup used in making sodas, increasing the cost of soda by 4 cents per can, but the tax was overturned by a citizen's veto.[104] California, Massachusetts, and Philadelphia, Kansas, Vermont, Mississippi, and Alaska among other municipalities, have all attempted to pass taxes on sugary beverages and met with broad resistance,[103] California most recently in the summer of 2013.

As of September 2014, nine states had sugar-sweetened beverage tax proposals filed in the current legislative session, some as sales tax, some as excise tax, and many with the idea of earmarking the proceeds to support health-related foods, activities, or obesity prevention.[105]

The federal government may also be considering a soda tax. On July 30, 2014, a bill was introduced in the House of Representatives to place an excise tax of one cent per 4.2 g (about a teaspoon) of caloric sweeteners used in certain beverages, with exemptions for milk, milk-alternative beverages (soy, almond, etc.), 100% fruit juice, alcoholic beverages, nutritional therapy beverages, and infant formula, with the revenue going to the fund initiatives addressing health conditions related to the consumption of sugar. The exemption on milk is described as "any liquid the primary ingredients of which are milk." In August 2014, this bill had been referred to the Subcommittee on Health.[106] The arguments to exempt milk-based products are sometimes couched in the idea that the milk itself is nutritious. *Do we need to balance the risk outcomes of osteoporosis with the risk outcome of diabetes and obesity?* Refer to Chapter 7, "Milk," for more details about milk nutrition.

Coffee versus cola

Sweetened blended coffee drinks may be considered to be primarily milk. *If a sugar-sweetened beverage tax applies to soda but not to blended drinks served at a coffee shop, is the tax evocative of a "sin" tax, manifesting social disapproval over soda but not coffee drinks?*

According to the USDA nutrient database, a 12 ounce carbonated cola beverage contains just about 40 grams of sugar and a 16 ounce carbonated cola beverage contains just about 53 g of sugar.[107] Conversely, a 12 ounce blended sweetened flavored coffee drink (a mocha) may contain just over 40 g of sugar and 60 g of sugar in the 16 ounce version.[108]

Similar legislation is being voted on in November 2014 in California, San Francisco, and Berkeley, and interestingly, San Francisco's version retains the exemption for café-served coffee beverages, but Berkeley's does not.[109] Berkeley's measure passed and San Francisco's did not, which may provide an opportunity for a natural experiment. By the time you read this chapter, you will be able to find out what impact Berkeley's soda tax may have had.

The Berkeley tax is based on the amount of sugar added, instead of the size of the beverages. A June 2014 study funded by the Robert Wood Johnson Foundation found that a calorie-based tax on sugar-sweetened beverages was more effective than a volume-based tax as it would achieve a greater reduction in calories with a smaller effect on the amount that consumers would be willing to pay for a product, and generally people were more inclined to purchase the cheaper drink.[110]

The calorie based tax serves to highlight the questionable attribute (a subtle nudge) while preserving consumer choice. The consumer can use cost, volume, and taste preference to evaluate if they will choose a 12 ounce sweetened beverage even though it is taxed more than a larger beverage with fewer calories. *Would this type of tax influence product reformulation by companies?* In October 2013, Mexico passed a soda tax that went into effect January 2014 that may sway adoption of soda taxes elsewhere. Self-regulation by companies through promoting alternative drinks and smaller sizes is likely.[111]

Health impacts of taxation

However, results of studies of taxation on sugar sweetened beverage taxes and a potential impact on obesity have been mixed. One observational study found no significant association between state level grocery and vending soda tax rates and adolescent BMI, though a weak inverse association was reported between teens at risk of overweight and vending machine taxes. The authors hypothesize that the current soda tax rates are so low that any relationship with BMI would be weak.[112] However, another study found that states without a sales tax on soft drinks or snack foods were four times as likely to experience a relative increase in obesity between 1991 and 1998 as states that had such taxes. Further, states that had repealed a soft drink or snack food tax during that time were 13 times more likely to experience a high relative increase in obesity.[113]

Studies have also attempted to project the impact of raising taxes on sugar-sweetened beverages. Using the economic theory of price elasticity, one study estimated that a penny per ounce tax on sugar-sweetened beverages would result in a 24% decrease in consumption of the beverages.[114] Another study estimated that a 20% price increase on sugar-sweetened beverages would result in an average reduction of 37 calories per day for adults and 43 calories per day for children.[115]

Focusing a tax too narrowly might not have the intended effect as food policy, as another study found that even large price increases of 20–40% in carbonated sugar-sweetened beverage would result in only a reduction of 4.2 calories per day at 20% and 7.8 calories per day at 40% due to the effect of consumers switching to other caloric beverages. Expanding the price increase to include all sugar-sweetened beverages resulted in reductions of 7 calories per person per day at an increase of 20% and 12.4 calories per person per day at 40%, with estimated weight losses of 7 and 1.3 lbs per person per year, respectively.[116] It is also possible that these studies have not increased taxes enough to significantly sway consumption significantly. Another projection found a greater potential weight loss of 3.3 pounds per year for adults based on a 20% tax on sugar-sweetened beverages.[117]

Is a potential weight loss of 3.3 pounds per year enough to use the obesity epidemic as a justification for this tax?

Or is there a legitimate argument that this type of tax is indeed a "sin" tax, that is, taxation of behavior that some find objectionable? A policy does not have to be grounded in nutrition. A policy could be based on social disapproval of poor people spending their money on soda.

It is also possible that we need to consider other health outcomes besides obesity. A 2011 study estimated that a tax of one cent per ounce would decrease consumption of sugar-sweetened beverages by 15%, and that would be enough to reduce new cases of type-2 diabetes by 2.6% and the prevalence of obesity by 1.5%. Although these effects are small, it would make a difference over a 10-year period, with an estimate that it would result in 95,000 less coronary heart events, and by delaying the onset of type-2 diabetes through improved diet, prevent 2.4 million person-years of diabetes.[118] These potential effects exemplify the possible impact of improving risk in small amounts across large numbers discussed at the beginning of this chapter.

Though not as widely debated as sugar-sweetened beverages, taxation of snack foods and other high energy foods has also been discussed as a potential obesity-prevention or reduction strategy because diets high in these foods are also often nutrient poor and tend to have a low satiating power, leading to overeating and weight gain.[120] Since tobacco work finds that adolescents are particularly responsive to price changes in cigarettes,[98] it is possible that adolescents may be similarly responsive to price increases of desirable but unhealthy sugary beverages or foods.[113] Many states already impose low taxes, ranging from 1% to 7% on candy and snack items sold in stores or vending machines.[120] European countries have introduced taxes on unhealthy foods. Denmark began taxing food based on its saturated fat content in 2011.[121] Hungary introduced a tax on prepackaged food based containing high levels of sugar and salt in 2011, while Norway taxes sugar and chocolate, and Finland taxes sugary products including soft drinks.[122] Romania taxes fast food, cake and candy, soda, and snack foods.[123]

However, studies assessing the potential impact of taxing unhealthy foods have shown mixed results. An experiment studied the purchases of groceries from a web-based service by comparing the purchases made from one group of people in the absence of added taxes to the purchases made by a second group of people when exposed to a 50% added tax on high energy density foods. The participants with the added tax bought fewer calories and 16% less energy dense foods than the group with no added taxes.[124] An analysis of the effectiveness and cost-effectiveness of a 10% price increase on soft drinks and snack foods found a possible weight reduction of 1.6 kg (3.5 lbs) and categorized such an intervention as effective and cost saving.[125] Furthermore, some food items may be more or less resilient to a contemplated tax, as a study to predict the effect of a tax placed on dairy products (a source of dietary saturated fat) found that a 10% tax would result in a less than 1% reduction in fat consumption.[126]

Response to taxation strategies

Recent proposals to tax sugar sweetened beverages and/or unhealthy foods have caused a flurry of both positive and negative responses in the United States. Aside from possible health benefits, supporters of taxation of sugar-sweetened beverages and other unhealthy foods also cite the potential revenue that could be raised through such a tax, and suggest the use of this revenue exclusively for health promotion programs. One group estimated that a tax of 1 cent per 12-ounce soft drink would generate $1.5 billion per year.[127] Another group estimated that a penny per ounce sugar-sweetened beverage tax would raise $79 billion over 5 years.[114]

Some, however, express concern that a tax on sugary beverages alone is not an effective method to reduce intake of excess calories, because these beverages can be substituted by other calorie sources. The alternate solution could be a tax on sugar itself, to increase the price of all products containing sugar, or a repeal of the current agricultural subsidies for sugar.[128] Of course, a tax on sugar (or the removal of prices support) would make sugar more expensive for food manufacturers. Manufacturers may simply seek a less expensive sweetener. Other concerns include difficulties implementation of such a tax would pose, particularly regarding categorization of foods into healthy or unhealthy categories.[127] Are "diet" sodas healthy?

It is also important to point out that a tax on either sugary beverages or unhealthy foods would be a regressive tax toward low-income populations, who spend a greater proportion of their income on food.[113] And, that for both sugar-sweetened beverages and unhealthy food tax projects, the evidence appears to indicate that relatively large taxes will be needed to produce health results. While some public opinion polls have shown that people may be willing to pay small increases in order to support better health,[129] large tax increases are unlikely to be as well received.

Further, the beverage industry has made a strong showing against proposed sugar-sweetened beverage taxes. In New York State, for example, the trade spent $9.4 million to oppose it and went so far as to set up the advocacy group, "New Yorkers Against Unfair Taxes." Created under the guise of a consumer coalition, it received strategic guidance from a high profile public affairs company. Political support for the bill was weak as well, particularly among representatives with beverage distribution centers in their districts.[130]

Even though the NYC Board of Public Health proposed rule restricting soda size did not apply in all venues, did not restrict how many 16 ounce purchases an individual could make (and did not restrict refills), much of the controversy was over if the government should act as a "nanny state" by telling citizens what is best for them through circumscribing individual choices that were being presented by the market.

However, the soda industry, aware of the recommendations in the 2010 Dietary Guidelines and seeking to avoid a loss of sales in a more health-conscious market, also recently began producing smaller-sized containers. The reduction of sizes by industry was not met with the controversy that the New York City proposal garnered. The rest of this chapter discusses how such actions may be self-regulation by industry.

SELF-REGULATION IN THE FOOD SYSTEM

Self-regulation is a voluntary, self-policing approach used by industry as part of changing business practices, and provides an alternative to government enforced regulation. Self-regulation interests public health and food policy professionals because of its potential role in responding to food-related health issues, as there are some 20 different chronic diseases connected to food and food consumption practices, including obesity.[131] However, public health professionals and opponents of "Big Food" have been critical of the ability of self-regulating organizations to adequately respond to the obesity epidemic in particular.

This chapter section discusses self-regulation in the food system (beyond that of advertising to children). After defining the theory of self-regulation, it explores examples of self-regulation in the food system, and investigates the incentives for self-regulating activities. Finally, it discusses the food system's successes and failures in addressing public health issues related to food using self-regulation and offers suggestions for future application and research.

Self-Regulation Theory: Definitions and Terms

Industry self-regulation has been studied in economics, organizations and industries, and defined as "a regulatory process whereby an industry-level, as opposed to a governmental- or firm-level, organization (such as a trade association or professional society) sets and enforces rules and standards relating to the conduct of firms in the industry."[132]

Self-regulation has also been studied within psychology for its use in cognitive and behavior change. This breadth of study reveals some important components of self-regulation which may lead to its success or failure: (1) standards, (2) monitoring for comparing and tracking progress, and (3) an implementation process for change.[133] The self-regulating activities identified in the food system reflect these components. Moreover, the term "self-regulation" is synonymous with terms like self-policing and voluntary regulation. These terms all intend here to convey the idea of an industry enacting measures by which it holds industry members accountable.

Arguments against Self-Regulation

What has our self-regulated food system brought upon us? As of 2014, adult obesity is above 20% in all states with rates exceeding 35% for the first time in two states (Mississippi and West Virginia).[134] The situation is equally serious for the nation's child and adolescent population. The most recent data report from the Center for Disease (CDC) as of the writing of this book indicates that in 2012, 18% of children aged 6–11 are obese, and 21% of adolescents, ages 12–19, are obese.[135] More than one-third of US adults are obese, and some of the leading causes of preventable death, heart disease, stroke, type-2 diabetes, and some types of cancer, are conditions related to obesity.[136] The health status of the American people serves as the lens through which public health professionals argue against food industry self-regulation.

Critics suggest that motivations for self-regulation develop from within the corporation. Under this scenario, a corporate conscience, rather than external forces, inspire self-regulation. As a result, critics argue that industry adopts weak and inadequate standards when they are independently responsible. When comparing the social responsibility of the corporation to humankind, it is said that "no one would seriously suggest that individuals should regulate themselves, that laws against murder, assault, and theft are unnecessary because people are socially responsible."[137,138]

It is also argued that industry generated standards fail to go deep and wide to cover the full scope of the industry with their self-policing measures.[139] Yet, this criticism overlooks the system intricacies that field experts have the knowledge to address. At the core, critics believe that corporations simply cannot be trusted to function responsibly when their business goal is profit, and suggest profit goals conflict with notions of self-regulation.[137] Promises can be retracted if sales wane and blame cast on the self-imposed voluntary regulations. Enforcing regulations and generating compliance is difficult to achieve among peer competitors. With no leverage to enforce agreements, industry members lack incentive to follow through on promises.[140]

Self-regulation may not improve consumer education beyond the product specific encouragement to purchase. In their flexibility, voluntary regulations allow industry to be inconsistent and regress or change course at any time, as well as to develop and promote standards favoring their own products. Under company-based product labeling standards, consumers may become confused by contrasting information and the variety of health labels, and federally approved standards undermined.[137]

Can self-regulation foster solutions for the obesity epidemic and other food-related health goals?

The American Beverage Association, in conjunction with soft-drink companies, has attracted public praise over their plans to remove sodas from schools. Although laudable in intent, the scope of the promise falls short. While soft-drink vending machines have been removed from some elementary schools, they remain in many middle and high schools. The promises restrict product sales and promotions during school hours but anticipate sales at school events.[137] Public health professionals feel that self-regulation is too weak, too narrow and failing as a solution to the obesity epidemic.

Arguments for Self-Regulation

The United States government promotes self-regulation as an opportunity to limit its own intrusion into the marketplace. It is believed that this bottom up approach will empower industry to do the right thing and that voluntary regulation is a "low-cost, flexible, incentive-driven mechanism" that motivates honesty and integrity in business practices.[139] The flexibility of self-regulation gives industry latitude to choose its own standards, which in turn determine the quality of the product.

As a result, it is believed that industries choose standards that reduce their liability and their costs. It is theorized that the cost of adhering to a self-imposed standard is less than regulation from an external entity. Self-regulation can create a ripple effect that forces other players to also adopt stringent standards to stay competitive in the market. Because this mimics imposed standards, it appears to be costlier for industry rivals to compete for this quality standard. It is also reported that perceived weak public regulations entice industries to navigate routes to a premium market.[141]

Successful regulation of a product requires a thorough knowledge of the product. Self-regulation by industry organizations is favored because industries have this expertize within their organization. The approach does not rely on outside officials or policymakers to conjecture what regulatory measure would be adequate. Field experts offer in-depth understanding of issues, can address issues in greater detail and can act swiftly. Voluntary regulation can touch ethical and moral issues that may be outside of the scope of government-crafted regulation. Self-regulation is also favored because the financial burden is at least initially assumed by industry, rather than by taxpayers.[139,142]

Although in the end, the cost would most likely be passed on to the individual purchaser (who is still a taxpayer), the cost is born by those directly participating in the exchange, rather than being subsidized by all taxpayers.

The Food System and Self-Regulation

Recall the discussion of defining food policy and food systems in the introduction of this book and consider the numerous factors that touch on producing, processing, and consuming food. Our food system is surrounded by external factors.

The influence of these external forces varies with the length of the supply chain. The chain can be very short when the farmer directly sells to the consumer, as in a community supported agriculture program. The food supply chain is typically much longer because the food system extending from producer through processor to consumer largely operates in an international market. For example, the corn that makes any breakfast cereal may be grown in Iowa, refined in Indiana, processed into cereal in China, packaged in North Carolina, and retailed for purchase in Colorado.

Self-regulating activities can be found throughout the food system. Regulations, public or private, are a response to a problems like food safety, new scientific information, or consumer demand. Country of Origin Labeling (COOL) is an example of regulation influenced by consumer demand for more information about where the food they were purchased was produced.[143] However, federal regulations are challenged to keep pace with a developing food industry, as is clear in the discussions over health claims and labeling in Chapter 9, "Grains." Federal regulations can be either too broad or too specific to adequately regulate the system, leaving latitude for voluntary regulation to develop.

Privately enforced regulations are now widely used in the food system. Some sectors articulate them as "codes of practice" or "codes of conduct." Examples of these codes include good agricultural practice (GAP) and good manufacturing practice (GMP).[143] Industries promise to participate by making a pledge or by entering into an agreement, and self-regulation occurs as either private or public nonbinding agreements.

Private standards are growing rapidly in the food system to distinguish the quality characteristics of food products. While food quality and food safety are related, they are regulated differently. Food safety is more rigorously addressed in public regulation. "Quality assurance schemes" are generated by industry but with no public standards and may be defined by a wide range of characteristics: production methods, animal treatment, product composition, worker treatment, environmental impact, freshness, and location. These quality attributes are accentuated for the consumer quite strategically with voluntary labeling. Quality standards are in fact, driving competition in the food market. More than ever before, consumers are making quality-based decisions about food.[143]

Grower/Producer

At the point of production, self-regulation lives in farming practices. Farmer producers are self-regulating for quality as well as safety. An example of producer self-regulation was gleaned from a conversation with dairy farmer R. VanOord in November 2008. At this time, some dairy farmers were voluntarily starting to test their herds for Johne's disease (*Mycobaterium paratuberculosis*, or MAP). Dairy farmers have a growing concern that this disease, sometimes found in dairy herds, has an association with Crohn's disease in humans, and could be transported in milk. Researchers have found evidence that MAP is found more often in people with Crohn's disease than in people without Crohn's disease, but the direction of this association is unclear, that is, it is unknown which occurs first—Crohn's disease or MAP.[144] This potential association remains hotly contested. While researchers have not yet proven the link between the two diseases, dairy farmers are nonetheless concerned about the impact, such a finding would have on the dairy market and began to work out best practices. Research has furthered this effort and the USDA is helping to promote education and best practices, however, these measures are not required by law, but rather are voluntarily implemented.[145]

Not all farmers observe the same standards in food production. Organic and naturally grown products may be the result of a private standard observed by a producer or an organic growers organization such as the Northeast Organic Farmer's Association. The difference in production methods is expressed in organic/natural versus conventional practices, free range or grass-fed operations, or humane treatment versus confined animal feeding operations. Withholding the use of chemicals, antibiotics, or hormones in food production is a voluntary, self-regulating choice a farmer is free to make. Rejecting the use of genetically modified (GMO) seeds, plants, or animals is also an expression of self-regulation by the farmer.

In addition to a firm's adherence to a voluntary private standard, these firms may go one step further to ensure the credibility of their compliance. In the case of organics, the growing third-party certification of private standards is an example of private standards and enforcement operating in tandem with public regulations.[142] Third-party certifications also cover other quality areas that public regulations do not and consumers may seek out, such as evidence of humane treatment. Nongovernmental certifying bodies are providing "certified" humane stamps that consumers may rely on when selecting their purchases.

Processor/Packager

Firms that purchase, process, and package raw foods may also subscribe to private standards. Often such standards reverberate back to the production phase or are a continuation of a standard observed in the production phase. These voluntary standards are often made known to the public through voluntary labeling. These labels distinguish foods that are non-GMO foods, humanely grown or acquired through fair-trade, processed, additive and preservative free, hormone-free, or made with all-natural ingredients.

Distributor/Retailer

Once products are processed and packaged they are distributed to retailers. Supermarkets have been labeled as "gatekeepers of the food supply." As such, supermarkets possess the ability to control food availability, promotion, pricing, and sales. Supermarket strategies respond to not only consumer demands but also influence consumer choice.[146] Food stores can regulate sales, and therefore consumption, of unhealthy food products with their own selling policies. Food retailers, like the Park Slope Food Coop in Brooklyn, New York, may articulate self-regulating policies in their mission.

We offer a diversity of products with an emphasis on organic, minimally processed and healthful foods. We seek to avoid products that depend on the exploitation of others. We support nontoxic, sustainable agriculture …We prefer to buy from local, earth-friendly producers.[147]

Although the Park Slope Food Coop is a small, niche retailer, large food retailers may also self-regulate, with candy-free check out lanes and increased promotion of fruits and vegetables.[148] Leading chains are also working on front-of-pack nutrition labels and salt reduction in store-brand products, and some supermarkets are working on environmental and sustainability programs that may influence business decisions.[149]

Eater/Consumer

Consumers regulate through their foods purchasing choices. Researchers have termed individuals who are actively engaged in self-regulation "food citizens." "Food citizenship" describes the set of principles the consumer applies to food purchases.[150] These standards might include local or seasonal foods, sustainable production, just conditions for farm workers, humane animal treatment, limited packaging and processing of foods as well as a direct producer-consumer relationship.

Are food citizens simply responding to premeditated plans of the food industry, which limits consumer ability to influence self-regulation? Can self-regulation happen across the entire food system, or is it limited by scale? When thinking about these questions, consider that food citizens and consumers are not just individuals. Consumers are also institutions such as schools, universities or hospitals, restaurants or corporate cafeteria. Chapter 8, "Meat," describes the impact of organizations signing up for "Meatless Mondays." In 2008, over 100 colleges and universities were voluntarily regulating their food purchases around the idea of local foods farm-to-college,[151] and by now that number has certainly increased. Hospitals, too, have invoked private standards for the foods they purchase and serve. As of 2014, more than 400 care facilities across the United States have committed to The Healthier Food Challenge by signing the Healthy Food in Health Care pledge or by formally adopting a sustainable food policy.[152]

Incentives for Self-Regulation in the Food System

The food system is shaped by external systems as described above. These forces can operate as incentives to participation. Understanding the incentives may help in assessing the ability of these measures to mitigate food-related health issues.

Profit is the obvious and overarching incentive for industries to voluntarily regulate a product. Self-regulation creates a ripple effect throughout the industry. When one company sets a standard that reaps positive attention, competitors adjust to join. Expanding a market is part of the increased profit incentive. A corporation may embrace tighter standards if it yields a higher-quality product that can be sold at a premium price in a niche market. Food technology, product innovation, consumer demand, economic forces, and cultural system stimulate new and niche markets. For example, religious laws concerning food regulation demonstrate how cultural systems motivate self-regulation. Kosher foods are regulated by private religious standards. The choice to produce Kosher foods is voluntary; however, to label and market foods as Kosher requires compliance with a publicly enforced standard intended to protect consumers from fraud.

The growth of organic products (and the development of federal standards for organics) was inspired in part by consumer demand. Although some farmers switched to organic methods due to ecological and moral beliefs, the high-value, premium priced organic market motivated other farmers to convert to organic growing methods as well. Self-regulation in the form of private standards also occurs when public regulations are missing or deemed inadequate, in particular for these niche markets. As an example, standards regarding humane animal practices are promoted through a variety of third-party certifiers.

A market advantage created through self-regulation may increase profits but can also save money long-term. By operating offensively through self-regulation and product diversification, the firm chooses when its budget can bear the financial burden of stricter standards. Research indicates that it costs more to comply with a publicly imposed regulation than a privately imposed standard.[141,143]

Furthermore, market opportunities exist both domestically and internationally. Export opportunities motivate industries to self-regulate their products in order to access international markets, as product quality and characteristics are not always measured by the same standards globally. The nonacceptance of GMO foods in international markets serves as an excellent example. European markets ban GMO foods, but US markets allow it. If a food producer wants to access an international market with GMO restrictions, self-regulation would be required.

Motivation for Public Health–Related Self-Regulation

Product endorsements emphasizing nutrition and health goals can entice industry to produce foods that meet certain desired criteria without the specific requirements of government regulation. For example, nutrition criteria recommended by the government may be promoted by nongovernmental organizations through health promotions.

An example of a health promotion is a "seal of approval" program. When New Zealand's National Heart Foundation offered a "seal of approval" program for food with prescribed nutritional standards, the food industry self-regulated the nutritional content of their products in order to receive the heart healthy–approved "Pick the Tick" label. A study of the intervention showed that the label incentive reduced the sodium content of 390 processed foods by more than 12 tons over the course of a year, and one-quarter of the margarines sold during the study period were reformulated through self-regulation to acquire the Pick the Tick label.[153]

The growth of food related health issues such as obesity and diabetes has also motivated self-regulation. Investigators of food-related health issues are pointing fingers at the food industry. In developing countries, it has been suggested that the growth of supermarkets—and the processed foods they sell—is a driver of the "nutrition transition" to diets high in high-calorie, nutrient-poor foods.[154] The fear of negative publicity motivates producers, processors, and retailers to adopt more stringent standards, which in turn yields positive publicity.[155]

An industry's response to regulation, whether public or private, poses some similarities. Firms make calculated decisions about the degree to which they will comply with public regulations, and choose to be laggards or leaders. The laggards choose to fight against the laws or comply at the lowest level possible. Some industries evade compliance while aiming to minimize the perceived negative impact of doing so by garnering positive media attention associated with some other act. Soft drink and fast food industries are noted for using this strategy. However, some industries operate offensively. These industry leaders are early adopters of stricter standards because the leading edge offers a market advantage, and provides the opportunity to potentially preempt public regulation.[155]

Evidence of self-regulation in the form of private standards occurs in the grower, processor, retailer, and consumer phases of the food system. Researchers who have examined self-regulation in the light of public health concerns have made recommendations for its future application to food-related diseases such as obesity. First, it has been determined that if self-regulation is to address food-related health issues, self-regulating activities must embody public health initiatives.[157] This suggestion implies that definitions and standards for food safety and quality must develop to include nutrition. It also suggests that self-regulation organizations should not simply be composed of food industry representatives. Rather, to gain the support of public health officials, as well as to pass muster for an increasingly critically aware consumer base, food system-wide stakeholders must be involved in self-regulating activities.

Second, industries should incorporate measurable outcomes into self-regulating agreements.[140] Concrete data will help in assessing compliance and can be used to make future recommendations. These steps are predictors of successful self-regulation, and reflect the third component of self-regulation, namely that successful self-regulation must have a process for change.[133] Measurements reinforce accountability, and combining science-based measurement with an implementation process can bolster enforcement, the weakest area of self-regulation.

Impact of Self-Regulation on Food-Related Health Conditions

Nutrition standards continue to be vigorously debated by researchers, nutrition experts, and food industry representatives, presenting an on-going challenge for food policy in all arenas, but particularly so for self-regulation. As noted in the previous section, the flexibility of self-regulation allows greater speed of implementation as a potential tool to address food-related conditions. However, with evolving nutritional standards, there is a challenge of discerning whether or not the effects of self-regulation are meeting the expectations of public health officials. It is unlikely that voluntary regulation will yield an effect that meets the expectations of public health representatives. It is also equally unlikely that public regulations would deliver stringent enough laws to minister to the problem. However, a solution in the middle ground may still benefit consumer health.

In other words, self-regulation may inherently create change in our food environment, but the deliberate promotion of industry self-regulation as a partnership strategy to create optimal change remains an on-going challenge for public health. Achieving public health goals via industry self-regulation is likely not to succeed if consumer favored companies do not participate, the credibility and process of participating companies is poor, or if the standards are weak, variably applied, and difficult to evaluate.[156] If standards are weak, even strong enforcement will lead to weak results.[157]

CONCLUSION

Local, regional, and federal governmental regulatory powers such as taxation, zoning, and ordinances can influence "who eats what, when, how and with what impact," and conversely, so can self-regulation in the food system in the absence of such public rules. We have accepted the role of public health in environment modifications that limit our choices, and we have already extended this role to include food modification, such as the pasteurization of milk. We have accepted the role of public health in changing social norms in areas such as tobacco use. Modifying our food environment to nudge personal food consumption behavior is an appropriate sphere for public health intervention. As the Founding Fathers themselves supported and passed a law taxing private beverage consumption, it becomes difficult to argue that food policy nudging a private decision of consumption toward behavior to the common public good or social goal is "un-American." Nonetheless, corporate self-regulation may be more acceptable to an American public sensitive to infringement of individual liberties by government, and may contribute to desired health goals. Looking forward, public health officials face the challenges of using local ordinances, taxes, and regulations in manners that foster public health outcomes without raising the specter of "sin taxes" or the "nanny state."

REFERENCES

1. Johnson LB. Commencement address in New London at the United States Coast Guard Academy. June 3, 1964. Available at http://www.presidency.ucsb.edu/ws/?pid=26290, accessed November 20, 2014.
2. Upshur REG. Principles for the justification of public health intervention. *Canadian J. Public Health.* 2002;93:101–103.
3. Kersh R, Stroup DF, Taylor WC. Childhood obesity: A framework for policy approaches and ethical considerations. *Preventing Chronic Disease*, 8, 2011: A93. Available at http://www.cdc.gov/pcd/issues/2011/sep/10_0273.htm, accessed September 2, 2014.

4. Mill JS, *On Liberty*. New York: Dover Thrift Editions, 2002.

5. McKinnon RA, A rational for policy intervention in reducing obesity. *American Med. Assoc. J. Ethics.* 2010;12:309–315.

6. Peck A. Revisiting the original "Tea Party:" The historical roots of regulating food consumption in America. *University of Missouri-Kansas City Law Rev.* 2011;80.

7. George Washington's Mount Vernon. Whiskey Rebellion. Available at http://www.mountvernon.org/research-collections/digital-encyclopedia/article/whiskey-rebellion/, accessed November 20, 2014.

8. Whiskey, AG Jr. Margarine, and newspapers: A tale of three taxes. In: *Taxing Choice*, Shughart II, WF (ed.), 1997. Oakland, CA: The Independent Institute, pp. 57–77.

9. Rose G. Sick individuals and sick populations. *Int. J. Epidemiol.* 1985;14:32–38.

10. Esterl M. Coke tailors its soda sizes. *Wall Street J.* September 19, 2011, Available at http://online.wsj.com/article/SB10001424053111903374004576578980270401662.html, accessed September 2, 2014.

11. New York City Department of Health and Mental Hygiene, Notice of Adoption of Amendment to Article 81 of the New York City Health Code. Available at http://www.nyc.gov/html/doh/downloads/pdf/notice/2012/notice-adoption-amend-article81.pdf, accessed September 2, 2014.

12. Bennett D. New York City's giant soda ban is defeated once again. *The Atlantic Wire*, July 30, 2013. Available at http://www.theatlanticwire.com/national/2013/07/new-york-citys-soda-ban-defeated-once-again/67777, accessed September 2, 2014.

13. Dumanovsky T et al. Changes in energy content of lunchtime purchases from fast food restaurants after introduction of calorie labeling: Cross sectional customer surveys. *BMJ.* 343, 2011. Available at http://dx.doi.org/10.1136/bmj.d4464, published July 26, 2011.

14. Horovitz B. Starbucks, other marketers go huge with healthier offerings. *USA Today*, updated July 7, 2011. Available at http://usatoday30.usatoday.com/money/industries/food/2011-07-11-small-portions-lower-calories_n.htm, accessed September 2, 2014.

15. Letter from Dawn Sweeney, President and CEO, National Restaurant Association, to the Honorable Margaret A. Hamburg, MD, Commissioner, US Food and Drug Administration, July 5, 2011. Available at http://www.restaurant.org/Downloads/PDFs/advocacy/20110705_ml_fda_sweeney.pdf, accessed September 2, 2014.

16. Garb M. Health, morality, and housing: The "Tenement Problem" in Chicago. *American J. Public Health.* 2003;93:1420–1430. Available at http://www.ncbi.nlm.nih.gov/pmc/articles/PMC1447986, accessed September 2, 2014.

17. American Heart Association, A History of Trans Fat. http://www.heart.org/HEARTORG/GettingHealthy/FatsAndOils/Fats101/A-History-of-Trans-Fat_UCM_301463_Article.jsp, updated January 11, 2014.

18. New York City Department of Health and Mental Hygiene, press release #114-06, December 5, 2006, http://www.nyc.gov/html/doh/html/pr2006/pr114-06.shtml, accessed September 2, 2014.

19. Resnik D. Trans fat bans and human freedom. *Am. J. Bioethics.* 2010;10:27–32.

20. Angell SY et al. Change in trans fatty acid content of fast-food purchases associated with New York City's restaurant regulation. *Ann. Internal Med.* 2012;157:81–86.

21. Fuchs AW, Frank LC. *Milk Supplies and their Control in American Urban Communities of Over 1000 Population in 1936.* Washington, DC: United States Government Printing Office, 1939.

22. United States Food and Drug Administration, Food Safety and Raw Milk, November 1, 2011. Available at http://www.fda.gov/Food/FoodborneIllnessContaminants/BuyStoreServeSafeFood/ucm277854.htm, last updated March 22, 2013, accessed September 2, 2014.

23. Hartke K. *CDC Cherry Picks Data to Make Case Against Raw Milk.* The Weston A. Price Foundation, February 22, 2012. Available at http://www.westonaprice.org/press/cdc-cherry-picks-data-to-make-case-against-raw-milk, accessed September 2, 2014.

24. Ryan CA et al. Massive outbreak of antimicrobial-resistant salmonellasis traced to pasteurised milk. *J. Am. Med. Assoc.* 1987;258:3269–3274.
25. Olmstead AL, Rhode PW. Impossible undertaking. The eradication of Bovine tuberculosis in the United States. *J. Econ. History.* 2004;64:734–772.
26. Koh HK. A 2020 vision for healthy people. *New England J. Med.* 2010;362:1653–1656.
27. Levi J. et al. *F as in Fat How Obesity Threatens America's Future.* 2013. Washington, DC: Trust for America's Health, August 2013. Available at http://healthyamericans.org/report/108, accessed September 2, 2014.
28. Ogden CL et al. Obesity and socioeconomic status in adults: United States, 2005–2008. *National Center for Health Stat. Data Brief.* 2010;50:1–8.
29. Flegal KM et al. Prevalence and trends in obesity among US adults, 1999–2008. *J. Am. Med. Assoc.* 2010;203:235–241.
30. Giskes K et al. A systematic review of environmental factors and obesogenic dietary intakes among adults: Are we getting closer to understanding obesogenic environments? *Obesity Rev.* 2011;12:e95–e106.
31. Drewnowski A, Obesity, diets, and social inequalities. *Nutr. Rev.* 2009;67:S36–S39.
32. Ford PB, Dzewaltowski DA. Disparities in obesity prevalence due to variation in the retail food environment: Three testable hypotheses. *Nutr. Rev.* 2008;66:216–228.
33. Kumanyika SK et al. Population-based prevention of obesity: The need for comprehensive promotion of healthful eating, physical activity, and energy balance: A scientific statement from American Heart Association Council on Epidemiology and Prevention, Interdisciplinary Committee for Prevention. *Circulation.* 2008;4:428–464.
34. Beaulac J, Kristjansson E, Cummins S. A systematic review of food deserts, 1966–2007. *Preventing Chronic Dis.* 2009;6:A105.
35. Rundle A et al. Neighborhood food environment and walkability predict obesity in New York City. *Environ. Health Perspect.* 2009;117:442–447.
36. Morton LW, Blanchard TC. Starved for access: Life in rural America's food deserts. *Rural Realities.* 2007;1:4.
37. Treuhaft S, Karpyn A. The grocery gap: Who has access to healthy food and why it matters. *Food Trust and Policy Link,* 2010. Available at http://www.policylink.org/sites/default/files/FINALGroceryGap.pdf, accessed September 2, 2014.
38. Walker RE, Keane CR, Burke JG. Disparities and access to healthy food in the United States: A review of food deserts literature. *Health Place.* 2010;16(5):876–884.
39. Lovasi GS. et al. Built environments and obesity in disadvantaged populations. *Epidemiol. Rev.* 2009;31(1):7–20.
40. Gibson DM. The neighborhood food environment and adult weight status: Estimates from longitudinal data. *Am. J. Public Health.* 2011;101(1):71–78.
41. Morland K, Diez Roux AV, Wing S. Supermarkets, other food stores, and obesity: The atherosclerosis risk in communities study. *Am. J. Preven. Med.* 2006;30(4):333–339.
42. Morland K, Wing S, Diez Roux AV. The contextual effect of the local food environment on residents' diets: The atherosclerosis risk in communities study. *Am J. Public Health.* 2002;92(11):1761–1767.
43. Sharkey JR, Johnson CM, Dean WR. Food access and perceptions of the community and household food environment as correlates of fruit and vegetable intake among rural seniors. *BMC Geriatrics.* 2010;10:32.
44. Powell LM, Han E, Chaloupka FJ. Economic contextual factors, food consumption, and obesity among US adolescents. *J. Nutr.* 2010;140(6):1175–1180.
45. Rose D, Richards R. Food store access and household fruit and vegetable use among participants in the US food stamp program. *Public Health Nutr.* 2004;7(8):1081–1088.
46. Zenk SN et al. Neighborhood retail food environment and fruit and vegetable intake in a multiethnic urban population. *Am. J. Health Promotion.* 2009;23(4):255–264.

47. Pearson T et al. Do "food deserts" influence fruit and vegetable consumption? a cross-sectional study. *Appetite.* 45(2),2005:195–197.
48. Pearce J et al. The contextual effects of neighbourhood access to supermarkets and convenience stores on individual fruit and vegetable consumption. *J. Epidemiol. Community Health.* 2008;62(3):198–201.
49. Boone-Heinonen J et al. Fast food restaurants and food stores: Longitudinal associations with diet in young to middle-aged adults: The CARDIA study. *Arch. Internal Med.* 2011;171(13):1162–1170.
50. New York City Department of City Planning, The FRESH Food Store Area Program, May 2009. Available at http://www.nyc.gov/html/dcp/pdf/fresh/presentation_may 2009.pdf, accessed December September 2, 2014.
51. Chen SE, Florax RJGM. Zoning for health: The obesity epidemic and opportunities for local policy intervention. *J. Nutr.* 2010;140(6):1181–1184.
52. Giang T et al. Closing the grocery gap in underserved communities: The creation of the Pennsylvania fresh food financing initiative. *J. Public Health Management Practice.* 2008;14(3):272–279.
53. Evans D. Democratic Chairman Budget Briefing, The Pennsylvania Fresh Food Financing Initiative, March 2010. Available at http://www.ncsl.org/documents/labor/workingfamilies/PA_FFFI.pdf, accessed September 2, 2014.
54. Karpyn A et al. Policy solutions to the "grocery gap." *Health Affairs.* 2010;29(3):473–480.
55. New York City Department of City Planning, FRESH Stores. Available at http://www.nyc.gov/html/dcp/html/fresh/index.shtml, accessed September 2, 2014.
56. New York City Department of City Planning, *Going to Market: New York City's Neighborhood Grocery Store and Supermarket Shortage.* April, 2008. Available at http://www.nyc.gov/html/dcp/html/supermarket/index.shtml, accessed September 2, 2014.
57. Low Income Investment Fund, *New York Healthy Food and Healthy Communities Program.* Available at http://www.liifund.org/products/community-capital/capital-for-healthy-food/new-york-healthy-food-healthy-communities-fund/, accessed September 2, 2014.
58. Food Retail Expansion to Support Health. FRESH Projects. Available at http://www.nyc.gov/html/misc/html/2009/fresh_projects.shtml, accessed September 2, 2014.
59. Healthy Food Access, healthyfoodaccess.org/sites/default/files/updated-hffi-fact-sheet.pdf, accessed August 29, 2014.
60. Holzman DC. White House proposes healthy food financing initiative. *Environ. Health Perspec.* 2010;118(4):A156.
61. NYC Food, Shop Healthy. Available at http://www.nyc.gov/html/nycfood/html/shop/shop.shtml, accessed September 2, 2014.
62. Healthy Corner Stores Network. Available at http://www.healthycornerstores.org/, accessed September 2, 2014.
63. New York City Department of Health and Mental Hygiene, Green Carts. Available at http://www.nyc.gov/html/doh/html/living/greencarts.shtml, accessed September 2, 2014.
64. Tester JM et al. An analysis of public health policy and legal issues relevant to mobile food vending. *Am. J. Public Health.* 2010;100(11):2038–2046.
65. Stewart H, Blisard N, Jolliffe D. *Let's Eat Out: Americans Weigh Taste, Convenience, and Nutrition.* United States Department of Agriculture Economic Research Service, Washington, DC; 2006, Economic Information Bulletin no. 19.
66. United States Department of Agriculture, Economic Research Service. Available at http://www.ers.usda.gov/data-products/food-expenditures.aspx#.UjAIPb9Rxok, accessed September 2, 2014.
67. Nielsen SJ, Popkin BM, Patterns and trends in food portion sizes, 1977–1998. *J. Am. Med. Assoc.* 2003;289(4)450–453.
68. Prentice AM, Jebb SA. Fast foods, energy density and obesity: A possible mechanistic link. *Obes. Rev.* 2003;4(4):187–194.

69. Pereira MA et al. Fast-food habits, weight gain, and insulin resistance (the CARDIA study): 15-year prospective analysis. *The Lancet.* 2005;365(9453):36–42.

70. Fleischhacker SE et al. A systematic review of fast food access studies. *Obes. Rev.* 2011;12(5):e460–e471.

71. Larson NI, Story MT, Nelson MC. Neighborhood environments: Disparities in access to healthy foods in the US. *Am. J. Preven. Med.* 2009;6(1):7 4–81.

72. Fraser LK et al. The geography of fast food outlets: A review. *Int. J. Environ. Res. Public Health.* 2010;7(5):2290–2308.

73. Hennessy-Fiske M, Zahniser D. Council bans new fast-food outlets. *Los Angeles Times,* July 30, 2008. Available at http://articles.latimes.com/2008/jul/30/local/me-fastfood30, accessed September 4, 2014.

74. Medina J, In South Los Angeles, New Fast-Food Spots Get a "No, Thanks," *The New York Times,* January 15, 2011. Available at http://www.nytimes.com/2011/01/16/us/16fastfood. html, accessed September 4, 2014.

75. Now, Council limits new fast food outlets in South LA, *Los Angeles Times,* December 8, 2010. Available at http://latimesblogs.latimes.com/lanow/2010/12/council-limits-new-fast-food-outlests-in-south-los-angeles.html, accessed September 4, 2014.

76. City of Los Angeles, draft community plan. Available at https://sites.google.com/site/southlaplan/, accessed September 4, 2014.

77. Sturm R, Cohen D. Zoning for health? The year-old ban on new fast-food restaurants in South LA. *Health Affairs.* 2009;28(6):w1088–w1097.

78. Richardson AS et al. Neighborhood fast food restaurants and fast food consumption: A national study. *BMC Public Health.* 2011;11:543.

79. Creighton R. Cheeseburgers, race, and paternalism. Los Angeles' ban on fast food restaurants. *J. Legal Med.* 2009;30(2):249–267.

80. Strategic Alliance for Healthy Food and Activity Environments, ENACT Local Policy Database. Available at http://eatbettermovemore.org/sa/policies/policy_detail.php?s_Search=&issue=1&env=2&keyword=6&s_State=&jurisdiction=&year=&policyID=221, accessed September 4, 2014.

81. Ashe M et al. Land use planning and the control of alcohol, tobacco, firearms, and fast food restaurants. *Am. J. Public Health.* 2003;93(9):1404–1408.

82. Ahmed H. Obesity, fast food manufacture, and regulation: Revisiting opportunities for reform. *Food Drug Law J.* 2009;64(3):565–575.

83. McAlister AR, Bettina Cornwell T. Collectible toys as marketing tools: Understanding preschool children's responses to foods paired with premiums. *J. Public Policy Marketing.* 2011;31:30.

84. Strom S. Toys stay in San Francisco Happy Meals, for a charge. *The New York Times,* November 30, 2011. Available at http://www.nytimes.com/2011/12/01/business/toys-to-cost-extra-in-san-francisco-happy-meals.html, accessed September 4, 2014.

85. Madison Park, Happy Meal toys no longer free in San Francisco.*CNN,* December 1, 2011. Available at http://www.cnn.com/2011/11/30/health/california-mcdonalds-happy-meals/index.html?hpt=hp_bn10, accessed September 4, 2014.

86. Wootan MG. *McDonald's Seeks to Circumvent San Francisco Law on Fast-Food Toys.* Center for Science in the Public Interest, November 30, 2011. Available at http://cspinet.org/new/201111302.html, accessed September 4, 2014.

87. ChangeLab Solutions, Model Ordinance for Toy Giveaways at Restaurants. Available at http://changelabsolutions.org/publications/healthier-toy-giveaway-meals, accessed September 4, 2014.

88. Otten JJ et al. Food marketing to children through toys, response of restaurants to the first US Toy Ordinance. *Am. J. Preven. Med.* 2012;42(1):56–60.

89. Ashe M et al. Local venues for change: Legal strategies for healthy environments. *J. Law, Med. Ethics.* 2007;35(1):138–147.

90. Godoy M. Taco Bell says adios to kids' meals and toys. *NPR The Salt*, July 23, 2013. Available at http://www.npr.org/blogs/thesalt/2013/07/23/204899615/taco-bell-says-adios-to-kids-meals-and-toys, accessed September 4, 2014.

91. Viard AD. Should groceries be exempt from sales tax? July 25, 2011, American Enterprise Institute,State Tax Notes. Available at http://www.aei.org/article/economics/fiscal-policy/taxes/should-groceries-be-exempt-from-sales-tax, accessed September 4, 2014.

92. Idaho State Tax Commission, Idaho Grocery Credit. Available at http://tax.idaho.gov/i-1043.cfm, updated Jan 8, 2013, accessed September 4, 2014.

93. Wistron BD. House passes plan to cut income tax, abolish sales tax on groceries. *The Wichita Eagle*, March 25, 2014, updated August 8, 2014. Available at http://www.kansas.com/news/politics-government/article1088134.html, accessed September 4, 2014.

94. Centers for Disease Control and Prevention, Morbidity and Mortality Weekly Report, March 30, 2012;61(12):201–204, http://www.cdc.gov/mmwr/preview/mmwrhtml/mm6112a1.htm, accessed September 4, 2014.

95. Farrelly MC, Nonnemaker JM, Watson K. The Consequences of High Cigarette Excise Taxes for Low-Income Smokers. September 12, 2012. Available at http://www.plosone.org/article/info%3Adoi%2F10.1371%2Fjournal.pone.0043838, accessed September 4, 2014.

96. Altman A. A brief history of: Sin taxes. *Time*, April 2, 2009. Available at http://www.time.com/time/magazine/article/0,9171,1889187,00.html, accessed September 4, 2014.

97. Sadowsky J. The economics of sin taxes. *The Action Institute, Religion & Liberty*, 4(2), 1994. Available at http://www.acton.org/pub/religion-liberty/volume-4-number-2/economics-sin-taxes, accessed September 4, 2014.

98. Davis K et al. *Cigarette Purchasing Patterns Among New York Smokers: Implications for Health, Price, and Revenue.* New York State Department of Health Tobacco Control Program, March 2006.

99. Olsen N, Heitmann BL. Intake of calorically sweetened beverages and obesity. *Obes. Rev.* 2009;10(1):68–75.

100. Brownell KD, Frieden TR. Ounces of prevention—The public policy case for taxes on sugared beverages. *New England J. Med.* 2009;360:1805–1808.

101. Claire Wang Y et al. Tax policy measure: A penny-per-ounce tax on sugar-sweetened beverages would cut health and cost burdens of diabetes. *Health Affairs*, January 31, 2012;31(1):1199–1207; doi: 10.1377/hlthaff.2011.0410.

102. Congressional Budget Office. Budget Options, Vol. 1. Health Care. December 2008.

103. Hartocollis A. Failure of state soda tax plan reflects power of an anti-tax message. *The New York Times,* July 2, 2010. Available at http://www.nytimes.com/2010/07/03/nyregion/03sodatax.html?pagewanted=all, accessed September 4, 2014.

104. Chan S. A tax on many soft drinks sets off a spirited debate. *The New York Times*, December 16, 2008. Available at http://www.nytimes.com/2008/12/17/nyregion/17sugartax.html, accessed September 4, 2014.

105. Yale Rudd Center for Food Policy and Obesity, Legislation database. Available at http://www.yaleruddcenter.org/legislation/legislation_trends.aspx, accessed September 4, 2014.

106. Congress.gov, https://beta.congress.gov/bill/113th-congress/house-bill/5279/text, accessed September 4, 2014.

107. United States Department of Agriculture. Agriculture Research Service. United States Department of Agriculture. National Nutrient Database for Standard Reference. Release 27. Basic Report: 1418, Carbonated beverage, cola. Available at www.ndb.nal.usda.gov/ndb/foods/show/4258, accessed October 28, 2014.

108. Starbucks. Explore Our Menu. Nutrition Info. Available at http://www.starbucks.com/menu/drinks/frappuccino-blended-beverages/mocha-frappuccino-blended-beverage#size=11002668&milk=61&whip=NA, accessed October 28, 2014.

109. Aliferis L. Here's what would be taxed—or not- in SF, Berkeley Soda Tax Measures. *KQED*. October 29, 2014. Available at http://blogs.kqed.org/stateofhealth/2014/10/29/

heres-what-would-be-taxed-or-not-in-sf-berkeley-soda-tax-measures/, accessed October 28, 2014.

110. Chen Z, Brissette IF, Ruff RR. By ounce or by calorie: The differential effects of alternative sugar-sweetened beverage tax strategies. *Am. J. Agri. Econ.* 2014;96(4):1070–1083; first published online June 2, 2014.

111. Comlay E. Coke Femsa shares fall as Mexico passed food, drink taxes, October 31, 2013. *Am. J. Agri. Econ.* Available at http://www.reuters.com/article/2013/10/31/us-mexico-sodatax-idUSBRE99U16120131031, accessed September 4, 2014.

112. Powell LM, Chriqui JF, Chaloupka FJ. Associations between state-level soda taxes and adolescent body mass index. *J. Adolescent Health.* 2009;45:S57–S63.

113. Kim D, Kawachi I. Food taxation and pricing strategies to "thin out" the obesity epidemic. *Am. J. Preven. Med.* 2006;30(5):430–437.

114. Andreyeva T, Chaloupka FJ, Brownell KD. Estimating the potential of taxes on sugar-sweetened beverages to reduce consumption and generate revenue. *Preven. Med.* 2011;52:52413–5416.

115. Smith TA, Lin B-H, Lee J-Y. *Taxing Caloric Sweetened Beverages: Potential Effects on Beverage Consumption, Calorie Intake, and Obesity.* ERR-100, US Department of Agriculture, Economic Research Service, Washington, DC. July, 2010.

116. Finkelstein EA et al. Impact of targeted beverage taxes on higher- and lower-income households. *Arch. Internal Med.* 2010;170(22):2028–2034.

117. Novak NL, Brownell KD. Taxation as prevention and as a treatment for obesity: The case of sugar-sweetened beverages. *Curr. Pharm. Des.* 2011;17(12):1218–1222.

118. Claire Wang Y. et al. A penny-per-ounce tax on sugar-sweetened beverages would cut health and cost burdens of diabetes. *Health Affairs.* 2012;31(1):199–207.

119. Drewnowski A, Darmon N. The economics of obesity: Dietary energy density and energy cost. *Am. J. Clinical Nutr.* 2005;82(1 Suppl):265S–273S.

120. Chriqui JF et al. State sales tax rates for soft drinks and snacks sold through grocery stores and vending machines, 2007. *J. Public Health Policy.* 2008;29(2):226–249.

121. Bittman M. How about a little Danish? *The New York Times*, October 4, 2011, http://opinionator.blogs.nytimes.com/2011/10/04/how-about-a-little-danish, accessed September 4, 2014.

122. Holt E. Hungary to introduce broad range of fat taxes. *Lancet.* 2011;378:755.

123. Romania becomes first country to introduce junk-food tax. Euractiv. January 7, 2010, updated September 4, 2014. Available at http://www.euractiv.com/health/romania-country-introduce-junk-food-tax/article-188647.

124. Nederkoorn C et al. High tax on high energy dense foods and its effects on the purchase of calories in a supermarket. An experiment. *Appetite.* 2011;56(3):760–765.

125. Sacks G et al. "Traffic-light" nutrition labeling and "junk-food" tax: A modeled comparison of cost-effectiveness for obesity prevention. *Int. J. Obes.* 2011;35(7):1001–1009.

126. Chouinard HH et al. Fat taxes: Big money for small change. *Forum for Health Econ. Policy.* 2007;10:2.

127. Jacobson MF, Brownell KD. Small taxes on soft drinks and snack foods to promote health. *Am. J. Public Health.* 2000;90:854–857.

128. Fletcher JM, Frisvold DE, Tefft N. Are soft drink taxes an effective mechanism for reducing obesity? *J. Policy Anal. Manag.* 2011;30(3):655–662.

129. Powell LM, Chaloupka FJ. Food prices and obesity: Evidence and policy implications for taxes and subsidies. *Milbank Quart.* 2009;87(1):229–257.

130. Hartocollis A. Failure of state soda tax plan reflects power of an anti-tax message. *The New York Times*, July 2, 2010. Available at http://www.nytimes.com/2010/07/03/nyregion/03sodatax.html?pagewanted=all, accessed September 4, 2014.

131. Levi J et al. *F as in Fat How Obesity Threatens America's Future.* 2013. Washington, DC: Trust for America's Health, August 2013. Available at http://www.healthyamericans.org/report/108/, accessed September 4, 2014.

132. Gupta AK, Lad. Industry self-regulation: An economic, organizational, and political analysis. *Acad. Manag. Rev.* 1983;8(3):416–425.
133. Carver CS, Scheier M. *Attention and Self-Regulation: A Control Theory Approach to Human Behavior.* New York: Springer-Verlag, 1981.
134. The State of Obesity: Better Policies for a Healthier America—Trust for America's Health, September 2014, http://www.healthyamericans.org/report/115/, accessed September 4, 2014.
135. Centers for Disease Control and Prevent, Childhood Obesity Facts. Available at http://www.cdc.gov/healthyyouth/obesity/facts.htm, updated August 12, 2014, accessed September 4, 2014.
136. Centers for Disease Control and Prevention, Adult Obesity Facts. Available at http://www.cdc.gov/obesity/data/adult.html, updated September 3, 2014, accessed September 4, 2014.
137. Simon M. *Appetite For Profit: How the Food Industry Undermines our Health and How to Fight Back.* New York: Nation Books, 2006.
138. Bakan J. *The Corporation: The Pathological Pursuit of Profit and Power.* New York: Free Press, 2004, pp. 85–110.
139. Hawkes C. Self-regulation of food advertising: What it can, could and cannot do to discourage unhealthy eating habits among children. *Nutr. Bull.* 2005;30(4):372–382.
140. Kline R et al. Beyond advertising controls: Influencing junk-food marketing and consumption with policy innovations developed in tobacco control. *Loyola Law Rev.* 2006;39(1):603–646.
141. McCluskey JJ. Pre-empting Public Regulation with Private Food Quality Standards, (paper presented at IATRC Summer Symposium, Bonn, Germany, May 28–30, 2006).
142. Richards L. US Securities and Exchange Commission, "Self-Regulation in the New Era" (presented at NRS Fall 2000 Compliance Conference Scottsdale, Arizona September 11, 2000). Available at http://www.sec.gov/news/speech/spch398.htm, accessed September 4, 2014.
143. Henson S, Reardon T. Private agri-food standards: Implications for food policy and the agri-food system. *Food Policy.* 2005;30(3):241–253.
144. University of Wisconsin, School of Veterinary Medicine, Johne's Information Center, FAQS. Available at http://www.johnes.org/zoonotic/index.html, updated March, 2010, accessed September 4, 2014.
145. Collins MT, Eggleston V. *Healthy Cows for a Healthy Industry.* Johne's Information Central. Available at www.johnesdisease.org, accessed September 4, 2013.
146. Hawkes C. Dietary implications of supermarket development: A global perspective. *Development Policy Rev.* 2008;26(6):657–692.
147. Park Slope Food Coop. Mission Statement. Available at http://foodcoop.com/go.php?id=38&PHPSESSID=e3e617620762ffd315cce258fb74fb42, accessed September 4, 2014.
148. Alexander E. Tesco takes a stand: Candy-free checkout counters, updated May 22, 2014. Vitality Institute. Available at http://thevitality institute.org/tesco-takes-a-stand-candy-free-checkout-counters, accessed May 10, 2015.
149. Dibb S et al. Healthy and fair: A review of the Government's role in supporting sustainable supermarket food. *Sustainable Development Commission.* Available at http://research-repository.st-andrews.ac.uk/bitstream/10023/2369/1/sdc-2008-green-healthy-fair.pdf, accessed September 4, 2014.
150. Baker LE. Tending Cultural Landscapes and food citizenship in Toronto's community gardens. *Geo. Rev.* 2004;94(3):305–325.
151. Community Food Security Coalition, Farm to college. Available at http://www.farmtocollege.org, accessed November 22, 2008.
152. Healthier Hospitals, The Healthier Food Challenge. Available at http://www.healthyfoodinhealthcare.org/, accessed September 4, 2014.

153. Young L, Swinburn B. Impact of the Pick the Tick food information programme on the salt content of food in New Zealand. *Health Promotion Int.* 2002;17(1):13–19.
154. Popkin BM. Global nutrition dynamics: The world is shifting rapidly toward a diet linked with noncommunicable diseases. *Am. J. Clin. Nutr.* 2006;84(2):289–298.
155. Henson SJ, Hooker NH. Private sector management of food safety: Public regulation and the role of private controls. *Int. Food Agri. Manag. Rev.* 2001;4(1):7–17.
156. Sharma LL, Teret SP, Brownell KD. The food industry and self-regulation: Standards to promote success and to avoid public health failure. *Am. J. Public Health.* 2010;100(2):240–246.
157. Hawkes C. Self-regulation of food advertising: What it can, could and cannot do to discourage unhealthy eating habits among children. *Nutr. Bull.* 2005;30(4):374–382.
158. California Almond Board, Pasteurization, http://www.almondboard.com/HANDLERS/FOODQUALITYSAFETY/PASTEURIZATION/Pages/Default.aspx, accessed September 2, 2014.
159. Institute of Medicine of the National Academies, Action Taken, http://www.iom.edu/Reports/2009/School-Meals-Building-Blocks-for-Healthy-Children/Action-Taken.aspx, accessed September 2, 2014.
160. The White House, President Barack Obama, Office of the Press Secretary, December 13, 2010, http://www.whitehouse.gov/the-press-office/2010/12/13/president-obama-signs-healthyhunger-free-kids-act-2010-law, accessed September 2, 2014.
161. Congressman Steve King, http://steveking.house.gov/media-center/press-releases/king-supports-the-utritionnanny-challenge-to-usda, accessed September 2, 2014.
162. United States Department of Agriculture, School Meals Nutrition Standards, Policy Memos, February 25, 2013, http://www.fns.usda.gov/sites/default/files/SP26-2013os.pdf, accessed September 2, 2014.

Food Protection
Defense and Safety

13

> *I aimed for the public's heart and by accident*
> *I hit it in the stomach.*
> Upton Sinclair

INTRODUCTION

Food protection covers both food defense and food safety. Food defense concerns intentional contamination and will be touched on briefly at the end of this chapter. Food safety concerns ensure that food has not accidentally become unsafe to consume. Foodborne illness, pesticides, chemicals used in production, reactions between the food and packaging, or other possible hazards, such as food additives or food allergens are all concerns of food safety.

Thus, this chapter covers labeling as a component of food safety as well as the safety of food production and processing. We describe federal involvement in food safety; we discuss pathogens and risk; we review the transition toward emphasizing prevention over testing at the end of the twentieth century; and we examine food safety through labeling, concluding with other concerns and unintended consequences.

This chapter generally emphasizes the Food and Drug Administration (FDA), as the FDA is responsible for the safety of 80% of the US food supply, as well as the final word on the safety of food products, additives, and the necessities of details provided on ingredient labels. Furthermore, the FDA is being thrust to the forefront of US food safety policy with the Food Safety Modernization Act (FSMA) of 2011, the largest expansion of FDA's food safety authority since the 1930s.[1] The FSMA reflects the trend of favoring government review of Hazard Analysis and Critical Control Point (HACCP) or similar systems designed to identify critical food safety hazard points rather than emphasizing inspections; thus making industry responsible for safety and the government only the enforcer. Food cannot be made completely risk-free, but we can acknowledge concern over risk, and provide information so that individuals can make their own evaluation of safety. Even though we cannot achieve fully risk-free food, we can seek to achieve an overall culture of food safety.

FEDERAL INVOLVEMENT IN FOOD SAFETY

Although numerous federal agencies, as many as 15,[1] are involved in food safety, the most prominent are the FDA, the Food Safety Inspection Service (FSIS) of the United States Department of Agriculture (USDA), the Environmental Protection Agency (EPA), and the Centers for Disease Control and Prevention (CDC). The EPA approves which pesticides may be used on food and sets acceptable tolerances of pesticide residue that are then enforced by the FDA.[2] The information gathered by the CDC during research and outbreak surveillance informs new food safety regulation and guidance.[3]

The USDA, via the FSIS, has responsibility over about 20% of our food supply, which includes most meats and poultry. This area of responsibility covers to a large extent our animal food sources during slaughter and processing, but also to a lesser extent the resulting processed food products. The FSIS is "the public health agency in the US Department of Agriculture responsible for ensuring that the nation's commercial supply of meat, poultry, and egg products is safe, wholesome, and correctly labeled and packaged."[4]

The FDA protects consumers against impure, unsafe, and fraudulently labeled products and through the Center for Food Safety and Applied Nutrition (CFSAN), regulates foods other than the meat, poultry, and egg products regulated by FSIS. Although the FDA's covered food products are circumscribed, it is the only agency with "farm-to-table" regulatory authority for those foods, as unlike the FSIS, the FDA can assert jurisdiction on farm property. Its primary role is to set and enforce standards regarding food safety.[5]

The great emphasis on the animal industry and inspection from the earliest years of food safety regulation has left a legacy of responsibility assigned to a certain extent by food source and product, a complexity compounded further over the years by shifting political alliances and directives. Some of the intricacies are due to the issue that federal agencies are granted their authority from Congress. They can only do what they are authorized to do through statutes; and different agencies have been tasked over the years with various (sometimes overlapping) aspects of protecting food. Differing committees of the Senate and the House, with differing emphases, are concerned with food safety. Committees in the Senate include Agriculture, Nutrition and Forestry; Homeland Security and Governmental Affairs; and Health, Education, Labor and Pensions. House committees include Agriculture; Energy and Commerce; Oversight and Government Reform; and Science.[1]

Most significantly, food safety responsibility still carries the legacy of the 1940 removal of the FDA from its original home as a department of the USDA; apparently, this move was considered a means to prevent recurring conflicts between producer interests and consumer interests.[6]

As a result, we now have a complex system, with overlapping areas of responsibility resulting in absurdities such as the USDA regulating pizza with meat toppings but leaving the FDA in charge of cheese pizza.[1] Trading with other countries can be complicated by differences in regulations promulgated by different agencies for the same pathogen.

In addition to the challenges of integrating agency responsibilities that overlap in food safety, each agency has multiple priorities. As John Bailar, Professor Emeritus of the University of Chicago, comments: "each federal agency with food safety responsibilities has other missions that are generally regarded as more important."[7]

Historically, the FDA has always been underfunded in comparison to the USDA. In FY2012 the FDA's food safety program activities were funded at 866 million while the USDA FSIS were funded at 1.004 billion. In other words, the FSIS covers 10%–20% of the US food supply with 60% of the funds available allotted to the two agencies; the FDA covers 80%–90% of the food supply with 40% of the funds; though slowly the FDA is being allocated more funds.[1] A contributing factor to both the lopsided allocation in funding and the perception that food safety is not necessarily the FDA's primary mission is that FDA funding is still appropriated from congressional agricultural committees—not those focused on health. The legacy of the FDA's initial home as part of the USDA continues to influence food policy even today.[8]

Focus on the Food and Drug Administration

The FDA takes responsibility for "… protecting the public health by assuring the safety, efficacy, and security of human and veterinary drugs, biological products, medical devices, our nation's food supply, cosmetics, and products that emit radiation."[9]

The FDA is responsible for 80% of the food supply, as well as with the final word on the safety of food products and additives and the necessities of details provided on ingredient labels. The FDA's food safety authority has recently been further expanded with the FSMA of 2011.[1] However, although the FSMA greatly expands the food safety purvey of the FDA, even a casual glance at the FDA's website and mission statement speaks to the extent of other important tasks. "Food" is first in the name of the agency, but "drugs" are first on the described responsibilities noted above.

Within the FDA, four directorates report oversee the core functions of the agency: medical products and tobacco, foods, global regulatory operations and policy, and operations. The Office of Foods was created in 2009 with the goal to lead a "functionally unified FDA Foods Program," and in 2012 renamed "Office of Foods and Veterinary Medicine." It houses the CFSAN and the Center for Veterinary Medicine.[10]

The FDA assures the safety of our food supply as follows:

- Ensuring the safety of foods for humans, including food additives and dietary supplements, by setting science-based standards for preventing foodborne illness and ensuring compliance with these standards.
- Ensuring the safety of animal feed and the safety and effectiveness of animal drugs, including the safety of drug residues in human food derived from animals.

- Protecting the food and feed supply from intentional contamination.
- Ensuring that food labels are truthful and contain reliable information consumers can use to choose healthy diets.[11]

Federal Regulation of Food Safety

As discussed in the introduction of this book, by the review of the Commerce Clause, federal authority to regulate food, food production, and food safety was a topic of much debate. Congress accepted federal responsibility over food safety through enacting two pieces of legislation in 1906, the Pure Food and Drug Act and the Meat Inspection Act, but has historically been reactive. This is attributable in part, to Congressional tendency to act upon political pressure caused by public reaction to either information or a disastrous outbreak of illness. For example, although food safety legislation had been in the works for some time prior to 1906, it took the public outrage over *The Jungle*, Upton Sinclair's expose of insanitary and inhumane working conditions in meatpacking plants, to push Congress into motion. Our response cycle continued to be mostly reactive thereafter. The 1938 Food, Drug and Cosmetic Act had been languishing for five years but was quickly signed into law as a response when more than 100 people died, many of whom were children, after a drug company marketed an untested form of the new sulfa wonder drug that contained a highly toxic chemical analog of antifreeze as a solvent.[12] The fatal Jack in the Box outbreak prompted the 1996 implementation of HACCP in meat and poultry processors.[13] Even the recent FSMA of 2011 was prompted by a series of outbreaks in FDA-regulated foods, such as peanut butter and spinach.[14]

1906: Inspection and adulteration

Oversight of the 1906 Pure Food and Drug Act and the Meat Inspection Act was given entirely to the USDA as the prevailing pressure concerned meat inspections and keeping sick animals out of the food supply. (Chapter 7, "Milk," describes a similar concern with dairy inspections.) Internally, the USDA split this new responsibility over food safety between two Bureaus, giving Meat Inspection to the Bureau of Animal Industry and Food and Drugs to the Bureau of Chemistry.[8] The Bureau of Chemistry was renamed as the FDA in 1931, and as noted above, transferred out of the USDA in 1940[6], resulting in a division in authority and responsibilities over the nation's food safety that continues today.

The 1906 Meat Inspection Act required every animal and carcass produced for interstate commerce to be inspected. Meat "found to be not adulterated shall be marked, stamped, tagged, or labeled as 'Inspected and passed'."[15] Meat found to be adulterated could not legally be sold in interstate commerce. The inspection of slaughterhouses that only produced meat for sale within its state borders was left under state authority. This Act and amendments set inspection standards to determine adulteration which made sense in 1906, but under which later the USDA's authority regarding deadly pathogens in food would be greatly circumscribed.

Under the Meat Act, slaughterhouses could only be operated while the inspector was present. Through the inspection stamp, the inspector passed approval of the safety of the meat, and the packer and producer had no further safety requirements. In effect, under the Meat Act, the government was responsible for the safety of meat inspected and stamped as approved, even though authority of safety was limited to the slaughterhouse premises exactly, as the inspector had no ability to recall meat.

The Pure Food and Drug Act also defined what would cause a food to be considered adulterated,[16] but required sampling versus continuous inspection, and made producers responsible for safety. The government was only the enforcer.[8] Thus, the two originating laws, although each passed in 1906, manifested two different policy approaches, one approach in which the government was responsible, and the other approach in which industry was responsible.

However, the 1906 Pure Food and Drug Act struggled with the lack of means of enforcement, and this was further exacerbated by the acceleration of new production technologies for food and drugs. Amendments were cobbled onto the law, but by 1933, it was clear that a complete overhaul was needed. Nonetheless, the revisions languished for another five years. By 1938, increased federal powers supported by the commerce clause were seen in a more favorable light.[17]

Even more significantly for Congress, a precipitating event in the form of a medical drug disaster occurred in 1937. A new liquid formulation of medication had been devised and distributed by a pharmaceutical company, and, in absence of any required regulatory toxicity testing for new products, turned out to be lethal. The drug caused over 100 deaths in two months, including the chemist by suicide upon learning of his error. Under the 1906 law, the only authority where the FDA had to recover the product was through misbranding, and luckily the product happened to be "misbranded" because it coincidentally happened to be labeled an elixir when it was actually technically a solution. The languishing Federal Food, Drug, and Cosmetic Act became law a year later in 1938.[18] This 1938 Act, in conjunction with the FSMA of 2011, provides the basis of FDA's authority in food safety today.

FOOD SAFETY: PATHOGENS AND RISK

Thirty-one pathogens are known to cause foodborne illness in humans, resulting in an estimated amount of 48 million cases a year in the United States.[19] Healthy People 2020 (HP 2020) has objectives to reduce illnesses caused by six of these pathogens, thus our food safety policy in the United States is most concerned with these six: salmonella, campylobacter, vibrio, yersinia, shiga toxin producing *Escherichia coli*, and listeria monocytogenes. Box 13.1 has more information about these pathogens.

Foodborne illnesses can be viral, parasitical, and bacterial in nature. As bacteria multiple rapidly, food that started out only lightly contaminated can become highly dangerous.[20] This is why commingling of ground beef from many cows can become highly problematic. Likewise, salmonella has been found within chicken eggs, and although infected eggs are rare, if one such infected egg is commingled with others

BOX 13.1   THE SIX PATHOGENS

Salmonella. Salmonella bacteria are often found on poultry and pork. It exists environmentally, such as in chicken feed and bedding.[21] One strain has been found which penetrates the egg-shell. However, it can be controlled. Some countries and/or producers have gone to great lengths to control salmonella, and chicken flocks have become salmonella free.[22] Fresh produce outbreaks have occurred as well, notably in sprouts, cantaloupe, papayas, and tomatoes. Cantaloupes are problematic, as low acidity (pH 5.2–6.7) and high water activity (0.97–0.99) supports the growth of pathogens. Salmonella and other pathogens can exists on seed sprouts through the process of seed-to-sprout, multiplying to high levels during the sprouting process due to favorable conditions and then survive the typical shelf life of refrigeration.[23] Generally, symptoms of diarrhea, fever, and cramps occur 12–72 h after consumption, lasting 4–7 days. Most people do not need treatment, but severe diarrhea symptoms may need to be treated in the hospital. About 42,000 cases are reported yearly, but many milder cases are not reported, perhaps as many as 29 times more.[24]

Escherichia coli. Known as "*E. coli*," most strains are harmless and can contribute to a healthy human digestive tract. Other strains cause a variety of illnesses, including, perhaps surprisingly, pneumonia. The form of *E. coli* that is associated with "travelers' diarrhea" is usually found on raw vegetables and garden salads. Strains that produce the shigella toxin, such as *E. coli* O157:H7 can be very dangerous. The infectious dose is very low; it can develop acid-resistance and can grow rapidly in some types of raw fruit and vegetables. The consequences can include bloody diarrhea and hemolytic uremic syndrome. The vast majority of outbreaks have been associated with consuming undercooked beef and dairy products, but outbreaks have also been linked to lettuce, unpasteurized apple cider, cantaloupe, and sprouts. The CDC estimates that about 265, 000 infections occur yearly by shigella-producing *E. coli*, and about 36% of these are *E. coli* O157:H7.[25]

Campylobacter. Most human illness is caused by one species, *Campylobacter jejuni*, which can be carried by birds without the birds themselves becoming ill. Most cases are from eating undercooked poultry. Generally, these bacteria are not hard and can be killed by drying or oxygen; and freezing can reduce the number of bacteria on raw meat, however it can take very little (less than 500 bacteria) to infect a person. That could be just a drop of raw chicken "juice" which could then easily cross-contaminate produce and other items. An estimated 1.3 million persons are affected yearly. Most people recover within five days, but there are more seriously, albeit rare, consequences, including arthritis, Guillain–Barré syndrome, and death.[26] *Campylobacterenteritis* has also been associated with lettuce or salads.[23]

Yersinia enterocolitica. Animals, usually swine, are the natural reservoir, but it can also be found on raw vegetables, tending toward root and leafy produce.[23] Preparing raw chitterlings (pork intestine) can cause cross-contamination; handling infants or their toys, bottles or pacifiers could result in transmission and infection of the infant.[27]

Listeria monocytogenes. For the most part, listeria induces only mild symptoms in healthy adults, but poses a significant problem to elderly, immunocompromised, and pregnant women. Outbreaks can be small in number, with only about 1600 cases a year, but high in mortality, with about 1 in 5 deaths. It is the third leading case of deaths from foodborne illness in the United States.[28] It can grow on fresh produce in the refrigerator.[23]

Vibrio. Vibrio vulnificus bacteria live in warm seawater. It can be contracted from eating contaminated seafood, or from the seawater itself through open wounds. If healthy, a person could suffer gastro-intestinal symptoms, but if immunocompromised, it can infect the bloodstream with a 50% likelihood of fatality. It is rare but under-reported, with as many as 95 nationally reported cases annually, and an average of 50 culture-confirmed cases reported from the Gulf Coast region, with most occurring between May and October.[29]

such as in a restaurant preparation it can lead to numerous illnesses. The slaughtering process can cause bacteria from the animal's stomach or manure to spread. Further processing of meat, such as grinding or mechanical tenderizing, can also be risky. Mechanical tenderizing is done at the processor with a machine that presses blades or needles into the meat, increasing the risk of cross-contamination of fecal matter entering below the surface of the meat, and thus less likely to be destroyed in the cooking process.[30]

Although becoming a vegetarian sounds like one way to ensure safety, avoiding the consumption of animal products does not eliminate risk. Fruits and vegetables, if washed or irrigated with water contaminated with manure or human sewage can pick up pathogens, and peanut butter can be contaminated with salmonella.

Furthermore, foodborne illnesses are not all bacterial in origin. Although norovirus is not one of the six pathogens targeted for reduction in HP 2020, noroviruses are the leading cause of illness from contaminated food in the United States, causing about 50% of all food-related illness outbreaks,[31] and new strains tend to appear every two–three years.[32]

The CDC estimates that 3000 people die annually from foodborne disease, and 128,000 are hospitalized. Over the year, about 48 million, or one out of every six Americans are likely to experience foodborne illness.[33] In 2011, an outbreak of listeria killed 33 people.[34] Thus, even just one outbreak of illness can contribute significantly to mortality. And with modern distribution methods, illness and mortality borne by unsafe food can cross even international borders. However, although relatively widely experienced among Americans, foodborne illness is low on the list of causes of death in the United States at 3000 deaths. The 15th leading cause of death in 2011 was pneumonitis due to solids and liquids at just over 18,000 deaths.[35]

Mortality aside, we tend to think of food poisoning as a temporary illness. Unfortunately, the consequences can be far-reaching, and health statistics that speak only to mortality miss the impact of a less than full recovery. Cases of certain *E. coli* strains can result in kidney disease and hemolytic uremic syndrome. Salmonella and shigella infections can induce long-term effects such as reactive arthritis, urinary tract problems and damage to the eye. Campylobacter infection can lead to Guillain–Barré syndrome and ulcerative colitis (a chronic bowel inflammation).[36]

The combination of relatively low mortality but widespread exposure and chance of contracting illness means that the risk—and cost—of foodborne disease to our society is largely experienced in sick-time loss of productivity, cost of health care and treatment, and economically by the food industry through the cost of recalling food or through reduced purchasing of food and products associated with outbreaks.

"Summer Complaint," or How Food Poisoning Turned Deadly

Prior to refrigeration, an illness known as "summer complaint" in young children was common, and consequently, a common topic in the *Journal of the American Medical Association* (*JAMA*) from 1883–1938. This illness of diarrhea, stomach pain, and sometimes fever and vomiting occurred during warm weather and primarily during the first summer after being weaned from breast milk; and at an even younger age to babies who were bottle-fed. Not unexpectedly, given the lack of refrigeration, the "summer complaint" was mostly foodborne diseases.

Although "summer complaint" sounds fairly innocuous, it was often fatal. By the end of the 1930s, with increasing utilization of home refrigeration and a better understanding of how to prevent summer complaint, the infant mortality rate, generally and from diarrhea specifically, had radically dropped. Articles on summer complaint had vanished from JAMA.[37] Foodborne diseases are no longer a leading cause of death among infants; not even in the top 10.[35] Just as "summer-complaint" faded from our social memory as we began to enjoy home refrigeration, so did the potential fatality of food poisoning. As Marion Nestle describes it, food poisoning just happened when egg salad was left too long in the sun at a picnic, or maybe just from random bad luck. Like getting a cold, you just put up with it till the discomfort passed.[8]

In December 1992, this social paradigm that food poisoning was mostly harmlessness was shattered when a food-poisoning outbreak transmitted through fast food chain Jack in the Box hamburgers turned deadly. American households that previously just talked of "food poisoning" learned a dreadful new specific term: *E. coli*. By 1993, after four children died, the American public learned that *E. coli* was only "mostly harmless" and that a specific strain of *E. coli* (O157:H7) had somehow acquired the ability to produce the deadly shigella toxin.

Even worse, this new deadly pathogen was not limited to being carried by food—it could be passed from person to person. A 16-month-old toddler died without having ever eaten a hamburger, having contracted the disease through exposure at a day care to another child who had eaten a contaminated burger but whom had only a mild bout of illness.[38]

Serious food-related illness is the hardest contradiction for us to bear—something that nourishes and sustains us, often given to us as comfort, with sincere love, and accepted in gratitude—could kill us. In 1993, Jack in the Box took the brunt of the blame, but that virulent strain of *E. coli* was in fact first recognized by scientists in 1982 due to an outbreak at McDonald's that was not highly publicized. And, in fact, 22 other outbreaks of *E. coli* O157:H7, causing 35 deaths, had been documented over the 10 or so years between the 1982 McDonald's outbreak and the 1993 Jack in the Box outbreak.[39]

However, with the deaths of children caused by the iconic and child-friendly American fast-food burger, and even worse, the death of a child who had not even eaten a burger, the 1993 *E. coli* outbreak caused a paradigm shift. It transformed food poisoning from a passing discomfort caused by individual's lapses in keeping food chilled during summer picnics into a serious public health problem.[20] The risk of illness was no longer under the individual consumer's control.

The Jack in the Box *E. coli* outbreak also alerted Americans to the industrialized shift in our food production system. An industry goal of keeping costs down in order provide inexpensive food and yet still make a profit lead the US food production industry to develop economies of scale. Economies of scale meant numerous small processing plants turned into fewer, larger, ones, with a greater number of cattle crowded together in feedlots for fattening prior to slaughter—standing on their own feces—increasing the likelihood of spreading pathogens. Postslaughter commingling ground beef can then spread contamination widely through the beef processed that day.[39] Even prior to the slaughter process, larger farms, holding facilities, and markets bring together many animals in close contact with each other—and with humans—and our food production itself becomes a fertile ground for disease.[40]

Prevalence of Foodborne Illnesses

Although there may be a perception that foodborne outbreaks have been on the rise due to attention in the media, incidences for the foodborne illnesses tracked by the CDC from 1996–2011 have generally been on the decline or flat since 1996, with the exception of a salmonella uptick in 2010, and an overall increase over time in vibrio.[41] The vibrio increase may be attributable in part simply to our increased appetite for seafood and global imports.

The perception that foodborne illness outbreaks have been increasing may have contributed to the adoption of the FSMA. However, there has not been in fact such an increase in total numbers of outbreaks. Instead, outbreaks have increased among foods previously considered unlikely culprits, such as produce. Between 2006 and 2008, the CDC identified produce as either the first or second leading source of foodborne illness outbreaks in the United States. Produce items viewed as healthy, whole, nonprocessed foods—spinach, lettuce, cantaloupe—have become significant carriers of pathogens. Commodity produce as a group—fruits, nuts, and five vegetables is the leading cause of foodborne illness in the United States, and of all items, more illnesses were attributed to leafy greens than any other item.[42] We have now learned that fast-food hamburgers are not the only risky food—foods that are also "positive" markers of health and socioeconomic status can now also be risky.

Contamination of produce can occur through various means, but one particular concern is animal waste. Animal waste is often used as a fertilizer, and without any required treatment for reduction of pathogens, there is a risk of contamination if time between manure application and harvest is too short. The recent outbreaks have in fact caused leafy green producers to create more stringent requirements regarding manure than those promulgated by the USDA's National Organic Program. And, as part of the FSMA, the FDA will include a produce safety rule addressing the application of manure.[43]

Individual Risk

In short, food cannot be made completely and totally risk-free. Food safety policy may be viewed as a risk analysis consisting of assessment (a scientific evaluation), management (a political evaluation), and communication.[44] We evaluate the risk, manage the risk, and communicate the risk. In order to effectively evaluate and manage, we must also critically understand the nature of the risk. In other words, when considering specific risks, we have to ask: *Are individuals in fact feasibly able to take responsibility for their own safety?*

For example, consumer food safety education disseminated by the United States government—the clean, separate, cook, and chill campaign—informs individuals that their actions can help to keep themselves and their household safe from food-poisoning.[45] This communication is a very important component of food safety. Individuals can take these steps and assert agency and responsibility toward preparing safe food. However, these actions are not equally effective for all foods or all pathogens. Although safe storage and preparation of foods is an important part of consumer awareness, some of these associated risks may be beyond individual control.

In the case of hamburger, an individual can in fact choose to balance risk and personal taste through selecting cooking method and temperature, even at restaurants in most states. But ordering a steak at a restaurant may be problematic. Mechanically tenderized beef may not be labeled as such, and even the restaurant may not be able to warn or advise as to appropriate cooking temperature. The USDA's FSIS has proposed a rule for labeling mechanically tenderized beef, the final rule was released May 2015, to be effective May 2016.[46]

In the case of produce that is normally consumed raw, cleaning may not be enough to kill bacteria.[47] In fact the United States government does elsewhere inform us that washing is ineffective on raw sprouts.[48] And what is one to do about peanut butter, which is normally eaten uncooked and cannot feasibly be washed?

In situations where we cannot feasibly impact food safety through our own actions, we may be asked to accept an unknown burden of risk. Faced with this, one choice could be to simply stop eating the item. But food retailers and producer may also have to make similar decisions. Some food retailers have announced that they would stop selling sprouts due to the high likelihood of contamination.[49] Farmers may choose to stop growing items such as cantaloupes, or conversely, may choose to organize as a group to create a safer (and branded) product.[50] Consider these actions in light of the self-regulation discussion in Chapter 12, "Nanny State."

Even though an act of federal government to ban specific foods might be bitterly contested as an infringement on individual liberty, a combination of actions by consumers and industry may without uproar naturally cause a default change in our food environment, resulting in some foods simply not being as widely available.

Economic Risk: Epidemiology versus Testing

Protecting the safety of our food protects the health of our own citizens, but also achieves the goal of protecting the economic value of trade. In 1996, the link between "mad-cow disease" found in UK cattle and a particularly horrible and fatal human variation caused the market for UK beef, domestically and internationally, to collapse. The economic loss in just the following year was estimated to range from 740 to 980 million pounds, including the cost of UK government efforts to regain consumer confidence globally and at home.[51]

Food safety policies could also be used as a means to protect domestic trade from imports. The tolerated level of any particular adulterant or pathogen may be a policy that varies from country to country. In the case of the pathogen listeria, the European Union tolerates a small amount of listeria (an amount considered to be less than an infectious dose), but the FDA has a zero-tolerance policy for listeria in foods that are normally eaten without being further heated. If listeria is found in cheese, the FDA deems the product to be adulterated, a policy that has been affirmed in federal court. The FSIS follows the FDA on the zero-tolerance policy, and if listeria is found in meat or poultry products, the products are judged adulterated.[52] This policy may contribute to some suppliers dropping the US market for items such as smoked salmon.[53] It is also a standard that other countries have questioned as to underlying intent—that is, if this is actually a policy of protectionism against imports rather than a policy of protecting health.[54]

Likewise, the United States has complained about Russia deciding to ban US pork and beef imports over the use of a veterinary drug, saying that the decision is retaliatory as well as unscientific, and not appropriate for a member of the World Trade Organization (WTO).[55] The WTO has adopted the Sanitary and Phytosanitary Agreement setting out a basis for distinguishing legitimate food safety conflicts from veiled protectionist policies.[56]

Domestically, food safety is just as significant of an economic issue; playing out on a smaller, but no less significant stage. When an outbreak occurs and can be traced to a specific source, shutting down the food production or distribution from that source can incur significant economic loss to a small company. People lose jobs. The economic loss can ripple through a community dependent on local industry.

Balancing public health and safety and economic interests is a process of policy in action, that, perhaps not surprisingly, the FDA has historically been reticent as to the details.[57] One of the issues at play in this balance is how the FDA tracks down the source of an outbreak. Epidemiology is a significant tool, which may essentially rely on detective work using deductive and inductive reasoning without physical evidence. Proof based on testing for the presence of pathogens is often not possible, given the perishability of food and the chance that testing would occur after any associated product would have either been consumed or thrown away due to perishability. Furthermore, pathogens may not have been equally distributed, such that sampling of products or of the production environment may not then find pathogens. A business that is being asked to voluntarily recall food or shut down production based on epidemiological assessment may find it hard to swallow without the "proof" of a test showing the presence of a pathogen in their product or plant.[58]

Although consumers may feel that inspection is a valuable component of outbreak prevention strategy, as noted above, pathogen testing is not necessarily feasible as a prevention strategy. The USDA's Microbiological Data program, a pathogen-testing program for fruits and vegetables that was inexplicably housed in the USDA's Agriculture Marketing department, provides an excellent example. Although the program was perceived by consumers to be a valuable component of prevention strategy, it was not intended to prevent outbreaks but instead to collect data on prevalence of pathogens. Upon testing, if pathogens were found, the data was sent to the FDA, and the FDA issued a recall, which has happened about 30 times since 2009.

In 2012, when the Obama administration cut funding to the Microbiological Data Program, the USDA pointed out that the Microbiological Data Program could be moved to the FDA, as the FDA in fact has responsibility over fruits and vegetables and already conducts testing, albeit far less than the USDA program. However, the FDA did not take over the program, and has not indicated if it will increase testing to fill the void.[59]

Because recalls based on inspection appear to the consumer as an important piece of foodborne illness outbreak prevention, there was some struggle in the media and congress over the closure of this relatively small testing program reflecting balancing producer and consumer interests.

The food producers, risking reputations and sales when recalls happen, pointed out that the testing, and subsequent recall was done too late for true prevention; by the time of the recall the produce has either been consumed or has been discarded due to spoilage.[60] However, consumers, after hearing increasing news that fruits and vegetables are causing serious foodborne pathogen disease outbreaks, may find it easy to account industry lobbying as responsible for the cancelation of the program.[61]

HACCP, FSMA, AND FOOD SAFETY CULTURE: EMPHASIS ON PREVENTION

By the 1970s, the belief that regulation of food was an appropriate federal power was ensconced well enough that health officials began arguing that the government was failing to do the job required by law. Health officials, recognizing the limitations of physical inspection, pointed out that the USDA inspection stamp was a misleading reassurance as to the safety of the meat when it came to pathogens, and begin pushing the federal government to provide warning labels on meats. The resulting court case opinion (*American Public Health Association v. Butz*, 1974) gave USDA the choice of providing consumer education or applying warning labels. The USDA chose consumer education. More significantly, the court determined that salmonella did not fall under the statute's definition of adulterant. Salmonella was not an adulterant because it was not added to the meat, but instead inherently contained in meat.[62]

But even physical meat inspections of each carcass became difficult for the USDA to manage as meat processing speed increased. In the early 1980s, the Government

Accountability Office (GAO) recommended that the USDA reform meat inspection and institute Hazard Analysis and Critical Control Point (HACCP), a process developed by Pillsbury in 1959 aimed at preventing contamination at every stage of food production and processing. This process was developed to create a food system for NASA to provide critically safe food for astronauts.[8]

In 1988, a joint committee advised the USDA and the FDA to institute HACCP, but industry reaction was cool, as the principles of HACCP would place the burden on the producers. Although the USDA had begun warming up to HACCP and warning labels, neither had been implemented when the deadly Jack in the Box outbreak occurred. It took a lawsuit from an advocacy group to mandate the USDA to place warning labels on ground meat and poultry, and even in the face of continued deaths from *E. coli* O157:H7, industry complained.[8]

In 1994, with the appointment of Michael Taylor to run the FSIS, the USDA announced the proposal of rules requiring HACCP systems in all ground beef and poultry processing plants. More significantly, the USDA also reviewed the *APHA v. Butz* decision in light of *E. coli* and determined the deadly shiga-toxin producing strain *E. coli* O157:H7 would now be considered an adulterant. It would now be illegal to sell meat that contained any amount of this pathogen.

Industry continued to attempt to block the idea of testing for *E. coli*, emphasizing the importance of consistent messages to consumers about proper cooking. However, the courts ruled in favor of the USDA's decision that *E. coli* O157:H7 was in fact an adulterant on the grounds that it was not killed by the customary cooking temperatures for beef.[8] In other words, it was inappropriate for the individual consumer to bear the burden of this particular food safety risk.

However, industry had successfully pushed back on salmonella, leaving the USDA to conduct only limited sampling for salmonella as part of HACCP. Meanwhile, the FDA was also attempting to develop HACCP controls, starting with foods considered most hazardous—seafood in 1995, with raw sprouts and eggs to follow effective 1999 and fresh juice in 2001,[8] but this leisurely rollout was interrupted in fall of 1996 by an outbreak of *E. coli* O157:H7 in unpasteurized apple juice. Like the Jack in the Box incident, this outbreak was particularly shocking, with fatalities among children caused by apple juice—apple juice!—a food commonly given to children as healthy and nourishing. Investigators believe that the source of contamination was windfall apples (apples that were harvested after they had already fallen to the ground) that may have come into contact with animal manure. It had become clear that all foods should be produced under a HACCP system. And, although it has been well known for some time that six other *E. coli* strains also produce the deadly shiga-toxin, those six other strains were just finally flagged in 2011 as adulterants.[63]

HACCP was a significant change in regulatory philosophy, shifting the burden of responsibility from government to industry,[64] highlighted by the description of US food safety policy on "FoodSafety.gov, Your Gateway to Federal Food Safety Information:"

> The food industry is responsible for producing safe food. Government agencies are responsible for setting food safety standards, conducting inspections, ensuring that standards are met, and maintaining a strong enforcement program to deal with those who do not comply with standards.[65]

The New Law: The Food Safety Modernization Act of 2011

The FSMA of 2011 focuses on the FDA and the areas of regulations under the FDA's control, with no direct change to food safety efforts in other agencies, such as the USDA. FSMA amended the FDA's existing structure and statutory authorities.[1] The new law is aimed at emphasizing prevention versus outbreak response.[66]

The FSMA's policy of prevention versus outbreak management is implemented through the idea that improved process and oversight of food production and processing is a better prevention tool than increasing inspection. Early twentieth century food safety law, as manifested in the 1906 Acts, naturally focused on the then current industrial management practice. Early twenty-first century food safety law, heralded by the mid 1990s HACCP regulation and manifested in the 2011 FSMA, broadens the perspective up and down the supply chain. Although the FSMA does expand government powers, in turn, it may provide for more flexibility within industry to reach safety goals, through systematic review of process and performance goals rather than through inspection standards.[67]

Some very critical and expanded powers were available immediately to the FDA upon the signing of the FSMA:

- Access to documents at food companies tied to outbreak causing illness/death.
- Increased frequency of inspection.
- Power to order a mandatory recall and the ability to suspend operations.[68]

Prior to the FSMA, the FDA did not have the power to do a mandatory recall. Even though from a consumer perspective, the recalls may have appeared authoritative, they were in fact voluntary. The FDA has not been aggressive with this new power, instead continuing to seek voluntary compliance. In November 2012, the FDA suspended operation of a business for the first time.[69] As of December 2013, only one mandatory recall process had been initiated with a notification to a pet-food manufacturer that had been slow to do a full voluntary recall, but the manufacturer complied voluntarily post the notification.[70]

Although Congress mandated implementation deadlines for the FDA, the complexity has given some pause to implementation and the Congressional Research Service also found that Congress did not provide enough funding.[1] In early January 2013, a year behind schedule, the FDA released two proposed rules.[71] Some of the delay was due to the fact that the rules go though the Office of Management and Budget (OMB) before they can even be released for public review, and the delays at OMB have stretched beyond the normal 90-day limit.[72] The FDA had been sued in 2012 by the Center of Food Safety for failure to meet several deadlines, and the lawsuit sought not only that the FDA met the deadlines but also to prevent the OMB from delaying the FDA's compliance with the deadlines. In February 2014, a settlement was reached that extends and staggers the final rule deadlines out to 2016.[73]

As these rules are not finalized at the time of writing this chapter, our intent is to take a look at the first two rules proposed by the FDA to highlight some of the change in thoughts and trends regarding food safety. The two proposed rules published in January 2013 concerned preventive controls for human food (hazard analysis and risk-based preventative controls, similar to HAACP plans), and standards for the growing, harvesting, packing, and holding of produce for human consumption.[74]

Domestic and foreign firms that manufacture, process, pack or hold human food (with some exceptions) will be required to have written plans that identify hazards, specify the steps to minimize or prevent those hazards, identify monitoring procedures, and record monitoring results and specify what actions will be taken to correct problems that arise. The FDA is in charge of evaluating plans and inspecting facilities for plan implementation.[75]

The burden on industry could be significant one-time and recurring costs; as this would include not only adopting new plans, but also training workers, implementing new equipment and techniques, auditing suppliers, and documenting the process. The cost to industry is balanced against avoiding the economic cost of an estimated 1 million illnesses, costing $2 billion a year.[75] This calculation assumes each illness costs about $2000 a year. But what we are really hoping for, of course, is to prevent the very costly cases, such as the *E. coli* infections that cause hemolytic uremic syndrome and put children on lifetime dialysis.

The second rule involves more stringent safety guidelines for growing and harvesting fresh produce that is commonly eaten raw. Section 105 of the FSMA directs FDA to set science-based standards for the safe production and harvesting of fruits and vegetables. The proposed standards cover known means of microbial contamination of products, namely: (1) agricultural water; (2) biological soil amendments of animal origin; (3) health and hygiene; (4) animals in the growing area; and (5) equipment, tools and buildings, with additional provisions related to sprouts. It would not apply to produce rarely consumed raw, those produced for personal or on-farm consumption, and produce with a documented commercial processing destination.[76] Regulating foods through the defining characteristic of "commonly eaten raw" might have some unexpected results. Under the proposed rule, kale is on the exemption list as a vegetable almost always consumed only after cooking. Apparently, the recent raw kale salad and smoothie trend was not common enough to get kale on the "commonly eaten raw" short list.[77]

Furthermore, farms under certain criteria are exempted, including size and local nature of sales (such as within state). The exemptions, due to a concern that the new regulations could be too costly for smaller farmers, mean that about 8 in 10 growers will be exempt from the rule. Even with these exemptions, the FDA believes that 90% of acreage used to grow US produce will either be covered by the regulations or the regulations will not be applicable due to the produce being processed or cooked. The estimated number of prevented illnesses is 1.75 million, at a savings of 1.04 billion.[78] Unaccountably, the diseases prevented by the produce safety rule appear on average to cost less than those prevented by the production hazard analysis rule.

The broad scope of the exemptions is due in part to the Tester amendment, a provision sponsored by a Democrat Senator from Montana. The intention was to exempt small farms that sell mostly locally, which is operationalized as an exemption for farms making less than $500,000 annually, and who sell more than half of produce directly to consumers (which could be through direct sales online), or to stores and restaurants within 275 miles radius or within the same state.[79]

Although this exemption does help in supporting small, local, farming by avoiding the burdens of additional cost and regulations, local produce or farm-to-consumer sold produce is not necessarily risk free simply because we can look the grower (or apparent grower) in the eye.

In August 2011, Oregon health officials confirmed that an *E. coli* O157:H7 outbreak that sickened and killed 14 people was traced to strawberries, many of which had been sold at roadsides, some of which had been resold several times and were not in fact being sold by the grower, contrary to appearances.[80]

The benefit of a smaller distribution network is that if an outbreak occurs from local produce, the burden of illness is hopefully smaller since less people are exposed. But the local farmer still has to manage the risk. And we still might be the one to get sick. Local, sustainable, small farmed, humane can be a better choice for many reasons, but it is not automatically the safer choice.

Nonetheless, a number of states are considering or passing "cottage food" laws that encourage smaller food production; sometimes pitched on the idea that food safety is encouraged or more likely to happen when fostered by a sense of community, and that failing to sell safe food would put a local producer out of business. Some are clearly pitched as a response to the increased federal authority of the FSMA, as evident through names such as "Food Freedom Laws." Different states have different "cottage food" regulations and requirements for the produce, including restrictions on "higher risk" types of food. California, for example, does not allow cottage food producers to make foods with cream or meat-based fillings.[81]

Food Safety Modernization Act Cost

A significant cost of the FSMA will be borne by farmers and producers, with small farms chipping in about $13,000 a year and large farms about $30,000. For producers, the total cost is expected to be $320–$475 million.[82] As of FY2012, the Health and Human Services (HHS) secretary reported that FDA needed an increase of $400–$450 million over the existing FY2012 food safety budget, and a request to implement user fees was proposed.[83]

Recall the above section, regarding the estimates by the FDA as to the number of illnesses the new rules should prevent—almost 2 million is a lot of cases—but not very many of the total estimated cases of 48 million a year. We can only hope that these new rules will be highly cost-effective by preventing the worst possible 2 million cases.

Food Safety at the State Level and the Food Model Code

The roles and responsibilities of the states, however, are not completely subsumed by this federal expansion into food safety. The federal government needs help from the states in inspecting and enforcing, and may provide financial assistance to the states to do so. One of the interesting differences between agencies working on food protection is how the agencies address this issue. The USDA relies on federal employees, whereas the FDA on state cooperation.[84]

Food protection in the United States is greatly implemented at the state and local level, which naturally reduces chances of uniformity.[66] At the local level, there are over 3000 public health agencies involved in food safety. The retail and food service sector (grocery stores, restaurants, etc.) tends to be governed by state and local health codes, although the FDA has an increased advisory presence through the development of a Model Food Code. Local and state level agencies are involved in outbreak response and recalls, surveillance through collaboration with the CDC and using tools such as FoodNet, PulseNet, Outbreak Net; laboratory testing, technical training and assistance, education, and retail, processing and farm inspections. In fact, states have primary jurisdiction over enforcement pesticide regulations and they conduct the majority of "FDA" inspections, under contract with the FDA.[85]

Furthermore, as noted above, the state's independent powers allow food regulation in areas not preempted by federal law; thus the states may have differing rules related to food sold within the state borders. The differing state rules concerning raw milk and "cottage foods" are examples.

The diversity of state approaches, although irksome to food producers or retailers operating in multiple states, also allows for experiments that may become more widely adopted once shown to be successful, similar to the discussion in Chapter 12, "Nanny State."

The FDA has created advisory guidance in the form of a model Food Code, and has encouraged state adoption to help in increasing uniformity throughout the states regarding regulation of the retail and restaurant sectors of food provision. Published from 1993–2005 in a two-year cycle, full updates are now scheduled for every four years, with supplements being published between cycles. The most recent full Code was published in 2013.[86]

Although the goal of uniformity among the states is laudable, the feasible extent of uniformity is modified by several factors. For example, the states have used various versions of the Food Code as models for regulation, so inherent differences between the versions will remain as differences among the states, particularly if the states lack automatic means to update their regulations as the model code changes. Also, the regulations that states adopt may be based on the Code, but not exactly the same as the Code.

An analysis of how the Code comprehensively treats the allocation of the burden of risk across the food system (individual vs. retailer vs. producer) is beyond the scope of this chapter. However, it is important to realize that this Code, or indeed, variations of the Code, can convey significant policy aspects regarding individual choice and allocation of risk. For example, the Code currently recommends that restaurants only serve well-done (cooked to 155°) hamburgers to children. For adults, however, the risk of choosing to purchase and consume an undercooked burger is allocated to the individual. The restaurant's responsibility is to have held it correctly at appropriate temperatures prior to cooking, and to inform the diner (often via the menu) of the risk of acquiring a foodborne illness if they order an undercooked burger.[86]

Some states do not allow the consumer to accept any of the risk, and mandate that burgers must be cooked to 155°. Until September 2012, North Carolina was one of those states, having passed the "well-done" burger law in 1993 after the Jack in the Box deadly *E. coli* outbreak. However, most states follow the FDA model code; restaurants may prepare rare burgers for their customers as long as risk information is provided, but restaurants can still choose to only prepare well-done burgers.[87] Since restaurants, particularly high volume burger chain restaurants, may nonetheless still choose to only cook well-done burgers,[88] people may associate the inability to choose the preparation of a burger with their liking as "nanny-state" interference by the government, but in most states that could in fact be a choice limited by the restaurant's policies.

Food Safety Culture: Inspection, Testing, and Audits

Inspection, testing, and audits are all potential components of a food safety system. Each tool has limitations, however, and a regulatory or industry emphasis on any one tool will shape related policy and allocate burden of risk.

As discussed above, physical inspection, particularly in the animal industry, is a well-established food safety component in the United States. Physical inspection, however, has become increasingly more difficult as processors automate work lines at increasing speeds. Moreover, although physical inspection can catch animals and carcasses with evident disease and defects, some of which are transmissible to humans, such as bovine tuberculosis,[89] it does not catch the pathogens such as shiga-producing *E. coli* that have now become familiar causes of illness.

Both the USDA and the FDA conduct inspections. And both agencies have been reducing or attempting to reduce inspections. Furthermore, as the number of inspections performed decreased, facilities subject to inspection have increased in number; such that between 2004 and 2008, 56% of the relevant facilities were not inspected even once.[1]

Program certification is another means in which the FDA can reduce inspection efforts but if the certification program is sloppy, the burden of the risk may fall on the unwitting consumer. In 2011, because of a salmonella outbreak in Mexican papaya, the FDA hired a private lab to test every Mexican papaya load at the US border. One company's load of papayas was rejected nine times, at which point the company turned to a Mexican certification program. Once certified, the FDA then exempted the company from further testing, although there was no immediate apparent improvement in the desired food safety outcome.[90]

A 2012 USDA proposal regarding poultry inspection has drawn criticism from consumers and other groups. Instead of having USDA inspectors examine birds prior to processing, this role would be turned over to poultry plant employees. The USDA inspector would instead spend more time on evaluating process and programs. The inherent conflict of interest and pressure on the employee is one issue at hand. Another is that the proposal allows the assembly line to speed up from 140 birds a minute, already a challenging task, to 200 a minute.[91] HACCP is not a panacea. The USDA is right to propose expansion of authority to include critical review of a company's HACCP or other safety and testing procedures, as a plan can be developed, and even perfectly followed, yet may not ensure the production of safe food if the plan is faulty.[92] However, emphasizing review of process over inspection and worker safety is problematic.

At the local level, foodservice inspections are also an established centerpiece for local public health efforts. However, although inspection scores published in restaurant windows may inspire confidence in diners, they do not necessarily predict foodborne illness.[93]

Third-party auditing

Like inspection, the effectiveness and quality of an audit requires not only skilled observation and judgment, but also consistent and appropriate standards. Some trade groups, recognizing a mutual benefit, have created guidelines for audits. However, the for-profit inspectors themselves are not required to meet any federal standards. And, of course, they often may work as prescribed by the company—thereby potentially missing critical review of troublesome spots.

Furthermore, third-party auditors may be more-or-less third-party; with industry executives on the corporate board.[92] Another relevant issue with third-party audits is that the report is of course, given to the paying party—namely, the company. Third-party audits have no obligation to make their findings public.[90]

In August 2010, a diagnostics lab in Iowa found salmonella in eggs and reported it to the company, a month prior to a massive outbreak that likely sickened over 60,000 people and triggered the recall of 550 million eggs. The lab stated it did not have legal or ethical obligation to alert regulators or consumers of the particular salmonella strain found—even though the lab is partially funded by state taxpayers.[94]

Foodborne illnesses have also been associated with farms, retailers, and others, who had just passed third-party audits, such as the 2009 outbreak of salmonella linked to the Peanut Corporation of American (PCA) and a deadly outbreak of listeria at a cantaloupe farm. In the case of PCA, the company knew in advance when the auditors were arriving. In the cantaloupe outbreak, the third-party auditor had outsourced the audit to an independent contractor.[90]

Indeed, government inspections have also passed plants for satisfactorily complying with regulations right before an outbreak occurred. A 2008 listeria outbreak in Canada occurred in a plant that appeared to be doing everything right, including employees taking action, management keeping records, training staff and following a quality assurance program, and even testing beyond regulatory requirements. Unfortunately, the company failed to recognize a pattern of listeria that was revealed through that testing; and was not obliged to report their own testing to an external party; thus eliminating the chance of someone else spotting the listeria pattern. Overall, an investigation found that the company's management (as well as some of the government organizations) lacked a sense of urgency regarding the importance of the issues; and thus failed to be sufficiently focused on food safety although apparently doing the right thing. Companies must understand the snap-shot-in-time and other evaluation limitations of a commissioned audit and not rely on the findings without understanding these limitations.[92]

Testing

Inspection and auditing can uncover potential sources of contamination, but are not able to pinpoint the presence of pathogens. Microbial testing can do that; however, one problem with testing is that the proportion of pathogen to product can be very low.[95] Sample testing could still miss the problem. Nonetheless, this is a step that producers, particularly those who have had an outbreak and need to regain reputation, could add to their process. The cost to do so for a bagged lettuce/salad provider as of 2012 was an extra three cents per bag. Although not all growers and retailers support such extra costs, some do—the retailer Costco is willing to pay the extra three cents to the grower Earthbound.[90] Unfortunately, safety information is not frequently conveyed to the consumer at the market place; and thus the impact of safety on supply and demand is not easily discernable.[96]

Food safety culture

Some food safety experts argue that the best use of audits is toward the creation of a genuine food safety culture. All businesses have a goal of making money. However, if profit takes precedence in a food business without regard to food safety principles, such as through encouraging ill employees to keep working, it reveals a lack of food safety culture. Training employees in food-safety principles is important, but not sufficient; training and culture needs to extend through to the supply chain and manifested by second-party audits.[92]

Food safety culture could be considered a behavioral norm that causes employees at any and all levels to recognize urgency as to potential illness of others and an understanding of why processes are important for safety. Such standards and customs must be collectively recognized and positively reinforced throughout the organization. Developing shared values and understandings however, might be made more difficult if the workforce is drawn from a diverse group, with differing languages and social backgrounds.[97]

In 2002, a survey found that two-thirds of a decrease in salmonella found in testing products was due to management decisions rather than proscriptive requirements. Investments in human and physical capital, technology, and changes in organizational structure made this difference. Nearly half of the reduction was caused through contractual arrangements with suppliers that gave incentives, including price premiums.[98]

FOOD SAFETY THROUGH LABELING

The FDA is in charge of regulating food product labeling. Labeling may not seem directly connected to food safety, as the label itself cannot make the food inherently safe or unsafe when it comes to pathogens, a very much currently emphasized issue of food safety.[8] However, although pathogen contamination is the most critical issue for food safety at a population level, individually we are also highly concerned about risks that are "invisible, involuntary, imposed, and uncontrollable."[8] Consider the risk

allocated to individuals, whether purchasing and preparing food for themselves or consuming food prepared by others. *What information about food and how food is processed is important for individuals to understand and accept risk?*

For example, knowingly accepting a risk such as drinking raw milk may not be scientifically advisable, but it is a choice made under an individual's autonomous sphere.
Unknowing consumption can result in shouldering an inappropriate burden of risk.

The idea of safe food and the original mandate for the FDA was grounded in what
we value as to the nature of food, namely food as "pure and wholesome." In order for
any consumer to be confidant in the purity or wholesomeness of a purchased food product, some basic information must be provided to the consumer. Although this sounds
obvious to us today as we read ingredient and nutrition information on labels, this level
of regulation and access to this level of information is comparatively recent. With standardized label requirements, followed by regulations regarding ingredients, nutritional
information and allergens, food labeling began to act more concertedly as protection for
the consumer. Regulation of labeling also serves to create a more efficient marketplace
for business; it allows for greater consumer acceptance of food products that were previously commonly self-prepared or from a different region.

A Brief History of Labeling Regulation Relevant to Food Safety

The 1906 Pure Food and Drug Act granted authority to the FDA over misbranding and
deception on food labels. The expanded regulatory powers of the Food and Drug Act
of 1938 removed the requirement that intent to defraud must be present in misbranding
cases, and added the ability to regulate standards of identity and quality, which was an
important tool for regulating enrichment and fortification (see Chapter 11, "Enrichment
and Fortification"). In 1958, the FDA was tasked with reviewing the safety of food additives under the Food Additives Amendment of the Federal Food, Drug, and Cosmetic
Act, which also gave the authority to the FDA to establish the conditions of usage for
food additives (to ensure safety) including labeling. Any food additive shown to cause
cancer in animals or human was prohibited. In 1965, interstate commerce is once again
used to establish and strengthen federal oversight over food with the enactment of the
Fair Packaging and Labeling Act (FPLA), which regulates appropriate labeling of consumer products in interstate commerce. In 1990, Congress passed the Nutrition Labeling
and Education Act (NLEA), which preempted state requirements and standardized the
food ingredient panel, serving size, and some important descriptive terms such as "low
fat" and "light."[99] See Chapter 4, "Nutrition Information Policies," for more detail on
the 1990 NLEA. In 2004, Congress passed the Food Allergen Labeling and Consumer
Protection Act (FALCPA), which is applicable to all food regulated by the FDA. Under
FALCPA, the label must clearly identify if any of the ingredients are sourced from the
eight most common allergens. Moreover, the 2013 regulation issued by the FDA concerning "gluten-free" is made under the authority of FALCPA. (See Chapter 9, "Grains.")

For purposes of our discussion in this chapter about the allocation of risk to individuals, we turn now to review in more detail two of these laws, namely the 1958 Food
Additive Amendment and the 2004 FALCPA.

1958 Food Additive Amendment

This amendment provides a legal definition of food additives for the purpose of imposing a premarket approval requirement by the FDA. Food additives are defined as "all substances, the intended use of which results or may reasonably be expected to result, directly or indirectly, either in their becoming a component of food, or otherwise affecting the characteristics of food." The definition specifically excludes items whose use is "generally recognized as safe;" government approval is not needed for such items to be added to food.[100] Pesticide chemicals, pesticide chemical residue, and color additives are also exempted because those items are subject to different legal regulations for premarket approval.

If a substance was used in food prior to January 1, 1958, an additive can be considered "generally recognized as safe" (GRAS) "through experience based on common use in food," defined as a history of consumption by a substantial number of users. The other option is to have qualified scientific experts evaluate the safety of the item.[101]

The FDA developed an initial list of food substances considered to be GRAS, however, much of what industry considered as GRAS was not on this list and manufacturers asked for opinion letters. In the late 1960s, the FDA banned an artificial sweetener, cyclamate, and President Nixon ordered the FDA to review the GRAS list. Accordingly, the authority of the letters were revoked and the FDA contracted a scientific review of presumed GRAS substances and established rulemaking procedures to affirm the GRAS status of substances that were either on the GRAS list or the subject of a petition from manufacturers.[102]

In 1997, to eliminate the resource-intensive rulemaking procedures, the FDA proposed to replace the GRAS affirmation petition process with a notification procedure.[103] FDA started accepting GRAS notices in 1998 as if the rule was final, however, the FDA has not (as of writing of this chapter) finalized the rule. The GRAS notification program provides a mechanism whereby the FDA is notified of a determination by a food manufacturer that the use of a substance is GRAS, rather than a petition asking the FDA to affirm that the use of a substance is GRAS.[104]

Under the GRAS notification process, the FDA evaluates whether the submitted notice provides a sufficient basis for the GRAS determination made by the submitter and responds with one of three letters:

1. The FDA does not question the basis for the GRAS determination.
2. The agency concludes that the notice does not provide a sufficient basis for a GRAS determination, either because information is lacking or because the information available raises safety questions.
3. The agency has, at the request of the notifier, ceased to evaluate the GRAS notice.[104]

The FDA makes it clear that if a manufacturer determines a substance to be GRAS, it can be marketed without notifying the FDA, and if the FDA happens to be notified, the product can still be marketed while waiting for a response. It is important to realize that this process means that there is no all-inclusive list of substances considered to be GRAS. Although the intent of the law in 1958 may have been to ensure that common food ingredients or items were not subject to food additive regulations, the change to a notification process allows the GRAS carve-out to cover new or innovative ingredients. The FDA has several lists of GRAS substances but also states that a substance is not required to be on these lists in order to be considered GRAS, as "the use of a substance is GRAS because of widespread knowledge among the community of qualified experts, not because of a listing or other administrative activity." We have to wonder how widespread of a community of qualified experts a company relies on to make the self-determination that a substance is GRAS. The FDA notes in the proposed rule that publication in a peer-reviewed scientific journal may be supplemented by secondary scientific literature, including the opinion of an "expert panel" specifically convened for this purpose, and that consensus does not mean unanimity.[105]

The FDA lists consist of the remnants of the list established by FDA regulation after the 1958 Food Additives Amendment, the items that were reaffirmed as GRAS in the 1970s by the FDA, the petitions received by the FDA during that time to affirm GRAS for particular ingredients, a list of substances the FDA has affirmed as GRAS for indirect food uses, and a list of the notices sent to the FDA by manufacturers post 1998, the date of the publication of the proposed rule.

You can search the post 1998 GRAS notices, along with FDA's response, from this link: http://www.accessdata.fda.gov/scripts/fdcc/?set=GRASNotices. As of August 31, 2014, 539 notices have been submitted to the FDA. The very first notice on this list is for soy isoflavone extract. The FDA closed this notice on November 3, 1998, ceasing evaluation at the request of the notifying company, Archer Daniels Midland (ADM). ADM requested that evaluation of the notice cease while updating the file with more information, to which the FDA advised that if ADM desired evaluation to continue, a new notice would need to be filed. As August 31, 2014, a new notice for soy isoflavone has not been filed.[106]

Finally, note that the FDA evaluation of the GRAS notification does not entail an independent scientific analysis. For a food additive, FDA determines the safety of the ingredient; whereas a determination that an ingredient is GRAS can be made by qualified experts outside of government and submitted to the FDA.[104]

Voluntary notification is a policy that favors corporate interest. An April 2014 report from the National Resources Defense Council (NRDC) spells GRAS out as "Generally Recognized as Secret" and comments that one problem with a voluntary notification program is, well, that it is voluntary. The NRDC report identified 275 chemicals from 56 companies that appear to have been marketed in the United States for use in food since 1997 by companies that made their own GRAS safety determinations without in fact, notifying the FDA; which means that a chemical's identity, chemical composition, and safety determination are not publicly disclosed. The report also notes that a manufacturer may continue to sell a product even the GRAS notification is voluntarily withdrawn.[107]

The criteria of "common use in food" leading to GRAS includes the amount of an additive commonly used and consumed; described by the FDA as "intended conditions of use of the substance" and "in consideration of the population that will consume the substance." This point is illustrated in a brief discussion on high-fructose corn syrup in Chapter 9, "Grains," and is manifesting in other controversies, such as caffeine.

Caffeine has been considered GRAS since the initial FDA list, but rising volumes in energy drinks and the use in alcohol have raised concerns. In 2010, the FDA declared that caffeine mixed with alcohol is not GRAS, citing in part, the lack of scientific data to support the idea that it would be safe for young adults, the intended market.[108] The idea of GRAS in context of historical consumption patterns evokes the idea that any risk in consumption is fairly known to the consumer or negligible, but when this is stretched by increasing amounts or in new combination, the consumer may be unknowingly accepting an unfair burden of risk.

2004 Food Allergen Labeling and Consumer Protection Act

Some consumers must avoid specific foods due to allergies. For a consumer with life-threatening allergies, the consequence of unknowing consumption could be a very high risk indeed. Although the 1990 NLEA required ingredients to be printed under their common or usual name, it may be difficult to discern if such an ingredient was sourced from an allergen. In 2004, Congress passed FALCPA which is applicable to all food regulated by the FDA. Under FALCPA, the label must clearly identify if any of the ingredients are sourced from the eight most common allergens. The label can either make a "contains" statement (contains soy, wheat, and eggs), or list the source in parentheses next to the ingredient lecithin (soy), whey (milk). If it fails to do either but contains any of the eight major allergens, the food product is considered misbranded.[109] See Box 13.2, FALCPA Allergens.

Although this information helps to highlight these specific allergens, consumers with allergies should still read the ingredient list on the label. For example, some people who are allergic to peanuts are also allergic to sweet lupin, a legume in the same family

---

BOX 13.2   FALCPA EIGHT ALLERGENS

1. Milk
2. Eggs
3. Fish (e.g., bass, flounder, and cod)
4. Crustacean shellfish (e.g., crab, lobster, and shrimp)
5. Tree nuts (e.g., almonds, walnuts, and pecans)
6. Peanuts
7. Wheat
8. Soybeans

These eight foods, and any ingredient that contains protein derived from one or more of them, are designated as "major food allergens" by FALCPA, and account for 90% of all food allergies.

From United States Food and Drug Administration, Food Allergies: What You Need to Know. Available at http://www .fda.gov/Food/ResourcesForYou/Consumers/ucm079311.htm, updated April May 15, 2014, accessed August 30, 2014.

as peanuts. Lupin flour and protein are becoming more commonly used in the United States, such as in gluten-free products. Lupin will not be highlighted as an allergen, but it must be declared in the list of ingredients.[110]

Furthermore, advisory statements such as "may contain" or cross-contamination of allergens are not required under FALCPA. This means that a manufacturer who uses tree nuts in one product on a manufacturing line does not need to make a statement on a different product that does not contain nuts but is produced on the same equipment. The FDA states that such precautionary statements as "may contain" or "produced in facilities that also processes tree nuts," must be truthful, not misleading, and not a substitute for good manufacturing practices.[109]

At first glance, a precautionary statement would seem to be helpful to the individual consumer. However, the manufacturer may have a greater incentive to simply make a broad, sweeping general statement rather than take on tighter control of the food-processing environment. In a scenario in which a manufacturer is relying on words over action, an advisory statement may not be an accurate expression of the risk to the consumer. An advisory statement may also keep the manufacturer safe from tort liability actions as well as prevent the FDA from declaring the product to be misbranded or adulterated.[111] Thus there is an on-going battle over what information is required, as industry believes, rightfully in some cases, that consumers will reject their product once they know what it contains, even if considered scientifically "safe."

Genetically Modified Foods

One of the most, if not the most, contentious issue regarding food labeling and food safety in the United States is genetically modified/genetically-engineered foods, often referred to as GMOs. As of 2014, 80%–89% of corn (depending on variety) and 94% of all soybeans grown in the United States were genetically engineered varieties.[112] Even if consumers do not choose to eat soy in a recognizable whole form, such as tofu or soy beverage, soy is present in many processed foods in the United States. Thus, mostly all Americans have been exposed to some amount of genetically modified soy.

The FDA first approved genetically modified crops in 1994, having earlier decided in 1992 that labeling was not required. The FDA did not see a distinction between the normal practices of hybridization as conducted by farmers and genetic modification and felt the important issue was the end-result. The 1992 policy statement pointed out that food product labeling is required to "reveal all facts that are material in light of representations made or suggested by labeling or with respect to consequences which may result from use." Thus, the FDA concluded in 1992 that consumers must be appropriately informed through labeling if an engineered product differed from a traditional counterpart such that the common name was no longer applicable or if there was a safety or usage issue.[113] However, the label would refer to that difference, not necessarily to the genetic engineering process that caused the difference. For example, in December 1996, when DuPont's genetically modified soybean oil proved to have a significantly different fatty acid profile than conventional soybeans with a much higher level of oleic acid, it no longer met the standard of composition of soybean oil. DuPont proposed it to be labeled accordingly as "high oleic soybean oil" to which the FDA agreed.[114]

Under the FDA's 1992 policy, as far as food safety is concerned, labeling of a genetically engineered food would be required if known food allergy issues existed, if for example, a peanut protein was introduced to a tomato. A peanut-protein-modified tomato, even if looked and tasted like a regular tomato, would require an informative label as to the peanut protein, but not necessarily to the genetic engineering process that introduced the protein.[115]

In 2001, the FDA issued draft guidance for industry regarding labeling bioengineered foods, which reaffirmed its decision not to require special labeling of all bioengineered foods, noting that the comments received had been mainly expressions of concern about the unknown, and that there was still no basis for concluding the fact that a food or its ingredients was produced using bioengineering is a material fact that must be disclosed under sections 403(a) and 201(n) of the Federal Food Drug and Cosmetic Act.[116] However, in the draft guidance, the FDA also curtailed label descriptions from food makers who do not use genetically modified or engineered ingredients and wish to promote their product accordingly. Since the FDA has concluded that the use or absence of bioengineering does not in and of itself mean there is a material difference in the food, the label must not somehow suggest to the consumer that genetically modified food is inferior to nonmodified foods, or less safe, and thus falsely alarm consumers as that would be misleading. The FDA noted in the draft guidance that using the term "GMO free" could be misleading on most foods, because most foods do not contain organisms.[116] Although technically accurate, it seems the FDA may have a different concept of "misleading" than consumers might.

In the meantime, many food manufacturers who wish to promote their nonGMO products have used a "NonGMO Project" association stamp.[117] Publishing final guidance for manufacturers who wish to voluntarily label their foods as being made with without the use of bioengineered ingredients is on the FDA's CFSAN's agenda of program priorities for 2013.[118] Significantly, final guidance on this topic could preempt laws made by states, such as the GMO labeling law passed by Vermont in 2014. The Vermont law requires that food produced entirely or in part from genetic engineering include the phrase, "Produced with Genetic Engineering" on the label. Connecticut and Maine passed labeling laws as well during summer 2013, but the laws will take effect only if additional states sign up.[119]

The FDA has made science-based decisions without consistently addressing larger social concerns held by consumers; not all of which are health-related concerns. Environmental and social justice concerns are also relevant issues that manifest in arguments over food policy.[44] In the absence of informative labeling, the consumer must accept the risk of unknowing consumption, and this is a risk that is hard to swallow when there are very little perceived benefits being offered in return. As noted in Chapter 12, "Nanny State," even if people are wrong about risk, the ethical principle that they are entitled to the autonomous ability to make the choice is generally an American belief. Thus, arguments in support of labeling are often based on "a right to know" argument, rather than safety.[120]

One specific concern regarding genetic engineering/modification and food safety is the potential of creating unrecognized new food allergens. All food allergens are proteins, and the technology of introducing a gene into another plant can carry along an allergenic protein.[113] The good news is that we can test for known allergens. For example, one company used a Brazil nut gene to improve the nutritional quality of soybeans, but then had to pull the soybean from the market after testing revealed that people who were allergic to Brazil nuts were also now allergic to the genetically engineered soybean.[121] To this extent, the risks of allergies can be mitigated through testing. However, the allergic potentiality of a new protein cannot be tested. The potential misuse of biotechnology, in a world with a highly globalized and centralized food supply, is also an issue of food safety.

OTHER FOOD SAFETY CONCERNS AND UNINTENDED CONSEQUENCES

Food Defense: Agroterrorism/Bioterrorism

Food safety includes food defense, namely, the protection of our food supply from intentional tampering or contamination. Although a widespread act of political nature is a deeply held concern in post-9/11 United States, there has been to date only one instance of food terrorism in the United States. In 1984, a religious cult in Oregon sprayed salad bars with salmonella; the goal was to sicken people to keep them from voting in a local election.[122] Nearly 751 people were sickened. However, it took a year for an investigation to determine that the sources was not an "ordinary" outbreak, but instead a deliberate contamination.[123]

Given only one known incidence of bioterrorism in the United States, the risk of another feels low. However, it raises the question as to if we would even be able to discern intentional contamination of our food supply. Farm animals and livestock, plant crops, and the food processing, distribution and retailing system could all provide potential terrorist targets.[124] As urban growth has occurred, agricultural operations, including farms, packinghouses, and processing plants have become larger, more centralized, and more intensive. With such consolidation, a targeted agroterrorism event could have a serious, adverse impact.

A poison or contaminant may be our first concern over food-related terrorism, but a livestock epidemic could also be disastrous. For example, foot-and-mouth disease (FMD) confined to a very small geographically distinct herd is a vastly different situation than FMD spreading through a large cattle operation.[125] Although FMD does not infect humans, it is highly transmissible among animals. A 2002 estimate calculated the costs of even a 10-farm outbreak in the United States at $2 billion dollars.[126] Furthermore, the biological agents that could inflict damage to livestock may be more readily available then those that could harm humans.[127]

The Public Health Security and Bioterrorism Preparedness and Response Act was enacted in 2002, a response to the September 11, 2001 terrorist attacks. This act expanded the FDA's authority over food manufacturing and imports, authorized expanded security at USDA facilities, and provided for criminal penalties for terrorism against animal enterprises. The 2006 Animal Enterprise Terrorism Act expanded these penalties and prescribed acts include threats, act of vandalism, property damage, criminal trespass, harassment, or intimidation. Thus broadly defined, terrorism includes domestic groups conducting animal rights activism.[126]

Viewed as overbroad, and clashing with first-amendment speech rights, the 2006 Act also created the possibility of a federal prison sentence if someone felt threatened, even if the acts were not intended to be threatening.[128] A preenforcement challenge lawsuit was filed at the end of the 2011 in the US District Court of Massachusetts challenging the constitutionality of the act; in March 2013, the District Court decided the plaintiffs did not have the standing to sue and dismissed the case. The First Circuit of Appeals affirmed the dismissal in March of 2014, relying on a 2013 case opinion that plaintiffs in a preenforcement challenge to a criminal challenge must show that their prosecution is "certainly impending." The plaintiffs have filed a petition with the Supreme Court pointing out a potential conflict with a 2014 Supreme Court decision.[129]

Deliberate food contamination most often is not political or widespread, however, but instead simply the act of a disgruntled employee. Technology can provide solutions, such as the creation of zones in factories in which entry by a staff member not wearing the appropriate authorized radio tag would set off security alerts.[130] However, HACCP plans regarding food safety could also include critical check points that are relevant to employees—not just employee training, but employee satisfaction, such as living wages and humane working conditions.

Unintended Consequences, Dietary and Otherwise

In the introduction to this book, we discussed how food policy affects who eats what, when, how, and with what impact. Food safety also impacts how and where we eat food. Food safety concerns may end up with regulations prohibiting foods rather than incorporating specific standards. Thus, cart vending may more often be regulated to allow ice cream, candy, and other highly processed foods, unintentionally limiting access to healthier foods. The New York City Green Cart program specifically increased the number of permits available for mobile vendors selling fruit and vegetables, but does not allow these carts to sell cut up fruit.[131]

Food safety concerns may also be used as arguments to achieve goals that may not be related to food safety. For example, some argue that a Philadelphia city rule banning volunteer group meal feeding of the homeless in parks and instead requiring that such meal service to be conducted in approved kitchens and by food safety trained workers is not really intended to curb foodborne illness outbreaks among the homeless, but instead to present a cleaner face to tourists.[132]

Food policies are not always consistent and coherent. Objectives can sometimes come into conflict. Increasing opportunities for farmers to sell directly to the public at outdoor farmers' markets may bump into local regulations regarding food safety.

Promotion of fruit and vegetable consumption

Likewise, the new role of produce as a potential significant food-safety risk highlights how food safety policies have to navigate not only philosophical issues of choice and individual responsibility but also goals promulgated by other policies. For example, as discussed in Chapter 3, "Dietary Guidance," government public health messages promote the increased consumption of fruits and vegetables. Promoting the consumption of fruits and vegetables in the face of increasing foodborne illnesses related to produce is a balance of risk.

HP 2020's food safety objectives are to reduce infections caused by six key pathogens transmitted through food, to reduce postdiarrheal hemolytic uremic syndrome (HUS) in children under five years of age, to reduce the number of outbreak-associated infections caused by Shiga-toxin producing *E. coli* O157, Campylobacter, Listeria, or Salmonella associated with the commodity food groups of beef, dairy, fruits and nuts, leafy vegetables, and poultry; to prevent increase in the proportions of certain pathogens found in humans that are resistant to certain antimicrobial treatments, to reduce severe allergic reactions in adults with diagnosed food allergies, and to increase the number of consumers who follow governmental food safety guidance. Another objective, considered development as there is no means yet to track data, is to increase safe food preparation in the retail and food service sectors.[133]

HP 2020 nutrition goals include increasing consumption of fruit and vegetables. The importance of consuming vegetables is underscored by the identification of the objective to increase the contribution of total vegetables to the diets of the population ages two or over as a Leading Health Indicator, which are a subset of HP 2020 objectives selected to communicate high priority health issues. Increasing the variety of vegetables consumed is also an HP 2020 objective with a specific focus on dark green vegetables, such as spinach and kale, orange vegetables, and legumes.[134] Sprouts are often sprouted legumes. People who turn to sprouts as part of their vegetable and legume consumption must be wary, as sprouts are germinated seeds that are at high risk for contamination due to the conditions required for germination. The 2010 *Dietary Guidelines for Americans* specifically recommend avoiding raw sprouts due to the high risk of contamination.[135]

The HP 2020 objectives do not prescribe the means of increasing fruit and vegetable consumption; and the objective is to generally increase the consumption of vegetables and other produce in any form—cooked, canned, processed, fresh, or raw. However, promoting the convenience of having raw fruits and vegetables available as snacks is an often-used tool to encourage consumption of produce, and is suggested by the 2010 Dietary Guidelines.[136]

The last paragraph of Appendix 3 of the 2010 *Dietary Guidelines for Americans*, entitled "Risky Eating Behaviors," emphasizes the dangers of consuming raw or undercooked animal products and the risk of foodborne illness. The recommendation to avoid raw sprouts is repeated, specifically for populations at high risk, but there is no other mention of risk in relation to consuming any other type of raw produce.[137]

Yet, as noted above, recently more foodborne illnesses have been attributed to leafy greens than any other item, in part because leafy greens are often prepared raw, and in part to norovirus being a likely culprit. We make the consumption of vegetables more convenient through incorporation into mixed dishes, often prepared for us, and the high risk attributed to leafy greens is attributed in part by the CDC to this dietary pattern.[42] When food is produced at a great distance, we may have less opportunity to learn about the methods and sanitation conditions of production, harvesting, processing, and distribution. And then, of course preparation of food, even fresh food, may happen at a distance, perhaps ahead of time. It might be sliced, diced, chopped, combined, and sometimes—but not always—cooked. As more people handle food, the risk of inadequate cooking or cross-contamination increases.

Norovirus

Norovirus is the leading cause of foodborne disease outbreaks tracked by the CDC that have a known cause, and foods that are normally consumed raw, such as leafy greens, fruit, and shellfish are the ones most commonly involved in norovirus outbreaks. Norovirus contamination can occur at any point through the farm-to-table cycle through a number of sources, including contaminated water used for irrigation. However, most norovirus food outbreaks are caused through transmission from food workers. Norovirus is highly contagious, and can persist on surfaces due to a resistance to many disinfectants. This is a particular problem for cruise ships, schools, and other institutional settings.[138]

Norovirus is not one of the 60 key pathogens targeted in food safety objectives of HP 2020. In the "social determinants" section of the food safety section overview, however, it is noted that the processing and retail food industries continue to be challenged by two issues related to food-service workers; namely a large employee population with high rates of turnover and a need for appropriate training.[139] Although HP 2020 recognizes that a living wage is a social determinant of health for individuals receiving the wage,[140] it fails to take the next step of recognizing that in the case of foodborne disease outbreaks transmitted by people who cannot afford to stay home when ill, a living wage might actually be a social determinant of health not just for the worker, but also for those who receive the services of the food-worker.

In the case of food-service and food production workers, a living wage and appropriate work conditions could potentially reduce employee turnover and allow for better food safety training and education as part of an employee participating in an overall food safety culture.

Upton Sinclair's intended message of *The Jungle* is still lost on us. He had not meant to singlehandedly launch a food safety movement. His goal had been to stir public outrage over the condition and plight of the workers.[141]

CONCLUSION

Food safety policy balances the allocation of risk and responsibilities among government, industry, and individual and must balance the costs of avoiding the risks, as well as the cost of the consequences of the risks. And, although our current risk-analysis focuses on significant specific pathogens, surveillance, and control of past problems, such as trichinosis, must continue in order to ensure our health.[142]

Food cannot be made completely risk-free, and producers may not be able to measure the risk or guarantee safety. We must evaluate the risk, manage the risk, and communicate the risk, as well as understand if individuals are in fact feasibly able to take responsibility for their own safety, especially in the absence of information. Food labeling is an important component of food safety that provides information on risk.

Due to market and economic interests, the balance of risk allocation between individual and industry can be affected by inputs outside of government regulation. Where we allocate the benefit of economic interest is an important component of food policy. Looking forward, if our food policy promotes inexpensive food for the consumer, the consumer may gain economically only in the short run, considering the health impacts of diet and food safety.

Furthermore, the emphasis on individual actions, in particular, cooking to a certain temperature reveals an important underlying component to our food safety policy culture, namely that the consumer has been allocated the burden of risk in cooking as a trade-off for the enjoyment of inexpensive food. Yes, we can cook chicken to safe temperature. And also be careful not to cross-contaminate our kitchen prior to cooking the chicken. *But should we have to?*

Salmonella can be drastically reduced in chicken flocks.[143] Campylobacter is highly prevalent, but some processors are clearly better at minimizing outbreaks than others.[144] Armed with knowledge, a consumer could choose to pay more—not just for perceived quality standards (organic), but also for verifiable standards of safety. How we think of and treat our food, our food animals, our food production, and our production workers in farm, plant, and kitchen is coming back to bite us in the gut.

REFERENCES

1. Johnson R. *The Federal Food Safety System: A Primer*, Congressional Research Service, December 15, 2010.
2. United States Environmental Protection Agency, *The EPA and Food Security*. Available at http://www.epa.gov/pesticides/factsheets/securty.htm, updated August 1, 2014, accessed August 30, 2014.
3. Centers for Disease Control and Prevention, *CDC and Food Safety*. Available at http://www.cdc.gov/foodsafety/cdc-and-food-safety.html, updated April 19, 2014, accessed August 30, 2014.
4. United States Department of Agriculture, *Food Safety and Inspection Service*. Available at http://www.fsis.usda.gov/About_FSIS/index.asp, updated April 29, 2014, accessed August 30, 2014.
5. United States Government, http://www.foodsafety.gov/about/federal/index.html, accessed August 30, 2014.
6. United States Food and Drug Administration, *About FDA*. Available at http://www.fda.gov/AboutFDA/WhatWeDo/History/Overviews/ucm056044. Updated March 11, 2014, accessed August 30, 2014.
7. Institute of Medicine, Forum on Microbial Threats. *Addressing Foodborne Threats to Health: Policies, Practices, and Global Coordination: Workshop Summary*, 2006. Available at http://www.ncbi.nlm.nih.gov/books/NBK57085/, accessed August 30, 2014.
8. Nestle M. *Safe Food: The Politics of Food Safety*. California: University of California, 2010.

9. United States Food and Drug Administration, *About FDA*. Available at http://www.fda. gov/aboutfda/whatwedo/default.htm, updated August 5, 2014.

10. United States Food and Drug Administration, *About FDA* Available at http://www.fda.gov/ AboutFDA/CentersOffices/OfficeofFoods/default.htm, updated June 14, 2013, accessed August 30, 2014.

11. United States Food and Drug Administration, *About FDA*. Available at http://www.fda. gov/AboutFDA/Transparency/Basics/ucm242648.htm, updated April 10, 2014, accessed August 30, 2014.

12. United States Food and Drug Administration, *FDA History*. Available at http://www.fda. gov/aboutFDA/WhatWeDo/History/origin/ucm054826.htm, last updated September 24, 2012, accessed August 30, 2014.

13. Glenn Morris J Jr, Potter ME (eds.). *Foodborne Infections and Intoxications*. Waltham, MA: Academic Press, 2013, p. 501.

14. Banerjee N, Semuels A. FDA proposes sweeping new food safety rules. *The Los Angeles Times*, January 4, 2013. Available at http://articles.latimes.com/2013/jan/04/business/la-fi-fda-food-safety-20130105, accessed August 30, 2014.

15. 21 U.S.C. § 604, United States Food and Drug Administration. *Regulatory Information.* Available at http://www.fda.gov/RegulatoryInformation/Legislation/ucm148693.htm, updated May 25, 2009, accessed August 30, 2014.

16. Federal Food and Drugs Act of 1906 (The "Wiley Act") P.L. Number 59-384, 34 Stat. 786, 1906. Available at http://www.fda.gov/RegulatoryInformation/Legislation/ucm148690. htm, accessed November 20, 2014.

17. Commerce Clause, *Legal Information Institute*, Cornell University Law School. Available at http://www.law.cornell.edu/wex/commerce_clause, accessed August 30, 2014.

18. United States Food and Drug Administration, *About FDA*. Available at http://www.fda. gov/AboutFDA/WhatWeDo/History/ProductRegulation/SulfanilamideDisaster/default. htm, updated October 7, 2010, accessed August 30, 2014.

19. Centers for Disease Control and Prevention, *Estimates of Foodborne Illness*. Available at http://www.cdc.gov/foodborneburden/2011-foodborne-estimates.html, updated January 8, 2014, accessed August 30, 2014.

20. Ackerman J. Food, how safe? *National Geographic*, 2012;201(5):2–31.

21. Jones FT. A review of practical *Salmonella* control measures in animal feed, *J. Appl. Poultry Res.* 2011;20(1):102–113. Available at http://japr.oxfordjournals.org/content/20/1/102.full, accessed September 2, 2014.

22. Scandi Standard, Why Scandinavian Chicken? Available at http://www.scandistandard. com/en/About-us/Why-Scandinavian-chickenisk-kyckling/, accessed September 2, 2014.

23. United States Food and Drug Administration, Safe Practices for Food Processes. Available at http://www.fda.gov/Food/FoodScienceResearch/SafePracticesforFoodProcesses/ucm091265.htm, updated April 24, 2013, accessed September 2, 2014.

24. Centers for Disease Control and Prevention, Salmonella. Available at http://www.cdc.gov/salmonella/general/index.html, updated April 5, 2012, accessed September 2, 2014.

25. Centers for Disease Control and Prevention, *E. coli*. Available at http://www.cdc.gov/ecoli/general/index.html, updated August 3, 2012, accessed September 2, 2014.

26. Centers for Disease Control and Prevention, National Center for Emerging and Zoonotic Infectious Diseases, Camplyobacter. Available at http://www.cdc.gov/nczved/divisions/dfbmd/diseases/campylobacter/, updated June 3, 2014.

27. Centers for Disease Control and Prevention, Disease Listing, Yersinia enterocolitica. Available at http://www.cdc.gov/ncidod/dbmd/diseaseinfo/yersinia_g.htm, updated October 25, 2005, accessed September 2, 2014.

28. Centers for Disease Control and Prevention, CDC Vital Signs, Recipe for Food Safety. Available at http://www.cdc.gov/vitalSigns/listeria/index.html, updated June 4, 2013, accessed September 2, 2014.

29. Centers for Disease Control and Prevention, Vibrio Illness. Available at http://www.cdc. gov/vibrio/vibriov.html, updated October 21, 2013, accessed September 2, 2014.
30. McGraw M. Beef's raw edges. *The Kansas City Star*, December 8, 2012.
31. Centers for Disease Control and Prevention, Norovirus trends and outbreaks. Available at http://www.cdc.gov/norovirus/trends-outbreaks.html, updated July 8, 2014, accessed August 30, 2014.
32. Centers for Disease Control and Prevention, Notes from the field: Emergence of new norovirus strain gii.4 sydney, United States 2010. *Morbidity and Mortality Weekly Report*, January 25, 2013. Available at http://www.cdc.gov/mmwr/preview/mmwrhtml/mm6203a4. htm?s_cid=mm6203a4_x, updated January 25, 2013, accessed August 30, 2014.
33. United States Department of Health and Human Services, Centers for Disease Control and Prevention, Estimates of Foodborne Illness in the United States. Available at http:// www.cdc.gov/foodborneburden/2011-foodborne-estimates.html, updated January 8, 2014, accessed August 30, 2014.
34. United States Department of Health and Human Services, Centers for Disease Control and Prevention, Multistate outbreak of listeriosis linked to whole cantaloupes from Jensen farms, Colorado. Available at http://www.cdc.gov/listeria/outbreaks/cantaloupes-jensen-farms/index.html, updated September 4, 2012, accessed August 30, 2014.
35. Hoyert D, Xu J. Deaths: Preliminary Data for 2011, National Vital Statistics Report, October 10, 2012;61(6), United States Department of Health and Human Services, Centers for Disease Control and Prevention, National Center for Health Statistics. Available at http://www.cdc.gov/nchs/data/nvsr/nvsr61/nvsr61_06.pdf, accessed August 30, 2014.
36. McKenna M. Food poisoning's hidden legacy. *Scientific American,* 2012;306(4):26–27.
37. Sliter-Hays SM. *Narrative and Rhetoric: Persuasion in Doctor's Writings about the Summer Complaint, 1883–1939*, PhD Dissertation, University of Texas, Austin, 2008.
38. 17-month-old is 3rd child to die of illness linked to tainted meat, *The New York Times*, February 22, 1993. Available at http://www.nytimes.com/1993/02/22/us/17-month-old-is-3d-child-to-die-of-illness-linked-to-tainted-meat.html, accessed August 30, 2014.
39. Drexler M. *Secret Agents: The Menace of Emerging Infections.* Washington, DC: The National Academies Press, 2002.
40. World Health Organization, 10 facts on food safety. Available at http://www.who.int/ features/factfiles/food_safety/facts/en/index3.html, accessed August 30, 2014.
41. United States Department of Health and Human Services, Centers for Disease Control and Prevention, Foodborne Diseases Active Surveillance Network (FoodNet). Available at http://www.cdc.gov/foodnet/data/trends/trends-2011.html, updated January 30, 2013, accessed August 30, 2014.
42. Painter JA et al. Attribution of foodborne illnesses, hospitalizations, and deaths to food commodities by using outbreak data, United Sates, 1998–2008. *Emerg. Infect. Dis.* 2013;19:3. Available at http://wwwnc.cdc.gov/eid/article/19/3/11-1866_article.htm, updated February 25, 2013, accessed August 30, 2014.
43. Erickson MC, Doyle MP. Plant food safety issues: Linking production agriculture with one health. *Improving Food Safety Through a One Health Approach, Workshop Summary*, Institute of Medicine, 2012. Available at http://www.ncbi.nlm.nih.gov/books/NBK114507/, accessed August 30, 2014.
44. Sperling D. Food law, ethics, and food safety regulation: Roles, justifications and expected limits. *J. Agri. Environ. Ethics.* 2010;23:267–278.
45. United States Government. Available at http://www.foodsafety.gov/keep/basics/index. html, accessed August 30, 2014.
46. Regulations.gov. Available at http://www.regulations.gov/#!docketDetail;D=FSIS-2008-0017, accessed September 2, 2014.
47. Jensen DA et al. Cross contamination of *Escherichia coli* O157:H7 between lettuce and wash water during home-scale washing. *Food Microbiology,* 2015;46:428–433.

48. United States Food and Drug Administration, Raw Produce. *Selecting and Serving it Safely.* Available at http://www.fda.gov/food/resourcesforyou/consumers/ucm114299, updated June 2, 2014, accessed August 30, 2014.

49. Weise E. Kroger stores stops selling sprouts as too dangerous. *USA TODAY,* October 20, 2012. Available at http://www.usatoday.com/story/news/2012/10/19/kroger-bans-sprouts-too-dangerous/1645147/, accessed August 30, 2014.

50. Wyatt K. Colorado cantaloupes return; growers push safety. *The Associated Press,* July 12, 2012. Available at http://bigstory.ap.org/article/colorado-cantaloupes-return-growers-push-safety, accessed August 30, 2014.

51. Atkinson N. The impact of BSE on the UK economy. Available at http://www.veterinaria.org/revistas/vetenfinf/bse/14Atkinson.html, accessed August 30, 2014.

52. US Food and Drug Administration, *Listeria monocytogenes Risk Assessment.* Available at http://www.fda.gov/Food/FoodScienceResearch/RiskSafetyAssessment/ucm184052.htm, updated March 29, 2013, accessed August 30, 2014.

53. Rothschild M. Listeria in smoked Salmon: Examining the risk. *Food Safety News,* September 11, 2013, http://www.foodsafetynews.com/2013/09/listeria-in-smoked-salmon-examining-the-risk, accessed August 30, 2014.

54. Buzby JC. Effect of food-safety perceptions on food demand and global trade, in: *Changing Structure of Global Food Consumption and Trade,* Economic Research Service, Washington, DC, pp. 55–66. Available at http://www.ers.usda.gov/media/293613/wrs011i_1_.pdf, accessed August 30, 2014.

55. Bottemiller H. Russia to ban USmeat over ractopamine residues this month. *Food Safety News,* February 1, 2013. Available at http://www.foodsafetynews.com/2013/02/russia-to-ban-u-s-meat-over-ractopamine-residues-this-month/accessed August 30, 2014.

56. World Trade Organization, *Understanding the WTO Agreement on Sanitary and Phytosanitary Measures.* Available at http://www.wto.org/english/tratop_e/sps_e/spsund_e.htm, accessed August 30, 2014.

57. United States Government Accountability Office, Food Safety: FDA's food advisory and recall process needs strengthening, GAO-12-589, July 26, 2012. Available at http://www.gao.gov/products/GAO-12-589, accessed August 30, 2014.

58. Pignolet J. Year after FDA blamed outbreak on Evergreen Produce, founders still baffled. *The Spokesman-Review,* October 7, 2012. Available at http://www.spokesman.com/stories/2012/oct/07/seeds-of-doubt/, accessed August 30, 2014.

59. Bottemiller H. MDP shuts down, USDA testing of produce for pathogens halted. *Food Safety News,* January 3, 2013. Available at http://www.foodsafetynews.com/2013/01/mdp-officially-shut-down-pathogen-testing-for-produce-halted/#.UQbi9-gt0fE, accessed August 30, 2014.

60. ElBoghdady D. Produce-safety testing program on chopping block. *The Washington Post,* July 12, 2012. Available at http://www.washingtonpost.com/business/economy/produce-safety-testing-program-on-chopping-block/2012/07/12/gJQAHdHWgW_story.html, accessed August 30, 2014.

61. Bottemiller H. USDA budget cut could slash 80 percent of produce testing. *Food Safety News,* July 10, 2012. Available at http://www.foodsafetynews.com/2012/07/usda-budget-cut-could-slash-80-percent-of-produce-testing/#.UlXYfL9Rw1g, accessed August 30, 2014.

62. 511 F.2d 331, 1974. *American Public Health Association et al. Appellants, v. EarlButz, Secretary of Department of Agriculture et al.*

63. United States Department of Agriculture, Food Safety and Inspection Service. Available at http://www.fsis.usda.gov/wps/portal/fsis/newsroom/meetings/newsletters/small-plant-news/small-plant-news-archive/spn-vol5-no9, updated July 22, 2013, accessed August 30, 2014.

64. United States Department of Agriculture, Food Safety and Inspection Service, FSIS history. Available at http://www.fsis.usda.gov/wps/portal/informational/aboutfsis/history, updated May 25, 2013, accessed August 30, 2014.

65. United States Government. Available at http://www.foodsafety.gov/keep/government/index.html, accessed August 30, 2014.

66. How does federal food safety legislation protect the nation's food supply? *Robert Wood Johnson Foundation*, Health Policy Snapshot Series, September 2011. Available at http://www.rwjf.org/en/research-publications/find-rwjf-research/2011/09/how-does-federal-food-safety-legislation-protect-the-nation-s-fo.html, accessed August 30, 2014.

67. Hoffman S. Food safety policy and economics: A review of the literature. Resources for the Future, July 2010. Available at http://www.rff.org/Publications/Pages/PublicationDetails.aspx?PublicationID=21310, accessed August 30, 2014.

68. United Stated Food and Drug Administration, Food Safety Modernization Act. Available at http://www.fda.gov/Food/GuidanceRegulation/FSMA/ucm257978.htm, updated August 5. 2014.

69. Satran J. Sunland: FDA didn't warn of peanut butter recall-related suspension. *Huffington Post*. Available at http://www.huffingtonpost.com/2012/11/28/sunland-fda-peanut-butter_n_2206353.html, November 28, 2012.

70. United States Food and Drug Administration, Annual report to Congress on the Use of Mandatory Recall Authority, December 2013. Available at http://www.fda.gov/Food/GuidanceRegulation/FSMA/ucm382490.htm, last updated August 5, 2014.

71. United States Food and Drug Administration, FSMA Progress Reports. Available at http://www.fda.gov/Food/GuidanceRegulation/FSMA/ucm255893.htm#progress_oct_dec, accessed August 30, 2014.

72. Bottemiller H. Key FSMA rules continue to languish at OMB, months after deadline. *Food Safety News*, April 23, 2012. Available at http://www.foodsafetynews.com/2012/04/key-fsma-rules-continue-to-languish-at-omb-months-after-deadline/#.U14QWr9Ry1s, accessed August 30, 2014.

73. Center for Food Safety, February 20, 2014. Available at http://www.centerforfoodsafety.org/files/2014-2-20-dkt-82-1—joint—consent-decree_26503.pdf, accessed August 30, 2014.

74. United States Food and Drug Administration, FSMA. Available at http://www.fda.gov/Food/GuidanceRegulation/FSMA/ucm334120.htm, accessed August 30, 2014.

75. United States Food and Drug Administration, FSMA Proposed Rule for Preventative Controls for Human Food. Available at http://www.fda.gov/food/guidanceregulation/fsma/ucm334115.htm, updated August 4, 2014, accessed August 30, 2014.

76. United States Food and Drug Administration, FSMA Proposed Rule for Produce Safety. Available at http://www.fda.gov/Food/GuidanceRegulation/FSMA/ucm334114.htm, updated August 4, 2014, accessed August 30, 2014.

77. Regulations.gov, February 12, 2013, Proposed Rule. Available at http://www.regulations.gov/#!documentDetail;D=FDA-2011-N-0921-0001, accessed August 30, 2014.

78. United States Food and Drug Administration, FSMA Proposed Rule for Produce Safety. Available at http://www.fda.gov/food/guidanceregulation/fsma/ucm334114.htm, updated September 26, 2013.

79. Bottemiller H. Nearly 80 percent of produce growers to be exempt from food safety rule. *Food Safety News*, January 11, 2013. Available at http://www.foodsafetynews.com/2013/01/nearly-80-percent-of-produce-growers-to-be-exempt-from-new-food-safety-rule, accessed October 14, 2013.

80. United States Food and Drug Administration, Safety, Recalls. Available at http://www.fda.gov/safety/recalls/ucm267667.htm, updated February 12, 2014, accessed August 30, 2014.

81. California Department of Public Health, Cottage Food Operations, Approved Food Products List, July 30, 2014. Available at http://www.cdph.ca.gov/programs/Documents/fdbCFOfoodslist.pdf,accessed August 30, 2014.

82. United States Food and Drug Administration, FSMA preliminary regulatory impact analysis. Available at http://www.fda.gov/downloads/Food/FoodSafety/FSMA/UCM334117.pdf, accessed August 30, 2014.

83. United States Food and Drug Administration, Implementing the FDA Food Safety Modernization Act, February 5, 2014. Available at http://www.fda.gov/NewsEvents/Testimony/ucm384687.htm, accessed August 30, 2014.
84. Fortin ND. *Food Regulation: Law, Science, Policy and Practice.* New Jersey: John Wiley & Sons, 2011.
85. David S et al. The essential role of state and local agencies in food safety and reform. The Food Safety Resource Consortium, www.thefsrc.org/State_Local/StateLocal_June17_background.pdf, accessed October 15, 2013, no longer available online as of August 30, 2014.
86. United States Food and Drug Administration, FDA Food Code. Available at http://www.fda.gov/food/guidanceregulation/retailfoodprotection/foodcode/default.htm, updated April 21, 2014, accessed August 30, 2014.
87. Purvis K. Want a rare burger? You're a step closer in NC. *News Observer*, May 19, 2012. Available at http://www.newsobserver.com/2012/05/19/2070999/want-a-rare-burger-youre-a-step.html, accessed August 30, 2014.
88. Five Guys. Available at http://www.fiveguys.com/about-us/faq.aspx, accessed August 30, 2014.
89. United States Department of Agriculture, Animal and Plant Healthy Inspection Service, Animal Health. Available at http://www.aphis.usda.gov/wps/portal/footer/topicsofinterest/applyingforpermit?1dmy&urile=wcm%3apath%3a%2Faphis_content_library%2Fsa_our_focus%2Fsa_animal_health%2Fsa_animal_disease_information%2Fsa_cattle_health%2Fsa_tuberculosis%2Fct_bovine_tuberculosis_disease_information, updated September 2, 2014.
90. Armour S, Lippert J, Smith M. Food sickens millions as company-paid checks find it safe. *Bloomberg Markets Magazine*, October 10, 2012. Available at http://www.bloomberg.com/news/2012-10-11/food-sickens-millions-as-industry-paid-inspectors-find-it-safe.html, accessed September 2, 2014.
91. Nixon R. U.S.D.A. poultry inspection plan sets off dispute. *The New York Times*, April 5, 2012. Available at http://www.nytimes.com/2012/04/05/us/usda-poultry-plan-sets-off-dispute.html, accessed September 2, 2014.
92. Powell DA et al. Audits and inspections are never enough: A critique to enhance food safety. *Food Control*, 2013;30:686–691.
93. Jones TF et al. Restaurant inspection scores and foodborne disease. *Emerging Infectious Diseases*, 2004;10(4):688–692. Available at http://wwwnc.cdc.gov/eid/article/10/4/03-0343_article.htm.
94. Foley RJ. Lab warned for salmonella risk before massive egg recall. *Associated Press*, June 4, 2012. Available at http://www.nbcnews.com/id/47677729/ns/health-food_safety/t/lab-warned-salmonella-risk-massive-egg-recall, accessed September 2, 2014.
95. Institute of Medicine, Forum on Microbial Threats, Addressing Foodborne Threats to Health: Policies, Practices, and Global Coordination: Workshop summary, 2006. Available at http://www.ncbi.nlm.nih.gov/books/NBK57085/, accessed September 2, 2014.
96. Hoffman S. Food safety policy and economics: A review of the literature. Resources for the Future, July 2010. Available at http://www.rff.org/Publications/Pages/PublicationDetails.aspx?PublicationID=21310, accessed August 30, 2014.
97. Griffith CJ, Livesey KM, Clayton DA. Food safety culture: The evolution of an emerging risk factor? *British Food Journal*, 2010;112(4):426–438.
98. Hoffman S, US Food safety policy enters a new era, *Amber Waves*, December 1, 2011. Available at http://www.ers.usda.gov/amber-waves/2011-december/us-food-safety-policy.aspx, accessed September 2, 2014.
99. Significant Dates in US Food and Drug Law History. US Food and Drug Administration, accessed August 17, 2013, http://www.fda.gov/AboutFDA/WhatWeDo/History/Milestones/ucm128305.htm

100. Code of Federal Regulations, 21 CFR 170.3, http://www.gpo.gov/fdsys/pkg/CFR-2012-title21-vol3/xml/CFR-2012-title21-vol3-sec170-3.xml, accessed May 13, 2014.

101. US Food and Drug Administration, Generally Recognized as Safe (GRAS), http://www.fda.gov/food/ingredientspackaginglabeling/gras/default.htm, last updated February 12, 2014.

102. United States Food and Drug Administration, Significant Dates in US Food and Drug Law History, last updated March 25, 2014. Available at http://www.fda.gov/AboutFDA/WhatWeDo/History/Milestones/ucm128305.htm, accessed September 20, 2014.

103. United States Food and Drug Administration, FDA's approach to the GRAS Provision: A History of Processes, last updated April 23, 2013. Available at http://www.fda.gov/Food/IngredientsPackagingLabeling/GRAS/ucm094040.htm, accessed September 20, 2014.

104. United States Food and Drug Administration, How US FDA's GRAS notification program works, January 2006, http://www.fda.gov/Food/IngredientsPackagingLabeling/GRAS/ucm083022.htm, accessed September 20, 2014.

105. Food and Drug Administration, HHS. Substances Generally Recognized as Safe. Proposed Rule. Federal Register Volume 62, Number 74. April 17, 1997. http://www.gpo.gov/fdsys/pkg/FR-1997-04-17/html/97-9706.htm, accessed November 20, 2014.

106. United States Food and Drug Administration, Guidance for Industry: Frequently Asked Questions about GRAS, December 2004. Available at http://www.fda.gov/Food/GuidanceRegulation/GuidanceDocumentsRegulatoryInformation/IngredientsAdditivesGRASPackaging/ucm061846.htm#Q16, accessed November 20, 2014.

107. Neltner T, Maffini M. Generally recognized as secret: Chemicals added to food in the United States, NRDC Report, April 2014. National Resources Defense Council. Available at http://www.nrdc.org/food/files/safety-loophole-for-chemicals-in-food-report.pdf, accessed September 20, 2014.

108. Unites States Food and Drug Administration, Inspections, Compliance, Enforcement, and Criminal Investigations. Warning letter, Phusion Projects Inc. November 17, 2010. Available at http://www.fda.gov/ICECI/EnforcementActions/WarningLetters/ucm234023.htm, accessed September 20, 2014.

109. United States Food and Drug Administration, Food Allergies: What you need to know. Available at http://www.fda.gov/Food/ResourcesForYou/Consumers/ucm079311.htm, updated April May 15, 2014, accessed August 30, 2014.

110. United States Food and Drug Administration, Frequently Asked Questions on Lupin and Allergenicity, last updated August 15, 2014. Available at http://www.fda.gov/Food/IngredientsPackagingLabeling/FoodAdditivesIngredients/ucm410111.htm, accessed November 21, 2014.

111. Roses JB. Food allergen law and the food allergen labeling and consumer protection act of 2004: falling short of true protection for allergy sufferers. *Food Drug Law J.* 2011;66(2):225.

112. United States Department of Agriculture, Economic Research Service. Available at http://www.ers.usda.gov/data-products/adoption-of-genetically-engineered-crops-in-the-us/recent-trends-in-ge-adoption.aspx, updated July 14, 2014, accessed August 30, 2014.

113. United States Food and Drug Administration, Guidance & Regulation. Available at http://www.fda.gov/food/guidanceregulation/guidancedocumentsregulatoryinformation/biotechnology/ucm096095.htm, updated August 15, 2013, accessed August 30, 2014.

114. United States Food and Drug Administration, Biotechnology Consultation Note. Available at http://www.fda.gov/food/foodscienceresearch/biotechnology/submissions/ucm161157.htm, updated December 26, 2013, accessed August 30, 2014.

115. United States Department of Agriculture, Guidance and Regulation. Available at http://www.fda.gov/food/guidanceregulation/guidancedocumentsregulatoryinformation/labelingnutrition/ucm059098.htm, updated July 7, 2015, accessed August 30, 2014.

116. United States Food and Drug Administration, Draft Guidance for Industry, January 2001. Available at http://www.fda.gov/food/guidanceregulation/guidancedocumentsregulatory information/labelingnutrition/ucm059098.htm, updated July 7, 2014, accessed August 30, 2014.

117. Non GMO Project. Available at http://www.nongmoproject.org, accessed November 20, 2014.

118. United States Food and Drug Administration; Center for Food Safety and Applied Nutrition Plan for Program Priorities, 2013–2014, updated September 4, 2013. Available at http://www.fda.gov/AboutFDA/CentersOffices/OfficeofFoods/CFSAN/WhatWeDo/ucm366279.htm?source=govdelivery&utm_medium=email&utm_source=govdelivery, accessed August 30, 2014.

119. Kaste M. So what happens if the movement to label GMOS succeeds? *npr*, October 16, 2013. Available at http://www.npr.org/blogs/thesalt/2013/10/16/235525984/so-what-happens-if- the-movement-to-label-gmos-succeeds.

120. Raab C, Grobe D. Labeling genetically engineered food: The consumer's right to know. *J. Agrobiotechnol. Manage. Econ.* 2003;6(4):155–161. Available at http://www.agbioforum.org/v6n4/v6n4a02-raab.htm, accessed April 9, 2014.

121. Nordlee JA et al. Identification of a brazil-nut allergen in transgenic soybeans. *New England J. Med.* 1996;334:688–692.

122. Zaitz L. Rajneeshee leaders take revenge on the Dalles' with poison, homeless—part 3 of 5. *The Oregonian*, April 14, 2011. Available at http://www.oregonlive.com/rajneesh/index.ssf/2011/04/part_three_mystery_sickness_su.html, accessed August 30, 2014.

123. Torok TJ et al. A large community outbreak of salmonellosis caused by intentional contamination of restaurant salad bars. *JAMA.* 1997;278:389–395.

124. Goodrich Schneider R et al. Agroterrorism in the US: An overview. Institute of Food and Agricultural Sciences, University of Florida, August 2005. Available at http://edis.ifas.ufl.edu/fs126, accessed September 2, 2014.

125. Olson D. Threats to America's economy and food supply. The Federal Bureau of Investigation, Agroterrorism, February 2012. Available at http://www.fbi.gov/stats-services/publications/law-enforcement-bulletin/february-2012/agroterrorism, accessed September 2, 2014.

126. Monke J. Agroterrorism: Threats and preparedness. Congressional Research Service Report for Congress, March 12, 2007.

127. Yeh JY et al. Livestock agroterrorism: The deliberate introduction of a highly infectious animal pathogen. *Foodborne Pathogens Dis.* 2012;9(10):869–877. Available at http://www.ncbi.nlm.nih.gov/pubmed/23035724, accessed September 2, 2014.

128. Fields G, Emshwiller JR. 'Mens Rea' takes hit from the animal enterprise terrorism act. *The Wall Street J.* September 27, 2011. Available at http://online.wsj.com/article/SB10001424053111903791504576586790205241376.html, accessed September 2, 2014.

129. Center For Constitutional Rights, *Blum v. Holder.* Available at http://ccrjustice.org/our-cases/Blum, accessed September 2, 2014.

130. Watson E. How vulnerable is your supply chain? Food defense, FSMA, big brother and virtual 'hot zones.' Food Navigator-USA.com, December 18, 2010. Available at http://www.foodnavigator-usa.com/Business/How-vulnerable-is-your-supply-chain-Food-defense-FSMA-Big-Brother-and-virtual-hot-zones.

131. New York City Department of Health and Mental Hygiene, Green Carts. Available at http://www.nyc.gov/html/doh/html/diseases/green-carts.shtml, accessed October 20, 2013.

132. Armour S. Philadelphia regulates brotherly love to curb homeless picnics. *Bloomberg Businessweek*, March 22, 2012. Available at http://www.businessweek.com/news/2012-03-22/philadelphia-regulates-brotherly-love-to-curb-homeless-picnics, accessed September 2, 2014.

133. Healthy People 2020, Food Safety. Available at http://www.healthypeople.gov/2020/topicsobjectives2020/objectiveslist.aspx?topicId =14, updated August 31, 2014.

134. Healthy People 2020, Nutrition and Weight Status. Available at http://www.healthypeople.gov/2020/topicsobjectives2020/objectiveslist.aspx?topicId=29, updated August 31, 2014.

135. United States Department of Agriculture and United States Department of Health and Human Services, *Dietary Guidelines for Americans*, 2010, 7th edition, 48. Washington, DC: Government Printing Office, 2010. Available at http://www.health.gov/dietaryguidelines/2010.asp, accessed August 30, 2014.

136. United States Department of Agriculture and United States Department of Health and Human Services, *Dietary Guidelines for Americans*, 2010, 7th edition, Appendix 2 Key Consumer Behaviors and Potential Strategies for Professionals, Table A2 1,65. Washington, DC: Government Printing Office, 2010. Available at http://www.health.gov/dietaryguidelines/2010.asp, accessed August 30, 2014.

137. United States Department of Agriculture and United States Department of Health and Human Services, *Dietary Guidelines for Americans,* 2010, 7th edition, 72. Washington, DC: Government Printing Office, 2010. Available at http://www.health.gov/dietaryguide lines/2010.asp, accessed August 30, 2014.

138. Centers for Disease Control and Prevention, Surveillance for Norovirus Outbreaks. Available at http://www.cdc.gov/features/dsnorovirus/, updated June 3, 2014.

139. Healthy People 2020, Food Safety. Available at http://www.healthypeople.gov/2020/topic sobjectives2020/overview.aspx?topicid=14, updated August 31, 2014.

140. Healthy People 2020, About Healthy People. Available at http://healthypeople.gov/2020/about/DOHAbout.aspx, updated August 31, 2014.

141. Arthur A. Upton Sinclair, *The New York Times*. Available at http://www.nytimes.com/ref/timestopics/topics_uptonsinclair.html, accessed August 30, 2014.

142. Centers for Disease Control and Prevention, Parasites. Available at http://www.cdc.gov/parasites/trichinellosis/epi.html, updated August 8, 2012, accessed September 2, 2014.

143. Wegener HC et al. Salmonella control programs in Denmark. *Emerging Infectious Diseases*, July 2003. Available at http://wwwnc.cdc.gov/eid/article/9/7/03-0024.htm, accessed August 30, 2014.

144. Consumer Reports, Lax rules, risky food. January 2010. Available at http://www.consumerreports.org/cro/magazine-archive/2010/january/viewpoint/overview/lax-rules-risky-food-ov.htm, accessed August 30, 2014.

Looking Forward

14

Food is embedded water.
Tim Lang[1]

This chapter is intended to briefly introduce topics of greater current interest and, as such, suffice as a conclusion. All of these topics deserve much more detailed exploration than we can give here; and in fact, we hope that you are already informed on these topics.

In the first section we reference topics covering labeling and activism, in the second section we examine food waste, and in the third section we discuss water. Food waste and water each deserve their own chapter and more attention and detail than we can provide here. Our research was limited, but in neither case of food waste or water did we locate an overarching federal policy. The federal government does seem to present more consumer facing material, information, and programs about food waste than about water.

Although this chapter is titled "Looking Forward," water, like waste, is being addressed today. The Environmental Protection Agency (EPA), US Department of Agriculture (USDA), and Food and Drug Administration (FDA) all have a role in the current efforts on managing food waste. The USDA is addressing food recovery at a programmatic level, but does not seem as focused on an overall food recovery strategy as we might expect. Water policy as related to food seems more diffuse, with less consumer-oriented information readily accessible. The USDA and the Department of the Interior (DOI) seem to be the major players in food-related water policy, except for the EPA's role in ensuring the waters from which we fish are safe. Although we are not able to enter into a full examination of water use policies, it seems that the current nexus of our food–water policy is aimed at improving technology and efficiency of water use in order to sustain agricultural production in type and quantity that we have grown accustomed to, rather than adjusting what and where we farm. This leads us to the conclusion of this chapter, in which we touch briefly on sustainability and food security, a topic not yet discussed in this book.

LABELING AND ACTIVISM

Nutrition

A USDA Economic Research Service (ERS) assessment of a food-away-from-home nutrition labeling policy indicates that even if consumers do not directly modify their choices, an ancillary benefit may occur through producer-initiated reformulation of products.[2] Section 4205 of the Affordable Care Act, signed into law on March 23, 2010, directs the FDA to establish labeling requirements for restaurants, similar retail food establishments and vending machines. The FDA issued proposed regulations on July 23, 2010 for vending machines and on April 6, 2011 for restaurants, and the final rules for each were released on November 25, 2014.[3] Recall from reading Chapter 11, "Enrichment and Fortification," that Congress has a habit of pushing back on the FDA, and we now have an opportunity to see if, under pressure from the food industry, Congress will attempt to roll back these two labeling rules.

Natural

We are back to definitions again. What does "natural" mean? The FDA has not defined natural, but will object if use of the term is misleading. The FDA has not objected to the use of "natural" if the food does not contain added color, artificial flavors, or synthetic substances.[4] In 2013, several federal judges requested the FDA's opinion as to the definition of natural and if a food product labeled "100% natural" could contain genetically modified organisms (GMOs). In Chapter 5, "Activism," we discussed some state-led initiatives around GMO labeling. Some of these initiatives are rejecting the use of "natural" on GMO products.

In January 2014, the FDA replied, declining to provide an opinion in this manner, pointing out that the proper course of action is to go through the formal regulatory process, and advised that it has been notified that the Grocery Manufacturers Association (GMA) intends to file a citizen petition asking the FDA to authorize the use of the term "natural" on foods with ingredients derived from biotechnology.[5]

Right to Know: Standard of Identity

Standard of Identity is not new, and we have touched on the topic in earlier chapters. However, the backbone of Standard of Identity is somewhat like "generally recognized as safe" (GRAS), in the sense that the FDA has condensed characteristics of foods considered to be nonlethal into a short-cut of "you don't need to know." As the fact that "we do want to know" might be the undercurrent that can take the activism around GMO labeling and surge to coalesce the food movement, we are taking this opportunity to highlight the research Alissa Hamilton did on the standard of identity for orange juice in the 2009 book *Squeezed*.[6]

"Standard of Identity" was meant to be a shortcut for consumers to be assured the item they bought was what they expected, a necessity when very little information was required on food labels. The FDA initially modeled Standards of Identity on the idea of recipes; defining the relevant characteristics for commonly known food items which were then staples that might be prepared in our own home kitchen, such as ketchup, tomato sauce, or jam. As our kitchens began to increasingly contain foods in increasingly processed forms, however, determining consumer expectations of staple foods became more complex. For orange juice, the struggle between the evocative label descriptions of manufacturers and the FDA's concerns over misleading labeling came to a head in the early 1960s. What did we expect orange juice to be? Chilled orange juice was only half fresh, the other half was reconstituted from concentrate, and it had all been previously heat-treated. Frozen concentrate was not just evaporated frozen orange juice, but instead, a blend of oranges from different harvest dates and orchards, possibly even different countries. Orange oil and orange essence could be stripped out of juice, and then added back in postheat treatment (and not necessarily to the same batch); all to achieve a taste as close as possible to fresh squeezed.[6]

The concerns which the FDA voiced that orange oil and essence were additives were assuaged by arguments that these items being added back into orange juice were no different than the oil and essence found in orange juice's natural state. The FDA concurred and allowed these as additions and also determined that these items did not need to be revealed as additions on the label.[6]

At the time, one of the benefits of a standard of identity for a food was that the actual contents did not need to be listed on the label. The 1990 Nutrition Labeling and Education Act preempted this exemption enjoyed by standardized foods, and today, even foods with a standard of identity must list the ingredients. Hence, you will find that the ingredients of milk include milk and then any added vitamins. However, if we look at a container of 100% orange juice, we will likely find the word pasteurized and the phrase "not from concentrate," but oil and essence will still not be listed as ingredients, even though according to Hamilton, they are most likely, if not assuredly, there.[6]

As Hamilton summarizes it, "when constituents of the juice are restored rather than modify a juice's flavor, the FDA does not believe they should be labeled," but the modern processes of making orange juice that she describes, requiring flavor scientists and flavor packets, has moved far beyond mere restoration. As she points out, we pay a premium now for the "not from concentrate" evocation of fresh orange juice, when in reality, the product is heat-treated, heavily handled, sits in storage for lengthy periods of time, and requires the addition of a flavor pack to be made tasty.[6]

Drinking orange juice that is less than fresh and less than flavorful without the addition of flavor packets would not harm us. Just because it is apparently safe, does that mean that the production process is irrelevant to us? She believes that it is a mistake to ignore the right to know in situations that are harmless. Even though the outcome may be harmless, consumers are nonetheless basing their purchase decisions on false information. Without information, we cannot make meaningful choices.[6] *Do we have the right to know how our food is produced?*

In particular, without information about the elements and characteristics of food and food production that we may personally value, we cannot make meaningful choices that also allow us to contribute to shaping our community and environment. Hamilton argues that right-to-know surpasses the idea of freedom from harm; right-to-know encompasses the freedom to choose.[6]

As a child growing up in Minnesota, the younger author drank only orange juice from frozen concentrate. As an adult, she was able to choose to purchase fluid orange juice, and lacking expectations as to the taste of the "real thing," found fluid orange juice delicious. Always a label reader, she has based her purchases of orange juice since then on ingredients and other available label information, and is completely dismayed to learn about "flavor packets."

In *Looking Forward From the Past*, we have raised similar comments and questions throughout defining food, about labeling, about our values, and how we make choices. Consumers are interested in knowing how their food is produced, and look to certifications on labels (government or otherwise). Consumer interest and demand resulted in the COOL labeling regulations referenced in Chapter 8, "Meat," and the GMO legislation referenced in Chapter 5, "Activism."

Looking forward, with the possible coalescence around GMO labeling, we may see continued pressure toward the consumer right to know. Or, conversely, we may see consumers today simply characterizing most food as processed anyway, and lacking expectations about the taste and characteristics of food in a more "original" state may shrug the "right to know" about food production process off as irrelevant.

Ingredients in Food Packaging

Concern over food packaging is not new, as evidenced by the highlight on Bisphenol A (BPA) in Chapter 4, "Nutrition Information Policies." However, the extent of our of lack of knowledge about food packaging or other materials that come into contact with our food through the manufacturing and distribution process may be surprising. The 1997 Food and Drug Administration Modernization Act (FDAMA) streamlined procedures used by the FDA, such as the notification process commented on above and elsewhere in this book for GRAS substances. A similar notification-instead-of-petition process was also introduced for packaging and other food-contact substances, such as the linings in cans featured by the discussions over BPA. The FDA's definition of food-contact materials includes the characterization of intended use by the manufacturer; "such use is not intended to have a technical effect in such food."[7]

Unlike GRAS, the notification process for packaging and food-contact substance is not voluntary, but like GRAS, industry provides the data.[7] There are exemptions to the notification process; mostly regarding materials considered GRAS or previously approved, but there is also a threshold exemption if the use of such a contact material results in a very low concentration of material passed to the food.[8]

Under the "intended use" definition, since the packaging is not intended to have a technical effect on the food, it is not considered to be an actual food additive. Thus, the package is exempt from food ingredient labeling requirements. ("Front of the label" information such as "BPA-free" is marketing information presented by the manufacturer.) One concern about the regulatory process for food packaging is that the FDA is assessing safety under a "high exposure-acute effect" paradigm and failing to address modern scientific knowledge regarding the potential harm of low-level, cumulative exposure, particularly to hormone-influencing chemicals.[9]

Safety aside, in the absence of labeling, consumers cannot raise questions or make purchasing choices that may be based on personal values about food.

GMOs

The promise of Golden Rice (rice that has been bioengineered to include a gene giving it both a yellow color and beta-carotene, providing the rice with a nutrition boost of vitamin A) is finally almost upon us. As one activist points out in an NPR story, "the poor have always been at the center of each and every assertion about the importance of GMOs to mankind," and the rice may help the poor, but in turn, it would no doubt help the image of biotechnology.[10] As of 2014, Golden Rice is apparently in trials in Bangladesh and the Philippines, but no news was available at the time of writing this chapter on the results. *Will Golden Rice be more or less expensive and effective than traditional food assistance programs?*

In November 2014, the USDA approved a genetically engineered potato for commercial planning that has been altered to resist bruising when the potato is fried, to produce less acrylamide, a suspected carcinogen.[11] Perhaps this is "GMO 2.0," in which biotechnology is aimed at providing consumer benefits rather than benefits to commodity crop farmers. If the consumer perceives benefits, they may be more willing to accept any possible perceived risk. The benefits of this potato could be touted not only as means to justify our continued enjoyment of French fries, but also as incurring less food waste due to less bruising loss. This sounds like a win-win for the food service industry, since if other states follow Massachusetts lead on food waste (see discussion below in this chapter), the food service industry may be facing escalating economic pressure to reduce food waste. Regardless of any particular "right to know" state laws, we would imagine that because the benefits of this potato are relatively consumer oriented, the genetic engineering would likely be advertised in some manner.

Many food activists concerned with GMOs are also concerned with the escalating level of seed and plant patenting that may prevent farmers from harvesting seeds for use the next year. One group has decided to distribute new varieties with an "open-source seed pledge" stating that any new crop varieties developed must remain free for everyone to use under the idea that the genetic resources of seeds are a common resource that anyone can use.[12] *Do you think seeds are a social good?*

GRAS

If GMOs fail to coalesce the food movement into a unified rally, GRAS should be the next candidate; with similar concerns over the allocation of risk becoming an unknown burden shouldered by the consumer. In Chapter 13, "Food Protection," we discuss how the FDA began accepting GRAS notices from manufacturers in 1998 under a proposed rule as if the rule was indeed final. In February 2013, the Center for Food Safety sued the FDA for violation of the Administrative Procedure Act (APA), which governs the process of agency rulemaking. The outcome of the settlement agreement with the Center for Food Safety is that the FDA must finalize the GRAS rule by August 2016, and it is expected that there would be the opportunity for public comment.[13] Perhaps unsurprisingly, a food industry association is preparing an initiative to further modernize the process.[14]

Nanotechnology

The foods we consume already naturally contain components at the nanosize scale in the form of globular proteins, casein micelles, and ribbon-shaped polymer polysaccharides.

We also have already long-adopted food processing methods that produce nanoparticles; namely the homogenization of milk.[15] As we discussed in Chapter 7, "Milk," homogenization of milk results in a richer mouth feel due to the increased surface area of the smaller but more numerous fat globules.

The increased surface area of materials that are deliberately engineered into nanosized versions can change how the substance is capable of interacting with other substances, and at nano-sizes, substance may be able to cross biological membranes. A number of novel applications exist, including nutrient delivery and food safety, but safety must also be assessed. In 2010, the Government Accountability Office (GAO) issued a report calling for the FDA to strengthen its oversight of the GRAS notification process, and further commented that the FDA encouraged, but did not require, companies to consult with the FDA as to whether engineered nanomaterials in food would be GRAS. However, in Canada and the European Union at that time, food ingredients that incorporate engineered nanomaterials were required to be submitted to regulators prior to marketing.[16]

In June 2014, the FDA issued final guidance for industry, which is nonbinding. Generally, the FDA is not making any judgment on the relative merit as to the overall safety of nanotechnology, is not making any definitions specific to nanotechnology, and affirms an interest in products "deliberately manipulated by the application of nanotechnology," rather than materials naturally occurring in that size range. (In other words, they are not concerned about homogenized milk.)

The FDA commented that they are not aware of any intentionally engineered nanoscale food in which there is enough data available for the foundation of a GRAS determination. A formal premarket review and approval by the FDA would likely be warranted, and this would be such for any significant manufacturing process change, not limited to the application of nanotechnology.[17] These comments, however, are nonbinding recommendations.

Micronutrient delivery is among potential uses of nanotechnology in food, and, like the promise of Golden Rice noted above, it could be a useful tool for refining public health nutrition programs.

WASTE

Waste is nothing new. As consumers, we are well aware of the marketing and the incentives splashed around us encouraging the purchase of new things. This often entails the disposal of old things; sometimes as hand-me-downs or gifts to friends and family, sometimes as donations to nonprofit organizations, sometimes sold, sometimes stored for future use, and sometimes sent to the landfill.

We do this with our food as well, when we have purchased in excess, bought something we realize later that we cannot use, acquired something new that we did not like, or have grown too much zuchinni in our gardens. However, the perishable nature of food presents some difficulties. Once we have "used" food by cooking or preparing it, our options for sharing portions left uneaten or that have spoiled in storage are limited. For the most part, unless we have the ability to compost, or animals that can eat table scraps, this food goes to the landfill. But until spoiled or inedible, it remains food.

Do you think there is any distinction between "food waste" and "food that is wasted?"

The USDA's ERS tracks food loss in the United States. The idea of food loss starts with characterizing "food" as what is available for human consumption. For example, we do not normally consume the peel of a banana. When we buy a pound of bananas, we are buying somewhat less than a pound of "food," due to the contributing weight of the nonedible peel. After all the nonedible parts of food are removed, ERS defines food loss as the amount of food available for human consumption that is then not consumed for any reason. Food loss includes factors like cooking shrinkage, or loss due to spoiling or pests, in storage or in the field due to weather. Food waste is a subcategory of food loss, which covers edible food that is discarded, such as plate waste or produce items rejected by a supermarket.[18]

Some amount of food loss is unavoidable. But does the same hold true for food waste? We are taking this opportunity to highlight the research on the waste presented in Jonathan Bloom's 2010 book *American Wasteland*. Jonathan Bloom argues that food waste is not unavoidable, and that food is wasted when "an edible item goes unconsumed as a result of human action or inaction." We are culpable for the decisions made from the farm to the garbage or compost bin (see Figure 14.1).[19]

Food recovery generally covers the redistribution for consumption of all types of edible food that would otherwise go unconsumed, such as field gleaning, perishable

FIGURE 14.1 This war-time poster reminding us not to waste food may be just as timely today. (With permission from Special Collections, National Agricultural Library.)

produce from wholesale and retail sources, the rescue of excess foods prepared by restaurants and institutions (except plate waste) and nonperishable foods. Composting food waste is better than food going to the landfill, but it is still a waste.

Food as a Social Good: Water and Energy

The USDA's ERS estimates that 31% or 133 billion pounds of food available for consumption in the United States in 2010 went uneaten. The value of the uneaten food, at retail prices, is over 160 billion dollars.[18] Although this number reflects the amount of money a consumer would spend, it may not fully reflect the external costs of the food production.

As Bloom points out in *American Wasteland*, food is a social good. Food may not always be included in the classic examples of social good resources, such as clean air and clean water. In the introduction of this book, we briefly discussed how the Supreme Court used the Commerce Clause to find authority for the federal regulation of the production of food. The Supreme Court's decision was grounded in the idea that even small, private consumption decisions by individuals could aggregate into an excessive cost to the public as a whole. Given the impact of small private decisions, and considering Professor Lang's argument that food is embedded water, we must clearly include food in the category of "social goods." Time, energy, and resources of fossil fuel and water all contribute to our food production, and the waste of food squanders those social goods.

As discussed in Chapter 8, "Meat": agriculture in general is the largest user of water resources in the United States, and raising animals to eat for meat is resource-intensive. By satisfying our hunger for meat without bounds, we are risking becoming thirsty.

Although calculations of water use vary, in general, water is used to grow the feed, to provide drinking water to the animal, and to wash and maintain equipment and the facility. By one estimate, a hamburger is worth over 634 gallons of water (10,000 eight ounce glasses of water) and that does not include the bun.[20]

Can we eat our meat and have it too? Food products using pea-protein to simulate the texture of meat are one example of innovative solutions that may address our taste for meat while hopefully using up less of our resources.

As individuals, we also carry a cost in food that we have purchased and are storing for future use. For consumers, refrigerators and freezers can be cost-saving if the use of these appliances allow us to purchase in bulk (and make fewer trips using less gas to go shopping), and of course, if we consume the food. Nonetheless, we also incur energy cost. The US Energy Information Administration (EIA) reports the average residential cost of electricity in the United States. As of October 7, 2014, the average cost in 2014 was 12.48 cents per kilowatt-hour, with a rise to 12.69 cents forecasted for 2015. Collectively, we incur refrigeration costs for fresh meat, milk, and produce through the entire chain from farm to consumer. One research study estimates the energy embedded in wasted food to be about 2% of annual energy consumption in the United States.[21]

American Wasteland provides a detailed examination of food waste. Here are just a few examples of how food waste can happen along the entire food system chain: food is left in the field, losses are incurred between harvest and distribution, produce is culled (as only the best can survive long-range transportation), stores are overstocked to compete for customers and excess discarded, customers purchase items but fail to use them prior to spoilage. As Bloom comments, our excess and abundance itself leads us to devalue food, which in turn fosters waste.[19]

In general, we have seen our food production consolidate and specialize, and this too contributes to wastage. Egg producers that hatch their own chicks cull the male chicks instead of raising them for food. Produce farms are also specialized, with a tendency for farms to grow vegetables either for the fresh consumption market or for the canned market. Thus, in the case of growers selling to the fresh market, when vegetables do not meet the standards for consumer purchase, the vegetables may simply be discarded rather than diverted to the processed market. Furthermore, the efficiency of our packaging and distribution system may require vegetables of a just-so length and size. With awareness and creative attention to the issue, these less than ideal but perfectly edible foods can be recovered instead of dumped.[19]

As noted in the introduction of this book, our transportation system with refrigerated trucks and train cars allows for national distribution of fresh produce. Although this provides a diet that is perceived to be healthier (particularly providing greater access to fresh produce during the winter season), this system also incurs loss. If the end goal of distribution is produce that remains fresh and appealing to consumers, produce may be culled and discarded long prior to spoilage in order to incorporate travel time into the predicted life-span of the product. Bloom also points out that there are less "garbage" markets for produce; supermarkets want the best, freshest looking produce in order to sell us on patronizing their store. With the competition and increasing scale of retail outlets, we have less access to stores that specialize in produce; such stores might offer lower prices and a greater range of quality by leaving some blemishes for the consumer to deal with.[19]

Chapter 7, "Milk," discusses the social context behind how milk became the iconic "nature's perfect food"; but with our recent greater emphasis on locavore diets, connecting with farmers, and eating "mostly plants," perhaps vegetables may have jostled milk off of the pedestal of "nature's perfect food." Viewed in this context, and realizing that we are immersed in media showcasing the beauty of nature's bounty, it is perhaps not surprising that consumers seek perfect fruits and vegetables. Consumer education is critical, and media participation in seeking "ugly" vegetable photos could be helpful.

Another means in which supermarkets are competing for shoppers is by providing ready-to-eat salad bars and buffet bars for meals and entrees. But Bloom points out that because these items are self-serve (unlike the prepared foods behind the deli counter), excess food cannot be recovered; items left on the steam table at the end of the day are considered the equivalent of plate waste.[19] *Can we regulate self-service in grocery environments, or even at buffet restaurants? Can we give up the convenience of serving ourselves?*

Food Waste and Dietary Guidance

Recognizing that our dietary guidance (and the further dissemination of such guidance through education and public discussion) influence our food purchases, it seems logical that our dietary guidance could also influence food waste. References to fruits and vegetables in the 2010 Dietary Guidelines stress that fresh, frozen, and canned are equal sources of nutrition, and that to achieve enough fruit intake we should use canned, frozen, and dried fruits, as well as fresh fruits.

A preference for fresh produce consumption is only insinuated twice; namely with the following statements: "Enjoy a wide variety of fruits, and maximize taste and freshness by adapting your choices to what is in season" and "Consume more fresh foods and fewer processed foods that are high in sodium." We mentioned in Chapter 4, "Nutrition Information Policies," the FDA's attempt to decrease the use of sodium in canned goods.

Health-seeking consumers may be attempting to limit consumption of canned goods due to concerns over sodium (or other issues, such as BPA lining). Culturally, many people may equate healthful foods with fresh produce. Thus, even though fresh and canned fruits and vegetables are treated equally in the 2010 Dietary Guidelines when it comes to nutrition, consumers may still perceive fresh produce as the healthier choice over canned, at least until sodium levels are reduced across the board in canned foods. When it comes to food waste, however, canned foods are a much better choice than fresh produce, as there is much less risk of the contents spoiling prior to consumption, so perhaps the value-added argument of sustainability may increase the esteem of canned vegetables. (The equation of healthy with fresh produce may also be contributing to the perception that fresh produce should look good.) In any case, we are seeking fresh produce to be healthy, yet fresh produce seems likely often to be the largest culprit in food waste.

Food waste is not mentioned in the 2010 Dietary Guidelines. *Should food waste be mentioned in the 2015 Dietary Guideline, and if so, how? Could the use of seasonality also be used to help limit food waste?* By searching the public comments for the next edition, we can review the number and substance of comments regarding food waste.

Likewise, food waste is not mentioned in Healthy People 2020. There is one environmental objective, EH-12, to increase recycling of municipal solid waste (MSW).[22] However, after accounting for recovery of MSW through recycling and composting, food is the single largest component of MSW hitting the landfill, at 21%.[23] In addition to the sheer waste of food entering the landfill instead of feeding people, food waste rotting in landfills becomes a significant source of methane emissions (more than 20%) in the United States.[24] *Should food waste be mentioned in the next Healthy People?*

Food recovery is not mentioned in the 2014 Farm Bill. Food waste is mentioned once as an excluded term in the Biomass Crop Assistance Program provision. Eligible material for the Biomass Crop Assistance Program is defined as renewable biomass harvested directly from the land, but not including "food waste and yard waste."[25] The USDA touts the 2014 Farm Bill as "enabl(ing) USDA to further expand markets for agricultural products at home and abroad, strengthen conservation efforts, create new opportunities for local and regional food systems, and grow the bio-based economy. It will provide a dependable safety net for America's farmers, ranchers, and growers. It will maintain important agricultural research, and ensure access to safe and nutritious food for all Americans."[26] *In light of ensuring access to safe and nutritious food, should food recovery be a topic of the next Farm Bill? How does the bio-based economy balance energy and hunger?*

Given the interest the EPA has in landfills, it is not surprising to see this agency promote the reduction of food waste and publish a Food Recovery hierarchy.

FIGURE 14.2 EPA food recovery hierarchy shows preventing waste as a first resort and feeding hungry people and animals prior to industrial usage (such as energy). (From Environmental Protection Agency, www.epa.gov/foodrecoverychallenge.)

Although we are just briefly touching on the role of the EPA, given an increased emphasis on sustainability of our resources and food waste, the EPA may become more publicly prominent in food policy. In addition to discussing the environmental, economic, and social benefits of diverting food waste from landfills, the EPA prioritizes actions that can be taken to minimize food waste and divert from landfills. At the top of the inverted pyramid shown in Figure 14.2 is the most important: preventing food waste before it is created.[24]

The EPA runs a Food Recovery Challenge as part of the Sustainable Materials Management Program,[27] and in June 2013, the USDA and EPA jointly announced the US Food Waste Challenge. The goal of the US Food Waste Challenge is to raise awareness about food waste and provide resources that help the organizations and individuals by doing the following:

- Reduce food waste by improving product development, storage, shopping/ordering, marketing, labeling, and cooking methods.

- Recover food waste by connecting potential food donors to hunger relief organizations like food banks and pantries.
- Recycle food waste to feed animals or to create compost, bioenergy and natural fertilizers.[28]

The US Food Waste Challenge website provides a great resource list of organizations working to recover food.[29]

The USDA is also addressing waste at the production level through specialty crop grants. As an example, funds have been allocated to the Minnesota Department of Agriculture to test the market for cosmetically imperfect fruits and vegetables.[30] And, plate waste is of particular issue in school food programs. One of the pushbacks at the efforts at improving nutrition is expressed as concern over school plate waste caused by students rejecting the food.

In fact, the School Nutrition Association, which supported the updated nutritional standards, is calling for a rollback of some of the standards.[31] (Keep in mind that another related issue concerns whether students are buying less school meals because of the new standards, which also is a financial issue for the school meal programs.) As of the writing of this chapter in the fall of 2014, there was plenty of media attention to this issue as students returned to school; but study data on plate waste has not been conducted on a widespread scale. And of course, plate waste itself was certainly not a new issue caused by the updated lunch standards; schools had plate waste and kitchen waste prior to the new standard. By the time we read this chapter, we may be able to discern if the magnitude of plate waste had actually changed due to the updated nutrition standards, and also if the USDA has compromised the nutritional standards and rollout schedule to accommodate the concerns of the School Nutrition Association.

As part of the US Food Waste Challenge, the USDA offers the Reduce Food Loss and Waste in The School Meals Program. Program activities include gathering new data on plate waste, encouraging "offer versus serve" and family-style service, as allowing children to select type and amount of foods will help in reducing plate waste, investigating the use of behavioral economic approaches, and providing training and support for food service personnel for reducing food waste in the production of meals and managing left over foods and donation efforts.[32]

Although the federal government is clearly making programmatic moves to address food waste, arguably, a greater overall federal policy is needed. For example, in the mid-1990s, the USDA had a dedicated food recovery and gleaning coordinator. In a 2012 report, the National Resources Defense Council (NRDC) recommends that the United States should make the reduction of food loss in the US a national priority, starting with the establishment of clear and specific food waste reduction targets.[33] It is reported that the fruits and vegetables incurring the greatest amount of loss are fresh apples, grapes, peaches and strawberries, and fresh and canned tomatoes and fresh and frozen potatoes and the authors reporting this data comment that focusing efforts on these foods may be beneficial to reduce food loss.[34] This research provides a possible starting point from which we could establish clear and specific food waste reduction. With clear data and objectives, it seems feasible to include food waste reduction objectives in Healthy People. Perhaps it is time again for a dedicated food recovery and gleaning coordinator. Perhaps we could incentivize our food system, or at least our produce production, to aim at seasonality, to minimize long-distance transit, and to encourage field-to-harvest or processing operations to maximize the use of all produce.

Food Waste and Labeling

And here we are again back to labeling. Another recommendation by the NRDC concerns addressing customer confusion over the variety of dates on food packages.[33] Such dates are included at the manufacturer's discretion, usually to indicate quality. Dates on packaging are not regulated by the FDA, and the law does not preclude the sale of any food past the expiration date, except for infant formula. Thus, although the dates may appear to indicate food safety to the consumer, they are not part of the FDA's food safety regulatory scheme. Even though a consumer may feel that a "use by" or similar date is at least a conservative safety estimate, the date does not account for actual food safety controls, such as field and processing conditions and temperature control throughout the product's lifecycle.[35] The FDA will take action if a food product is dangerous to consumers, independently of any date that may be on the label.[36]

Even though dates on food packages are not regulated at the federal level, states, and even some local governments, may have specific food packaging regulations concerning dates. On the industry side, manufacturers choose whether or not to provide information, use differing terms and have differing means of determining the relevant quality-based date. The NRDC argues that these variations in state laws and application by manufacturers mean that consumers cannot rely on package dates to consistently have the same meaning.[37] This is true, but this is most relevant for consumers when comparing choices from differing manufacturers. The variability between states seems less relevant to consumers given that most are likely to be purchasing a majority of their foods within one state. Nonetheless, as noted elsewhere in this book, smoothing out state variations in food packaging laws is naturally helpful to industry.

Dates on food packaging arose due to the increasing gap of knowledge between consumer and the history behind food source.[35] Also, as noted in Chapter 9, "Grains," the development of packaging allowed manufacturers to convey more information about quality to consumers. Open dating, meaning date information accessible to consumers, provides a perceived means of assessing the value of intended purchases in light of quality.

Open date labeling, using terms such as sell by, best if used by, or best before, freeze by, use by, baked on, and packed, provides information to retailers and consumers about optimum quality, shelf-life of a product, and enable easy stock rotation. Consumers may apply these terms and dates indiscriminately to assess food safety rather than food quality, and take such a date to mean that a product should not be eaten after such date. By discarding food when it is still edible (albeit perhaps of lessened flavor or texture), consumers are contributing to food waste. Products with a sell-by-or-freeze date will typically have a third of the product's shelf life remaining after that date.[35]

In the 2013 report "The Dating Game," the NRDC recommends making "sell-by" dates or any other retail stock control dates closed (invisible to the consumer), but in turn, creating an open uniform dating system that clearly distinguishes between safety dates and quality dates for consumers. The food industry can of course take their own action on dating conventions, (recall the discussion of corporate self-regulation in Chapter 12, "Nanny State"), and consumers can educate themselves and be educated. Moreover, this report supports the idea of date regulation by the FDA and USDA under the existing authority these agencies have regarding misleading package labeling, which would allow action in this area without new legislative direction by Congress.[35]

Food Recovery Policies

As noted in Chapter 13, "Food Protection," foodborne illness can be costly and fatal, and thus, the specter of liability may be a barrier for food recovery. The Bill Emerson Good Samaritan Food Donation Act, signed into law on October 1, 1996, has the express purpose of "encouraging the donation of food and grocery products to nonprofit organizations for distribution to needy individuals"[38] and relieves donors from liability lawsuits (short of gross negligence or misconduct). Prior to the passage of the Bill Emerson Act, businesses relied on state statutes limiting liability of food donors, but naturally these varied from state to state. The Act provides a floor of consistent legal protection, which helps in eliminating barriers to donations from businesses selling or operating in more than one state.

Other barriers still exist to full adoption of food recovery efforts by food producers and businesses. Depending on crop conditions and the market at the time of harvest, harvesting may not be economically feasible for food producers. Although unable to find updated general information on crop insurance and gleaning, it seems that farmers may be able to utilize crop insurance while still having the harvest gleaned.[39] Further examination of crop insurance is beyond the scope of this chapter, but an analysis as to potential impact of crop insurance on overplanting may be useful. As Bloom points out, plowing the crop rather than harvesting it is better than if the harvested crop just ends up in the landfill. In other words, if overplanting leads to a failure to harvest and the crops are plowed back under, we have failed in a moral obligation of providing access to that food for the hungry among us, but at least we have made the least egregious squandering of the embedded social goods of water and energy.

However, as NRDC recommends, regulatory measures could incentivize a complete harvest and tie it to food assistance efforts. In California, a tax credit is available to farmers who donate crops, and similar credits exist in Arizona, Colorado, North Carolina, and Oregon. The NRDC also calls for updating federal tax incentives, as a temporary measure that gave small companies enhanced tax deductions has expired, while commenting that without tax changes, the food will be lost, as it would remain cheaper for the company to landfill the food.[33]

But incentives to donate food do not necessarily save the food from the waste; in some cases, it may just push the waste further down the food chain. For example, as one author observed while volunteering at a food pantry, even though there were often plenty of French bread style loaves for each client to take two or more loaves, most clients did not even take one loaf. The food pantry clients actively rejected the bread. That bread was just not going to get eaten, and would end up in the landfill, tax break regardless. Studies have reported that in industrialized countries, retail practices increase the likelihood that fresh perishable items, such as bakery products, will reach their sell by dates before being sold and then likely to be wasted.[35] And because of the perishability of these products, even donation may not be the answer to keep them out of the landfill.

Are tax incentives for businesses the only means to recover food for the hungry? Could other incentives be more effective at reducing processed food waste?

We regret not being able to delve significantly into the idea of incentives to prevent food waste before it happens, as there must already be some fascinating work done on this topic. However, given that we are *Looking Forward From the Past*, we would like to highlight something from Chapter 11, "Enrichment and Fortification." During World War II, the "Order No. 1" that made the enrichment of commercially baked bread mandatory also regulated the baking, packaging, and distribution of bread in order to ration supplies of all kinds. Order No. 1 made it illegal for stores to return excess bread to the bakery.[40] Then, as now, stores overstock on some things in order to draw customers. Stores had to order bread more carefully, could not overstock their shelves, and the bakeries were more careful in production.

Instead of focusing on increasing incentives to donate food, what about creating incentives to better manage production to prevent food waste before it happens? Could we tailor food-specific tax or other incentives?

Food Waste to Animal Feed

Companies that produce commodity food staples are already well versed with finding markets for the by-products, such as selling for incorporation into animal feed. According to a Quaker Oats plant manager, in 2012, just 1% of the Quaker Oats plant's product went to landfill.[41] The by-products of commodity food production could consist of both biomass inedible for humans (soybean pods) as well as potentially edible food waste.

However, human food waste is indeed used for animal feed. The Food Waste Reduction Alliance (consisting of the GMA, the Food Marketing Institution, and the National Restaurant Association) have been tackling food waste since 2011.[42] (Perhaps not coincidentally, a year after *American Wasteland* was published.) The GMA has reported that 70% of manufacturers' food waste is going to animal feed and only 5% to landfills.[43]

Animal feed is regulated by the FDA, and for good reason. See Chapter 8, "Meat," for a discussion on bovine spongiform encephalopathy (BSE) and antibiotic residue in chicken feed due to the practice of adding feather to chicken meal. As just one other example of disease transmission through animal feed, pigs being fed raw garbage are susceptible to diseases, such as the vesicular exanthema of swine (VES) virus. An outbreak in California that started in the 1930s was contained until 1952, when a passenger train between San Francisco and Chicago served California pork and discarded the raw pork trimmings into the garbage in Cheyenne, Wyoming; that garbage was then fed to swine auctioned off and sent throughout the US. Within 14 months, VES had broken out in 41 states, leading to federal involvement, including enforcement of laws requiring garbage to be cooked before fed to swine.[44]

Under the Food Safety Modernization Act (FSMA) (see Chapter 13, "Food Protection"), the FDA was given a legislative mandate to establish preventative controls across the food supply, and that includes animal feed. In October 2013, the FDA proposed a rule regarding preventative controls for animal food that included new sanitation and record-keeping requirements, including food safety plans for all of the by-products going into feed. In an interesting corollary to the food-waste argument pushback against the updated school nutrition standards, the GMA has argued that the proposed safety rule is too costly, and that the cost and burden will result in more food waste in landfills rather than being diverted to animal feed.[45]

Based on public comments, the FDA is proposing some revisions more applicable to the animal food industry, including agreeing with industry comments that human food by-products or waste are subject to human preventive control regulations up to the point of diversion, and thus, do not need to be processed under the same rules that animal food manufacturers do. The final rule is expected in August 2015.[46] Compromises may be made that might affect the cost and burden of the rule and the according impact on the amount of food waste diverted out of landfill.

In other words, the idea of an overall arching federal priority for food waste is a complicated endeavor and may impinge on other government objectives, such as the developing biomass programs.

Local and Regional Solutions to Food Waste

Tax incentives are a carrot for businesses, but as discussed in Chapter 12, "Nanny State," state and local governments can create sticks as well for food policy issues, including food waste prevention programs. In the fall of 2014, the state of Massachusetts introduced a "stick" by banning high levels of commercial waste disposal. Any entity that disposes of at least one ton of organic material per week must donate or repurpose any useable food. Any remaining food waste must be shipped to an anaerobic digestion facility or sent to composting and animal-feed operations. There is also the "carrot" that the companies should be able to save on disposal costs.[47]

The synergy of the US Food Waste Challenge and this new legislation has resulted in a "Dairy Power Project," sponsored by the Innovation Center for US Dairy, located nearby. Twelve local companies are sending nearly 100% of their food waste (20,0000 tons) to dairy farm anaerobic digesters to mix with the manure of over 350 dairy cows. The result is $75,000 worth of crop fertilizer, enough energy to power over 300 homes, and amelioration of the methane gas that would have been produced in the landfill.[48]

According to a press release, the Massachusetts commercial food waste ban is intended to stimulate increased food donation, as well as the recycling and conversion of food waste into valuable products, including renewable energy and compost.[49] However, although the donation of food appears as a priority in the wording of the law, the suggested methods of implementation do not make food donation necessary. The Mass.gov website provides information about compliance assistance from Recycling Works, including a fact sheet that states "Delivering food waste to an off-site composting or anaerobic digestion facility through a hauler is a common strategy, but you can also reduce waste through purchasing controls and production changes, donating surplus food, or installing on-site technologies."[50] The ease of having all food waste hauled away might be an incentive for businesses to skip the time and cost of sorting out edible food. Donation of food for those who are hungry may end up being subject to the values of corporate self-regulation.

A new grocery store, the Daily Table, has been planning to open in Dorchester, Massachusetts with the aim of selling food rescued from the food waste stream at a deep discount. The Daily Table's description of food waste priorities matches that of the EPA hierarchy, with the first priority to prevent waste, the second to feed people, and the third to feed animals that we will be eating.[51] The use of the retail channel to move food waste is also evocative of how the Supplemental Nutrition Assistance Program (SNAP) as food assistance uses the regular retail channel, and in fact, a store like the Daily Table could provide a great benefit to SNAP users through maximizing benefit purchasing power and by increasing access to still healthy produce.

However, this novel solution to move edible food waste through the retail rather than charity channel might be derailed if perishable but edible local food waste is too greatly diverted to compost or energy solutions.

Food Recovery: Economy Solutions

Just as the Daily Table is finding a niche for food waste recovery within a retail channel, rather than for charity, other food recovery programs are seeking to support economic stimulation through food recovery, rather than relying solely on volunteer effort. Volunteers of course provide valuable food recovery support, but one concern is whether volunteer-based efforts can support truly sustainable food recovery and food access. Food recovery may provide opportunities for some work and income, and as Bloom notes, the poor need money.

There are various organizations nationally and globally that are focusing on food recovery as work; we will just use two as examples. Food Shift is using food recovery services to provide training and jobs for people who need them while reducing waste.[52] The Farm to Family program in California recovers more than 120 million pounds of produce per year from farms and packers for distribution to food banks, but does so through concurrent picking, whereby workers harvest unmarketable produce alongside the marketable grades. The associated excess costs to harvest are covered by the program, and the grower still can claim the tax deduction.[33]

Food Waste: People, Energy, and Pigs

Although the idea of turning food waste into energy is an exciting application of technology, Bloom points out that although food waste to energy is better than landfill, creating large-scale embedded food-waste-to-energy solutions are, in the long term, at odds with the overall priority of reduction of food waste, and at odds with an ethical priority to first address hunger.

It may be important to make a distinction between food waste and the umbrella term "organic waste." Organic waste that is diverted to fuel production is an excellent use of the waste—unless food waste is included. On September 28, 2014, California Governor Jerry Brown signed AB 1594 that would prohibit local governments from receiving recycling credits for landfilling green material, such as yard clipping, starting from January 2020. The recycling diversion credit for landfilling green waste was originally granted to encourage collection of green waste prior to having an end use market, but is now competing with the more efficient uses of this waste, such as composting and anaerobic digestion. He also signed AB 1826, which requires the state's commercial sector: supermarkets, large venues, and food processors, to separate food scraps and yard trimmings and arrange for organic recycling. Compliance will be phased in, starting in 2016 with businesses generating eight cubic yards or more a week of organic waste.[53]

Under AB 1826, "organic waste" means food waste, green waste, landscaping waste, nonhazardous wood waste, and food-soiled paper waste that is mixed in with food waste. "Food waste" is not defined in the bill, and a keyword search for the phrase food waste in the California Codes does not reveal any definition of "food waste" at this time.[54] We can only hope that "food waste" in this context is limited to "plate waste," and that other incentives will encourage these companies to divert any edible food waste away from this stream.

Clean World, a company in Sacramento, California (the state's capitol) has the largest commercial high-solids food waste anaerobic digester in the United States, which according to the company's website, is "especially engineered with food waste in mind, including agricultural residues, commercial food processing waste, and restaurant and supermarket food scraps."[55] Although the digester produces renewable energy, we can only hope that the food waste being sent truly does consist of "residues" and "scraps." *Is it possible that increased anaerobic digester solutions could contribute to continued subsidization of over-production of food?*

Existing sewage treatment facilities can also add on food scraps. In Oakland, CA, the Eat Bay Municipal Utility District has been recycling food scraps from restaurants and supermarkets into renewable energy for the last seven years by using the existing anaerobic digesters. The energy generated can power the wastewater plant, with some to spare to sell back to the grid. This localized "add-on" solution seems like an efficient way to deal with local plate waste while not diverting actual (edible) food waste from appropriate channels.[56]

This "neighborhood" solution to handle food service sector plate waste bring us to another potential neighborhood solution that (per the EPA hierarchy) is actually a higher priority usage than energy: feeding animals that feed us. Food waste did not always go into landfills. One reason why food waste started ending up in the landfill in such prodigious amounts is simply because we stopped feeding it to pigs. In 1914, hotel and resort waste in New Jersey was not going to landfill. It was feeding pigs. In the Secaucus area, there were more than 25,000 pigs that were fed entirely on hotel waste from the nearby metropolitan areas.[57]

However, as noted above, we learned that feeding uncooked garbage to pigs was not a good idea. The 1980 Swine Health Protection Act requires that any garbage which contains meat waste be heated (sterilized) before being fed to pigs. States may impose additional requirements such a licensing or restrictions completely banning the feeding of food waste to pigs. Also because the nutrients of a food waste stream will be inherently inconsistent due to the nature of the supply, garbage as a main dish is not efficient in industrial raising operations (see Chapter 8, "Meat").[58]

Nonetheless, feeding waste to pigs can be done in a manner both safe and appropriate for pigs. Even in states that ban feeding garbage to commercially raised pigs, there may be exemptions allowing household pigs to be fed household waste. Like the use of local restaurant scraps to feed local energy production, decentralizing some meat production in the form of local pigs may provide another source of avoiding the cost and energy of food waste transit, and supporting local food production. This may not be a solution for a very urban, dense, large city, but it is not inherently limited to rural areas.

In sum, if we over-emphasize keeping food waste out of the landfill in order to feed our need for energy (rather than ourselves or our pigs), we may lose sight of the real objective of preventing food waste to begin with; and this is relevant to our next discussion—that is of water.

WATER

Through the examination of food waste, we note the educational messages about the loss of water in connection with food waste. *As a society, are we backing slowly into water policy through addressing food waste?*

Water and Dietary Guidance

The IOM has derived DRIs for water that provides guidelines for Adequate Intake. There is no upper limit to water intake as normally functioning kidneys can handle more than 24 ounces an hour. For infants, the adequate intake (AI) is less than a liter, for children less than 2 L, for adolescents it ranges from 2.1 to 3.3L, for adult women who are not pregnant or lactating it is 2.7 L (those conditions bump up the need for water to 3.0 and 3.8 L, respectively) and for adult men, it is 3.7 L. One liter is almost 34 ounces, and a quart of water is 32 ounces. With that rough conversion, adult men would require just under a gallon of water a day. Luckily, this needed water intake can also be sourced from other liquids and high moisture foods, such as watermelon and meat; the moisture in food accounts for about 20% of total water intake. The IOM also comments that thirst and the consumption of beverages at meals are adequate to maintain hydration. In other words, if we drink beverages at meals and when we are thirsty, we will have achieved AI.

The 2010 Dietary Guidelines mention water as an example replacement for sweetened beverages, and makes the note that the AI was based on median total water intake estimated from US dietary surveys and should not be considered as a specific requirement level. *Do you think the Adequate Intake of water should be reflected in our dietary guidance?*

If any specific water requirement is included in our official dietary guidance, how will that impact, inform, or prioritize our food and water policy?

Recall the discussion in Chapter 6, "Federal Food Assistance," about how language used to describe the Thrifty Plan was revised due to concerns that it was evoking a guarantee that could be considered a "right."

Speaking of nutrients, water can be a source of calcium, as noted in Chapter 7, "Milk." The calcium content of water that we drink in the United Stated varies widely, considering that some people do still drink well water that is not otherwise treated. And even in the case of municipally treated water, the median calcium content could range from 20 to 105 mg per liter.[59] Assessments of dietary intake of calcium will be underestimated if calcium provided through drinking and cooking water is not included.

Depending on our water source, our water may also contain higher levels of fluoride due to fortification. As in the case of calcium noted above, our drinking water prior to municipal treatment already contains some variable amounts of fluoride. The practice of adding more fluoride came about as a recommendation by the US Public Health Service to combat tooth decay, and is considered by the CDC to be one of the 10 great public health achievements of the twentieth century.[60] The Department of Health and Human Services provides recommendations as to the level of fluoridation, most recently in January 2011 of 0.7 mg per liter of water.[61]

Although federally recommended, fortification of water with fluoride is regulated at the community level, and from time to time, communities will review fluoridation programs, and may decide to discontinue fluoridation. Unlike the examples discussed in Chapter 11, "Enrichment and Fortification," the costs of which are paid by the food industry and passed along to the consumer, the cost of fluoridation is borne directly by consumers through local taxes. And, given the wider availability of fluoride exposure through dental treatments, toothpaste, and even some bottled water, as well as recognizing that there is already some fluoride inherent in the water, some communities are stopping fluoridation because it seems like an unnecessary expense.

The history and ongoing controversy of fluoridation deserves more than this brief mention. For our purposes of examining water, however, it serves to highlight that many of us are already drinking fortified water; water that may be a rich source of desirable minerals, such as calcium; and water from which we may be filtering out other less desirable minerals, such as lead. Our water will affect our dietary intake nutrient profile and must be considered as part of "consumption epidemiology." Fluoridation also serves to highlight the issue of local community water control issues, which we discuss in the next section.

Water Rights versus Right to Water

Originally, this concluding chapter was meant to focus on water and the associated ethical and political complexities around how we navigate and prioritize the use of this public good when it comes to food consumption and public health, including the tension local communities are facing over the transformation of water as a public good to private profit. Although we unfortunately do not have the ability to fully address this question in this chapter, we must take this opportunity to ask: *Do we have a right to drinking water?*

Detroit's struggles are evoking this question about a human right to water. We know Detroit has problems, problems so dire that in 2013, the Michigan governor suspended democracy by appointing an emergency manager to take control of the city. Historically, enforcement of water bills was lax. A sudden and abrupt change of policies under the pressure of bankruptcy resulted in sudden and abrupt residential water shut offs for people who owed at least $150 or were two months behind. As reported by the National Geographic, in July 2014, the average past due residential bill was $1703, an amount literally dwarfed by the average commercial past due bill of $13,014, the average industrial past due bill of $67,487, and the average municipal past due bill of $276,961.[62]

However, the total number of residential past due bills added up to almost 50% of the total number past due accounts. We included this detail because if we were looking at this from the viewpoint of lowering risk in public health, we would see the solution that Detroit took of going after residential bills as being the same approach of lowering a health risk of the overall population by changing the behavior of the largest number of people, even if those people are individually at the smallest risk. One can only hope that this was not the form of justification used by Detroit's emergency manager.

Although we do not expect to hear the United Nations talking about human right violations here in the United States, the National Geographic article reports that a UN expert declared water to be a fundamental human right, and access to water should be denied only to those able to pay but whom refuse to pay.[62]

By the time you read this chapter, the crisis in Detroit may be over. Or not. Inform yourself at http://detroitwaterproject.org.

In other communities, we may not see water shut offs or an immediate impact on residents, but decisions are being made, or have been made, that may in the long run impact our expectations of a right to drinking water. In 2012, the Wall Street Journal (WSJ) reported that most Americans still get residential water through services that are publically owned and operated, but communities are increasingly shifting water utilities to the private sector such that nearly 73 million people get water through private companies. For arguments on both sides of this issue, read the WSJ article cited in the footnotes,[63] but you will surely also easily find more recent news on this topic. *Do you know if your water service is public or private?*

Water and Agriculture in the West

Detroit may be an extreme example of a local communities struggle over water; and is particularly ironic due to being sited on a river between two of the Great Lakes. Other cities, such as Phoenix, are instead sited in a water scarce basin from which agricultural consumption counts for more than 90% of the freshwater being drawn. When farmers of European descent settled in the area in the 1860s, they found an irrigation system that had been built hundreds of years earlier by the Hohokam people, a system that sustained a population of close to 50,000 for hundreds of years, and happily rebuilt and expanded it, along with planting water-intensive crops, such as wheat. Unfortunately, even with a population of just 1700 people in 1861, the water supply for the new City of Phoenix was known to be undependable in the summer heat.[64]

In Chapter 9, "Grains," we describe the Homestead Act and the impact on settlement that encouraged our nostalgic perception today of our history as agrarian, even though the American population had been decidedly becoming urban until that time. The settlers in the West struggled in developing the needed irrigation to farm in that climate.

John Wesley Powell's 1878 *Report on the Lands of the Arid Region of the United States* called for the Federal Government to control the West's rivers for irrigation and equal distribution for planned development in order for settlement in the west to succeed. The cause for federal involvement was greatly helped by the ascension of Theodore Roosevelt to the Presidency after the 1901 assassination of McKinley. It seemed perfectly reasonable to harness the otherwise un-used rivers into supporting farming, families, and economic stability. The Reclamation Act of 1902 committed the Federal Government to construct and maintain dams, reservoirs, and canals throughout the west to be funded by sale of public lands, with similar provisions of acreage, residence, and agricultural use, as had been promulgated earlier in the Homestead Act. Twenty years later, the Bureau of Reclamation had moved beyond irrigating farmland into providing electricity, which could then fund more, and larger, projects.[65]

By providing 600 dams and reservoirs later, the bureau has undeniably shaped the patterns of inhabitation and farming in the Western United States and had become the second largest hydroelectric power producer in the nation, with 53 plants producing more than 40 billion kilowatt hours annually and generating nearly a billion dollars in power revenues.[66]

The 1911 Roosevelt Dam, one of the Bureau of Reclamation projects, and the first to include hydroelectricity, brought agriculture to the Phoenix area. In the 1950s, the advent of air-conditioning brought people. Today, now with a population of over 4 million, the city is struggling with water management. Historically, urban planning tended to look for more water before trying to conserve. A case study report analyzing Phoenix and other similarly situated cities argues that increased urban–rural partnerships, such as ones in which cities may pay rural communities for providing water, are an important potential tool for urban water management.[64]

However, a note of caution is relevant. Los Angeles managed to acquire more water in a urban–rural partnership with the Imperial Irrigation District (the single largest water user in California) that started in 2003. The water was made available through increasing efficient irrigation and other methods that kept excess water from seeping into the ground. Unfortunately, it turned out that the excess water had only been lost to human use. The water had not been lost on the environment. Previously, the excess water drained through the soil into the Salton Sea, and the loss of that water was now problematic. Efficiency means less returns to the source, not necessarily less drawn from the source. Efficiency without behavior change is not the same as conservation.[67]

The water law in the west is complex, and does not favor the health of rivers. The DOI's Fish and Wildlife Service, Division of Water Resources explains the appropriation doctrine (the basis of water laws developed in the west) as essentially a rule of capture that awards a water right to a person actually using the water for a beneficial use. Priority is given on a first-come basis, so the earliest appropriator on a stream has the first right to use the water. The definition of beneficial is derived from state constitutions, statutes, or case law, and such uses may be different in each State, and the definition of what uses are beneficial may change over time.[68] Water rights are contentious.

Given such complexity, perhaps, it is not so surprising that there is apparently less consumer-facing information from the federal government on water policy related to food production.

Water: Changes in How, What, and Where We Produce Food

Agriculture is a heavy user of water. Research has been done to illustrate the differing water demands of our food production. Waterfootprint.org is an excellent resource on this topic, including a visual "product gallery" rotating slideshow with associated data on water usage.[69]

We have been producing more food because we have been watering more; global production of food doubled since 1950 but we tripled our use of water to do so, facilitated by government subsidization for water and the electricity to pump the water, and increasingly advanced pumping technology.[64]

Unfortunately, we are unable to delve fully into the extent of government subsidization for water and irrigation. However, it is worth noting that the USDA spent $1.1 billion between 2004 and 2013 for high-efficiency irrigation. The Environmental Quality Incentives Program (EQIP), first authorized in the 1996 farm bill, provided help to farmers to purchase more efficient irrigation equipment; but the goal according to a USDA water engineer was not to save water but to make agriculture as productive as possible. The EQIP contracts also did not restrict the farmers from then using the water conserved by the use of the high-efficiency equipment, thus making it clear that the end goal was not about saving water. Not surprisingly, research found that farmers were likely to do what they do best—grow more food using some of the water, rather than reduce consumption. Environmental activists want to keep the water conservation subsidy, just change it so that subsidized farmers would be restricted from increasing water consumption, but that idea failed to make it to the 2014 Farm Bill.[70]

But we already know that we need to change our water usage. Farmers and researchers have been working on it, just maybe not fast enough. The Ogallala Aquifer, which Kansas taps into, is being sucked dry. A 2013 study estimates that with no further change in our agricultural production technologies or in what we choose to grow or raise in Kansas, agricultural production in Kansas will peak around 2040 and decline, but immediate cuts in groundwater pumping, which appear to be feasible, would extend that to 2070, with a more gradual decline thereafter.[71] Keeping the aquifer from ever running out, however, would require far less corn and cattle production in Kansas.[72] If we produce less corn and cattle, we might eat less corn and cattle. Our dietary guidance may not much affected and our overall health may actually improve. However, considering our potentially fragile water systems, we must reflect upon the specialty crop production in the west. California alone produces 99% of the US almond and walnut crop, 97% of the plums, 95% of the celery crop, 90% of the strawberry crop, 90% of the broccoli, 90% of the lettuce, 80% of the carrots, much of the tomatoes, almost all of the cauliflower, and all of the artichoke.[73] If we had to dial back that much specialty crop production due to water policy management, then undoubtedly our dietary guidance, and possibly our health, would reflect those changes.

Cooking and Food Safety Methods

In fact, even our cooking methods might reflect a change, and not just in the sense that we might learn new methods to cook with less water, or emphasize the reuse of cooking water. As noted above, there is water content in our foods, and if we have a paucity of drinking water, we might need to physiologically count on the moisture that is trapped in some of our foods. What happens when you carmelize onions? The onions become sweeter, and the pieces, smaller. The water is sweated out and we have lost the opportunity to consume that (actually) embedded water. It is not hard to imagine; we might naturally change how we cook our foods if we were need to rely to a greater extent on the moisture captured in our foods.

Also, a good number of our food safety measures rely on water and some could use alternative solutions, such as the recommendation that ice cream scoops be stored with the serving end in a container with running water that removes food particles and washes them away, or thawing under running water for several hours.

Drink More

In 2013, First Lady Michelle Obama, as hononary chair, launched the "Drink Up" campaign to encourage the consumption of water, stating: "I've come to realize that if we were going to take just one step to make ourselves and our families healthier, probably the single best thing we could do is to simply drink more water. It's as simple as that. Drink more water."[74]

The "Drink Up" campaign is run by the Partnership for a Healthier America (PHA).[75]

By encouraging increased healthy behaviors, and making the "healthy choice the easy choice," PHA is using this campaign to help squeeze out less desired behaviors, such as drinking sugar-sweetened beverages, without directly stating as such.

Unsurprisingly, in November 2014, the International Bottled Water Association (IBWA) announced continued support of the "Drink Up" campaign,[76] noting that bottled water is projected to be the number one beverage sold in the United States by 2020. The IBWA commissioned a study about how much water goes into one liter. According to the IBWA, it takes 1.39 L of water to make one liter of water; less than it takes to make the same amount of soda. Although an improvement on soda, others point out that this still underestimates water usage, as it does not include the water used to make the packaging.[77]

Drinking water is healthy, and people enjoy bottled water. Can we continue to improve the sustainability of this choice?

Food: Can Water and Oil Mix?

Speaking of sustainability, looking forward, we may need to account for the impact oil production has on our ability to grow food. Producing oil may provide more energy that we can use to grow food. However, the wastewater produced by hydraulic fracturing ("fracking") may prevent us from growing food.

Although often the wastewater from fracking is returned deep underground, it may also be injected into more shallow aquifers that the EPA has designated as "exempt," meaning that the water is not appropriate for human use. In California, the state accidentally allowed oil and gas companies to dispose of wastewater into nine wells that tap into aquifers which are designated by the EPA as "nonexempt," or high-quality water meant to be protected for drinking water and irrigation. One of the wells is in the middle of a citrus grove. As 40 other water supply wells, many used for drinking water, are within a mile of that well, local residents and farmers fear that slow contamination may occur.[78]

In North Dakota, a similar conflict is arising from oil and food being produced cheek by jowl. Farmers are concerned about highly salinized wastewater spills. Water, so salty that it will kill current crops and prevent future ones from growing, has spilled into wetlands and on their fields from ruptured pipelines. One farmer can point to 80 acres that were contaminated by a wastewater spill 50 years ago, and has been fallow ever since, no matter what the family tried to plant on the acreage. Although oil-field practices have improved over the years, and regulation has tightened, wastewater spills do still occur. According to the Associated Press, 74 pipeline leaks occurred in 2013 in North Dakota, spilling a total of 22,000 barrels, with 17,000 attributed to one spill alone. In 2014, a million gallon spill (24,000 barrels) occurred when a pipeline burst on Fort Berthold Indian Reservation. Due to the lack of an alert system on the pipe, the leak was discovered through reviewing production loss reports.[79]

Even in the absence of nearby farming, we would still need tight regulations for wastewater production; would it make sense to attempt to separate our oil and food production a bit more?

In order to continue to invest in the agrarian social idea promoted by the Homestead Act, should we recognize the modern day impact of what—and where—those farmers grow?

Although we have acknowledged the impact of climate change and water resources will require improvements to our farming practices, it seems that we are mostly doing so *in situ. Should our farming subsidies and incentives become more highly localized? Could the USDA consider incentivizing farmers in Kansas to switch production away from corn and cows? Or is it more likely that we will continue to pursue technological solutions?*

Sustainability Is Food Security

We have not yet addressed federal involvement in food security policy short of describing the USDA's role in federal nutrition assistance programs. Food security refers to having consistent access to enough useable and nutritionally appropriate food.[80] However, this definition from the World Health Organization (WHO) does not specify affordability; the idea of affordability is perhaps assumed to be included in access.

The 2010 Dietary Guidelines "Call to Action" include ensuring that all Americans have access to nutritious foods and opportunities for physical activity. One cited means of working toward this goal is "to increase food security among at-risk populations by promoting nutrition assistance programs."[81]

Healthy People 2020 has two food insecurity objectives: NWS -12: Eliminate very low food security among children and NWS- 3: Reduce household food insecurity and in doing so reduce hunger.[82]

The September 2014 USDA report on Household Food Security reports that "a minority of American households experience food insecurity at times during the year, meaning that their access to adequate food is limited by a lack of money and other resources."[83]

Read together, these statements lead to the idea that in the United States, food security is considered very specifically a problem for only those individuals who lack money to purchase food. Because the phrase "other resources" is connected to a lack of money, which is a resource lacked by an individual, it is unclear if "other resources" are also meant to be limited to resources lacked by an individual, such as transit to get to food, as opposed to a lack of community resources such as a supermarket.

Yet individuals cannot control the ability of the food chain to ensure availability of food. In the beginning of this chapter, we made the argument that food is a social good. Lang expands this argument with the point that if food is a social good, it must be socially secure. Our perspective on food security needs to widen its focus from the individual. Looking forward, food security should include addressing the total sustainability of our food system. Lang argues that from an ecological public health perspective, food security *is* sustainability; only sustainable food systems can deliver meaningful security.[1] To this statement, we will simply add: to everyone.

CONCLUSION

Through examining food waste, and considering the great extent to which food really is embedded water, it becomes clear that our food policy is our water policy, and vice versa. The same goes for our agricultural policy.

Although the 2015 Dietary Guidelines are not yet available as we conclude this book, it seems certain that the Dietary Guidelines will address the sustainability of our diet in some manner. Perhaps the pendulum of dietary guidance will swing from eat more to eat less, particularly when it comes to eating animals. *But does sustainable mean we will have to eat less meat? Or does sustainable mean we will be eating less meat because it will be a luxury?*

We are making a fine distinction here between the idea of a forced change in behavior (such as rationing) versus preserving choice but nudging behavior by making meat more expensive (such as taxation).

During the rationing of World War II, most Americans viewed the imposition of rationing as a deprivation. The government campaigns and business ads countered by communicating messages around the idea of sharing, such as in Figure 14.3, so that everyone had "(...) an adequate supply of what they needed, rather than the greedy few hoarding all they could while the unfortunate many had to do without."[84]

Another author, in discussions over food waste, points out that it is unrealistic to assume that society today would accept the level of government interference required

FIGURE 14.3 This war-time poster uses a fairness argument that may be applicable again in our future. (With permission from Special Collections, National Agricultural Library.)

by rationing, but comments that an intelligent regulation across the food chain could work.[85] Perhaps those that would reject rationing perceive they would still be in the group which is able to afford meat if taxed. In any case, a "meat tax" or any similar nudging behavior at the consumer level would have to be supported with complementary regulation up and down the food chain.

We opened this chapter with the comment that for neither the case of food waste or water did we locate an overarching federal policy. In hindsight, however, we believe we did in fact discern an overall thread, but one emphasizing energy, not food waste or water.

In 1939, Congress gave the nod of approval to the Bureau of Reclamation to develop projects with the principal benefit of hydroelectric power.[65] Overall, the USDA is focused on making agriculture more productive, not on conserving water. Making agriculture more productive may be at odds with preventing food waste; and the USDA seems less focused on food recovery than might be expected. Overproduction of agriculture creates biomass, which can be used to make energy, so the incentives to prevent food waste at the level of production may not be clearly aligned; and the 2014 Farm Bill clearly emphasizes biomass over food recovery.

One study of the resources that go into food production phrased it in this manner: "Producing uneaten food requires a major investment of resources."[86] That phrasing, albeit unintentionally so, supports our point. In order for biomass energy production to continue, we may indeed literally need to produce "uneaten food."

In the past, government policies, programs, and education on the issue of food waste cropped up during times of economic or social strife. Today, the underlying issue is water. Food waste and water are indubitably and intimately connected. Expressing the issue of reducing food waste as one of conserving water gives us the means of easing into the more difficult conversation about water and our sustainability. Whether through action or inaction, our water policies will impact who eats what and when.

As noted in the introduction, the genesis of this book is the confluence of past and present food and nutrition policy. Our goal was to illustrate how the past gave rise to the present. Armed with this information, we leave it to you to determine the future.

REFERENCES

1. Lang T. Crisis? What Crisis? The normality of the current food crisis, *Journal of Agrarian Change*, 10(1);2010:87–97. Available at http://www.agroecology.wisc.edu/courses/agro-ecology-702/materials/8-poverty-and-hunger/hunger/lang-2010.pdf, accessed November 23, 2014.
2. Variyam JN. *Nutrition Labeling in the Food-Away-From-Home Sector: An Economic Assessment*. US Department of Agriculture, Economic Research Service. Economic Research Report No. ERR-4. April 2005. Available at http://www.ers.usda.gov/publications/err-economic-research-report/err4.aspx#.U58ihaiW8zw. Accessed June 16, 2014.
3. US Food and Drug Administration. Menu and Vending Machines Labeling Requirements. Available at http://www.fda.gov/Food/IngredientsPackagingLabeling/LabelingNutrition/ucm217762.htm, accessed November 26, 2014.
4. US Food and Drug Administration. FDA Basics. What is the meaning of "natural" on the label of food? Updated April 10, 2014. Available at http://www.fda.gov/aboutfda/transparency/basics/ucm214868.htm, accessed November 4, 2014.
5. Letter from Leslie Kux, Assistant Commissioner for Policy Food and Drug Administration, to Judges Gonzalez Rogers, White, and McNulty, January 6, 2014. Available at http://www.hpm.com/pdf/blog/FDA%20Lrt%201-2014%20re%20Natural.pdf, accessed November 4, 2014.
6. Squeezed HA. *What You Don't Know About Orange Juice*. New Haven: Yale University Press, 2009.
7. US Food and Drug Administration. Packaging & Food Contact Substances. Updated September 17, 2014. Available at http://www.fda.gov/Food/IngredientsPackagingLabeling/PackagingFCS/default.htm, accessed November 6, 2014.
8. US Food and Drug Administration. Determining the Regulatory Status of Components of a Food Contact Material. Updated March 24, 2014. Available at http://www.fda.gov/Food/IngredientsPackagingLabeling/PackagingFCS/RegulatoryStatusFoodContactMaterial/default.htm, accessed November 21, 2014.
9. Grossman E. When it comes to food packaging, what we don't know could hurt us. October 27, 2014. Available at http://ensia.com/features/when-it-comes-to-food-packaging-what-we-dont-know-could-hurt-us/, accessed November 21, 2014.
10. Charles D. In a Grain of Golden Rice, A World of Controversy over GMO Foods, March 7, 2013. *NPR*. Available at http://www.npr.org/blogs/thesalt/2013/03/07/173611461/in-a-grain-of-golden-rice-a-world-of-controversy-over-gmo-foods, accessed November 21, 2014.
11. Pollack A. U.S.D.A. approves modified potato. Next up: French fry fans. *The New York Times*. November 7, 2014. Available at http://www.nytimes.com/2014/11/08/business/

genetically-modified-potato-from-simplot-approved-by-usda.html?emc=eta1&_r=0, accessed November 15, 2014.

12. Open Source Seed Initiative. Available at http://www.opensourceseedinitiative.org, accessed November 14, 2014.

13. Center for Food Safety. Press Release. Victory! CFS Wins First Step in Major Legal Battle to Protect Food Safety. October 20, 2014. Available at http://www.centerforfoodsafety. org/press-releases/3550/victory-cfs-wins-first-step-in-major-legal-battle-to-protect-food-safety, accessed November 4, 2014.

14. Zuraw L. Food Industry Association Plans to make GRAS more transparent. August 29, 2014. Available at http://www.foodsafetynews.com/2014/08/gma-plans-to-make-gras-more-transparent, accessed November 4, 2014.

15. Padua GW, Wang Q (eds.). *Nanotechnology Research Methods for Foods and Bioproducts.* IA: Wiley-Blackwell Publishing, 2012.

16. United States Government Accountability Office. Food Safety: FDA Should Strengthen Its Oversight of Food Ingredients Determined to be Generally Recognized as Safe. February 2010. GAO report 10-246.

17. United States Food and Drug Administration. Guidance for Industry: Assessing the Effects of Significant Manufacturing Process Changes, Including Emerging Technologies, on the Safety and Regulatory Status of Food Ingredients and Food Contact Substances, Including Food Ingredients that Are Color Additives. June 2014. Available at http:// www.fda.gov/Food/GuidanceRegulation/GuidanceDocumentsRegulatoryInformation/ IngredientsAdditivesGRASPackaging/ucm300661.htm#introduction, accessed November 4, 2014.

18. Buzby JC, Wells HF, Hyman J. The Estimated Amount, Value, and Calories of Postharvest Food Losses at the Retail and Consumer Levels in the United States. Economic Information Bulletin No. (EIB-121) February 2014. Available at http://www.ers.usda.gov/publications/ eib-economic-information-bulletin/eib121.aspx, accessed November 4, 2014.

19. Bloom J. *American Wasteland. How America Throws Away Nearly Half of its Food (and What We Can Do About It).* Cambridge, MA: Da Capo, 2010.

20. Finley JW, Seiber JN. The nexus of food, energy, and water. *Journal of Agricultural and Food Chemistry.* 2014; 62(27):6255–6262. Available at http://www.silae.it/files/jf501496r. pdf, accessed November 4, 2014.

21. Cuéllar AD, Webber ME. Wasted food, wasted energy: The embedded energy in food waste in the United States. *Environ. Sci. Technol.* 2010;44(16):6464–6469. Available at http://pubs.acs.org/stoken/presspac/presspac/abs/10.1021/es100310d, accessed November 6, 2014.

22. HealthyPeople.gov. 2020 Topics & Objectives, Environmental Health. Available at https:// www.healthypeople.gov/2020/topics-objectives/topic/environmental-health/objectives, accessed November 6, 2014.

23. United States Environmental Protection Agency. Municipal Solid Waste Generation, Recycling, and Disposal in the United States: Facts and Figures for 2012. Available at http://www.epa.gov/solidwaste/nonhaz/municipal/pubs/2012_msw_fs.pdf,   accessed November 11, 2014.

24. United States Environmental Protection Agency. The Food Recovery Hierarchy. Last updated November 6, 2014. Available at http://www.epa.gov/foodrecovery/, accessed November 9, 2014.

25. Agricultural Act of 2014. H.R. 2642-286 Biomass Crop Assistance Program Section 9011(a)(6)(C)(iii). Available at http://www.gpo.gov/fdsys/pkg/BILLS-113hr2642enr/pdf/ BILLS-113hr2642enr.pdf, accessed November 11, 2014.

26. United States Department of Agriculture. Food, Farm and Jobs Bill. Last Updated September 25, 2014. Available at http://www.usda.gov/wps/portal/usda/usdahome?navid=farmbill, accessed November 11, 2014.

27. United States Environmental Protection Agency. Sustainable Materials Management, Food Recovery Challenge. Available at http://www.epa.gov/smm/foodrecovery/index.htm, accessed November 6, 2014.

28. United States Department of Agriculture. Office of the Chief Economist. US Food Waste Challenge. Available at http://www.usda.gov/oce/foodwaste/index.htm, accessed November 6, 2014.

29. United States Department of Agriculture. Office of the Chief Economist. US Food Waste Challenge. Available at http://www.usda.gov/oce/foodwaste/resources/donations.htm, accessed November 11, 2014.

30. Minnesota Department of Agriculture. Minnesota specialty crop grant recipients split $1.4 million from USDA. October 2, 2014. Available at http://www.mda.state.mn.us/en/news/releases/2014/nr20141002-speccrop.aspx, accessed November 6, 2014.

31. School Nutrition Association. Protect School Meal Programs: Provide commonsense flexibility so schools can prepare healthy meals that students will eat! July, 2014. Available at http://schoolnutrition.org/uploadedFiles/Legislation_and_Policy/SNA_Policy_Resources/July2014-Protect%20School%20Meal%20Programs.pdf, accessed November 11, 2014.

32. United States Department of Agriculture. Office of the Chief Economist. Reduce Food Loss and Waste in The School Meals Program. Available at http://www.usda.gov/oce/food-waste/commitments/fns/school.htm, accessed November 6, 2014.

33. Gunders D. Wasted: How America is losing up to 40 Percent of Its Food from Farm to Fork to landfill. NRDC Issue Paper. August 2012. Available at http://www.nrdc.org/food/files/wasted-food-ip.pdf, accessed November 11, 2014.

34. Buzby JC, Hyman J, Stewart H, Wells HF. The Value of Retail- and Consumer-level fruit and vegetable losses in the United States. *J. Consumer Affairs*. Fall 2011;45(3):492–515. Available at http://ucce.ucdavis.edu/files/datastore/234-2202.pdf, accessed November 6, 2014.

35. Newsome R et al. Applications and perceptions of date labeling of food. *Comprehensive Rev. Food Sci. Food Safety*. 2014;13:745–769.

36. United States Food and Drug Administration. FDA Basics. Did you know that a store can sell food past the expiration date? Last Updated April 10, 2014. Available at http://www.fda.gov/AboutFDA/Transparency/Basics/ucm210073.htm, accessed November 11, 2014.

37. National Resource Defense Council and Harvard Food Law and Policy Clinic. The Dating Game: How Confusing Food Date Labels Lead to Food Waste in America. September, 2013. Available at http://www.nrdc.org/food/files/dating-game-report.pdf, accessed November 11, 2014.

38. Bill Emerson Good Samaritan Food Donation Act. Report from the Committee on Economic and Educational Opportunities. July 9, 1996. Available at http://www.gpo.gov/fdsys/pkg/CRPT-104hrpt661/html/CRPT-104hrpt661.htm, accessed November 11, 2014.

39. United States Department of Agriculture Risk Management Agency. 2009 Insurance Fact Sheet. Gleaning. Available at http://www.rfhresourceguide.org/Content/cmsDocuments/USDA_RMA_GleaningFactSheet.pdf, accessed November 11, 2014.

40. Ed. Bread: The Staff of Life falls victim to the war. January 15, 1943. *St. Petersburg Times*.

41. Masterson K. Cows munch on recycled captain crunch. *Harvest Public Media*, 2012. Available at http://harvestpublicmedia.org/article/991/food-waste-into-livestock-feed/5, accessed November 21, 2014.

42. Food Waste Reduction Alliance. Welcome. Available at http://www.foodwastealliance.org, accessed November 11, 2014.

43. Smith E. Industry coalition raises flag on food waste regulations. *Associations Now*, 2014. Available at http://associationsnow.com/2014/04/industry-coalition-raises-flag-food-waste-regulations/, accessed November 11, 2014.

44. Smith AW et al. Calicivirus emergence from ocean reservoirs: Zoonotic and interspecies movements. *Emerg. Infect. Dis.* 1998;4(1):13. Available at http://wwwnc.cdc.gov/eid/article/4/1/98-0103, accessed November 11, 2014.

45. Bottemiller Evich H. Food companies unhappy with FDA feed proposal. *Politico*, 2014. Available at http://www.politico.com/story/2014/04/fda-trash-food-recycling-105726.html, accessed November 11, 2014.

46. Federal Register. Food and Drug Administration Proposed Rule. Current Good Manufacturing Practice and Preventive Controls for Food for Animals. September 29, 2014. Available at https://www.federalregister.gov/articles/2014/09/29/2014-22445/current-good-manufacturing-practice-and-hazard-analysis-and-risk-based-preventive-controls-for-food, accessed November 21, 2014.

47. Mass.gov. Energy and Environmental Affairs. Patrick Administration Finalizes Commercial Food Waste Disposal Ban. Press Release. January 31, 2014. Available at http://www.mass.gov/eea/pr-2014/food-waste-disposal.html, accessed November 21, 2014.

48. United States Department of Agriculture. Office of the Chief Economist. Dairy Power—Food Waste Repurposing. Available at http://www.usda.gov/oce/foodwaste/commitments/innovation/dairy.html, accessed November 21, 2014.

49. Mass.gov. Energy and Environmental Affairs. Patrick Administration Kicks off Statewide Commercial Food Waste Ban. Press Release. October 1, 2014. Available at http://www.mass.gov/eea/agencies/massdep/news/releases/patrick-administration-kicks-off-statewide-commercial-.html, accessed November 21, 2014.

50. Recycling works Massachusetts. Options for Complying with the Commercial Organics Waste Ban. Available at http://www.recyclingworksma.com/commercial-organics-waste-ban/, accessed November 21, 2014.

51. Leschin-Hoar C. Banning food waste: Companies in Massachusetts get ready to compost. *The Guardian*, 2014. Available at http://www.theguardian.com/sustainable-business/2014/sep/09/food-waste-ban-massachusetts-compost-landfill-food-banks, accessed November 17, 2014.

52. Food Shift. Our Approach. A new way to do business, shifting food waste into jobs. Available at http://foodshift.net/our-approach/, accessed November 11, 2014.

53. California's New Laws to Accelerate Organics Recycling. September 30, 2014. Biocycle. Available at http://www.biocycle.net/2014/09/30/californias-new-laws-to-accelerate-organics-recycling/, accessed November 11, 2014.

54. California Legislative Information. Text Search "Food Waste." Available at http://leginfo.legislature.ca.gov/faces/codesTextSearch.xhtml, accessed November 11, 2014.

55. Clean World. Our Technology. Available at http://www.cleanworld.com/technologies/, accessed November 11, 2014.

56. East Bay Municipal Utility District. Food Scraps Recycling. Available at https://www.ebmud.com/water-and-wastewater/environment/food-scraps-recycling, accessed November 11, 2014.

57. Westendorf ML (ed.). *Food Waste to Animal Feed.* Iowa: Iowa State University, 2000.

58. Ramirez A, Zaabel P. Swine Biological Risk Management. http://www.cfsph.iastate.edu/pdf/swine-biological-risk-management.

59. Park YK, Yetley EA, Calvo MS. Calcium Intake Levels in the United States: Issues and considerations. FAO Corporate Document Repository. Available at http://www.fao.org/docrep/w7336t/w7336t06.htm, accessed November 11, 2014.

60. Centers for Disease Control and Prevention. Community Water Fluoridation. Available at http://www.cdc.gov/fluoridation/, accessed November 14, 2014.

61. Federal Register. Proposed HHS Recommendation for Fluoride Concentration in Drinking Water for Prevention of Dental Caries. January 13, 2011. Available at www.federalregister.gov/articles/2011/01/13/2011-637/proposed-hhs-recommendation-for-fluoride-concentration-in-drinking-water-for-prevention-of-dental, accessed November 14, 2014.

62. Mitchell B. In Detroit, water crisis symbolizes decline and hope. *National Geographic*, 2014. Available at http://news.nationalgeographic.com/news/special-features/ 2014/08/ 140822-detroit-michigan-water-shutoffs-great-lakes/.

63. Little RG, Hauter W. Are we better off privatizing water? *Wall Street Journal*, 2012. Available at http://online.wsj.com/articles/SB1000087239639044381680457800228092625 3750, accessed November 14, 2014.

64. Richter BD et al. Tapped out: How can cities secure their water future? *Water Policy*, 2013;15(3):335–362. Available at http://www.iwaponline.com/wp/01503/0335/015030335. pdf, accessed November 15, 2014.

65. United States Department of the Interior. National Park Service. Water in the West. Available at http://www.nps.gov/nr/travel/ReclamationDamsIrrigationProjectsAndPowerplants/Water_ In_The_West.html, accessed May 10, 2015.

66. United States Department of the Interior, Bureau of Reclamation. About Us. Last updated September 9, 2014. Available at http://www.usbr.gov/main/about/, accessed November 21, 2014.

67. Walton B. Spending to conserve water on California farms will not increase supply. *Circle of Blue*, 2014. Available at http://www.circleofblue.org/waternews/2014/world/ conserve-water-california-not-increase-supply/.

68. US Fish and Wildlife Service. Water Rights Definitions. Available at http://www.fws.gov/ mountain-prairie/wtr/water_rights_def.htm, accessed November 15, 2014.

69. Water Footprint Network. Available at http://www.waterfootprint.org/?page=files/product-gallery, accessed November 11, 2014.

70. Nixon R. Farm subsidies leading to more water use. *The New York Times*, 2013. Available at http://www.nytimes.com/2013/06/07/us/irrigation-subsidies-leading-to-more-water-use. html?_r=0, accessed November 15, 2014.

71. Steward DR et al. Tapping unsustainable groundwater stores for agricultural production in the High Plains Aquifer of Kansas. *Proceedings of the National Academy of Sciences in the United States of America*. 110(37) E3477-E3486. September 10, 2013. Available at http://www.pnas.org/content/110/37/E3477.abstract, accessed November 15, 2014.

72. Plumer B. How long before the Great Plains runs out of water? 2013. Available at http://www.washingtonpost.com/blogs/wonkblog/wp/2013/09/12/how-long-before-the-midwest-runs-out-of-water/, accessed November 15, 2014.

73. Keep California Farming. Which Foods Come From California, accessed November 15, 2014.

74. The White House. Office of the First Lady. Remarks by the First Lady at the "Drink Up" Campaing Launch. September 12, 2013. Available at http://www.whitehouse.gov/the-press-office/2013/09/12/remarks-first-lady-drink-campaign-launch, accessed November 14, 2014.

75. Partnership for a Healthier America. http://ahealthieramerica.org, accessed November 14, 2014.

76. Water Technology. IBWA continues support of the Drink Up campaign. November 6, 2014. Available at www.watertechonline.com/articles/169053-ibwa-continues-support-of-the-drink-up-campaign, accessed November 14, 2014.

77. Gustafson TA. How much water actually goes into making a bottle of water? *NPR The Salt*, 2013. accessed November 14, 2014.

78. Stock S, Meak L, Villarreal M, Pham S. Waste Water from Oil Fracking Injected into Clean Aquifers. November 14, 2014. Available at http://www.nbcbayarea.com/investiga-tions/Waste-Water-from-Oil-Fracking-Injected-into-Clean-Aquifers-282733051.html, accessed November 15, 2014.

79. Wood J, Macpherson J. Associated press pipeline breach renews call for more monitor-ing. *San Jose Mercury News*, 2014. Available at http://www.mercurynews.com/science/ ci_26129866/officials-no-evidence-brine-bay-after-nd-spill, accessed November 15, 2014.

80. World Health Organization, Food Security. Available at http://www.who.int/trade/glossary/story028/en/, accessed August 30, 2014.

81. US Department of Agriculture and US Department of Health and Human Services. *Dietary Guidelines for Americans, 2010.* 7th Edition, Washington, DC: US Government Printing Office, December 2010.

82. US Department of Health and Human Services. Office of Disease Prevention and Health Promotion. Healthy People 2020. Washington, DC. Available at https://www.healthy people.gov/2020/topics-objectives/topic/nutrition-and-weight-status/objectives, accessed November 23, 2014.

83. Coleman-Jensen A, Gregory C, Singh A. Household food security in the United States in 2013.*Economic Research Service.* Report Number 173. Available at http://www.ers.usda.gov/media/1565415/err173.pdf, accessed November 23, 2014.

84. John Bush Jones. *All-out for Victory!: Magazine Advertising and the World War II Home Front.* Brandeis University Press: New Hampshire. 2009.

85. Hudson J, Donovan P. *Food Policy and the Environmental Credit Crunch: From Soup to Nuts.* Routledge: London and New York. 2014.

86. Reich AH, Foley JA. Food Loss and Waste in the US: The Science Behind the Supply Chain. April, 2014. Available at http://www.foodpolicy.umn.edu/policy-summaries-and-analyses/food-loss-and-waste-us-science-behind-supply-chain, accessed November 23, 2014.

Index

A

AAA, *see* Agricultural Adjustment Administration (AAA)
Acceptable Macronutrient Distribution Range (AMDR), 74
ACG, *see* American College of Genetics (ACG)
Activism, 114
    digital, 115–116
    food activist poster, 114
    food activists, 113
    food industry, 122–123
    food movement, 123–124
    labeling, 459–464
    social media, 115–116
    250 years of food activism, 116–120
Adequate intake (AI), 74, 154, 477
ADM, *see* Archer Daniels Midland (ADM)
Administrative Procedure Act (APA), 463
ADUFA, *see* Animal Drug User Fee Act (ADUFA)
Adult onset diabetes, *see* Type-2 diabetes
Affordability, 113, 483, 134, 391
Agrarian values, 37
    demographics and marketing outlets, 38–39
    federal organic regulations, 37–38
    federal support for organics, 39–40
    organic standards, 38
    people's department, 40–41
Agribusiness, 288, 291
    agricultural subsidy, 23
    consolidation of livestock, processors, and food production, 32–33
    development, 288
    food retailers, 33–34
    lobbying expenditures, 29, 248
Agricultural Act (1949), 157
Agricultural Act (1954), 288
Agricultural Act (2014), 5, 14, 168
Agricultural Adjustment Act (1933), 6, 19, 245
Agricultural Adjustment Act (1938), 20
Agricultural Adjustment Administration (AAA), 20
Agricultural excess reduction, 20
    commodity distribution, 20–22
    trade, 22
Agricultural policy/policies, 5, 37, 287
    agribusiness, 288
    demographics and marketing outlets, 38–39
    federal organic regulations, 37–38
    federal support for organics, 39–40

    grain surplus, 288–293
    organic standards, 38
    people's department, 40–41
Agricultural safety net, 22
    agribusiness, 32–34
    agriculture establishment, 28
    bees, 31–32
    commodities, 23–24
    crop insurance, 23–24
    Farm Bills, 28–29, 34–37
    farms consolidation, 30–31
    food industry lobbying, 28
    subsidies, 23–24
    impact of subsidies and crop insurance, 24–27
    U.S. agriculture production, 29–30
    writing rules, 27–28
Agricultural transition, 267–268, 294
Agriculture, 255, 481
    establishment, 28
    horticulture and, 304
    industrial, 255
    organic, 36, 37, 39, 125, 258
    promotion, 269
    sustainable, 129, 255
    urban, 129–132
    and water in west, 479–481
Agriculture Act (1935), 157
Agriculture and Consumer Protection Act (1973), 158, 168
Agroterrorism, 445–446
AHA, *see* American Heart Association (AHA)
AI, *see* Adequate intake (AI)
AICR, *see* American Institute for Cancer Research (AICR)
Air pollution, 254–255
ALA, *see* Alpha-linolenic acid (ALA)
Alcoholics, 351
Allen, Will, 130
Alpha-linolenic acid (ALA), 26
AMDR, *see* Acceptable Macronutrient Distribution Range (AMDR)
American values, 7–8
American Beverage Association, 403
American College of Genetics (ACG), 369
American diet, 88, 95, 133, 153, 162–163, 192, 208, 249, 273, 351, 386; *see also* Healthy Eating Index (HEI)
American Heart Association (AHA), 95
American Heritage Dictionary, 1–2